BARNYARD IN YOUR BACKYARD

BARNYARD IN YOUR BACKYARD

A Beginner's Guide to Raising Chickens, Ducks, Geese, Rabbits, Goats, Sheep, and Cattle

Edited by Gail Damerow

Contributing Authors: Gail Damerow, Darrell L. Salsbury, Nancy Searle, Paula Simmons, Heather Smith Thomas

The mission of Storey Publishing is to serve our customers by publishing practical information that encourages personal independence in harmony with the environment.

Text edited by Nancy W. Ringer and Carey L. Boucher
Art direction and cover design by Meredith Maker
Front cover photographs (clockwise from top left) ©Mark Turner/MIDWESTOCK, ©Linda Dufurrena/Grant Heilman Photography, ©Thomas Hoviand/Grant Heilman Photography, ©Cliff Riedinger/MIDWESTOCK, ©Mark Turner/MIDWESTOCK, ©Larry Lefever/Grant Heilman Photography. Back cover photographs ©Mark Turner/MIDWESTOCK (top), Artville (bottom and spine).
Illustration credits appear on page 408
Text design by Kathy Herlihy-Paoli, Inkstone Design, Inc., and Susan Bernier
Text layout and production by Susan Bernier

Printed in the United States by VonHoffmann Graphics
10 9 8 7 6 5 4 3 2 1

Library of Congress Cataloging-in-Publication Data

Barnyard in your backyard : a beginner's guide to raising chickens, ducks, geese, rabbits, goats, sheep, and cattle / Gail Damerow, general editor.

p. cm.
ISBN 1-58017-456-6 (alk. paper)
1. Domestic animals. I. Damerow, Gail.
SF61 .B348 2002
636—dc21

2002008708

CONTENTS

Introduction

One of the great things about living in the country is having enough space to keep livestock. The space requirements of various animals range from those of rabbits, which need so little room that just about anyone can keep them, to as much space as you wish to devote to a whole herd of cattle, sheep, or goats. When I started out with livestock, I lived on approximately one acre of land, on which I raised a variety of chickens, ducks, and geese. What attracted me to that property in the first place was its backyard population of colorful bantam chickens and its little pond on which floated a picturesque armada of mallard ducks.

I soon learned that each of those chickens and ducks, as well as the numerous barnyard animals I have since known, had its own personality. Of the many chickens in our barnyard today, two hens have found a way to escape when they see my husband and me coming to do the chores, and we now look forward to Henny and Penny greeting us halfway along the path.

Each animal has not only a unique personality but also a unique voice that lets you identify it just by sound. Even in those early days, it didn't take me long to recognize each rooster by the sound of its crow, and today I can recognize each goat in my herd by its bleat. Some goats make sounds so unlike all the others that we give them nicknames, such as The French Lady, whose call sounded like the French word *moi,* and The Brat, whose "Mah!" sounded like a spoiled child insisting on getting her way "now!"

WHY RAISE ANIMALS?

Taking time to observe the sight and sounds of the animals in your barnyard offers a refreshing change from the hurry and scurry of modern life. Provided you leave the cell phone behind, doing barnyard chores offers time out for relaxation and quiet enjoyment. Since livestock must be cared for daily, they pull you away from your indoor activities and force you to get out for a little exercise and fresh air. I often spend long days in the office and look forward to doing the evening chores that not only let me stretch my muscles but also help clear my head after a busy day.

As well as being pleasurable, barnyard animals are also purposeful. In exchange for your care, they give you unadulterated food products, such as eggs, meat, or milk, for your table; wonderful fiber products, such as wool, feathers, or hides, for use in a variety of crafts; and environmental benefits, such as manure to fertilize the land, automatic lawn mowing, and the joy of seeing active, healthy animals in your backyard. What a treat to pause and gaze out your window and see your animals peacefully going about their daily activities.

Having outdoor pets can be one reason to raise barnyard animals, which are not unlike house pets in some ways. The four-legged breeds tend to be somewhat like dogs in the companionship that can develop between them and their human keepers. The two-legged breeds tend to be more like cats, in that they tolerate and even appreciate your presence but have their own agenda and in general prefer each other's company to yours.

Recreational value is another reason for raising backyard livestock. Not only is caring for animals pleasurable in itself, but your chosen species may have a local or regional club that meets periodically to share information and ideas, giving you a chance to get acquainted with others who share your enthusiasm. Sometimes these groups are involved in breed exhibitions, privately or at county fairs, which offer the opportunity to share your enthusiasm with the public and encourage other people to become involved. Recreational opportunities include hitching your goat, cow, steer, or sheep to a cart for pleasure driving, entering local parades, and maybe using the animals to help with light chores.

Most people don't associate barnyard livestock with community service, but an often overlooked activity is taking well-trained and well-mannered animals to visit elderly people in nursing homes. Animals provide terrific therapy for drawing out unresponsive patients and will brighten the day of any shut-in. The animals, too, learn to enjoy all the attention lovingly showered on them.

Educational value is yet another reason for raising backyard livestock. Anyone who keeps animals knows that caring for them involves a constant learning process that is particularly valuable for children. If you enjoy working with children, you might set up an informal petting zoo for neighborhood kids, or take your animals to local schools.

Helping care for animals at home is a wonderful way to teach children responsibility, patience, dedication, and compassion. Raising barnyard animals also offers an excellent way for kids to learn about the natural processes of procreation, birth, and death.

Cautionary Considerations

All these marvelous rewards don't come without their price, however. For one thing, barnyard animals require constant care, day after day, week after week. No matter what else may be going on in your life on a particular day, or how tired you are at day's end, you must make time to take care of your livestock. These daily chores do not involve a lot of time or a lot of hard work, but they are an important responsibility. If you cannot find a substitute caretaker for times when you must be away, you may soon feel tied down by your animals.

But that's largely a matter of attitude. Many's the time I've not felt like doing barnyard chores, only to have my spirits lifted when I got out to the barn and was greeted by the animals eagerly waiting for my arrival. On the other hand, sometimes when I turn down an invitation from a friend or relative because I can't leave my animals for long, I'm exhorted to get rid of them so I can "be free." Of course, none of these people have, or have had, livestock of their own, and my friends who do have barnyard animals would never consider suggesting such a thing.

Other downsides that can be part and parcel of keeping backyard livestock include the need to deal with the manure, odor, noise, flies, and complaining neighbors. All of these potentially negative factors can be handily dealt with through proper management. I consider manure to be an advantage rather than a disadvantage because I am also a gardener, and manure makes an outstanding compost. My backyard barnyard full of animals provides me with a constant supply of manure. Odor-free composting techniques are covered in such books as *Let It Rot!*, by Stu Campbell. If you are not a gardener, surely you know someone who is and who would be delighted to have a source of free, natural fertilizer. And if you are enterprising, you might exchange the manure for help in cleaning out your barn, or you might sell the manure as a way for your livestock to help earn their keep.

Properly dealing with manure automatically solves the problems of odor and flies, which leaves us next to deal with noise. Barnyard noise is particularly problematic because not everyone considers it a problem. When I hear my neighbor's cow bellow, I know that her calf is being weaned or the cow is ready to be rebred. I once had a neighbor who, when she heard a cow bellow, became alarmed and called in a vet at her own expense. Now that can become pretty annoying if you are the cow's owner.

A crowing rooster is another noisemaker that not everyone considers to be a problem. I enjoy hearing the sound of a cock crow. I have had customers come to my place to buy a single rooster just to have the pleasure of hearing it crow. On the other hand, various of my chicken-keeping friends have been in constant battles, sometimes ending up in court, over their crowing roosters.

If barnyard noise is a potential problem, consider silent animals, such as rabbits or Muscovy ducks. The latter are sometimes called quackless ducks because their sound is so muted that it can be heard only at close range. Rabbits and Muscovies not only are quiet but also make great pets that even the crankiest neighbor might learn to appreciate.

Breeding and Offspring

If breeding is to be part of your backyard program, it must be well thought out in advance. Will you be breeding top-quality animals or run-of-the mill stock that only increase the population of mediocre animals? And what will become of the offspring? Hatching chicks in an incubator is a popular school project and is a super educational tool, provided plans are made in advance for the future of the resulting chicks. The worst possible scenario, as happens at some schools, is to send each student home with one or two chicks, which are more likely than not to meet some unpleasant fate.

School chicks and other offspring resulting from backyard breeding can be raised for meat, but you must be prepared to deal with the eventuality that one day the animals will be butchered. This event can be traumatic if you, or especially your child, has become attached to the animals. How well I remember the rabbits our family had when I was little. I had thought that they were my pets until the day I came home from school and found them hanging from the rafters to be skinned. I can't tell you how betrayed I felt. I eventually got over it, and today rabbit is one of my favorite meats. But as a child, I would have appreciated knowing that the rabbits our family was raising, and that I was allowed to play with, were destined to feed us.

When you raise animals for meat, you must be prepared to deal with butchering day. For starters, never give an affectionate name to a meat animal. If you name it at all, call it "Sir Loin Chop" or "Finger Licking" or something similar as a reminder of its purpose in life. Dealing with butchering involves not only overcoming the emotional aspects but following the prescribed procedures that result in safe, tender, tasty meat. Educate yourself by reading a book such as *Basic Butchering of Livestock & Game,* by John J. Mettler, Jr., and if the process sounds like something you'd rather not get involved with, find out ahead of time if you can count on someone else to do it for you.That someone might be a friend or neighbor involved with similar livestock, or perhaps a professional slaughterhouse.

Not all slaughterhouses accept all kinds of livestock. Some take only poultry, whereas others take only larger animals. Even a custom butcher who handles goats and steers might have a seasonal schedule: for example, taking in only game animals during the hunting season. When you find a slaughterhouse you plan to use, try to get endorsements from past customers. We once took in a goat to be butchered by a shop we had not used before. Despite our best instructions, the meat was wrapped in packages that were much too large for our small family and all the fat was included in the ground meat. As a result, the meat didn't store well in the freezer.

Perhaps you don't want to get involved in raising meat at all. Consider that right from the start in making your selection of livestock to raise. If you raise a calf, will it be a steer for meat or a heifer that will eventually become the family milk cow? If the latter, you must know that a cow gives milk only as a result of calving, so you will have another calf to deal with in the future. The same is true of milk goats. The question still remains of what you will do with the offspring.

Some breeds of goat and most breeds of sheep may be raised for fiber, which is used for spinning and other craft projects. If you find these animals appealing, you will have to contend with shearing. Learning to shear is a skill in itself, and may be more than you bargained for if you have only one or two animals. Custom shearers usually won't deal with small numbers of animals, but you may be able to take your sheep or goats to a larger operation at shearing time. And if you live in an area where fiber-bearing animals are not common, you may have a hard time finding a shearer at all. A fiber-bearing alternative to breeds that need shearing is cashmere goats or angora rabbits, from which the hair is harvested by combing rather than shearing.

Most rabbits are raised for meat and some are raised for their hides. Both activities involve breeding and butchering. You can, of course, keep rabbits strictly as pets. Similarly, chickens may be raised for meat or eggs or kept as pets, and ducks and geese may be meat animals or pets. Many people raise chickens or waterfowl solely for ornamental reasons — to enjoy the pleasure of having a couple of bantams pecking on the lawn or a pair of ducks or geese floating on the backyard pond. That's fine, too, although it's likely that one of the birds will eventually steal off into the underbrush to lay and hatch a batch of eggs, thus greatly increasing your backyard population. The offspring could be considered a bonus and given away or raised for meat.

An advantage to raising livestock solely for meat is that the project can be short-term. A batch of broiler chickens, for example, may be raised and butchered all within 6 weeks' time. A lamb may be ready to turn into chops in 6 or 7 months. These short time frames give you a chance to decide whether you like raising livestock at all. If the answer is yes, you then have the choice of doing another short-term meat project or engaging in a long-term project involving breeding more meat animals.

Education Is Key

Once you've made your livestock selection, begin to educate yourself about what's involved. Read not only this book, which provides an overview of each type of animal, but also some of the books mentioned in the Recommended Reading section on page 379, which offer more in-depth details on specific breeds. Also, subscribe to a periodical dealing with your chosen breed. Network with others who raise the breed by joining a local club, if one exists, and regional or national breed clubs. Visit the fair in your county, and perhaps in surrounding counties, to meet people who have the breed that interests you.

A super place to gather information is a 4-H show, where the kids involved are well educated about their animals and eager to share their knowledge. Nothing pleases children more than the opportunity to show a novice adult how smart they are. The people you meet during your networking will be invaluable later when you have questions about how to milk a cow, trim a goat's hooves, or assist a ewe giving birth.

FINDING STOCK

The same places that offer good networking opportunities are also excellent sources for locating livestock to purchase. If possible, try to buy animals from someone who lives nearby. Livestock purchased close to your home will be well adapted to your area, and you will have someone to turn to if you need help later on. When you buy from a local breeder, you can see for yourself whether the animals come from a clean, healthy environment and whether the breeding population has the proper conformation. If you are buying a female animal — a cow, ewe, or doe (goat or rabbit) — the seller may have a male animal to which you could breed her when the time comes.

An excellent place to find local sellers of livestock is the farm store. Many farm stores maintain a bulletin board where breeders may advertise livestock for sale, and the clerks often can tell you who buys feed for the species you are seeking. The county Extension office is another possible source of information, although some agencies are more active and knowledgeable than others. Larger livestock operations might advertise in the Yellow Pages of your phone book, in the newspaper classified ads, or in the freebie shopper newspapers that abound in every community. The farm store and Extension office can also tell you if your area has a club or other interest group dedicated to your breed. Also check with the national association that promotes your breed or species, most of which maintain a membership list that is available to the public. Some organizations publish their membership list on Web sites, where you can readily locate the members nearest you. If you are interested in a less common breed, contact the American Livestock Breeds Conservancy for their annually updated list of breeders.

If you have difficulty finding the animal you want locally, cast your net a little farther afield. Again, the national breed association may be helpful, as may advertisements in periodicals devoted to your chosen breed. Even when purchasing from a distance, it's best to try to travel to the seller's location to view the breeder stock and pick up your purchase. No matter how carefully animals are transported, shipping always involves certain risks. We have occasionally purchased a calf from a dairy in the next county and transported it home in our pickup camper, and we never had a problem except on one extremely hot day. During the 45-minute drive home, we stopped to offer the calf some water. It was too frightened to drink, so we decided the better plan was to get home fast and get it off the hot truck. We arrived home to find the calf nearly prostrate from heat and dehydration. After a good hosing down with cold water and several gallons of Gatorade, the calf was fine, but the incident gave us quite a scare. We vowed that from then on if we had to transport livestock in the summer, it would be only in the cool hours of early morning or late evening.

If you cannot pick up your purchase in person but must arrange to have it shipped, have a clear written understanding with the seller regarding who bears the risk if the animal gets sick or dies. The stress of long-distance travel compromises an animal's immune system, leaving it open to any infection that might come along during travel or on arrival at its new home.

If you decide to purchase from outside your area, try to get recommendations from satisfied customers. Every breed seems to have a few disreputable dealers. No matter whom you buy from, the supplier should be a reputable breeder rather than a dealer. Dealers, who buy and sell animals but do not raise them, often cannot attest to the breeding history or health of the animals they trade. A breeder, on the other hand, knows exactly what is being sold and will be able to answer your questions about the breed in general and about the particular animal you are acquiring.

Another place to avoid purchasing livestock is at an auction or sale barn, where animals are constantly coming and going. You will have no idea where your animal came from, and you can't tell how healthy or unhealthy it may have been to start with or what kind of diseases it may have picked up along the way. The last thing you want is for your first livestock experience to turn into a fiasco involving multiple expensive visits with the veterinarian, administering medications to a reluctant animal, and in the end possibly losing the animal despite your best efforts.

Don't be shy about closely examining your potential purchase for fear of insulting the seller. You will instead be showing that you are a well-informed buyer. If you appear to be thorough in your examination, the seller may tell you things that might otherwise not have been mentioned. Don't be shy either about asking questions, which is how most of us learn. The only dumb question is the one you fail to ask.

If the animal you are purchasing is mature, check it for such problems as crooked back or legs, or asymmetry in a cow's or goat's udder and in the pouches between a goose's legs. Ask the animal's age and be familiar with the prime production age and longevity of your chosen species, as well as the relative characteristics of young and old animals. Unfortunately, once an animal is fully mature, its exact age may be difficult to verify. To avoid getting stuck with a geriatric case, seek animals that are nearly but not quite full grown. That way, you can be sure of the age, you will get the advantage of the animal's full productivity while avoiding the uncertainties involved in raising a baby, and you will have a pretty good idea of what the animal will look like at full maturity.

If you are purchasing a registered animal, obtain all the necessary paperwork at the time that money changes hands. Stories abound of people who paid for a registered animal only to never receive the registration papers. Something similar happened to me when I purchased my second goat as a companion to my first goat. After we agreed on a price and loaded the goat into the truck, the seller went inside to get the registration papers. A short while later, he came back out to inform me that he couldn't find the papers but would mail them to me later. After several reminders, he finally did supply papers, but the name on the registration was not the name he had given me when I bought the goat. I wasn't too bothered, since many people give their animals both a registration name and a barn (or pet) name. Because I later had a question about the goat, I contacted the original owner, whose name was on the registration as the person who had sold it to the man who sold it to me. I described the goat and, after much discussion and confusion, learned that the animal I had bought was not the animal for which I had received papers. Fortunately, it was not an important issue for me, because I was not interested in breeding that goat as a registered animal, but I learned firsthand how easily a buyer can be duped by an unscrupulous seller.

Preparing a Home

Before bringing home your first animal, a little additional advance preparation will help smooth the way.

Ensure Family Support

Check with all members of your family to see how they feel about having barnyard livestock in your backyard. It's always best to have everyone's full support, especially when you may need a substitute to do your daily chores while you're away. If not all members are involved in maintaining the livestock, strife can result when the uninterested members feel that the others spend too much time at the barn. In contrast, relations in families in which everyone is involved in some phase of animal care are usually harmonious.

In our family, my husband and I normally do chores together; we each have certain responsibilities, but both pitch in for the other when need be. We enjoy our time together walking to and from the barn, but at the barn, we devote our full attention to the animals.

Establish Caretaking Responsibilities

Establish a caretaking schedule and decide who in your family will do what chores daily, weekly, monthly, and seasonally. Children, for instance, may be in charge of the daily routines of feeding, milking, and gathering eggs; these simple tasks will help them learn about responsibility. Adults or older teenagers should probably be involved in less frequent but more difficult tasks, such as vaccinating, cleaning stalls, or attending a birth.

Check Zoning Regulations

Check the zoning regulations for your area. Zoning laws may prohibit you from keeping certain species of livestock, limit the number of each species you may keep, regulate the distance animal housing must be from nearby human dwellings or your property line, or restrict the use of electric fencing.

I saw firsthand how zoning works on my first little farm, where I raised poultry and waterfowl. The area was rezoned after I moved in. Of course, my poultry activities were grandfathered in, meaning that the authorities could not make me get rid of the birds I had, but I was not allowed to increase the population. Now, the nature of raising chickens is that after the spring hatch, you have more, and as the year progresses and you sell or butcher some, you have fewer. Complying with the new law meant I would eventually be out of business. I managed to prevail as long as I lived there, but not without hassles from neighbors and occasional visits from the authorities.

One of my acquaintances experienced a zoning issue when he purchased some property on which he planned to breed chickens. Between the time he agreed to buy the land and the time the purchase was finalized, the area was rezoned to ban roosters, without which he could not breed his hens. If you plan to raise livestock on property you have yet to purchase, check not only existing zoning laws but also proposed changes.

Prepare Facilities

Once you learn of any zoning regulations that might influence where on your property you may keep animals, prepare their facilities. Most animals will require all-weather housing. If your area has particularly hot days or cold days, take those extremes into consideration right from the start, or you may never get around to providing proper housing. If you are starting out with babies, remember that they will grow; make sure your facilities are of sufficient size to handle fully mature animals. If you wish to breed your stock to raise future babies, chances are pretty good you'll want to keep one or more of them, so allow space for herd or flock expansion.

Have your facilities ready and waiting before you bring home your first animals. Things have a way of taking longer than expected, and you don't want your new animals to suffer the discomfort of staying too long in temporary and perhaps unsuitable housing.

Provide adequate feed and water stations, and lay in a supply of feed. If you will be using feed that is different from what your new animal has been eating, purchase some of its usual feed from the animal's seller. Gradually mix in greater quantities of new feed with the old. The goal is to avoid an abrupt switch that can cause stress or digestive upset.

Install Fencing

Secure the barnyard area with a sturdy fence that not only keeps in your livestock but also keeps out predators. When most people hear the word "predator," they think of wild animals such as foxes, raccoons, or coyotes. But the number one predators of domestic livestock are neighborhood dogs. I woke up one night to hear the squawking of some young chickens I was keeping in a raised hutch in my driveway. I ran to the window and saw what looked like a man reaching in to grab my chickens. When I snapped on the light, he crouched and ran. In my sleepy state, it took me a moment to realize I had just scared off a huge dog. Next morning, the animal control warden told me the surrounding neighbors were being terrorized by a marauder that killed not only countless chickens but a man's calf, a couple of sheep belonging to a little boy, and most of one little girl's rabbits. Based on my description, the dog was finally identified; it was determined a menace to the public and euthanized. Because the warden didn't believe that one dog could possibly do so much damage in such a short time, the animal's stomach contents were examined; they included a wide variety of fur, wool, and feathers. The irony is that the woman who owned the dog, and who had turned it out each night to run loose, was furious with the authorities for having deprived her children of their pet dog. Sadly, such tales are far too common.

Sometimes the predator dog is not the neighbor's but your own. I was once talked into getting a dog to guard my chickens. One morning, I went out to be greeted by a happy canine that had spent the night killing my fryers and neatly laying them out along the path with all the heads pointing in one direction and feet pointing in the other direction. The dog seemed immensely pleased with his gift to me. My gift to him was to find him a home with a nice couple who didn't raise chickens.

Inform the Neighbors

Let your neighbors know about your plan to raise livestock. Explain that you are taking great pains to keep your animals from getting into other people's yards and to keep other people's animals out of your yard. Describe what you are doing to maintain clean housing and minimize odors and flies. By letting the neighbors in on your plans, you are less likely to hear complaints from them later.

Depending on your relationship with your neighbors, you might get them involved by asking for their input and advice. Perhaps they'll be interested in fresh eggs from your chickens, fresh milk from your cow or goat, or barnyard compost for their garden. You might even interest them in helping with your livestock adventure, thereby developing an ally who can keep an eye on things when you must be away. You might even pique their interest enough that they'll end up having a backyard barnyard of their own.

1

Chickens

Introducing Chickens

If you've never raised livestock before, keeping chickens is a great start. They're easy to raise, they don't need a lot of space, and they don't cost a lot of money to buy or to feed. Everything you learn about feeding, housing, and caring for your chickens will help you later if you decide to raise some other kind of animal.

People have raised chickens for at least 5,000 years. All chickens belong to the genus *Gallus,* the Latin word for cock, or rooster. The English naturalist Charles Darwin traced all chickens back tens of thousands of years to a single extant breed, the wild red jungle fowl of southeast Asia *(Gallus gallus).* These fowl look like today's brown Leghorns, only smaller.

Wild jungle fowl are homebodies, preferring to live and forage in one place as long as possible. This trait made taming wild fowl an easy task. All people had to do was provide a suitable place for the chickens to live and make sure that the flock got plenty to eat. As a reward, they had ready access to fresh eggs and meat.

Early chickens didn't lay many eggs, though, and they made pitifully scrawny meat birds. Over time, chicken keepers selected breeders from those that laid best, grew fastest, and developed the most muscle — and thus came about today's domestic chickens. The Romans called household chickens *Gallus domesticus,* a term scientists still use.

Different people who have kept chickens over the years valued different traits, which led to the development of many different breeds. In 1868, Darwin took inventory of the world's chicken population and found only 13 breeds. Now we have many times that number. Most of today's breeds were developed during the 20th century, when chickens became the most popular domestic food animal.

Getting Started

How much it costs to get started depends on such factors as the kind of chickens you want and how common they are in your area, how simple or elaborate their housing will be, and whether you already have facilities you can use or modify.

Chickens must be housed to protect them from wind and harsh weather, but the housing need not be fancy. An unused toolshed, or the corner of a barn or other outbuilding, can provide comfortable quarters. If your yard isn't fenced, you'll need to put one up. A good fence keeps dogs and other predators away from your chickens and keeps your flock from bothering your garden or your neighbors' flower beds.

In deciding where to put your chicken yard, consider whether crowing may bother your neighbors. Male chickens — called roosters, or cocks — are well known for their inclination to crow at dawn. Ancient people believed they crow to scare away evil spirits lurking in the dark. Cocks occasionally crow during the day, and if two cocks are within hearing distance, they will periodically engage in an impromptu crowing contest. A rooster rarely crows during the dark of night, unless he is disturbed by a sound or a light.

If the sound of crowing might cause a problem in your area, consider keeping hens without a rooster. Although the rare persnickety neighbor may complain about hen sounds, the loudest noise a hen makes is a brief cackle after she's laid an egg. Contented hens "sing" to themselves by making a soft, pleasant sound that only a grouch could object to. Without a rooster, hens will still lay eggs. The rooster's function is not to make hens lay eggs but to fertilize the eggs so they can develop into chicks. Without a rooster, you won't be able to hatch the eggs your hens lay.

Comparing Benefits and Drawbacks

Raising chickens has some downsides. One is the dust they stir up, which can get pretty unpleasant if they are housed in an outbuilding where equipment is stored. Another is their propensity to scratch, which becomes a problem if they get into a bed of newly planted seedlings. Chickens also produce plenty of droppings that, if not properly managed, will smell bad and attract flies.

Before you set up a chicken farm, make sure that you and your family are not allergic to chickens. You can find this out ahead of time by visiting a poultry show at your county fair or spending a few hours helping a friend or neighbor care for their chickens. If you have an allergic reaction, you will have avoided the expense and heartache of setting up a flock you immediately have to get rid of.

Until you raise your own chickens, it may be hard to believe that people become attached to their chickens and have difficulty letting them go when it's time to butcher meat birds or replace old layers with younger, more efficient hens. The only alternative, though, is to run a retirement home for chickens, which gets pretty expensive, and the birds will still get old and die eventually. You'll have to come to grips with the loss.

For many people, the upside of raising chickens far outweighs the downside:

- Chickens provide wholesome eggs and meat for your family, and you can take pride in knowing that the flock that puts food on your table lives under pleasant conditions.
- Raising chickens is educational. By watching chickens interact, you will learn something about how all birds live and behave.
- Chickens are attractive. They come in all sizes, shapes, and colors. You can find a breed that appeals to your aesthetic sensibilities.

Caring for a home flock takes a few minutes each day, to provide feed and water and to collect eggs. In hot or cold weather, these jobs must be done twice daily, seven days a week. If you raise chickens for meat, the project will be finished in 2 to 3 months. If you raise hens for eggs, you must care for them year-round. As long as you keep in mind that your flock relies on you for its survival, raising chickens is a breeze.

Choosing the Right Breed

No one knows for sure how many breeds of chicken may be found throughout the world. Some breeds that once existed have become extinct, new ones have been developed, and forgotten ones have been rediscovered. Only a fraction of the breeds known throughout the world are found in North America.

The American Standard of Perfection, published by the American Poultry Association, contains descriptions and pictures of the many breeds and varieties officially recognized by that organization. These breeds are organized according to whether they are large chickens or bantam (miniature) chickens, and each group is divided by class.

Large chickens are classified according to their place of origin: American, Asiatic, English, Mediterranean, Continental, and Other (including Oriental). Bantams are classified according to characteristics such as whether they have feathers on their legs. Each class is further broken down into breeds and varieties.

Chickens of the same breed all have the same general type, or conformation. A chicken that looks similar to the ideal for its breed, as depicted in the *Standard,* is true to type, or typey.

The feathering of some breeds distinguishes them from others. A Polish chicken, for example, has a crest of feathers on its head. The name Polish does not come from this chicken's place of origin — which is the Netherlands — but from its poll or feathery crown.

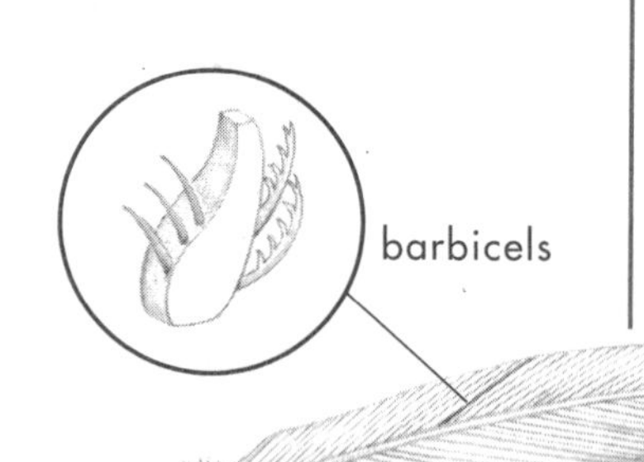

The smooth web of the feather is held together by barbicels.

Most feathers have a smooth, satinlike surface called a web. This web is created by barbicels, tiny hooks that hold the web together. If you run your fingers along a feather from top to bottom, the barbicels let go, causing the web to separate. Rub the feather the other way, and the barbicels hook onto each other, bringing the web back together.

A feather with no barbicels looks fuzzy. The feathers of a Silkie have no barbicels, making the bird look like it has fur. Silkies differ from other chickens in another way: Most chickens have white or yellow skin, while Silkies have black skin. If you are raising chickens for meat, the skin color may make a difference to you. People of Asian cultures tend to prefer chickens with darker skin, Europeans prefer white skin, and most Americans prefer yellow skin.

Hen feathering also distinguishes one breed from another. A cock usually has pointed feathers on his neck (hackle) and lower back (saddle), whereas hens have rounded feathers in those places. In a hen-feathered breed, however, the cock has round feathers like a hen's. Sebright bantams are an example of a hen-feathered breed.

Some breeds come in more than one color; each color constitutes a variety. The colors may be plain — such as red, white, or blue — or they may have a pattern, such as speckled, laced, or barred. Two varieties of the Plymouth Rock are white and barred.

Most chickens are clean-legged, meaning that they have no feathers on their legs. Chickens that have feathers growing all the way down to their toes are called feather-legged. Some breeds, such as the Frizzle, come in a clean-legged variety and a feather-legged variety. (The Frizzle gets its name from its curly, or frizzled, feathers.)

Sometimes a chicken has a tuft of feathers growing under its chin. This trait is aptly called a beard. Some breeds, such as the Polish, come in a bearded variety and a nonbearded variety.

Another feature that may distinguish one variety from another is the style of comb. Most breeds sport the classic single comb, with its series of sawtooth zigzags. The Sicilian Buttercup, by contrast, has two rows of points that meet at the front and back, giving the comb a flowerlike look. Other comb styles are pea, cushion, strawberry, and rose. Rose comb and single comb are two varieties of Leghorn.

With all these different possibilities, how do you choose the breed and variety that is best for you? Narrow your choices by deciding what you want your chickens to do for you:

- Do you primarily want eggs?
- Do you want to raise your own chicken meat?
- Do you want to help preserve an endangered breed?
- Do you want to compete at shows?
- Do you simply want the pleasure of seeing chickens roam your yard as pets?

Comb Styles

Buttercup comb

Cushion comb

Pea comb

Rose comb

Rose comb (spiked)

Single comb

Strawberry comb

V comb

Walnut comb

Egg Breeds

All hens lay eggs, but some breeds lay more eggs than others. The best laying hen will yield about 24 dozen eggs per year. The best layers are smallish breeds that produce white-shelled eggs. These breeds originated near the Mediterranean Sea, hence their classification as Mediterranean. Examples are Minorca, Ancona, and Leghorn, respectively named after the Spanish island of Minorca and the Italian seaport towns of Ancona and Leghorn (Livorno).

Leghorn is the breed used commercially to produce white eggs for supermarkets. Leghorns are inherently nervous, flighty birds that are unlikely to calm down unless you spend a lot of time taming them. The most efficient layers are crosses between breeds or strains within a breed. The strains used to create commercial layers are often kept secret, but you can be sure a Leghorn is involved.

Most layers produce white eggs, but some lay brown eggs. Brown-egg layers are calmer than Leghorns and therefore more fun to raise.

After a year or two, the laying ability of hens decreases. Unless you choose to keep your spent hens as pets, the best place for them is the stew pot.

Meat Breeds

Good layers are scrawny, because they put all their energy into making eggs instead of meat. If you want chickens mainly to have homegrown meat, raise a meat breed.

The various terms for chicken meat depend on the stage at which the bird is butchered. Broilers and fryers weigh about 3½ pounds and are suitable for frying or barbecuing. Roasters weigh 4 to 6 pounds and are usually roasted in the oven.

For meat purposes, most people prefer to raise white-feathered breeds, because they look cleaner than dark-feathered birds after plucking. The best breeds for meat grow plump fast. The longer a chicken takes to get big enough to butcher, the more it eats. The more it eats, the more it costs to feed. A slow-growing broiler therefore costs more per pound than a fast grower.

Most meat breeds are in the English class, which includes Australorp, Orpington, and Cornish. Of these, the most popular is Cornish, which originated in Cornwall,

Strains

A strain consists of related chickens of the same breed and variety that have been selectively bred for emphasis on specific traits, making the strain readily identifiable to the trained eye. Most strains are named after their breeder, although sometimes a breeder will develop more than one strain and give each a code name. Two strains may be so different from each other that generalized comparisons about their breed become difficult. Indeed, sometimes more variation occurs among strains than among breeds. To further complicate matters, the farther a strain gets from its source the more it may change, because the new breeder may not share the vision of the original breeder. When a strain is taken up by several new breeders, chances are good that each will selectively breed in a different direction until the original strain is barely recognizable.

England. The ideal Cornish hen weighs exactly 1 pound dressed. The fastest-growing broilers result from a cross between Cornish and New Hampshire or White Plymouth Rock. The Rock-Cornish cross is the most popular meat hybrid. Those 1-pound Cornish hens you see in the supermarket are actually 4-week-old Rock-Cornish crosses, and may not be hens but cockerels (young cocks).

A Rock-Cornish eats just 2 pounds of feed for each pound of weight it gains. By comparison, a hybrid layer eats three times as much to gain the same weight. You can see, then, why it doesn't make much sense to raise a laying breed for meat or a meat breed for eggs. If you want both eggs and meat, you could keep a flock of layers and raise occasional fryers, or you could raise a dual-purpose breed.

Dual-Purpose Breeds

New Hampshire

Plymouth Rock

Rhode Island Red

Endangered Breeds

Chantecler

Dominique

Exhibition Breeds

Cochin

Sebright

Silkie

Dual-Purpose Breeds

Dual-purpose breeds kept for both meat and eggs don't lay quite as well as a laying breed and aren't quite as fast-growing as a meat breed, but they lay better than a meat breed and grow plumper faster than a laying breed. Most dual-purpose chickens are classified as American, because they originated in the United States. They have familiar names like Rhode Island Red, Plymouth Rock, Delaware, and New Hampshire. All American breeds lay brown-shelled eggs.

Some hybrids make good dual-purpose chickens. One is the Black Sex Link, a cross between a Rhode Island Red rooster and a Barred Plymouth Rock hen. Another is the Red Sex Link, a cross between a Rhode Island cock and a White Leghorn hen. (A sex link is a hybrid whose chicks may be sexed by their color or feather growth.) Red Sex Links lay better than Black Sex Links, but their eggs are smaller and dressed birds weigh nearly 1 pound less.

Each hatchery has its favorite hybrid. Although hybrids are generally more efficient at producing meat and eggs than a pure breed, they will not reproduce themselves. If you want more of the same, you have to buy new chicks from the hatchery.

Endangered Breeds

Many dual-purpose breeds once commonly found in backyards are now endangered. Because these chickens have not been bred for factory-like production, they've retained their ability to survive harsh conditions, desire to forage, and resistance to disease.

The American Livestock Breeds Conservancy keeps track of breeds and varieties it believes are in particular danger of becoming extinct and encourages breeders to join its poultry conservation project. Among the breeds on its list is the Dominique, sometimes incorrectly called "Dominecker," the oldest American breed. A few years ago, it almost disappeared, but it is now coming back thanks to the efforts of conservation breeders. Canada's oldest breed, the Chantecler, experienced a similar fortunate turn of fate.

While the American Livestock Breeds Conservancy specializes in dual-purpose breeds, the Society for the Preservation of Poultry Antiquities specializes in tracking and conserving endangered exhibition chickens.

Exhibition Breeds

Exhibition chickens are bred for beauty rather than their ability to efficiently produce meat or eggs. Some of the same breeds kept for meat and eggs are also popular for exhibition, although the strains are different. Commercial strains used for egg or meat production are often hybrids, but even the pure production breeds are not necessarily true to type. Exhibition strains, on the other hand, are more typy but less efficient at producing eggs or meat.

Even among exhibition breeds, some lay better than others. Exhibition breeds in the Mediterranean class lay better than most other show breeds, even though they don't lay as well as production flocks of the same breed. Similarly, among exhibition birds, the Cornish and Cochin are more suitable than many others for meat production.

Bantams are miniature exhibition chickens weighing 1 to 2 pounds. They are popular because they eat less and require less space than larger chickens. The shrill crowing of a bantam cock, however, is more likely to irritate neighbors than the lower-pitched crowing of a larger rooster.

Some bantams are small versions of bigger breeds, only one fifth to one fourth the size. Others, such as Sebrights and Silkies, come only in the miniature version. Bantam eggs are smaller than those laid by large chickens; three bantam eggs roughly equal two regular eggs. The American Bantam Association publishes the *Bantam Standard* describing all recognized bantam breeds, many of which are not listed in *The American Standard of Perfection.*

Exhibition chickens differ from other livestock in that they don't come with registration papers. Your only assurances that you get what you pay for are the seller's word and reputation.

Making the Purchase

After settling on a breed, variety, and strain, your next decision is whether to purchase newly hatched chicks or grown chickens. Starting with chicks is less expensive than buying the same number of mature birds, and the chicks will grow up knowing their home territory. Some breeds and hybrids are sold sexed, meaning you know when you buy them how many are cockerels and how many are pullets (young hens). Most chicks are sold straight-run, meaning as-hatched; in this case, about half will be cockerels and half pullets. Sexed chicks are more expensive than straight-run chicks.

Starting with grown birds carries certain risks, among them the greater likelihood of buying diseased or spent chickens. To make sure you are getting young chickens, look for legs that are smooth and clean and breastbones that are soft and flexible. Advantages to purchasing older chickens are that you can easily tell the roosters from the hens and, if you plan to show, you can determine how typey they are. If you will be raising the chickens for meat or eggs, you won't have to wait as long if you purchase started birds.

Where you buy your chickens depends on what kind you want. If you want a production breed or hybrid, get chicks from a commercial hatchery. If none is nearby, deal with a reputable firm that ships by mail. You may find chicks at a local feed store, although chances are you won't be able to learn much about what they are or where they came from. Avoid bargain chicks that come free with your first purchase of a sack of feed; they are likely to be excess cockerels of a laying breed.

Dual-purpose breeds are sold by hatcheries, as well as by individuals who advertise in local newspapers. Visiting the seller lets you see what the flock looks like and the conditions under which the birds are raised.

How Long Do Chickens Live?

A chicken may live 10 to 15 years. Few chickens live out their full, natural lives. Chickens raised for meat have a short life of only 8 to 12 weeks. Chickens raised for eggs or as breeders are usually kept for 2 or 3 years, until their productivity and fertility decline. Chickens raised for show are in their prime at 1 year of age. After that, prizewinners may be shown for several more years or kept as breeders. A chicken kept in a protective environment may survive as long as 25 years.

The best place to find sellers of an exhibition breed is at a poultry show. Many people enter shows just to advertise their chickens for sale. Even if you can't buy chickens at the show (most shows prohibit on-site sales), you will meet people who have chickens for sale at home, and you can connect with a local, regional, or national association that promotes the breed that interests you. Chickens sold by a hatchery are not likely to be show quality. Top-quality exhibition birds are quite expensive.

In making your final selection, the main thing to look for is good health. Chicks should be bright-eyed and perky. If they come by mail, open the box in front of the mail carrier, in case something has gone wrong and you need to file a claim. In grown chickens, signs of good health include the following:

- Feathers that are smooth and shiny, not dull or ruffled
- Eyes that are bright, not watery or sunken
- Legs that are smooth and clean, not rough and dirty
- Combs that are full and bright, not shrunken and dull
- Soundless breathing; no coughing, sneezing, or rattling sounds

When you visit a seller, whistle as you approach the flock. The chickens will pause to listen, letting you easily hear any unusual breathing sounds. Before taking home your selection, look under each chicken's wing and around the vent under its tail to make sure it isn't crawling with body parasites.

How many chickens you take home depends, again, on your purpose in keeping them. If you are interested in egg production, determine how many eggs you want per day, divide that number by two, and multiply by three. If you want, for example, six eggs per day, you'll need at least nine hens. Since hens don't lay at the same rate all year long, sometimes you'll have more eggs than you can use, and other times you'll have too few. It's the nature of the game.

If you want a dual-purpose breeding flock, 50 chickens is a good number to have. If you wish to raise chickens for show or as pets, a trio (one cock and two hens) makes a nice little family.

Head Shaking

A chicken that shakes its head from side to side is telling you that you have frightened it by moving too fast or being too loud.

Handling Chickens

A flock of calm chickens is a joy to be around. Chickens that squawk and fly every time you come near aren't much fun. Part of flighty behavior is genetic in that some breeds are more easily frightened than others, but all chickens may be taught to be calm. The secret is for you to be calm when you are around them.

Whenever you approach your chickens, whistle or sing so they'll hear you coming and won't be startled. Spend at least 5 minutes a day with your flock. The more time you spend, the less easily frightened and the friendlier your chickens will become. Walk slowly among them, talking or singing softly. Some folks pick up and pet their chickens. Your neighbors may think you odd, but your chickens will love it.

Peck Order

Chickens are social animals. They are happiest when they are with others of their own kind. In every group of chickens, one emerges as the leader. Scientists call this "establishing the peck order." The peck order reduces stress by ensuring that every chicken knows how to relate to every other chicken. Most of the time, roosters are

Sparring is normal when chicks are young; once the peck order has been established, fighting should subside.

higher in rank than hens. If a hen is higher in rank than a cock (in other words, the cock is "henpecked"), it is usually because the cock is young and inexperienced or old and weak.

Chicks start establishing their place in the peck order when they are only 3 weeks old. After a couple of weeks, most of the fighting stops, because the peck order has been established. If a new chicken is introduced, the fighting starts again until the new chicken has found its place in the peck order.

When two chickens meet for the first time, their first order of business is to establish the peck order. The farther a chicken is from home, however, the more timid it becomes. The home flock therefore has an advantage over any newcomer, which will get badly beaten up if you simply turn it loose.

Instead, keep the new chicken apart from the flock for a few days. After the new chicken becomes used to you, start bringing in your other chickens, one at a time. Introduce the lowest-ranking ones first. They are usually the youngest or the oldest birds. After the new chicken has met all the other chickens one by one, put her with the whole flock. She should get along fine, because she has already established her place in the peck order. She will probably rank somewhere in the middle.

If you are introducing several new birds, you can help ease the transition by placing a fence between the old flock and the new for a few days. When you remove the separating fence, leave each group its own feed and watering station. If both groups include a mature cock, you can expect the cocks to fight, even through the fence.

Aggression

Most roosters are not mean, but they will defend their flock from a threat. However, they can't always tell when a threat is real. Something as simple as floppy shoelaces may look like a real threat to a rooster. If you always wear jeans when you tend your flock and one day you wear overalls or a rain suit, the floppy legs might be perceived as a threat. Sometimes a rooster will feel threatened by any man, or any woman, or any child, or any person but the one who normally brings feed and water.

A rooster may learn to be mean if he gets teased by a person or another animal. Children sometimes poke sticks at chickens and dogs enjoy chasing them, in both cases scaring the chickens out of their meager wits. Keep dogs and misbehaved children away from your flock, or teach them how to behave properly.

Sometimes a rooster gets the notion that you are a big chicken he must outrank in the peck order. He may move toward you sideways, with his head down, pretending he is looking at something on the ground. The term *cockeyed* comes from this ability of a rooster, or cock, to look someplace else while he is really watching you. He may not actually attack, but he may threaten by rushing at you or bumping against your leg. Try to make friends with him by picking him up and rubbing his wattles.

If a rooster does attack, pay attention to what you are wearing or carrying, especially anything different from the usual. Maybe he doesn't like your rubber boots, flip-flop sandals, or bare legs or the way the new feed bucket swings in your hand. If you are doing or wearing something different, avoid it next time and see if it makes a difference.

When a rooster gets his hackles up — that is, raises the feathers on his neck to make himself look big and fierce — he means serious business. Occasionally a cock remains mean no matter what you do. Such a rooster is big trouble. Get rid of him.

When a rooster gets his hackles up, watch out — he's threatening to attack.

Chicken Talk

Chickens use at least 30 different sounds to communicate. If you pay attention to the sounds they make as they engage in different activities, you'll soon be able to close your eyes and know what they are doing by the sound. Recognizing these sounds lets you understand your flock's mood and helps you be a better manager.

Cackle is the sound a hen makes after laying an egg. Humans tend to think she's bragging, but she's really protecting her future offspring by drawing the attention of potential predators to herself as she moves away from the nest.

Growl is the best way to describe the sound a hen makes when she's on the nest and you reach underneath her to collect eggs. She's telling you to leave the eggs alone so she can hatch them.

Peep is a chick's way of communicating with its mother, even before it breaks out of the shell. Chicks peep to keep in touch with one another and with their mother. They have many ways of peeping, depending on whether they are content, uncomfortable, afraid, lost, or sleepy.

Cluck is a sound made by a mother hen so her chicks will know where she is. This loud, clipped sound is so distinct that a hen with chicks is often called a clucker or a cluck.

Brrr is a warning sound a mother hen makes when she senses danger. It tells her chicks to hide by flattening against the ground until they hear her cluck.

Hawk is the sound the highest-ranking chicken in the peck order makes when a bird flies overhead. It causes the other chickens to run for cover, even if the "bird" turns out to be a high-flying airplane or a low-flying butterfly.

Screech is what a chicken does when it is unexpectedly grabbed to call a rooster or a senior hen to the rescue. When you hear this sound, you know someone or something (perhaps a dog) has gotten hold of one of your chickens.

Crow is what a rooster calls for lots of reasons: to establish territory, to brag after winning a fight, and to keep the flock together. When a rooster is about to crow, he flaps his wings and stretches his neck. Nearby roosters who hear the crowing may answer. Crowing must be important to chickens; if no rooster is around, sometimes an old hen will crow.

Singing is the best way to describe the melodious sound made by happy hens. Hearing this sound will make you feel happy, too.

Chickens and Other Animals

Chickens get along well with other pets and livestock. You can pasture them with sheep, goats, cows, or horses, but if the chickens have access to roosting on the manger or in the hay storage area, they will foul the hay for the other livestock.

Keeping chickens with ducks or geese is not a good idea. Waterfowl like it wet, and chickens need to stay dry to be healthy. Chickens shouldn't be kept with pigs or turkeys, either, because of the possibility of exchanging diseases.

Chickens get along with dogs and cats, provided the pets are properly trained. The best way to train a pet is to introduce it to chickens as a puppy or kitten, to which chickens look big and scary — something to stay away from. Adult cats need to be watched around baby chicks, and all dogs must be watched around chickens. Discourage your dog from playing with your chickens, as dogs tend to play rougher than chickens can withstand. Many a chicken has been killed by a family dog that didn't mean it.

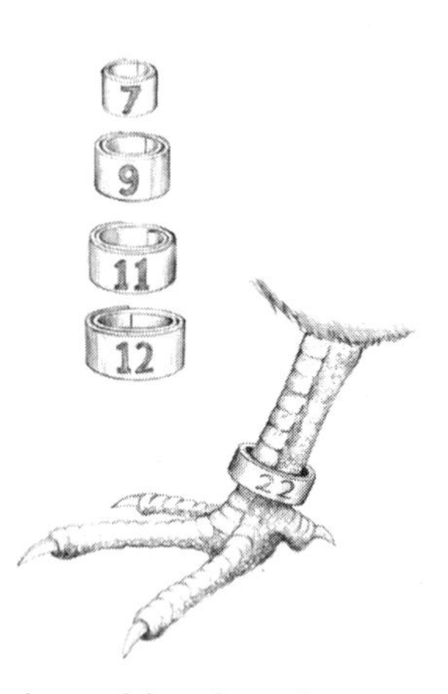

Numbered leg bands help you identify chickens in your flock.

Banding

When all the chickens in a flock look alike, it's a good idea to band them so you can tell them apart. Thin spiral leg bands may be used for whole flock identification. If, for example, you want to keep track of your chickens according to the year they were hatched, use a different color band each year.

To keep track of individual birds, such as for mating purposes, use numbered leg bands. Any chicken entered in an exhibition is required to wear a numbered band for identification.

Leg bands are sold through poultry supply catalogs and at some feed stores. Be sure to purchase the right size band for your breed:

- Number 7 (7/16") fits most bantams.
- Number 9 (9/16") fits hens of the light breeds, such as Leghorn and Ancona.
- Number 11 (11/16") fits cocks of the light breeds and most dual-purpose chickens, such as Wyandotte and Plymouth Rock.
- Number 12 (12/16") fits cocks of the heavy breeds, such as Jersey Giant and Cornish.

Catch and Carry

The easiest way to catch a chicken is to go out after dark, when the chicken is asleep, and pick it off its perch. If you need to catch a chicken during the day, you'll be happy if yours are tame enough for you to walk right over and pick them up. A chicken that isn't tame will stay just out of reach or run away from you. In that case, try to trap it in a corner, which is easier if you have help. Once you start to catch a chicken, don't give up or you'll only teach the bird to evade you in the future.

If you will have occasion to frequently catch chickens during the day, invest in a crook or a poultry net. The crook lets you snare a leg. With a net, you can snag the whole bird, if you're quick. Both devices are available from poultry suppliers.

After you have caught a chicken, hold it for a moment until it calms down. Stroke its neck and wattles to let it know you mean no harm. Cradle the chicken in one arm while you hold its legs with the other hand. When carrying more than one chicken, you can turn them upside down and use their legs as handles. Hold on to both legs or the chicken may churn around, scratching you with its claws until it gets away.

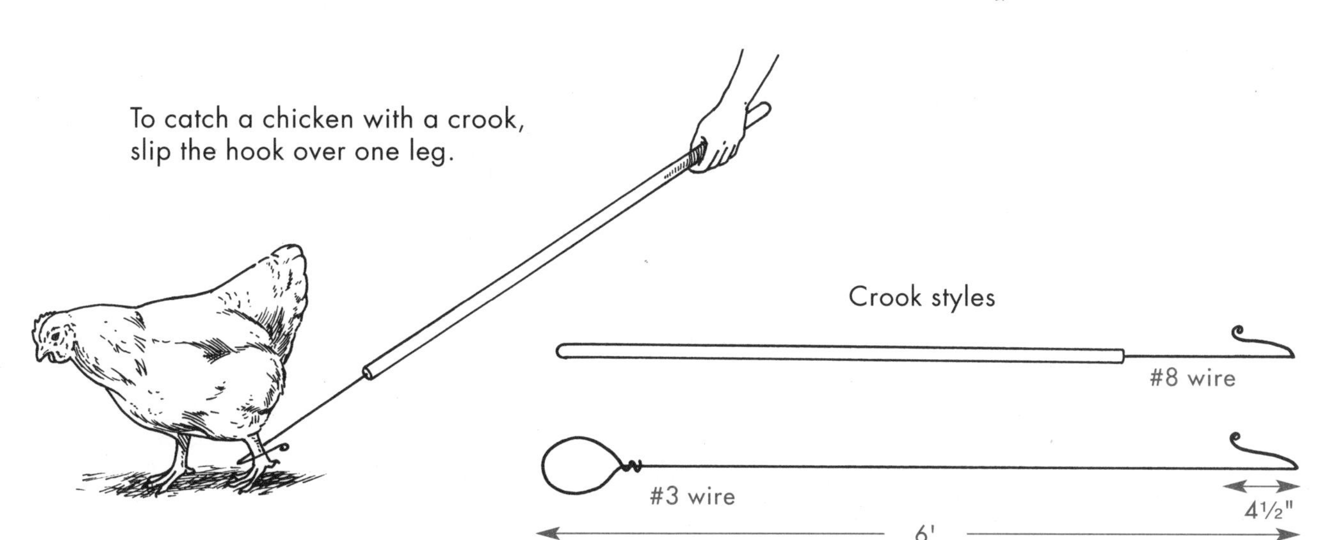

Transporting

Chickens may be safely transported in anything from a paper bag to a pet carrier, as long as the following conditions are met:

- The chicken can't get out.
- The container has no sharp edges or other injurious protrusions.
- The container is large enough for the bird to stand up and move around.
- The container is not so large that the bird can hurt itself flying about if it becomes frightened.
- The chicken gets enough air but is protected from drafts, cold, and rain. A stock rack or wire cage on the open bed of a pickup truck, for example, is not suitable for transporting chickens. A car trunk may not be suitable if it accumulates carbon monoxide fumes.
- The chicken has access to drinking water, if not in transit, at least during occasional stops.
- You don't leave the chickens in a car or truck during hot weather.

Training

All chickens should be trained to some extent. They should at least learn to remain calm when you move among them. Some people train their chickens to come when they are called and to stand calmly so they may be picked up and handled. This type of training takes a great deal of time. If you are interested in exhibiting your chickens at poultry shows, you must be willing to devote time to their training. Flighty chickens that are afraid of people and of being handled don't stand a chance of winning prizes.

Poultry shows are sponsored by fair boards and various local clubs and may be sanctioned by the American Poultry Association, the American Bantam Association, or both. The show's premium list tells you what organization, if any, has sanctioned the show, what prizes are being offered, what rules must be followed, and which classes will be shown.

Once you decide which of your chickens qualify for the show you intend to enter, they must be conditioned and trained. Conditioning involves making each bird look its best, and must be done anew before each show. Training teaches a bird to remain calm around people and to accept being handled. It may be started at any time during the bird's life, but the younger the chicken, the easier it will be to train.

Training involves handling each individual bird two or three times a day. Approach slowly, and if the chicken seems frightened, pause until it calms down. Reach across the bird's back and place one hand over its far wing, at the shoulder. Place your other hand under the breast, with one leg between your thumb and index finger and the other between your second and third fingers. Your index finger and second finger should be between the bird's legs. The bird's breastbone, or keel, will rest against your palm.

Gently lift the chicken and hold it quietly for a moment, then remove your hand from its wing. Let it sit in your hand another moment. Turn it to examine its comb and wattles. Open each wing. Initially keep the bird facing toward you; if you face it away, it may try to escape. After a few moments, gently put the bird down on its feet, pause before letting go, then slowly move away. Repeat this procedure often enough and your chickens will soon be crowding around you, eager for your attention.

Housing Chickens

Chickens are territorial. They rarely roam far during the day, and they come back at night to roost in a familiar place. In ancient times, that place would have been a tree. Even some modern chickens prefer to roost in trees, but trees give them no protection from inclement weather or from owls and other predators. For that reason, chickens should be provided with protective housing and encouraged to sleep there at night.

The type of housing you provide will be influenced to a great extent by suitable existing facilities, available space, and the amount of time you wish to devote to maintenance. Portable housing is ideal for chickens, because they are periodically moved to new ground, either in the garden or out in a pasture, but such a system works only if you are willing to take time to do the moving. An alternative is to divide the area around a stationary building into several separate yards and rotate the chickens among them to give vegetation a chance to regrow.

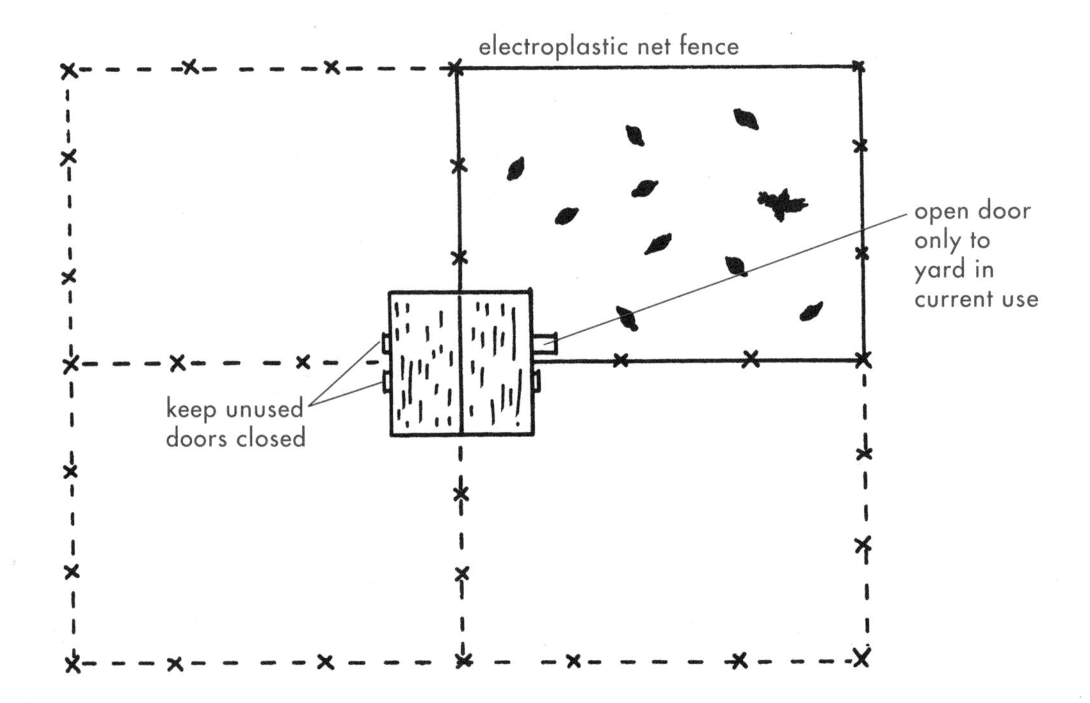

To rotate range without moving the housing, divide the yard around the coop with fencing. Install a chicken-sized door leading from the coop to each of the small lots. Keep the doors to the lots that aren't being used closed.

Because of the keeper's time and space constraints, the usual backyard setup involves a small stationary coop surrounded by a fenced yard. In short order, the chickens eat all the vegetation in the yard, which then develops a hard-packed barren surface. Provided such a yard is clean and dry, the chief disadvantage to such a situation is that the chickens have no access to forage and all their feed must be brought to them.

Coop Location and Design

The first thing to do before deciding where to house your chickens is to check your local zoning regulations. Since chickens generate a certain amount of odor, dust, and noise, laws may pertain to how far your chickens must be from your dwelling or property line. The ideal spot for a coop is on a hill or slope that offers good drainage in rainy weather.

Next, look around to see if you have a structure that might readily be converted into a chicken house, such as a playhouse the children have outgrown, an unused toolshed or other outbuilding, or a camper shell from a pickup truck.

You may prefer to buy a ready-made toolshed or to design and build your coop from scratch, making it as plain or as fancy as your heart desires. If you live in a mild climate, or if you raise an annual batch of broilers that won't be around all year, you need only a rudimentary shelter as protection from wind and rain. In a harsh climate where chickens are kept year-round, housing must be insulated and heated to keep combs and wattles from freezing.

The more room chickens have, the happier and healthier they are. Crowding leads to stress that can cause chickens to eat each other's feathers or flesh. Total room includes the coop and yard combined; the larger the yard, the less critical a large chicken house becomes. A coop measuring 8 feet by 12 feet is big enough for 30 regular-sized chickens or 50 bantams. For easy cleaning, the structure should be tall enough that you can stand inside without bumping your head.

The coop should have a door you can close and latch at night to protect your flock from roaming predators, and a few screened openings to provide ventilation when the door is shut. If you raise layers, you will need electricity so your hens have 14 hours of light during winter, when the days are short. Be sure the wiring is properly installed; hire an electrician if you are not experienced with wiring. It is not safe to run an extension cord from the house.

If you are interested in having a portable coop, these books will be helpful:

- *Chicken Tractor,* by Andy Lee and Patricia Foreman — small garden units
- *Pastured Poultry Profit$,* by Joel Salatin — small commercial units
- *Day Range Poultry,* by Andy Lee and Patricia Foreman — large commercial units

The Perch

A chicken's natural inclination is to roost in a tree at night. To satisfy this instinct and encourage chickens to roost indoors, proper chicken housing is fitted with a perch. An ideal chicken perch is about 2 inches in diameter for large breeds, 1 inch for bantams. Allow 8 inches of perching space for each chicken.

An old unused ladder makes a dandy perch. You can buy dowels from the hardware store, or use new lumber with the corners rounded off so the chickens can wrap their toes around it. Two things *not* to use as a perch are plastic and metal pipe, both of which are too smooth for chickens to grip well.

A ladder-style perch allows chickens to hop up to a high roosting spot without much trouble.

If your coop isn't big enough to have one long perch, install two or more shorter ones, spacing them 18 inches apart and at least 18 inches from the wall. Place the lowest perch 2 feet off the ground and the second one 12 inches higher than the first one. Your chickens will hop onto the first perch and use it as a step to get up to the second one, and so on until they reach the top. Attach the perches securely so they won't rotate or sag.

Nests

When a hen wants to lay an egg, she looks for a private place where she feels the egg will be safe. If you don't supply nests, your hens will lay on the ground, where their eggs will become soiled or cracked. Once chickens get a taste of a broken egg, they'll deliberately peck open fresh eggs. The bad habit of egg eating is hard to stop.

Allow one nest per four hens. You can buy ready-made nests from a store or catalog, make your own, or use found items. A covered cat litter box, for instance, makes a good nest. A reasonable nest size is 14" wide by 14" high by 12" deep.

Place nests on the ground until your pullets get used to laying in them, then firmly attach them to the wall 1½ to 2 feet off the ground. A two-tier system also works, with a lower tier for pullets and a higher tier your hens will prefer as they mature. Install a perch in front of each tier so a hen can check out a nest before entering. If one nest is occupied, she will wander down the perch until she finds a vacant nest.

Nail a 4-inch-wide board across the front of the nests to keep the eggs from rolling out and to hold in nesting material. A handy feature is to have a door at the back of the nests that lets you replace nesting material and collect eggs from the outside.

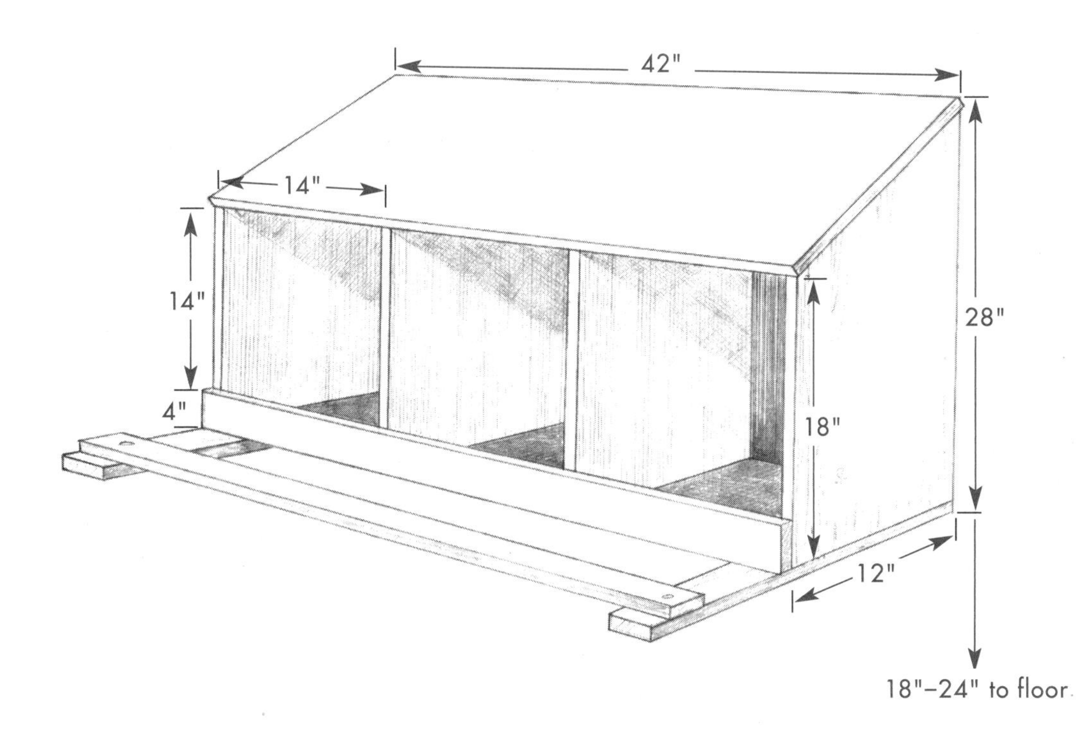

Three nests will accommodate 12 hens.

When pullets first start to lay, leaving an egg in each nest helps teach them what the nests are for. You don't have to use real eggs; use plastic or wooden ones from a poultry supply catalog or hobby shop.

Litter

Nesting material, also called bedding or litter, keeps eggs clean. Litter may consist of wood shavings, wood chips, sawdust, rice hulls, peanut hulls, chopped straw, soft hay, ground-up corncobs, shredded paper, or any other soft, absorbent material. Place 3 to 4 inches of litter in each nest.

Occasionally an egg will break, and sometimes a hen will leave her droppings in a nest, creating quite a mess. A handy way to clean things up is to keep a supply of squares cut from a cardboard box to line your nests. When you clean out a nest and remove the old cardboard, the nest floor underneath it will be clean. Put down a new piece of cardboard before adding fresh nesting material.

A thick layer of the same litter spread on the floor of your coop will help keep your chickens clean and healthy. If you don't use litter, manure will collect on the floor, particularly beneath the perches, and will smell unpleasant and attract flies. When you use litter, your chickens will scratch in it and stir in the manure.

Start with a layer of litter at least 4 inches deep. If the litter around the doorway or under the perch gets packed down, break it up with a shovel and rake. If any litter gets wet from a leaky waterer or poor drainage, remove the wet patch and replace it with dry litter. Be sure to fix the problem that caused the litter to get wet.

A chicken coop that smells like manure or has the pungent odor of ammonia is mismanaged. These problems are easily avoided by keeping the litter dry, adding fresh litter as needed to absorb droppings, and periodically replacing the old litter with a fresh batch.

A Chicken Fence

Chickens enjoy being in sunlight and fresh air. A fence keeps them from straying and protects them from dogs and other predators. The fence should be at least 4 feet high, or higher if you keep one of the lightweight breeds that tends to fly.

The best kind of fence for chickens is a wire mesh fence with small openings. Chicken wire, also called poultry netting, has 1-inch-wide openings woven in a honeycomb pattern. It makes a super chicken fence but doesn't hold up over the years.

A more durable fence is yard-and-garden wire with 1-inch spaces toward the bottom and wider spaces toward the top. The small openings at the bottom keep chickens from slipping out and small predators from getting in.

To further protect your chickens from predators, string electrified wire along the top and the outside bottom of your fence. Check first to make sure local regulations allow electric fence in your area.

If you plan to keep newly hatched chicks inside the fence, they will slip through 1-inch openings. Get a roll of 12-inch-wide aviary netting, which looks like chicken wire but has ½-inch openings. Fasten the aviary netting tightly along the bottom of your fence.

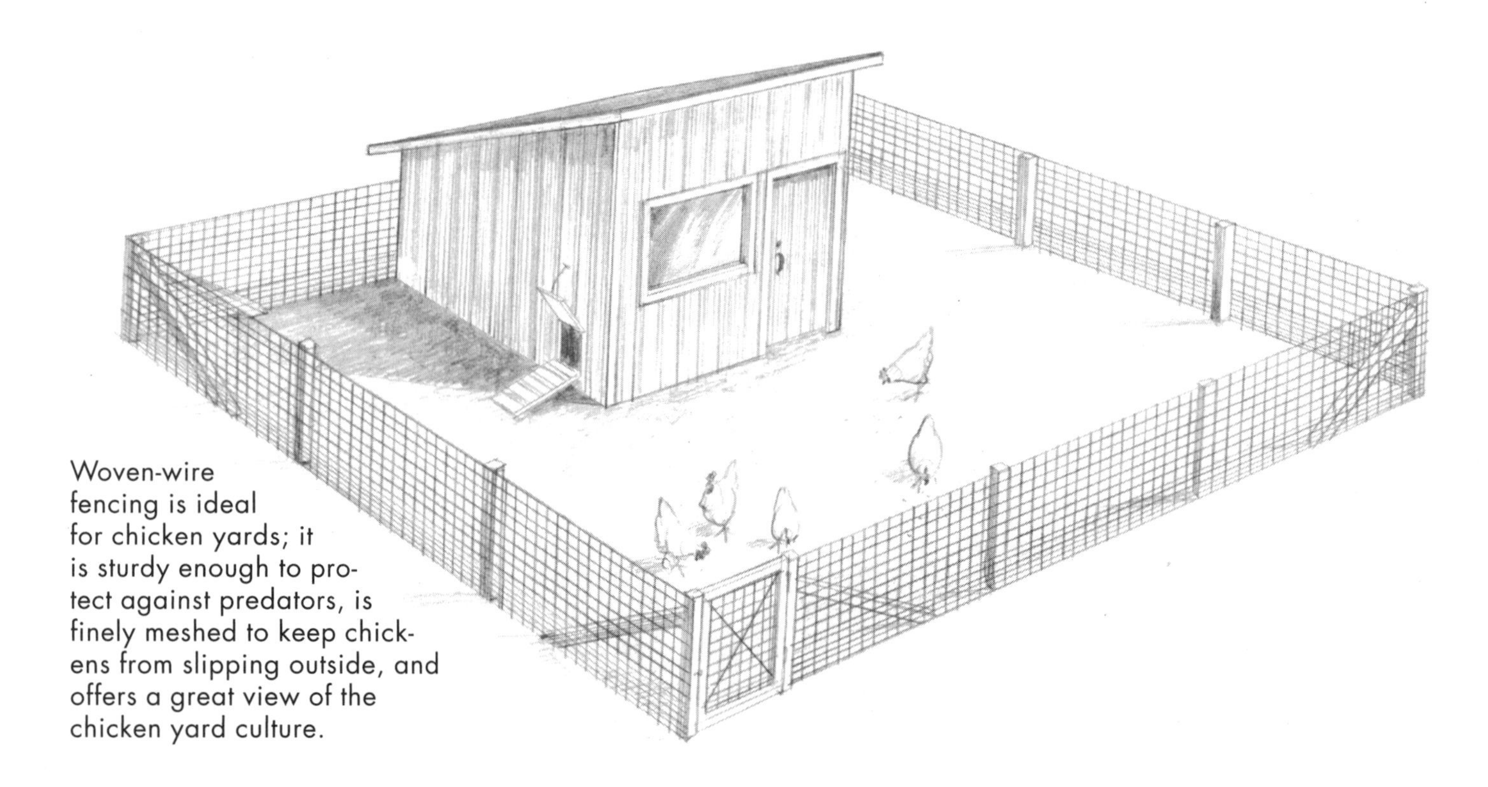

Woven-wire fencing is ideal for chicken yards; it is sturdy enough to protect against predators, is finely meshed to keep chickens from slipping outside, and offers a great view of the chicken yard culture.

Keeping House

How often a chicken coop needs to be cleaned depends on how big it is in relation to the number of chickens it houses, how well-kept the litter is, and the prevalence of disease. A properly maintained coop housing a healthy flock should not have to be cleaned more often than once a year.

The best time to clean a coop is on a warm, dry spring day. Wear a dust mask, or tie a bandanna over your nose and mouth, so you won't inhale the fine dust you'll be stirring up.

First, remove all fixtures, feeders, waterers, perches, and nests. Shovel out the used litter. With a hoe, scrape droppings from the perches, walls, and nests. Use an old broom to brush dust and cobwebs from the walls, especially in corners and cracks. If a source of electricity is nearby, this job is better handled with a shop vacuum.

Mix 2 tablespoons of chlorine bleach into 2 gallons of boiling water. Use an old broom to scrub the inside of the coop with the bleach-water. Leave the doors and windows open so the coop will dry fast. While the building is drying, scrub all the items you removed with fresh bleach-water and leave them in the sun to dry. When the structure and fixtures are dry, replace the fixtures and add fresh litter and nesting material.

While you are at it, check around outside and pick up any junk that may be lying around, including feed-bag strings and bits of broken glass that your chickens might be tempted to eat. Remove any scrap lumber or rolls of fencing that provide hiding places for rodents and snakes, and get rid of old tires and other items that can accumulate water where pesky insects might breed.

FEEDING CHICKENS

Wild jungle fowl, the ancestors of modern chickens, met their nutritional needs by consuming a variety of plants and insects. Given enough room to roam, some of today's breeds remain active foragers. In most backyard situations, though, chickens are confined to a small area and must be furnished everything they need to enjoy a balanced diet.

Chicken Feed

The easiest way to make sure your chickens eat right is to buy chicken feed or poultry ration at a farm store. Mature chickens should be fed a lay ration containing 16 percent protein plus all the other nutrients they need. Lay ration comes in crumbled or pelleted form. Less feed will be wasted if you use pellets. Since two-thirds of the cost of keeping chickens goes into feed, the less you waste, the lower your cost.

Each chicken eats about 2 pounds of ration per week. Dual-purpose hens eat a bit more, bantams a bit less. All chickens eat less in summer than in winter, when they need extra energy to stay warm.

Lay ration comes in 25- or 50-pound sacks. It's a good idea to keep a little extra on hand so you won't run out, but don't stock up too far ahead. Chicken feed goes stale, especially in warm weather, causing the nutritional value to decrease. Buy only as much as your flock will consume within 2 or 3 weeks.

Chickens love cracked corn or a mixture of grains called scratch, which should be viewed as chicken candy. You can tame a flock by throwing down a handful of scratch whenever you visit your chickens; pretty soon, they will learn to come to you. In cold weather, a little scratch fed late in the day helps keep chickens warm overnight. Too much scratch makes chickens fat, creating lazy layers.

After opening a sack of scratch or lay ration, store the feed in a plastic trash container with a tight-fitting lid. A 10-gallon container holds 50 pounds of feed. The closed container helps prevent feed from getting stale and keeps out mice. Place the container in a cool, dry area, out of the sun. Use up all the feed in the container before opening another sack.

If you buy more feed at a time than will fit into the container, store unopened bags away from moisture and off the floor or ground. Wooden or plastic pallets are ideal for this purpose. Avoid attracting rodents by keeping your feed storage area swept clean. To prevent mice from nibbling holes in your feed sacks and inviting their cousins to dinner, reduce the rodent population by frequently setting traps.

Reducing Feed Costs

You can reduce the cost of buying feed by treating your chickens to leftover table scraps and garden produce. Chickens love tomatoes, lettuce, apple parings, bits of toast, and other tasty treats from the kitchen. Take care, though, to avoid strong-tasting foods such as onions, garlic, or fish, which will give an off flavor to eggs and meat. Don't feed chickens anything that has spoiled or rotted. And avoid feeding raw potato peelings, which are not digestible. If you wish to feed potato peelings, cook them first. Do not rely on table scraps as your flock's sole source of food.

Another way to reduce the cost of feed is to let your chickens roam on a lawn or in a pasture for part of the day. By eating plants, seeds, and insects, they will balance their diet and eat less of the expensive commercial stuff. Take care not to put your chickens on grass or around buildings that have been sprayed with toxins.

How much you save on commercial rations by allowing your chickens to forage depends on the quantity and quality of the forage, which varies seasonally. Some layer flocks can survive entirely on forage. Broilers, on the other hand, would grow much more slowly and in the long run may end up costing more per pound.

Supplements

Eggshells are made of calcium. To lay eggs with hard shells, a hen needs adequate calcium in her diet. Lay ration contains enough calcium for pullets, and hens that forage freely find plenty of calcium in the form of hard-shelled insects. Older hens that do not forage need supplemental calcium to keep them from laying thin-shelled eggs that break easily.

Farm stores carry supplemental calcium in the form of ground oyster shell or limestone (but *do not* feed dolomitic limestone, which can harm egg production). Place the calcium supplement in a hopper where your hens can eat as much as they want. Fill a second hopper with granite grit, which chickens need to eat because they have no teeth. Everything the chicken eats passes through the gizzard; the grit lodges in the gizzard and grinds feed up finely enough to be digestible. Foraging chickens may find sufficient pebbles and sand to use as grit, but you can't count on their finding enough.

Granite grit is sometimes called insoluble grit, meaning it wears down over time rather than dissolving. Calcium supplements, by contrast, are sometimes called soluble grit, because the particles are hard and may temporarily serve the same purpose as grit until they dissolve. This distinction may be important in making sure you get the right thing when you ask for grit at the farm store.

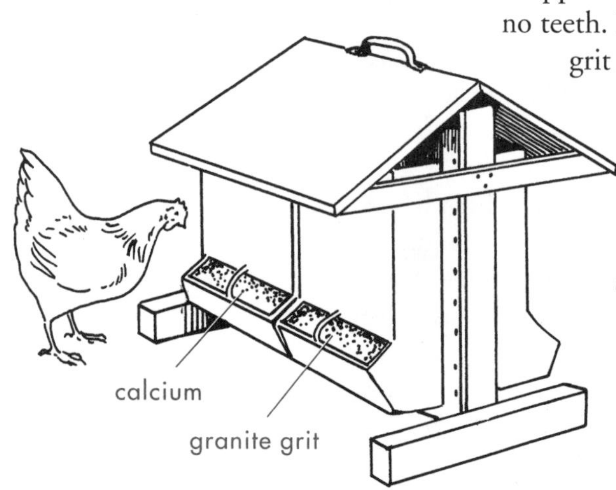

Two double-sided hoppers, one for calcium and one for granite grit, hang side by side.

Feeders

Eating is chickens' main activity. They like to peck a little here and a little there, eating all day long, which works fine for chickens that can forage. If chickens that have all their feed brought to them are allowed to go hungry, they soon start pecking on one another, which leads to trouble. Confined chickens should be fed free choice so they can eat whenever they want to. Free-choice feeding requires a large enough feeder to ensure that your flock won't run out of rations before the next feeding.

Feeders come in many designs, but good ones share important features:

- They are designed to discourage chickens from sitting on top and messing in the feed.
- They are designed to prevent billing out, a habit chickens have of using their beaks to scoop feed onto the ground, where it is wasted. Billing out is discouraged by a feeder with edges that roll or bend inward.
- They are easy to clean.
- Their height may be adjusted to suit the chickens' size. A hanging feeder works well because the chain from which it hangs may be shortened as the chickens grow. The ideal is to place feed at the height of the chickens' backs.

A typical hanging feeder consists of a bucket-shaped container, open at both ends, with a shallow dish attached at the bottom and a handle at the top from which the feeder is hung. Feed is poured in at the top, and as the chickens eat from around the dish, gravity causes more feed to replace that which has been eaten. Not all feeders have lids, but they should, to keep the droppings from chickens that roost on the feeder from getting into the feed. If your feeder lacks a lid, fashion one from the lid of a 5-gallon bucket by notching out two sides to fit under the handle.

As a rule of thumb, one hanging feeder is enough for up to 30 chickens. If you use a trough feeder, allow 4 inches in trough length for each bird; if chickens can eat from both sides of the trough, allow 2 inches for each bird.

Another rule of thumb is to furnish enough feeder space so all your chickens can eat at the same time. Otherwise, the lowest-ranking chickens in the peck order will get pushed aside and go hungry. You can't go wrong by providing more than the minimum feeder space, but you *can* go wrong by providing too little.

If your flock includes more than one cock, provide one feeder per cock and place the feeders at least 10 feet apart. Each cock will claim a feeder and entice some of the hens to join him, reducing competition between the males and thus reducing fights.

Avoid Moldy Rations

Wet feed soon goes moldy and can cause illness. If feed gets wet, dispose of it and scrub out the feeder before filling it with fresh ration.

This trough has adjustable legs to increase or decrease the trough's height as well as an antiroosting reel that will spin and dump any bird that tries to hop up on top of it.

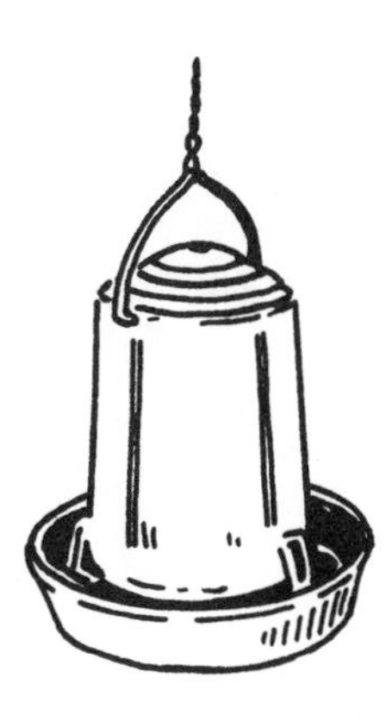

This hanging tube-type feeder has raised sides at the bottom to prevent billing out.

Water

The most important part of your flock's diet is water. Chickens must be able to drink whenever they desire. A chicken can't drink much at once, so it has to drink often. Depending on the weather and on the chicken's size, each bird will drink 1 to 2 cups of water per day.

When the temperature gets above 80°F, chickens may drink two to four times more than usual. Provide fresh cool water several times a day to encourage drinking.

In cold weather, chickens will drink less than normal, but they still need water. To make sure they drink enough when the weather is cold enough to freeze their water, provide warm water several times a day. If your coop has electricity, use a water-warming device from the farm store to keep the drinking water from freezing.

Like feeders, waterers come in many designs. The best waterer holds enough that your flock doesn't run out before you have a chance to supply more. It keeps the water clean by not allowing chickens to step in it or to roost over it.

The worst kind of waterer is a rain puddle. Chickens will walk in it, mess in it, drink from it, and get sick. Even if you provide plenty of clean drinking water, your chickens will drink from puddles if they can. Avoid puddles by filling areas of standing water with dirt, sand, or gravel. Since chickens love to dig dust holes in the yard, and dust holes become puddles in wet weather, filling puddles in the chicken yard is a never-ending job.

Automatic waterers are wonderful. Every time a chicken takes a sip, fresh clean water flows in. The water is always clean, and the waterer is always full. All you have to do is check every day to make sure nothing has clogged the waterer and clean out the water bowl or trough once a week. The farther your waterers are from the feeders, the less feed your chickens will get in their water and the less often you will have to clean the waterers. Farm stores and poultry supply catalogs carry automatic waterers. They are not expensive, but piping water to your chicken coop may be.

The least expensive kind of waterer is made of plastic and holds 1 gallon. Fill the container, screw on the base, and turn the waterer over. Each time a chicken drinks, water runs out of the container through a little hole in the base. Plastic waterers don't

Metal waterer: Fill the inner container with water, then slide the outer shell over it.

Plastic waterer: Fill the container with water, screw on the base, and flip it over.

hold up well, especially when left outside in the sun or in freezing weather, so be prepared to replace them often.

A bit more expensive is a metal waterer consisting of an inner container and an outer shell. The inner container has a basin at the bottom. When you fill the inner container and slip on the outer shell, the shell presses against a clip that lets water flow into the basin. This style comes in a 3-gallon size and a 5-gallon size. They eventually rust out, but sometimes you can keep one going a little longer by patching holes with epoxy.

In the summer, locate waterers in the shade where the water won't be warmed by the sun. In winter, put water in the sun to help prevent freezing. If your flock includes more than one cock, provide one waterer per cock, just as with feeders, creating a separate territory for each male.

Provide drainage beneath each waterer to prevent standing puddles in the event of spillage. A bed of sand or gravel beneath the waterer will improve drainage. Make a small platform by nailing together a wooden frame, 42 inches wide by 42 inches long by 12 inches high. Staple strong wire mesh to the top. Set the platform on the sand or gravel, and place the waterer on top of the platform. Any spills will fall through the wire, to be absorbed into the sand or gravel. Your chickens will stay healthier because they won't be able to drink dirty water from the ground.

Egg Production

A pullet starts laying when she is 20 to 24 weeks old. Her first eggs are quite small, and she will lay only one egg every 3 or 4 days. By the time she is 30 weeks old, her eggs will be normal in size and she will lay about two eggs every 3 days.

When a pullet is born, she carries in her body as many as 4,000 ova, or undeveloped yolks. When the pullet reaches laying age, one by one the ova grow into full-size yolks and drop into a 2-foot-long tube called the oviduct.

As a yolk travels through the oviduct, it becomes surrounded by egg white and encased in a shell. About 24 hours after it started its journey, it is a complete egg ready to be laid.

A hen cannot lay more eggs than the total number of ova inside her body. From the day she enters this world, each female chick carries with her the beginnings of all the eggs she can possibly lay during her lifetime. Few hens, however, live long enough to lay more than 1,000 of the possible 4,000 they start with.

The Life of a Layer

A good laying hen produces about 20 dozen eggs in her first year. At 18 months of age, she stops laying and goes into a molt, during which her old feathers gradually fall out and are replaced with new ones. Chickens molt once a year, usually in the fall, and the process generally takes 2 to 3 months. Because a hen needs all her energy to grow replacement feathers during the molt, she lays few eggs or none at all. Once her new feathers are in, she looks sleek and shiny, and she begins laying again.

After her first molt, a hen lays larger but fewer eggs. During her second year, she will lay 16 to 18 dozen eggs. Some hens may lay more, others fewer. Exactly how many eggs a hen lays depends on many factors, including breed and strain, how well the flock is managed, and the weather. Hens lay best when the temperature is between 45 and 80°F. When the weather is much colder or much warmer, hens lay fewer eggs than usual. In warm weather, hens lay smaller eggs with thinner shells.

All hens stop laying in winter, not because the weather is cold but because winter days have fewer hours than summer days. When the number of daylight hours falls below 14, hens stop laying. If your henhouse is wired for electricity, you can keep your hens laying year-round by installing a 60-watt lightbulb. Use the light in combination with daylight hours to provide at least 14 hours of light each day.

You can leave the light on all the time, or save electricity by plugging the light into a timer switch from an electrical supply or hardware store. If you use a timer, remember to adjust it occasionally as daylight hours change, as well as anytime the power goes out and throws off your lighting schedule.

Layers versus Lazy Hens

You can improve your flock's overall laying average by culling and slaughtering the lazy layers. The hens you cull can be used for stewing or making chicken soup. When your flock reaches peak production at about 30 weeks of age, you can easily tell by looking at your hens and by handling them which ones are candidates for culling:

- Look at their combs and wattles. Lazy layers have smaller combs and wattles than good layers.
- Pick up each hen and look at her vent. A good layer has a large, moist vent. A lazy layer has a tight, dry vent.
- Place your hand on the hen's abdomen. It should feel round, soft, and pliable, not small and hard.
- With your fingers, find the hen's two pubic bones, which are located between her keel (breastbone) and her vent. In a good layer, you can easily press two or three fingers between the pubic bones and three fingers between the keel and the pubic bones. If the pubic bones are close and tight, the hen is not a good layer.

The four-point examination of a good laying hen.

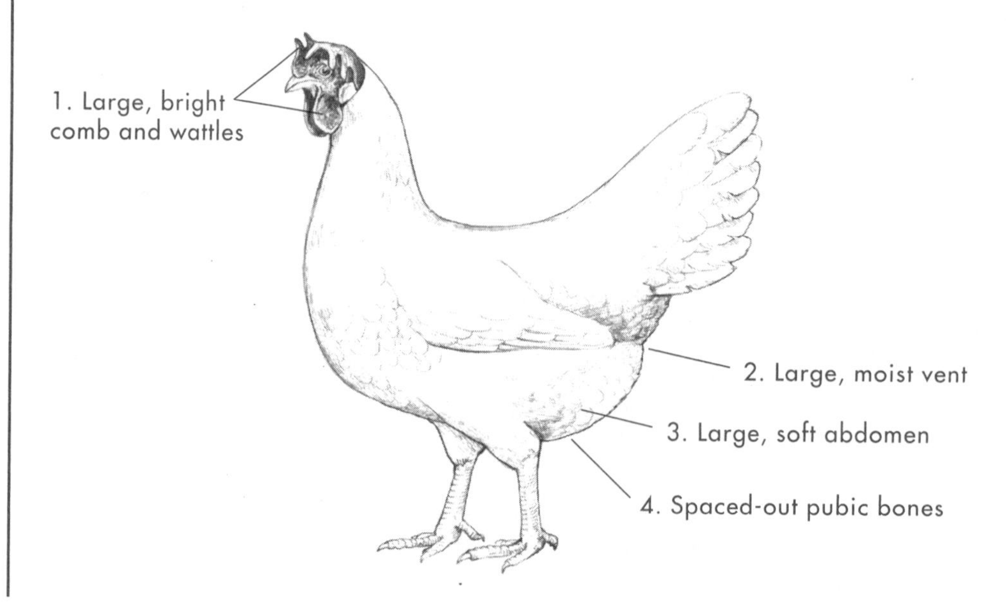

The Bleaching Sequence

If you raise a yellow-skinned breed, you can sort out the less productive hens by the color of their skin after they have been laying awhile. The same pigment that makes egg yolks yellow colors the skin of yellow-skinned breeds. When a hen starts laying, the skin of her various body parts bleaches out in a certain order. When she stops laying, the color returns in reverse order. You can therefore tell how long a yellow-skinned hen has been laying, or how long ago she stopped laying, by the color of the exposed skin on her beak and legs.

Bleaching Sequence

Body Part	Number of Eggs Required to Bleach	Approximate Weeks to Lay That Many Eggs
Vent	0–10	1–2
Eye ring	8–12	2–2½
Earlobe	8–10	2½–3
Beak	35	5–8
Bottom of feet	50–60	8
Front of shank	90–100	10

Replacement Pullets

A hen lays best during her first year. As she gets older, she lays fewer and fewer eggs. If you raise chickens primarily for eggs, you have the same concern as commercial producers — a time will come when the cost of feeding your hens is greater than the value of the eggs they lay. For this reason, commercial producers rarely keep hens more than 2 years.

To keep those eggs rolling in, buy or hatch a batch of chicks every year or two. As soon as the replacement pullets start laying, get rid of the hens. If you replace your hens every year, you might sell your old flock, which will still lay fairly well for at least another year. If you replace your hens every two years or more, you can sell or use them yourself as stewing hens.

Collecting and Storing Eggs

An egg is at its best quality the moment it is laid, after which its quality gradually declines. Properly collecting and storing eggs slows that decline.

Collect eggs often so they won't get dirty or cracked. Pullets sometimes lay their first few eggs on the floor. A floor egg is usually soiled and sometimes gets trampled and cracked. Well-managed pullets soon figure out what the nests are for. If they continue laying on the floor, perhaps not enough nests are available for the number of pullets in the flock.

Egg Aging

An egg stored at room temperature ages more in 1 day than an egg stored in the refrigerator ages in 1 week.

Eggs also get dirty when a hen with soiled or muddy feet enters the nest. Eggs crack when two hens try to lay in the same nest or when a hen accidentally kicks an egg as she leaves the nest. The more often you collect eggs, the less chance they will get dirty or cracked.

Frequent collection keeps eggs from starting to spoil in warm weather and from freezing in cold weather. Try to collect eggs at least twice a day. Since most eggs are laid in the morning, around noon is a good time for your first collection.

Discard eggs with dirty or cracked shells, which may contain harmful bacteria. If you are going to sell or hatch your eggs, sort out any that are larger or smaller than the rest or have weird shapes or wrinkled shells. You can keep the oddballs for your own culinary use. (And by the way, it's normal to occasionally find an extremely small egg with no yolk or an extra-large egg with more than one yolk.)

Store eggs in clean cartons, large end up so the yolk remains centered within the white. Where you store your eggs depends on whether they will be used for eating or hatching. If you plan to hatch them, store them in a cool, dry place, but not in the refrigerator. If you plan to eat them or sell them for eating, store them in the refrigerator as soon as possible after they are laid.

The egg rack on a refrigerator door is not a good place to store eggs. Every time you open the refrigerator, eggs on the door get blasted with warm air. When you shut the door, the eggs get jarred. The best place to keep eggs is on the lowest shelf of the refrigerator, where the temperature is coldest. Raw eggs in a carton on the lowest shelf keep well for 4 weeks.

Determining Freshness

Sometimes you'll find eggs in a place you haven't looked before, so you can't tell how long they've been there. One way to determine whether an egg is fresh is to put it in cold water. A fresh egg sinks, because it contains little air. As time goes by, moisture evaporates through the shell, creating an air space at the large end of the egg. The older the egg, the larger the air space. If the air space is big enough to make the egg float, the egg is too old to eat.

Nutritional Value

Eggs have been called the perfect food. One egg contains almost all the nutrients necessary for life. The only essential nutrient it lacks is vitamin C. Most of an egg's fat, and all the cholesterol, is in the yolk. To reduce the cholesterol in an egg recipe, such as scrambled eggs or omelettes, use 2 egg whites instead of 1 whole egg for half the eggs in the recipe. If the recipe calls for 4 eggs, for example, use 2 whole eggs plus 4 egg whites.

To eliminate cholesterol in a recipe for cakes, cookies, or muffins, substitute 2 egg whites and 1 teaspoon of vegetable oil for each whole egg in the recipe. In a recipe calling for 2 eggs, for example, use 2 egg whites plus 2 teaspoons of vegetable oil. If the recipe already has oil in it, you may omit the extra 2 teaspoons.

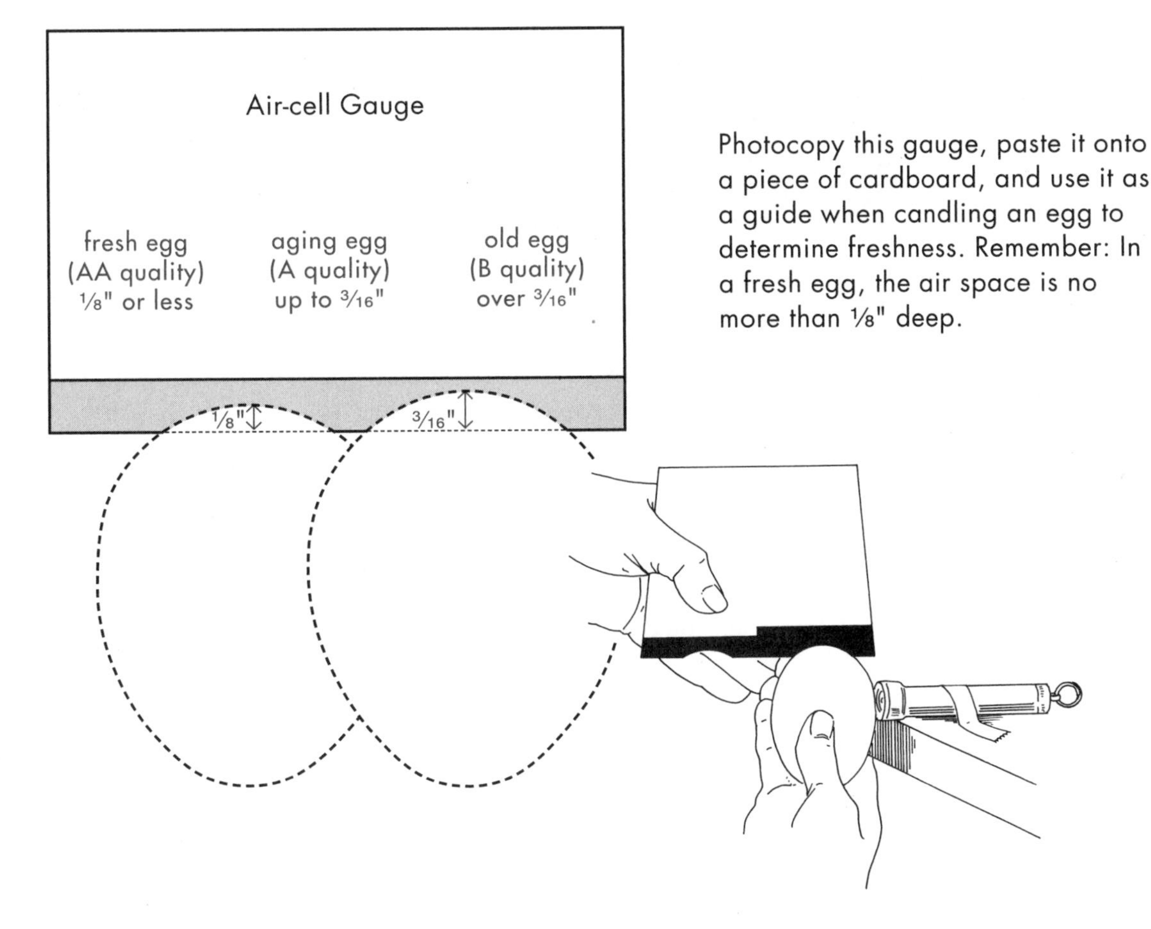

Photocopy this gauge, paste it onto a piece of cardboard, and use it as a guide when candling an egg to determine freshness. Remember: In a fresh egg, the air space is no more than 1/8" deep.

Another way to tell if an egg is fresh is to use a light to examine the air space, the yolk, and the white. This examination is called candling, since it was once done by using candles. These days an electric light is used, but the process is still called candling. A good penlight is ideal for candling.

Candle eggs in a dark room. Grasp each egg by its small end and hold it at a slant, large end against the light, so you can see its contents through the shell. In a fresh egg, the air space is no more than 1/8" deep. The yolk is a barely visible shadow that hardly moves when you give the egg a quick twist. In an old or stale egg, the air space is large and sometimes irregular in shape, and the yolk is a plainly visible shadow that moves freely when you give the egg a twist.

Occasionally, you may see a small, dark spot near the yolk or floating in the white. This spot is a bit of blood or flesh that got into the egg while it was being formed. Even though blood spots and meat spots are harmless, when you sort eggs for hatching or for sale, eliminate those with spots. Customers don't find them appetizing, and eggs with spots may hatch into pullets that lay eggs with spots.

If you aren't sure what you are seeing when you candle an egg, break the egg into a dish and examine it. Soon you will be able to correlate what you see through the shell with what you see in the dish.

When you break a fresh egg into a dish, the white is compact and firmly holds up the yolk. In an aging egg, the white is runny and the yolk flattens out. Try comparing one of your homegrown eggs with a store-bought egg — it's easy to tell which is fresher.

Shell Color

A hen's eggs have a specific shell color. All her eggs might be white, light brown, dark brown, speckled, blue, or green. The color of the shell has nothing to do with the nutritional value of the contents.

Hens of the Mediterranean breeds lay white-shelled eggs. Since Mediterraneans are the most efficient layers, they are preferred by commercial egg producers. Many consumers prefer eggs with white shells because that's what they're used to seeing.

Brown-shelled eggs are laid by American breeds. Since the Americans are dual purpose, they're popular in backyard flocks. Some consumers prefer brown eggs because they look homegrown. Brown eggs come in every shade from a dark reddish color to a light tan that appears almost pink.

Blue-shelled eggs are laid by a South American breed called Araucana and its relative, the Ameraucana. Because the eggs are so pretty, these breeds are sometimes called "Easter-egg chickens."

Green-shelled eggs are laid by hens bred from a cross between Araucana and a brown-egg breed. Unscrupulous sellers charge outrageous prices for blue or green eggs by falsely claiming that they're lower in cholesterol than white or brown eggs.

Hard-Cooked Eggs

Hard-cooked eggs are often called "hard-boiled," even though they shouldn't be boiled at all. This is the one dish for which you don't want to use fresh eggs. The contents of a fresh egg fill up the shell, making the shell difficult to peel. After some of the moisture evaporates from an egg, the contents shrink away from the shell. An egg that's 1 week old before being hard-cooked will be easier to peel.

The proper way to hard-cook eggs is to place cold eggs in a saucepan, cover them with cold water, bring the water to a boil, and turn off the heat. Cover the pan and leave the eggs in the hot water for 10 minutes. Remove the eggs from the hot water after 10 minutes and cool them quickly in ice water or under cold running water, to keep the yolks from turning green. A green yolk is safe to eat, but it doesn't look appetizing. Cooked eggs keep well in the refrigerator for 1 week.

Managing Breeders

If you're happy with hybrid layers or buying an occasional batch of broilers to raise, you don't need a breeder flock. But if you wish to hatch eggs and raise chicks, you'll need to maintain a breeder flock. Flock owners who hatch eggs include homesteaders who enjoy producing their own meat and eggs, exhibition breeders, and preservationists who work with endangered breeds.

Creating a Breeding Plan

People who collect and hatch eggs may be divided into two camps: breeders who emphasize quality and what you might call propagators or multipliers, who emphasize quantity. Both groups hatch a lot of chicks. For the propagator, getting lots of chicks is the goal. To the breeder, a large number of chicks is merely the means to an end. The more chicks you have, the more heavily you can cull; the more heavily you cull, the better the quality of your flock. The only way to improve your flock and maintain its quality is through a well-thought-out breeding plan that includes the following considerations:

- Begin with the best chickens you can find.
- Establish a long-range goal.
- Cull ruthlessly to meet that goal.

To have consistent results in your chicks year after year, your breeding flock must include only chickens of a single breed. Your results will be inconsistent if you hatch eggs from a flock containing different breeds or commercially crossbred chickens. The birds in your breeding flock should be healthy and free of deformities. They should be reasonably true to type, meaning that each bird is of the correct size, shape, and color for its breed. Cull any chicken from your breeding flock that doesn't measure up.

Culling

The quality of a small flock can degenerate rapidly if you make no effort to select in favor of health, vigor, hardiness, and good reproduction. In pursuit of your breeding goals, keep only your best offspring and get rid of the rest, even though "the rest" may represent a large percentage of each hatch.

How severely you need to cull will be influenced by the quality of your foundation flock. Unfortunately, you can't do much to ensure you're getting the stock you want, except to seek out someone who specializes in the strain you're interested in, has worked with it for a long time, and freely offers details about its background. Once you acquire your foundation stock, developing the birds to meet your goals is a matter of mating and culling.

Cull against birds that develop slowly, aren't energetic, or might otherwise be described as unthrifty. Cull in favor of birds that show some improvement over their parents. In addition to visually inspecting the birds, manually examine each one for skeletal irregularities or deformities. Any bird that doesn't measure up should go into the frying pan, the freezer, or the compost pile.

If you're raising exhibition birds, select breeders according to guidelines in the *Standard.* The best age to cull cocks for show is 8 months. Pullets are at their best when they're ready to lay their first eggs, at 5½ to 7 months. Cull a laying flock for quality and quantity of eggs and the size of the hens. Choose meat birds with short shanks and compact bodies.

In all cases, cull in favor of good temperament; it's no fun to raise chickens that are wild or downright mean. Although you'll hear all manner of advice on how to cure meanness, the only sure cure is to get rid of aggressive birds and breed for good disposition.

After the first generation, start culling problem breeders so your flock will include not only good birds but also those that transmit their good qualities to their offspring. Pay the same attention to cocks as to hens.

Optimizing Fertility

Although an egg must be fertilized in order to hatch, not all fertilized eggs hatch. Eggs fail to hatch for many reasons, not all of which are easy to determine. One possible reason is that the embryo died before incubation began. Commonly known as weak fertility, this event may have nothing to do with fertility at all but may be due to deficiencies within the egg.

Low fertility may result from inbreeding, which creates uniformity of size, color, and type but also brings out weaknesses, such as reduced rate of laying, low fertility, poor hatchability, and slow growth. Inbreeding doesn't cause these problems, but it concentrates them the same way it concentrates favorable traits.

More likely to play a significant role in low fertility are management factors, such as:

- The flock is too closely confined.
- The weather is too warm.
- The flock gets fewer than 14 light hours daily.
- The cock has an injured foot or leg.
- Breeders are infested with internal or external parasites.
- Breeders are too young or too old.
- Breeders are stressed.
- Breeders are undernourished.
- Breeders are diseased.
- The mating ratio is too high or too low.

Mating Ratio

The ideal cock-to-hen ratio is influenced by the cock's condition, health, age, and breed. On average, the optimum ratio for heavier breeds is 1 cock per 8 hens, although a cock in peak form can handle up to 12 hens. The optimum ratio for lightweight laying breeds is 1 cock for up to 12 hens, yet an agile cock may accommodate 15 to 20. The mating ratio for bantams is 1 cock for 18 hens, although an active cockerel might handle as many as 25. An older cock or an immature cockerel can manage only half the hens than can a virile yearling.

Too many cocks cause low fertility because they spend too much time fighting among themselves. Too few cocks cause low fertility because they can't get around to all the hens. If you have one cock and more than half a dozen hens, chances are he will favor some hens and ignore others. Because cocks play favorites, it pays to periodically switch them. The more cocks you have for switching, the less likely you are to experience inbreeding problems; it's far better to have chicks sired by several cocks than hatch the same number of chicks from one sire. In addition, by keeping extra cocks, you avoid the risk of losing your rooster and being without one.

If your flock is small enough for only one cock at a time, house the extras in separate pens or cages. Ironically, cocks are less likely to fight if you keep three or more together instead of just two. If you use several cocks at a time, rotate them as groups, rather than individuals, to avoid disturbing their peck order. If you lose one cock out of a group, it's better to risk a possible slight drop in fertility than have a drastic drop due to peck-order fighting caused by bringing in a replacement.

Flock Age

Maximum fertility and hatchability generally can be obtained from mature cockerels and pullets. Most cockerels reach maturity around 6 months of age. Early-maturing breeds may be ready to mate sooner, whereas late-maturing breeds may not be ready until 7 to 8 months of age. Comb development is the best indication of a cock's maturity.

You can start collecting hatching eggs when pullets are about 7 months old and have been laying for at least 6 weeks. Eggs laid earlier tend to be low in fertility, and those that hatch are likely to produce deformed chicks. As time goes by, fertility and hatchability improve and level out by the sixth week of laying.

After about 6 months of laying, fertility and hatchability begin to decline, gradually among bantams and more rapidly among the heavier breeds than among lighter breeds. Industrial broiler breeders are kept for 10 months or less, compared with 12 months or more for layer breeders.

Whether it's better to hatch eggs from hens that are more than or less than 2 years of age depends on your goal. If it is to improve the health and vigor of future generations, hatch from older hens. Two-year-old hens that are laying well must be relatively disease resistant and are likely to pass that resistance to their offspring. On the other hand, a slight but significant decline occurs in hatchability after a hen's first year and continues as the hen ages. After the second year, you'll see a greater percentage of early embryo deaths and failure of full-term embryos to hatch.

If you do breed old birds, take special care to avoid stress in the breeder flock. Chickens may become stressed by harassment from animals or people, changes in the feed or feeding schedule, and a shortage of or change in the taste of their drinking water. Hens are more easily stressed than cocks, and older birds of either sex are more strongly affected than younger ones.

Optimal Number of Hens per Cock

Breed	Optimum	Max.
Bantam	18	25
Light breeds	12	20
Heavy breeds	8	12

Breeder Housing

Breeder flock housing plays an important role in the fertility and hatchability of eggs. Facilities with environmental variety offer a cock privacy to mate undisturbed by other cocks. Excessive fighting among cocks may be a sign of poor facility design, since cocks that are lower in the peck order have no place to hide from dominant birds.

Floor space for the breeder flock should offer a minimum of 3 square feet per bird for large breeds, 2 square feet for smaller breeds, and 1½ square feet per bantam. Include one nest for every four hens, and frequently change nesting material so eggs won't get soiled or cracked.

Housing should protect the flock from extremes of climate, since sudden changes can cause a decrease in laying or fertility. During the early part of the season (late winter), provide 14 hours of continuous light to stimulate both egg and semen production.

Feeding Breeders

Assuming your breeder flock is healthy and free of both internal and external parasites, good nutrition is the most important factor in promoting fertility and hatchability. Poor nutrition may arise because rations are poorly balanced, insufficient, or nutritionally deficient.

The main cause of poorly balanced rations is feeding breeders too much scratch grain or other treats. To ensure the proper protein/carbohydrate balance, reduce your flock's grain ration about 1 month before the hatching season begins. Unbalanced rations, particularly those with excessive protein, can lead to gout, a condition in which uric acid crystals are deposited in joints or on interal organs.

See that your breeders get enough to eat for their size and level of activity, and for the time of year. Feed either free choice or often, so those lowest in peck order get their turn at the feed hopper. Examine each bag of feed to make sure it isn't dusty, moldy, or otherwise unpalatable, causing your chickens to eat less.

Periodically weigh a sample of cocks and hens. Weight loss of more than 10 percent can affect reproduction. Underfed cocks produce less semen, and underfed hens don't lay well.

A hen's diet affects the number and vitality of her chicks and the quality of carry-over nutrients they continue to absorb for several weeks after the hatch. Nutritional deficiencies that may not produce symptoms in a hen can be passed on to her chicks. What you feed the breeder hen therefore also affects her chicks.

The same ration that promotes good egg production doesn't necessarily provide embryos and newly hatched chicks with all the elements they need to thrive. Lay ration contains too little protein, vitamins, and minerals for the proper composition of hatching eggs and high hatchability. Feeding lay ration to a breeder hen can result in a poor hatch or in nutritional deficiencies in the offspring. The older the hen, the worse the problem becomes.

To improve your hatching success, feed your flock a breeder ration, starting 2 to 4 weeks before you start collecting eggs for hatching. If you're lucky, you'll find breeder ration at your local farm store. Be sure it's fresh, and use it within 2 weeks of the time it was mixed — even if you feed your flock the best breeder ration in the world, nutritional deficiencies will result if the feed is stored so long that fat-soluble vitamins are destroyed by oxidation.

In areas where breeder ration is not available, the closest alternative may be game bird ration. If you can't find either ration, 6 weeks before you begin collecting hatching eggs, toss your flock a handful of dry cat kibble daily, and add a vitamin/mineral supplement to its drinking water.

The problem of ensuring adequate breeder flock nutrition is compounded if you hatch early in the season, before your hens have access to fresh greens. If you live in a northern area, you can measurably improve your hatching success by delaying incubation until your breeder flock has access to plenty of forage and sunlight.

Hatching Eggs

The best months to hatch in most climates are February and March. If you live in a very cold climate, March and April are the best months. A breeding flock and its chicks are strongest and healthiest in spring.

Pullets hatched in spring will lay by fall and continue laying for 1 year. Pullets hatched in winter will lay by midsummer but may molt and stop laying in the fall. Chicks hatched in summer are not yet strong enough to ward off disease-causing organisms that flourish in warm weather. For best results, hatch in spring.

You can hatch eggs in two ways: Let a hen hatch them for you or hatch them in a mechanical device.

Natural Incubation

Letting a hen handle the hatching is called natural incubation. The hen doing the hatching is called a setting hen or a broody hen. The eggs in the hen's nest are called a setting or a clutch. The chicks that hatch are also sometimes called a clutch, but more often are called the hen's brood.

Since a hen stops laying when she starts setting, breeds developed primarily for their laying ability have been selectively bred against the instinct to brood. Luckily, the hens of some breeds are still good setters. The breeds most likely to brood are

Cochin, Orpington, Old English Game, Plymouth Rock, Rhode Island Red, Wyandotte, and Silkie.

When a hen is beginning to set, she stays in the nest long after she has laid her eggs. If you touch her or try to take her eggs away, she puffs up and growls and may peck your hand. If her eggs are left in the nest, within a few days she will settle down to serious business.

A setting hen typically gets off the nest for a few minutes each day to grab a bite to eat. While she is out, another hen may come along to lay an egg in her nest. When the broody hen comes back to find the other hen in her nest, she may get confused and go to a different nest. The other hen lays her egg and leaves. The uncovered, partially incubated eggs get cold, and the embryos die. For this reason, it's a good idea to separate a broody hen from the rest of the flock.

Silkie hens are such tenacious setters they are often used to hatch the eggs of rare and exotic birds.

Prepare a private place where the hen can brood without being bothered. Move the hen and her eggs at night. If you move her during the day, she may try to get back to the old nest. If you move her at night, she will wake up in the morning to find herself in a quiet place, and chances are good she will stay there. While you are moving her, check her body for lice and mites, and treat her if necessary.

Of course, chickens are like people — you can't always predict what they'll do. After you move a hen, she may stop setting. Even if you don't move her, she may stop setting before her eggs hatch. The best you can do is make sure your broody hen has a comfortable, quiet place and plenty to eat and drink, and then let nature take its course. If she sticks it out for 21 days, she will hatch out a nestful of downy chicks.

Artificial Incubation

If you hatch eggs in a mechanical device, the process is called mechanical or artificial incubation. The device in which the eggs are hatched is an incubator. Incubators are designed to imitate the temperature and humidity produced by a setting hen. They come in several styles and a wide range of prices and are sold through farm stores and poultry supply catalogs. Some incubators have a window in the cover so you can watch your eggs hatch. Some have a fan that circulates warm air to keep the temperature uniform throughout the incubator.

An automatic turner is a handy feature. An incubator without one costs less, but you have to turn the eggs by hand several times a day. Turning keeps the embryo floating within the egg white so it won't stick to the inside of the shell. When a hen hatches eggs, she constantly fidgets in the nest, causing her eggs to turn.

If you get an incubator with a window, automatic turning, *and* a fan, you'll have it made. Every incubator comes with a set of directions. For best hatching results, follow them carefully. (See page 95 for more information.)

Egg Selection

Select only eggs that are clean and of the proper size, shape, and color for your breed. Eliminate any that are cracked, unusually large or small, round, or oblong. Candle the eggs and eliminate those with blood spots, which can be hereditary.

Depending on how many eggs your incubator holds and on how well your hens lay, gathering enough eggs to fill your incubator will probably take several days. A hen also has to collect a clutch of eggs before she starts setting, in order for all the chicks to hatch at once. You can store the eggs for up to 10 days in a clean egg carton with their large ends oriented upward. Keep the eggs out of sunlight, in a cool place but not in the refrigerator. The best storage temperature for hatching eggs is about 55°F.

Hatching Rate

No matter how careful you are, not every egg will hatch. Some won't hatch because they are infertile. The cock may be too young or too old, or perhaps he favors some hens while ignoring others. Maybe he is ill. If you have more than one rooster, they may spend too much time fighting, rather than mating.

Assuming that all the eggs are fertile, even a setting hen is not 100 percent successful. A good hen, however, hatches a higher percentage of eggs than can be hatched in an incubator. The typical hatching rate in a good incubator is about 85 percent of all fertile eggs.

Fertile eggs fail to hatch for a variety of reasons, including the following:

- The breeding flock is unhealthy, weak, or improperly fed.
- The eggs are dirty.
- The eggs are improperly stored or have been stored too long.
- The eggs are not turned often enough.
- The incubation temperature is too high, too low, or not steady.
- The incubation humidity is too low or too high.
- The incubator is not properly ventilated.

Egg Anatomy

An egg consists of many parts. The first thing you see is the shell. If you look closely, you can see tiny pores, which allow oxygen and carbon dioxide to pass through the shell. The big end has more pores than the little end, causing more air to be trapped at the big end.

The shell is lined with a leathery outer membrane. Within the outer membrane is an inner membrane. These two membranes help keep bacteria from getting into the egg and slow the evaporation of moisture from the egg. Between the outer membrane and the inner membrane, at the larger end of the egg, is an air cell that holds oxygen for the chick to breathe.

The inner membrane surrounds the egg white, or albumen. Albumen is 88 percent water and 11 percent protein. One function of the albumen is to cushion the yolk floating in it. The yolk is made up of fats, carbohydrates, proteins, vitamins, and minerals that feed the growing embryo.

On two sides of the yolk are cords, or chalazae, that keep the yolk floating within the albumen. When you crack open an egg, the chalazae break and recoil against the yolk. They then look like white lumps on two sides of the yolk.

On top of the yolk is a round, whitish spot called the germinal disk or blastodisc. If the egg is infertile, the blastodisc is irregular in shape. In a fertile egg, this spot is called the blastoderm. A blastoderm looks like a set of tiny rings, one inside the other. During incubation, the blastoderm develops into a baby chick. If you break an egg into a dish and examine the yolk, you can tell whether the egg is fertile by whether it has an irregularly shaped blastodisc or a perfectly round blastoderm.

After the egg has been broken, of course, it can no longer be incubated. Unfortunately, there is no way to tell from the outside whether the egg is fertile.

You can, however, tell whether an egg is fertile after it has been incubated for a few days. Hold the egg against your candler and gently turn it. If the egg is infertile, it will look clear. If the egg is fertile, you will see a dark spot at the center — a newly formed heart! Watch for a moment, and you'll see the heart beat. Surrounding the heart are blood vessels that carry food and water from the yolk to the developing embryo.

Embryo Development

A blastoderm needs food, water, and oxygen to develop into an embryo. It gets food and water from the egg yolk. It gets oxygen from air coming through the pores in the shell.

Week 1. Within the shell, the embryo has everything it needs to develop into a chick. By the second day, it has a heart. By the third day, it has a head with eyes and little wings and legs. By the end of the first week, the embryo has all the body parts of a finished chicken, although they are still undeveloped.

Week 2. By the ninth day, the embryo begins to look like a chick. By the fourteenth day, it has feathers.

Week 3. At the beginning of the third week, the embryo turns its head toward the air-cell end of the egg. It continues to grow throughout the week, absorbing the remainder of the yolk.

By the twentieth day, the embryo occupies the entire space inside the shell, except the air cell. It needs more oxygen, so it uses its beak to break through the inner membrane into the air cell. There it takes its first real breath and starts to peep. You can hear it peeping inside the shell.

On the twenty-first day, the chick makes a tiny hole, or pip, in the shell by using its egg tooth — a sharp projection at the top of its upper beak. The egg tooth, having served its purpose, falls off soon after the chick hatches.

After the chick pips the shell, it rests for 3 to 8 hours. It then turns its head and breaks the shell all the way around. As it chips away at the air-cell end of the shell, the chick shoves against the small end with its feet. After about 40 minutes of hard work, it kicks free of the shell. Then, wet and exhausted, the chick rests.

The development of an embryo

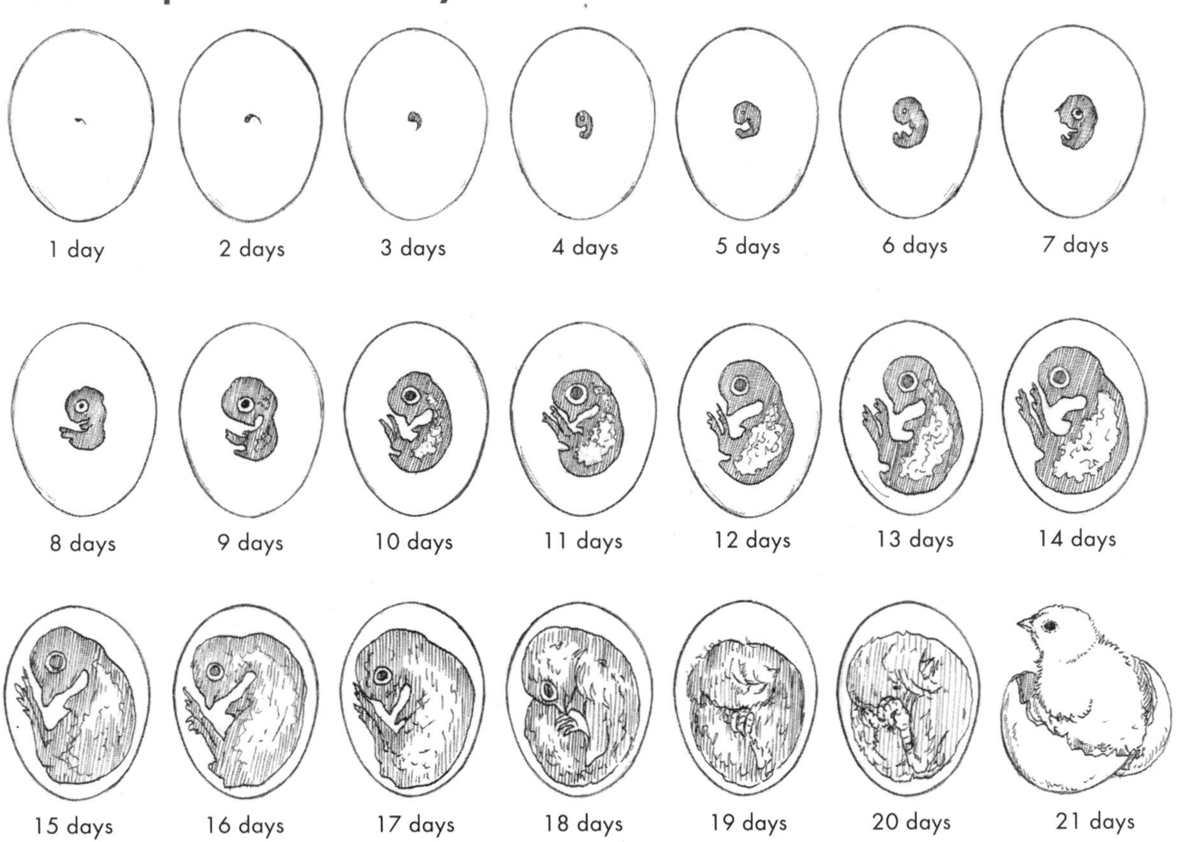

Raising Chicks

When a hen hatches out a brood, she keeps them in the nest until she feels they are ready to venture out into the world. Even after they leave the nest, she keeps them warm and helps them find food. A mother hen gathers her brood under her wings if it rains or she senses danger. She squawks and puffs up, making herself look as big as possible, to chase away any dog, cat, or human that might come near. When you hatch chicks in an incubator, you are responsible for providing them with warmth, food, and protection.

The Brooder Box

Remove newly hatched chicks from the incubator when they are completely dry and fluffy. By that time, they will be scrambling around and complaining loudly. When you open the incubator, you may find some chicks that are still wet. Leave them in the incubator for a few more hours to dry so they won't get a chill.

Chicks need a warm, dry, draft-free place where they are protected from dogs, cats, and other animals. Such a place is called a brooder. The simplest brooder is a sturdy cardboard box big enough to provide 76 square inches of space per chick.

Fasten a piece of chicken wire to the top of the box so air can get in but pets can't. Place a piece of cardboard or some newspapers over part of the top to keep out drafts.

At one end of the box, hang a lightbulb in a reflector; you can purchase this setup at a farm store or hardware store. The heat from the light will keep the chicks warm. To increase the warmth level, lower the light or increase the wattage; to decrease the warmth level, reduce the wattage, raise the light, or use a larger box.

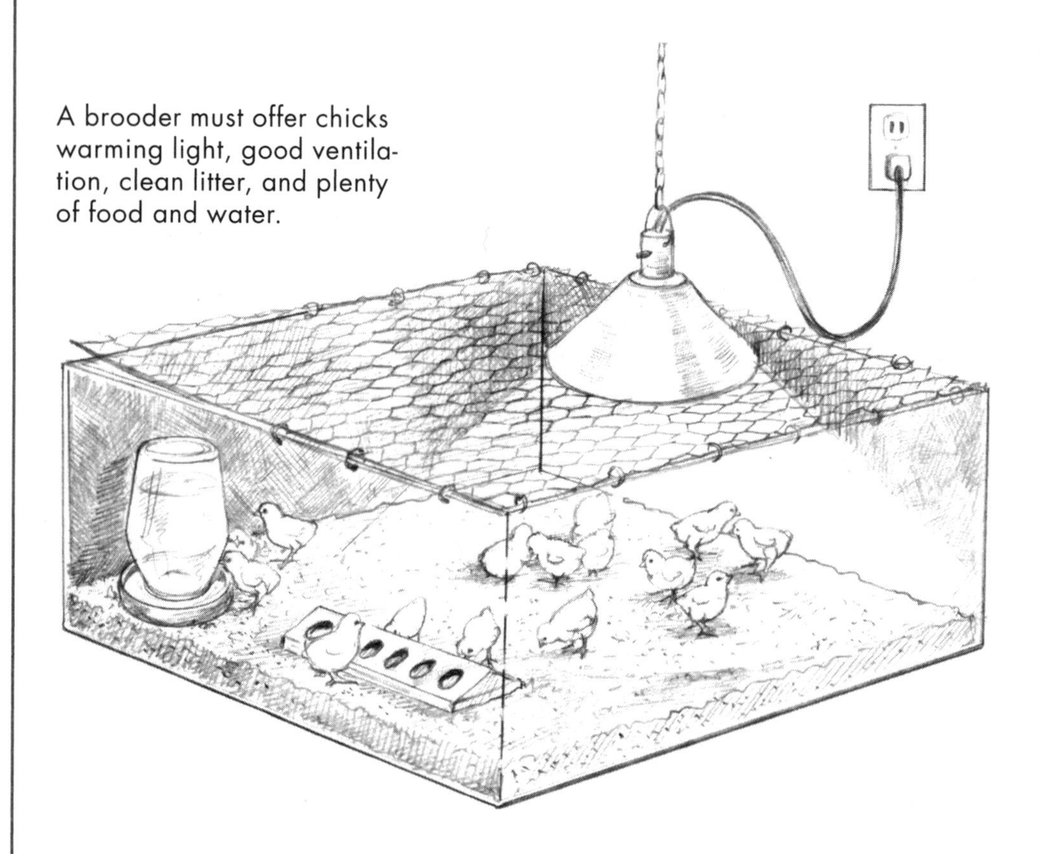

A brooder must offer chicks warming light, good ventilation, clean litter, and plenty of food and water.

Line the bottom of the box with several layers of newspaper topped with paper toweling. The paper toweling gives the chicks better footing than newspaper, which they would find slippery. After a few days, when the chicks are walking and eating well, you can use wood shavings, sand, or other litter at the bottom of the brooder. Litter absorbs droppings and helps keep the chicks warm and dry but should not be used until they learn to eat chick feed instead of litter. Each day, sprinkle a little clean litter over the old litter.

You can tell whether chicks are comfortable in the brooder by the way they act:

- ◆ If they are too warm, they will move as far away from the light as they can, crowding into the corners and possibly smothering one another.
- ◆ If they are not warm enough, they will complain loudly, crowd under the light, sleep in a pile, and possibly smother one another.
- ◆ When they are just right, they will move freely around the brooder, make contented sounds, and sleep nicely spread out.

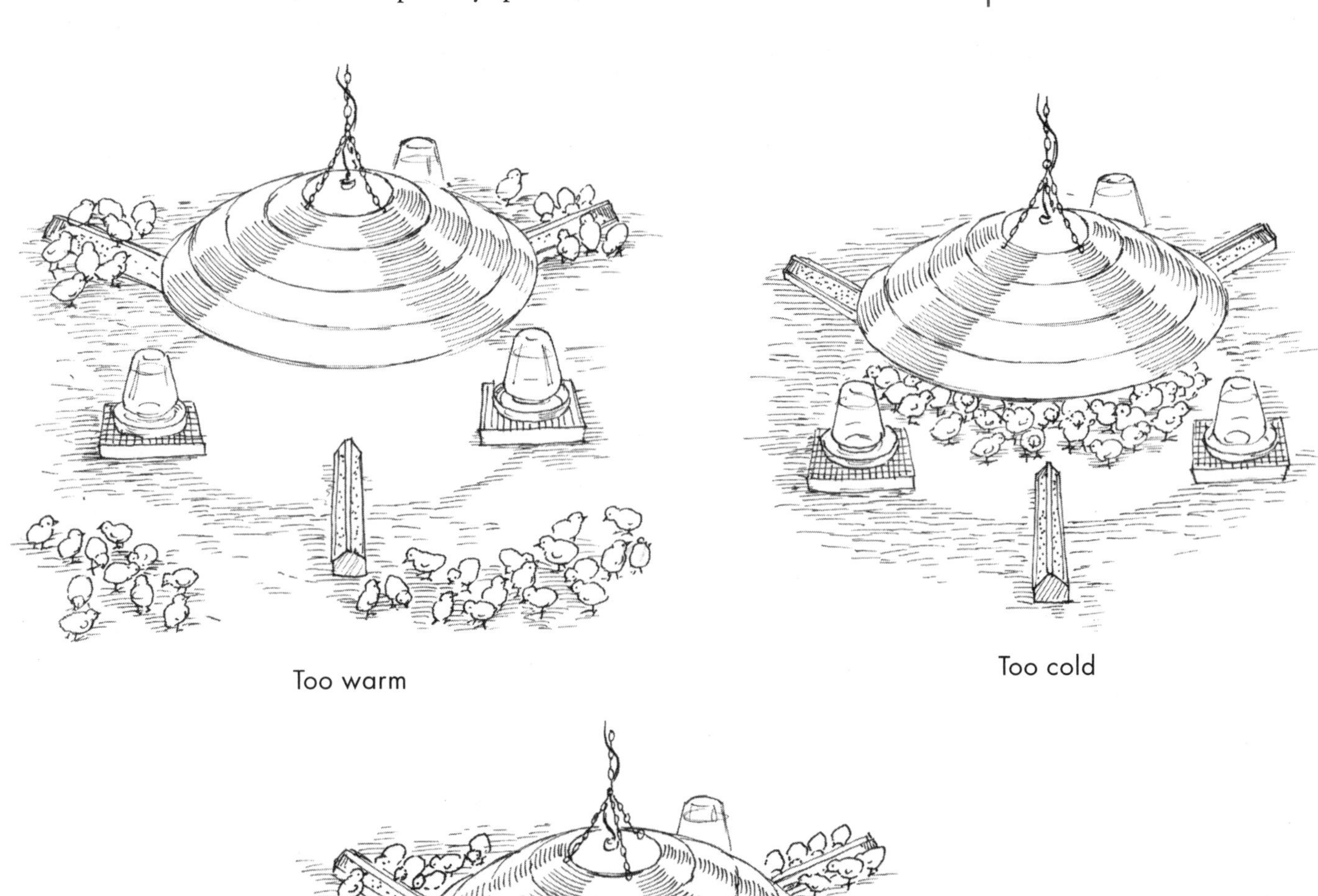

Feeding Chicks

A chick should drink its first water soon after it comes out of the incubator. Clean, fresh water should be available at all times thereafter. The easiest way to water chicks is to use a 1-quart glass jar fitted with a chick watering basin from the farm store or poultry supply catalog. Fill the jar with water, place the basin on top, and flip the jar over. Every time a chick takes a drink, water flows out of the jar into the basin. These devices are designed to keep chicks from getting the water dirty by walking in it or falling into the water and drowning. As the chicks get older and need more water and a larger basin, switch to the 1-gallon size.

Even though a chick should drink right away, it may not be ready to eat. It is still living on reserves supplied by the yolk it absorbed just before hatching. When you order chicks by mail, they are shipped right out of the incubator. Their yolk reserves let them survive many hours in the mail. By the time they arrive, though, they will be ready for food and water.

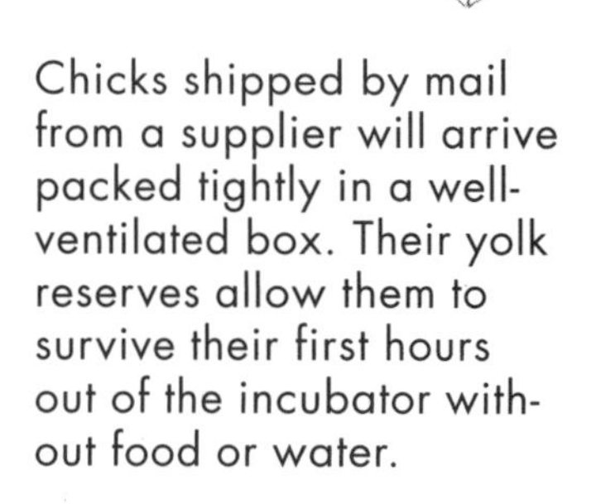

Chicks shipped by mail from a supplier will arrive packed tightly in a well-ventilated box. Their yolk reserves allow them to survive their first hours out of the incubator without food or water.

Feed chicks a starter ration purchased from the farm store. Starter is higher in protein and lower in calcium than lay ration, which should never be fed to young chickens. Some brands of starter are medicated, some are not. If you take proper care of your chicks and keep their housing clean and dry, medication is not necessary. Medicated feed is a poor substitute for good management.

In areas where chickens are big business, farm stores sell a variety of rations for chicks. You may find starter ration for newly hatched chicks and grower ration for older chicks. You may find one kind of grower ration for meat birds and another kind for layers. In most parts of the United States, though, you will find only one all-purpose starter or starter-grower.

Place the starter ration in a feeder designed especially for chicks. If the feeder has a cover with slots in it, allow one slot per chick. If the feeder is a trough type without a cover, allow 1 inch of trough length per chick, or half that if the chicks can eat from both sides.

To minimize waste, fill the feeder only two-thirds full. The top of the feeder should be as high as the chicks' backs. Raise the feeder as the chicks grow, either by using a hanging feeder with an adjustable chain or by fastening a wooden block to the bottom of the feeder. Once your chicks outgrow their baby feeder, switch to a chicken-sized model.

Feeding Layers versus Broilers

If you are raising layers, when they reach 18 weeks of age, gradually mix more lay ration into the starter until they are completely switched over to adult ration by the time they start laying. Of the laying breeds, each pullet will eat about 25 pounds of feed by laying age at 20 weeks. Dual-purpose breeds take a bit longer to reach laying age and eat about 27 pounds of feed.

If you raise chicks for eggs, you want them to grow slowly so they are fully developed by the time they start to lay. If you raise broilers, on the other hand, you want them to grow as fast as possible to keep them nice and tender until they get big enough to butcher. The younger the chicken, the more tender it will be.

Young birds convert feed into meat more efficiently than older ones. The most economical meat is a broiler or fryer weighing 2½ to 3½ pounds. Raising roasters that weigh 4 to 6 pounds costs more per pound. Feed meat chicks often to stimulate their appetite. The more they eat, the faster they grow.

An efficient broiler eats approximately 2 pounds of starter for every pound of weight it gains. If you raise an efficient meat breed or hybrid to 3½ pounds, each will eat 7 pounds of starter by the time it is ready to butcher at 7 to 8 weeks. If you raise a dual-purpose breed, your broilers won't grow as rapidly. Depending on their breed, by the time they weigh 3½ pounds, they may eat twice as much as a specialized meat breed.

Culling

As your chicks grow, cull any that are deformed. Watch especially for runts and chicks with crooked breasts or backs. Deformed chickens do not grow well, may not lay well, cannot be shown, and should not be kept for breeding. Except for runts and diseased birds, culls may be raised for home butchering, if you are so inclined. Otherwise, dispose of culls as they come to your attention. (See page 41 for more details.)

Health Problems

Different chick diseases prevail in different parts of the United States. Your county Extension agent or state poultry specialist can advise you about the need to vaccinate your chicks. Three common conditions that can affect nearly all chicks are pasting, coccidiosis, and brooder pneumonia.

Pasting occurs when droppings stick to the bird's rear end and clog the vent opening. Gently pick off the wad of hardened droppings, taking care not to tear the chick's tender skin. To prevent pasting, make sure that your chicks aren't getting a chill. If pasting persists, mix a little cornmeal or ground-up raw oatmeal with their starter. By the time your chicks are 1 week old, pasting should no longer be a problem.

Coccidiosis is a parasitic infection that causes droppings to be loose, watery, and sometimes bloody. Chicks raised in the cool weather of early spring are unlikely to get this disease unless they live in filthy conditions or are forced to drink dirty water. Coccidiosis occurs more often during warm humid weather, when the parasites naturally flourish. To prevent this disease, take measures to keep the drinking water free of droppings and scrub the waterer every time you refill it. Keep the brooder lined with clean litter and immediately replace dirty or wet litter.

If you regularly raise chicks, avoid coccidiosis by making or buying a brooder with a wire mesh floor that lets droppings fall below the mesh. Another way to prevent infection with coccidia is to feed medicated starter, which contains a coccidiostat. Medicated feed is designed to be a preventive measure and therefore won't help chicks that have coccidiosis. Treating the disease requires stronger medication that you can purchase from a farm store, poultry supplier, or veterinarian.

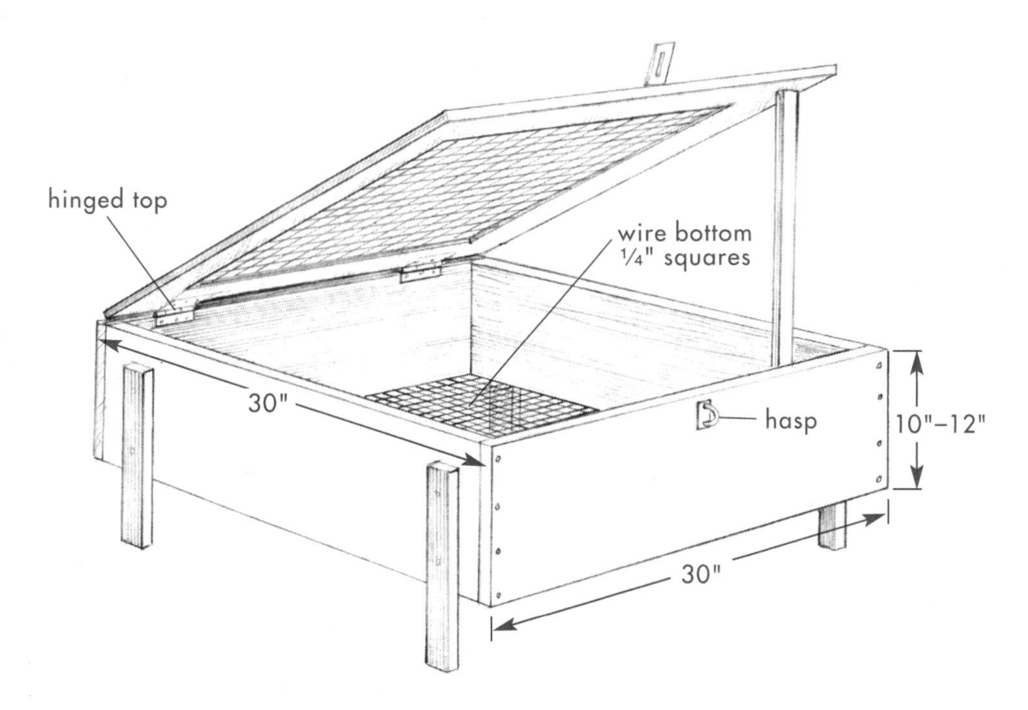

If you plan to raise chicks on a regular basis, a permanent brooder with a raised floor may be a wise investment to prevent infection by coccidiosis.

Brooder pneumonia is a fungal infection of the lungs. Affected chicks may have difficulty breathing, or they may just sicken and die. To prevent brooder pneumonia, be sure that the feed and litter are free from molds (fungi).

Keep your chicks warm, dry, and away from drafts and predators, feed them properly, and make sure they always have clean water, and chances are good they'll thrive.

As They Grow

Chicks start growing feathers on their wings within a day or two of hatching. Depending on their breed, they will be fully feathered by the time they are 4 to 6 weeks old. By then, they need at least 1 square foot of space each and no longer need artificial heat. They are ready to be moved out of the brooder and into the chicken house.

As the chicks grow, be sure to provide less heat and more space. Chicks that are kept too warm or crowded start pecking at one another, causing serious wounds that lead to cannibalism. Cannibalism is a learned habit that usually starts about the time feathers start growing on the lower back. Once cannibalism starts, it's difficult to stop. Preventive measures include increasing the available space, reducing heat and light, and using a red lightbulb that minimizes the attractiveness of blood and emerging blood-filled feathers.

Relieving boredom is another way to prevent cannibalism. Perches help relieve boredom by giving chicks something to play on. Given a chance, chicks will practice perching when only a few days old. Pretty soon, they'll be playing perching games. A chick may jump onto the perch and then jump off the other side, scaring the dickens out of the other chicks. Or one chick may follow another onto the perch, causing the first to lose its balance and hop down. After a few weeks, some chicks will roost on the perch overnight. By the time they are 4 or 5 weeks old, they will all roost. Allow 4 inches of roosting space per chick.

Don't use a perch, however, if you're raising broilers. A perch causes heavy meat birds to get blisters on their breasts. Blisters and calluses may also occur on heavy birds housed on a wire floor or on packed, damp litter. If you raise broilers, forget the perches and freshen the bedding every day.

Chicken Development

Development	Age in weeks
Chicks feather out	4–6
Cockerels crow	6–8
Pullets start laying	20–24

From Chicks to Chickens

As your chicks grow, you will soon be able to tell the cockerels from the pullets. At 3 to 8 weeks of age, depending on the breed, they will develop reddened combs and wattles. Cockerels have larger, more brightly colored combs and wattles than pullets.

Cockerels will develop spurs on their legs. The older a cock gets, the longer his spurs grow. Most hens have tiny spurs or little round knobs in place of spurs. (Game hens are one exception — they may have spurs as long as 1½ inches.)

In some breeds, the cock's feathers are a different color from the hen's. In most breeds, the cock's hackle and saddle feathers are pointed, whereas a hen's feathers are rounded. The cock also develops long sweeping tail feathers, called sickles.

And, of course, the cockerels are the ones that crow. Their first attempts will sound pretty funny, but soon enough they'll get the hang of it. When the cockerels start chasing the pullets, it's time to separate them. Select the best cockerels for breeding and fatten the rest for butchering.

Pullets start laying at 20 to 24 weeks, depending on the breed. The first eggs will be small and probably found on the floor. After a few weeks, you should find regular-sized eggs in the nests.

Chicken Health

Chickens can suffer many different health problems from many different sources. If you consult a book on poultry, you will find so many diseases listed you may be dissuaded from ever keeping chickens. Most disorders, though, are readily prevented through good management. Many infectious diseases can be prevented through vaccination and worming. How aggressive you need to be depends on many factors. The intensive husbandry of commercial poultry operations necessitates aggressive vaccination and worming. A well-maintained backyard flock, derived from healthy stock, will not require a lot of preventive medicine. Be aware, though, that backyard flocks that are not vaccinated or wormed can develop problems. If you show your chickens or regularly introduce new birds, you will need to vaccinate and worm more often than if yours is a closed flock, with no travel and no new additions. Consult your veterinarian or Extension agent for advice.

The longer you keep a chicken, the more likely it is to get a disease, which is why commercial growers and many experienced backyard flock owners won't keep chickens for more than a year or two — as soon as this year's replacement flock matures, last year's flock is out the door. Given a little extra care, though, your chickens can remain safe and healthy for many years.

How Diseases Spread

Disease-causing organisms are always present in the environment, but they may not cause problems unless a flock is stressed or kept in unclean conditions. Diseases are carried through the air, soil, and water. They may be spread through contact with other chickens or other animals, especially rodents and wild birds. They may be carried on the clothing, particularly the shoes, of the person who tends the flock.

Wild birds spread diseases by flying from one chicken flock to another, looking for spilled grain. If you live in an area where chickens are common, netting over your chicken yard will keep out freeloading birds.

Visiting other chicken yards is a good way to bring back diseases by way of manure clinging to your shoes. After such a visit, clean your shoes thoroughly before tending your own flock.

One chicken can get a disease from another, even if both birds appear to be healthy. After your flock is established, avoid introducing new chickens. Every time you bring home a new bird, you run the risk of bringing some disease with it. If you do acquire a new chicken, or bring one of your chickens back from a show, house it apart from the rest of your flock for at least 2 weeks, until you are certain the bird is healthy.

Chickens have a greater chance of remaining healthy if you take the following measures:

- Scrub out waterers and refill them with clean fresh water daily.
- Avoid feeding old or moldy rations.
- Clean out the coop at least once a year. In between, promptly remove wet litter and accumulated piles of droppings.
- Provide enough space at feeders and waterers for even the lowest chickens in the peck order.
- Vaccinate as recommended for your area by your county Extension agent, state poultry specialist, or veterinarian.
- Maintain a stress-free environment, and train your chickens to be calm.

Signs of Illness

Once you become familiar with how a healthy chicken looks and acts, you can easily detect illness by noticing changes. Each time you enter the coop, stand quietly for a few moments until your chickens get used to your presence and go on about their business. Then look for anything unusual.

Sound. The chickens in a healthy flock make pleasant, melodious sounds. Sick chickens may sneeze, gulp, or make whistling or rattling sounds, especially at night.

Smell. Notice how your chicken house usually smells. Any change in odor is a bad sign.

Appearance. A well chicken has a bright, full, waxy comb, shiny feathers, and bright, shiny eyes. A sick chicken's feathers may look dull, and its comb may shrink or change color. Its eyes may get dull and sunken, or swell shut. Sticky tears may ooze from the corners of its eyes. Its nostrils may drip or become caked.

Droppings. A chicken's droppings are normally gray with a white cap. The droppings of a sick chicken may turn white, green, yellow, or bloody or be loose. Occasional foamy droppings, however, are normal.

Behavior. A healthy chicken looks perky and alert, with its head and tail held high. A sick chicken hangs its head or hunches down, sometimes ruffling its feathers to get warm. It may drink more than usual, eat or drink less than usual, or lay fewer eggs. A mature bird may lose weight. A young bird may stop growing.

Dead chickens are, of course, one sign of disease, but don't jump to hasty conclusions if one chicken dies. The normal mortality rate for chickens is 5 percent per year. Naturally, you'll be upset when you find a dead chicken, but it's not a significant issue unless more chickens die or your flock shows other signs of disease.

Salmonella

Salmonella is a bacterial disease that can affect poultry. Eggs and meat contaminated by salmonella bacteria pose a significant human health risk. Infected birds may have diarrhea and obvious signs of illness, often leading to death. Carrier birds are those with no signs of illness that pass bacteria to their eggs and to other birds. Be sure to purchase stock from salmonella-free flocks.

Providing Treatment

Properly treating a sick chicken requires knowing what disease it has. Unfortunately, many chicken diseases mimic one another. If you do not know exactly what disease you are dealing with, administering the wrong medication can make things worse. Seek help from an experienced poultry person in your area, your county Extension agent, or your state poultry specialist, or consult a comprehensive chicken health manual (see Recommended Reading, page 379).

Definitely call your county agent or state specialist if several chickens suddenly get sick or die at the same time. Your flock may have a contagious disease that could easily spread to nearby flocks. You might take the dead chickens to a poultry pathologist. You won't get them back, but you will find out what ails them, from which you will learn how to treat the rest of your flock. Your county Extension agent can help you find the nearest poultry pathologist.

If one of your chickens appears sick, immediately isolate it well away from the rest of your flock. To avoid spreading disease, tend your well chickens before feeding and watering the sick one. For the same reason, always tend chicks or growing chickens before taking care of mature ones, even when they all appear to be perfectly healthy.

It's sad to say, but the best way to treat many diseases is to do away with the sick bird and burn or deeply bury the body. Usually, by the time you notice that a chicken isn't well, it's too sick to be cured. By getting rid of it, you may keep its disease from spreading. Even if you do cure the bird, it may remain a carrier and continue to infect other chickens in your flock, and it is almost certain not to fully recover its reproductive capabilities. Eliminating diseased chickens from your breeding flock helps make future generations more disease resistant.

Feather Loss

Chickens lose feathers for various reasons:

- The annual molt, during which chickens naturally renew their plumage
- Chicks picking newly emerging blood-filled feathers from one another, a form of cannibalism that is preventable through proper management
- Mating, generally among heavier breeds, in which the cock claws off a hen's back feathers; this feather loss is avoidable by rotating breeding cocks or housing cocks away from hens part of the time
- Lice and mites, which cause itching and result in chickens' pulling out their own feathers

Lice and Mites

Lice and mites may be brought to your flock by wild birds, rodents, and new chickens. They may be carried on used feeders, waterers, nests, and other equipment; if you recycle used equipment, scrub and disinfect it before putting it to use. Lice and mites bite or chew a chicken's skin and suck its blood, and a serious infestation can result in death. The habit chickens have of dusting themselves in dry soil, which is infuriating when they get into your garden, helps keep their bodies free of lice and mites.

Periodically check your chickens for lice and mites. Examine them at night, using a flashlight to look between the feathers around the head, under the wings, and around the vent. Also look carefully at the scales along the shanks. You won't need to check more than a few chickens, since these parasites spread rapidly from one chicken to another.

Lice leave strings of tiny light-colored eggs or clumps that look like miniature grains of rice clinging to feathers. You are unlikely to see the lice themselves, since they move and hide quickly, but you may see scabs they leave on the skin.

Body mites are tiny red or light brown insects that look like spiders crawling on the skin at night. During the day, they inhabit perches and nests. A setting hen on the nest is an easy target for mites.

If lice or body mites get into your flock, dust all chickens with an insecticidal powder. Use only products approved for chickens; these products are available through farm stores and poultry suppliers. Thoroughly clean out your coop and sprinkle insecticidal powder into all the cracks and crevices.

Leg mites get under the scales on a chicken's shanks, causing them to be raised instead of lying smoothly. A serious infestation is painful and causes the chicken to walk stiff-legged. To control leg mites, once a month coat perches and the legs of all your chickens with vegetable oil. Oral ivermectin is a very effective treatment for infected birds. Consult your veterinarian about the proper formulation and dosage, and be aware that there will be a withholding time after treatment for laying and meat birds.

Raised scales on a chicken's legs indicate leg mites.

Internal Parasites

Different kinds of internal parasites occur in different areas of the United States. The two most common are coccidia and worms.

Fecal Analysis

To find out for sure whether your chickens have worms, scoop some fresh feces into a plastic bag, seal it, and take it to your vet for analysis. If your chickens have worms, the vet can tell you what kind they are and recommend an appropriate treatment.

Coccidia are everywhere, but a properly managed flock develops a natural immunity to them. When coccidia get out of hand, they cause coccidiosis. Although many different animals are affected by coccidiosis, the coccidia that infect chickens do not infect other kinds of animals, and vice versa. Coccidiosis usually affects chicks, but adult chickens can also get it, especially when the weather is hot and humid. The first sign is loose droppings, sometimes tinged with blood. A medication for treating coccidiosis, sold through farm stores and poultry supply catalogs, must be used to treat the whole flock at once.

Worms in chickens are similar to those in dog and cats. A chicken gets roundworms by picking up worm eggs as it pecks for food on the ground. It gets tapeworms by eating an infected intermediary host, such as an earthworm, grasshopper, housefly, ant, snail, or slug. Confined chickens are more likely to have roundworms, whereas foraging chickens are more likely to have tapeworms. Signs of both kinds of worm are droopiness, decreased laying, and weight loss, or, in young birds, slow or no weight gain. Loose droppings or diarrhea are also possible. Sometimes you can see worms in the droppings.

To reduce the chances of your chickens' getting worms, prevent puddles from forming in their yard and keep the coop floor covered with clean, dry litter. Move pastured chickens often. Discourage wild birds and rodents from visiting, and worm any new chicken you bring into your flock.

Parasites in the environment are killed naturally by drying in the sun and, even more effectively, by being frozen in the winter. If your climate is mild, or if you have a particularly mild winter, you may need to perform aggressive parasite control.

Cancer

Only a few forms of cancer occur in chickens, but the ones that do occur with some frequency.

Marek's disease is caused by a herpes virus. It causes a cancer of lymphocytes, the circulating immune cells. Affected birds are usually less than 4 months old, and they often develop signs of weakness and leg paralysis. Some birds may just sicken and die. Vaccination is an effective preventive measure.

Lymphoid leucosis is also a cancer of lymphocytes, but it occurs in birds older than 4 months and is caused by a different virus. Affected birds often sicken and die. Pale combs, indicative of anemia, may be seen. There is no vaccine for lymphoid leucosis.

Cancers of the ovary and uterus occur in older hens. Any hen more than 2 years old can be affected. Birds with cancer of the reproductive tract may lay misshapen eggs or no eggs at all. They will develop gradual loss of condition, a prominent keel, and an enlarging soft, fluid-filled abdomen.

Summer Care

To keep themselves cool in warm weather, chickens breathe through their mouths. A mature chicken starts breathing through its mouth when the temperature reaches 85°F, a chick when the temperature reaches 100°F. As the temperature rises, chickens also breathe more rapidly and spread their wings away from their bodies. Both signs are indications of heat stress. At a temperature of 105°F or above, chickens may die.

To prevent deaths in hot weather, ensure continuous access to fresh water that remains as cool as possible, either by putting waterers in the shade or by periodically furnishing cool water. Since chickens drink less when their water isn't pure, avoid putting medications in drinking water during hot weather. Make sure your chickens can get into shade without crowding together. In a dry climate, you can cool off your chickens by spraying them lightly with a hose. In a humid climate, spraying doesn't help, because the air is already too full of moisture to allow evaporation.

During hot weather, hens lay smaller eggs with thinner shells, a problem you can mitigate somewhat by using a lay ration with a higher than usual protein level. If you live in a hot climate, your farm store probably carries high-protein ration during the summer months.

Winter Care

A chicken stays warm by ruffling its feathers to trap warm air next to its body. A cold draft blowing through the feathers removes the warm air and causes chilling. Check for drafts by quietly standing in your coop awhile, and then squatting to detect cold air moving at chicken level. Take any necessary measures to reduce coop draftiness.

Unlike a hen, a cock doesn't sleep with his head tucked under his wing. In cold weather, his comb may therefore freeze during the night, becoming quite painful and perhaps reducing his fertility. Insulation helps prevent frozen combs, as does mounting a small electric heater above the perch. To make sure your chickens won't get too warm, plug the heater into a thermostatic control that kicks on when the temperature falls below 35°F and kicks off above 35°F.

Chickens for Meat

Of all the forms of livestock, chickens put meat on your table with the least amount of time and effort. In a matter of weeks, your chicken-keeping chores are over and your freezer is full of tasty, healthful poultry. If, on the other hand, you keep a dual-purpose flock, the availability of poultry meat may be ongoing as you butcher surplus cockerels and spent hens throughout the year.

Meat Classes

Chicken meat may be divided into five basic classes:

Rock-Cornish game hen: not a game bird and not necessarily a hen, but a Cornish, Rock-Cornish, or any Cornish-cross bird, usually 5 to 6 weeks old, weighing less than 1 pound or about 1½ pounds.

Broiler or fryer: a young tender chicken, usually weighing 4 to 4½ pounds live weight, less than 13 weeks of age, with soft, pliable, smooth-textured skin and flexible breastbone; suitable for almost any kind of cooking.

Roaster: a young tender chicken, usually weighing 6 to 8 pounds live weight, 3 to 5 months of age, with soft, pliable smooth-textured skin and a breastbone that's less flexible than that of a broiler or fryer. This class of chicken is suitable for roasting whole.

Stewing hen, baking hen, or "fowl": a mature hen (10 months or older and usually a layer that is no longer economically productive) with a nonflexible breastbone and less tender meat; requires stewing or another moist cooking method.

Cock or rooster: any fully mature male chicken, the meat of which is dark, tough, strong tasting, and unfit to eat.

Managing Meat Birds

Methods for raising broilers fall into three categories:

1. Indoor confinement involves housing chickens indoors on litter and taking them everything they eat. This method is practical for small-flock owners who don't have much space. The goal of confinement is to get the most meat for the least cost by efficiently converting feed into meat. The standard feed conversion ratio is 2 to 1 — each bird averages 2 pounds of feed for every 1 pound of weight gain. To get a feed conversion ratio that high, you must raise Cornish-cross hybrids, which have been developed for their distinct ability to eat and grow. Efficient feed conversion means allowing birds only enough space to get to feeders and drinkers, and no more. If you don't like the idea of factory farming, you can give your meat birds more room than the minimums shown in the accompanying chart, but be prepared to feed them a bit longer to get them up to weight.

2. Range confinement, like indoor confinement, involves keeping broilers in a building, but this type of building is portable, is kept on pasture, and is moved daily. Range confinement reduces feed costs, especially if you move housing first thing each day to encourage hungry birds to forage for an hour before feeding them their morning ration. On the other hand, you need enough good pasture (or unsprayed lawn) to move the shelter daily, and you must do so each day without fail. As they reach harvest size, the birds will graze faster and deposit a higher concentration of droppings, and will have to be moved at least twice a day. Chickens raised by this method take longer to reach butchering size than do those confined indoors.

3. Free range lets chickens freely come and go from their range shelter. This method requires more land than either form of confinement, because you need enough space for both a shelter and pasture for grazing (and trampling), multiplied several times to allow for fresh forage. Figure at least one-quarter acre for 100 birds. This method requires less labor than range confinement (because you don't have to move the shelter daily) but more labor than indoor confinement (because you do have to move the shelter occasionally). Allowing the chickens to exercise creates darker, firmer, more flavorful meat but also causes them to eat more and grow more slowly — they don't reach fryer size until about 13 weeks. Because not everyone is willing to raise broilers for an extra 5 weeks and not everyone appreciates the full flavor and firm texture of naturally grown chicken, free ranging is less often used for meat birds than for laying hens.

Combination management involves raising birds in confinement for 8 weeks, butchering some as fryers, and putting the rest on pasture for another 4 to 5 weeks until they reach roaster size. Since feed conversion efficiency goes down as birds get older, raising roasters this way costs less than raising them entirely indoors but requires longer to achieve roaster weight.

Minimum Confinement Space

Age	Floor Space per Chick
0–2 weeks	0.5 sq ft
2–8 weeks	1 sq ft
8 weeks	2–3 sq ft

Feeding Meat Birds

Provide enough feeder space that your chicks can eat at will and that those lower in the peck order won't get pushed away by dominant birds. The general rule is to furnish enough feeder space so at least one third of your chicks can eat at a time.

A confinement-fed broiler eats approximately 2 pounds of feed for every pound of weight it gains. If you raise your birds to 4 pounds, each one will gobble up at least 8 pounds of feed during its lifetime. A nonhybrid may eat twice that amount.

The older a chicken is, the less efficiently it converts feed into meat and the costlier it becomes to raise. The conversion ratio starts out below 1 in newly hatched chicks and reaches 2 to 1 at about the fifth or sixth week. During the seventh or eighth week, the cumulative, or average, ratio reaches 2 to 1 — the point of diminishing returns. From then on, the cumulative ratio has nowhere to go but up, and the amount of feed the chicken eats (in terms of cost) can't be justified by the amount of weight the chicken gains.

Although the most economical meat comes from birds weighing 2½ to 3½ pounds, most folks prefer meatier broilers or fryers in the range of 4 to 4½ pounds. If you want nice plump roasters, be prepared to pay more per pound to raise them.

To estimate the minimum amount of feed 100 confinement-fed chicks should eat each day, double their age in weeks. For example, 100 four-week-old broilers should eat no less than 8 pounds of feed per day. (In metric, the age of the chicks in weeks roughly equals the minimum amount of feed, in kilograms, 100 chicks should eat each day). If feed use levels off or drops below this guideline, look for management or disease problems.

The Importance of Water

Regardless of your management method, provide free access to fresh water at all times. Meat birds that don't get enough to drink eat less and therefore grow more slowly.

Avoiding Drug Residues

To avoid drug residues in your homegrown meat, shun medicated rations and seek out sources of nonmedicated rations. Medications include low levels of antibiotics to improve feed conversion and coccidiostats to prevent coccidiosis, an intestinal disease that interferes with nutrient absorption and drastically reduces the growth rate of infected birds. If you choose to start your flock on a medicated ration, you must find a nonmedicated feed to use during the drug's withdrawal period, which represents the minimum number of days that must pass from the time drug use stops until drug residues dissipate from the birds' bodies. If the label does not specify a withdrawal period, ask your feed dealer to look it up for you in his spec book. Where nonmedicated rations are not available, scratch grains may be your only option during the drug withdrawal period.

If you use nonmedicated feed throughout the growing period, you'll have to be especially careful to prevent coccidiosis. This disease is especially problematic in areas where conditions are warm and humid. Keep litter clean and dry for indoor birds, move range-fed birds frequently to prevent build-up of droppings, and keep drinking water free of droppings.

Poultry Economics

Several methods are used to determine the economic efficiency of producing chicken meat. Since feed cost accounts for at least 55 percent of the cost of meat-bird production, most economic indicators factor in the amount of feed used.

Feed conversion ratio is the total amount of feed in pounds eaten by the flock divided by the flock's total live weight in pounds. Hybrid broilers raised under efficient commercial methods get by on as little as 1.85 pounds of feed per 1 pound of weight gained. At home, don't expect a conversion ratio much better than 2 pounds of feed per pound of live weight, or about 3 pounds of feed per pound of carcass. If your rate is considerably higher, take stock of your management methods.

Feed cost per pound is the feed conversion ratio multiplied by the average cost of feed per pound. Determine the average cost of feed per pound by dividing the total pounds of feed used into your total feed cost.

Performance efficiency factor is the average live weight divided by the feed conversion ratio, multiplied by 100. In factory-raised poultry, this index hovers around 200. The higher you get above 200, the better you're doing.

Livability is the total number of birds butchered or sold divided by the total number started. To convert that number to a percentage, multiply by 100. Good livability is 95 percent or better. If your livability is above 90 percent, you're doing as well as many commercial growers.

Average live weight is the total live weight divided by the total number of birds. The industry average for hybrids efficiently raised to 8 weeks of age is 4 to 4½ pounds. If, at the end of 8 weeks (13 weeks for nonhybrids), your birds aren't even close, look for ways to improve your management methods. One way to boost your average is to raise cockerels instead of straight-run chicks. As a general rule, cockerels weigh 1 pound more than pullets of the same age and on the same amount of feed.

Average weight per bird is the total dressed weight of all birds divided by the total number of birds. This index factors in weight lost to excess fat and inedible portions, such as intestines, feathers, heads, feet, and blood. A good average for the edible portion is approximately 75 percent of live weight.

Butchering

As butchering time draws near, seek out a fellow backyard chicken keeper willing to show you how to clean your broilers, or learn the procedure from a good book (see Recommended Reading, page 379). If you prefer not to butcher your own birds, perhaps you can find a custom slaughterer in your area who handles chickens.

2

Ducks & Geese

Introducing Ducks and Geese

Keeping ducks and geese is a relatively simple proposition. They prefer to forage for much of their own food in meadows and woodlands, and they require little in the way of housing; often, a fence to protect them from wildlife and marauding neighborhood pets and to keep them from waddling far afield will suffice. Ducks and geese do not need a pond to lead a happy life in your backyard, although keeping them on one simplifies their maintenance even further.

Talking about waterfowl can get a bit confusing. A male duck is a drake, but a female duck is a duck. A male goose is a gander, but a female goose is a goose. So a drake is a duck, but a duck isn't always a drake, and a gander is a goose, but a goose isn't always a gander. Got it? Dealing with these fowl in groups is much simpler: A bunch of ducks is a bevy, and a gang of geese is a gaggle.

A hatchling is a duck or goose fresh out of the egg. If the hatchling is a duck, it is a duckling; if it is a goose, it is a gosling. A hatchling that survives the first few critical days of life and begins growing feathers is called started. When it goes into its first molt (shedding of feathers), it's called junior or green. Until the hatchling has reached 1 year of age, it is young; after this age, it is old.

Waterfowl Families

The word *waterfowl* collectively refers to birds in the Anatidae family — ducks, geese, and swans. The Anatidae family comprises more than 140 species, all of which are web-footed swimming birds that have a row of toothlike serrations along the edges of their bills (which allows them to strain food out of water) and hatch precocial young. Precocial means that in contrast to the featherless and helpless hatchlings of many birds, waterfowl hatchlings can see, walk, eat, and vocalize within a few hours of coming into the world.

Waterfowl fall into two basic subfamilies. On one side are geese, along with swans and whistling ducks. These fowl share many common characteristics. Males and females have the same color pattern and a similar voice, molt annually, and mate permanently. The males in this group help incubate and care for the offspring. Being large birds, geese and swans have a hard time hiding while on the nest; thus, to protect themselves, they are instinctively aggressive. They nest after laying only a few eggs, and their hatchlings mature slowly. This group includes the 14 species of goose, which are classified into two groups. One of those groups includes the two species from which domestic geese were developed — the swan goose, which gave us African and Chinese geese, and the greylag, which gave us all other domestic geese.

The second basic subfamily of waterfowl includes all ducks except the whistlers. The shared characteristics of these ducks are that males and females differ in color pattern and voice, molt twice a year, and mate for a single season. When the female starts nesting, the male moves off by himself. These ducks are smaller than the birds in the first subfamily and therefore can hide themselves more easily. They hatch more eggs at a time, and their hatchlings mature more quickly. This subfamily includes three groups: diving ducks, dabbling or surface-feeding ducks, and perching or wood ducks. The Muscovy was derived from the perching ducks. All other domestic ducks were derived from the Mallard, which is one of the 11 species in the dabbling group.

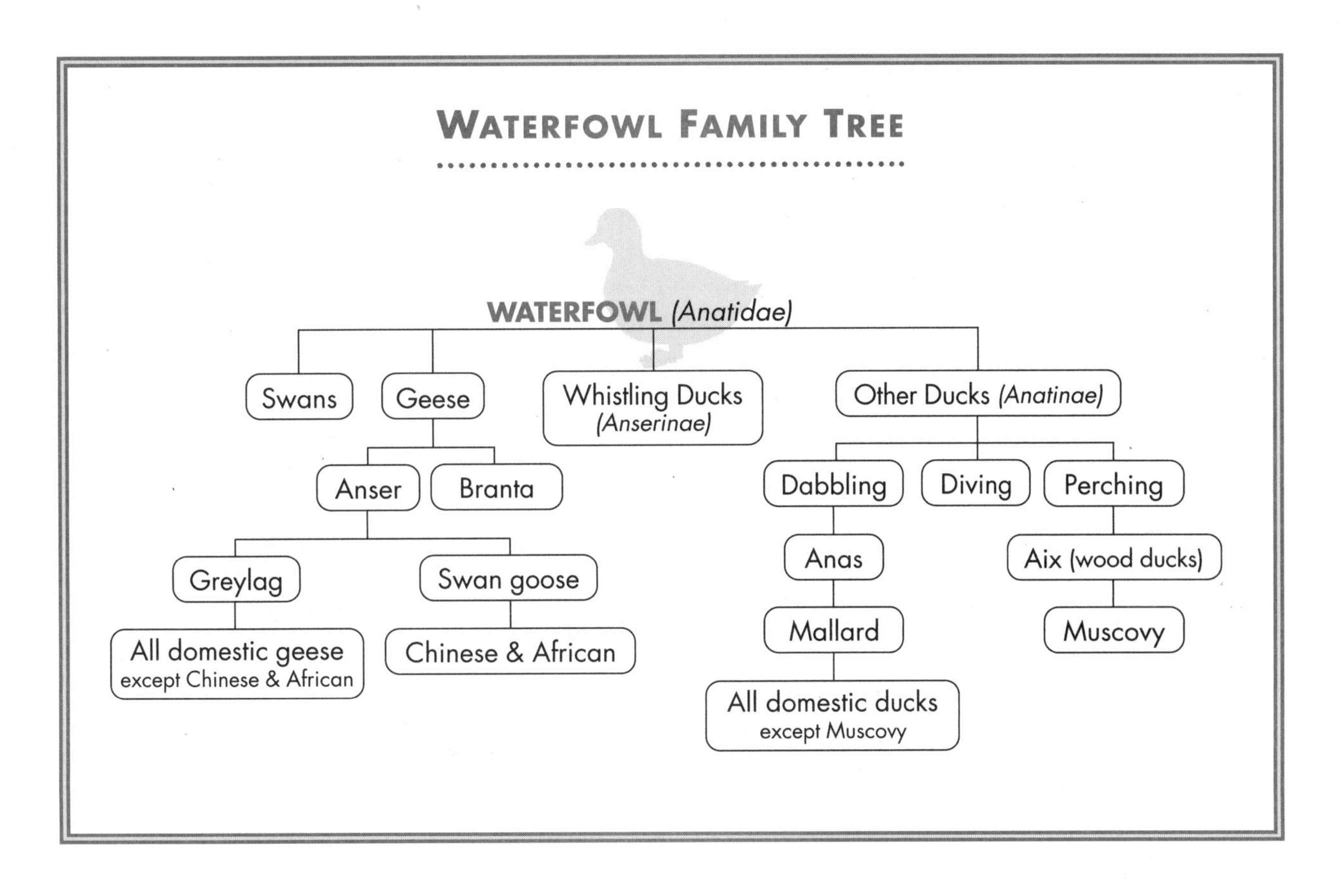

Waterfowl Traits

Ducks spend a lot of time in water, nibbling at plants, bugs, and other shoreline inhabitants. If you let them wander in your vegetable garden or flower bed, they will help control garden pests, although you have to take care they don't run out of other things to eat and start in on your pea vines or pansies. Muscovies in particular relish slugs, snails, and other crawly things. In fact, the San Francisco area once had a rent-a-duck service that loaned out Muscovies to local gardeners. Ducks also enjoy chasing flies, in the process offering not only fly control but also a great deal of entertainment. Ducks also keep mosquitoes from getting beyond the larval stage. Unfortunately, tadpoles will suffer the same fate.

Some breeds of duck are kept primarily for ornamental reasons, whether for their colorful plumage, their comical upright stance, or their diminutive profile. Other breeds are prized for their eggs. Khaki Campbells are particularly known for their laying ability, and their eggs make wonderful baked goods. Any breed may be raised for meat, and putting your excess ducks into the freezer both keeps meat on your family table and keeps down the population at the pond.

Some breeds like to fly around, and occasionally one will fly off into the sunset, but such wanderlust can readily be controlled by clipping a wing. The primary downside to ducks is their eternal quacking. In a neighborhood where the noise could become a nuisance, the answer is to keep Muscovies, also known as quackless ducks. Among the other breeds, the male makes little sound, but the female quacks loudly, and each bevy seems to appoint one particularly loud spokesduck to make all the announcements.

Geese also make a racket with their honking. Usually they holler only with good reason, but a less observant human might not detect that reason: for instance, the cat or weasel the geese have spotted slinking along the fence line. Besides announcing intruders, geese also have a tendency to run them off. A lot of people are more afraid of geese than of dogs, probably because they are less familiar with geese and feel intimidated by their flat-footed body charge, indignant feather ruffling, and snakelike hissing. Even as they fend off intruders, geese can become attached to their owners and are less likely to charge the family dog or cat than roaming pets and wildlife.

Geese are active grazers, preferring to glean much of their own sustenance from growing vegetation. They are often used as economical weeders for certain commercial crops; farmers take advantage of their propensity to favor tender shoots over established plants.

Geese lay enough eggs to reproduce, but no breed of goose lays as many eggs as the laying breeds of duck. The meat, however, is plentiful and delicious. Goose is a traditional holiday meal, and when roasted correctly, the meat defies its reputation for being greasy. Goose fat rendered in the roasting makes terrific shortening for baking (leaving your guests wondering what your secret ingredient might be) and in the old days was used as a flavorful replacement for butter as a spread. The feathers and down from plucked geese may be saved up and made into comforters, pillows, and warm vests.

Ducks and geese get along well together and may be kept in the same area. Given sufficient living space with water in which to wash themselves, they will remain spanking clean. When properly fed and protected from harm, your ducks and geese will give you many years of pleasure.

Choosing the Right Bird

Numerous breeds of duck and goose may be found worldwide, and more are created all the time. Others, however, are becoming endangered or extinct. Only a few of the developed breeds are commonly found in the United States. Other breeds with much lower populations are kept by fanciers or conservation breeders. Your purpose in keeping waterfowl will to some extent determine which breed is right for you, but in the long run, your best bet is to select the breed you find most visually appealing.

Each breed has been developed for specific characteristics valued by humans. Among ducks, some are efficient at laying eggs, whereas others grow rapidly to provide an economical source of meat. Still others are prized primarily for their unusual plumage, although most breeds are attractive in one way or another. Geese are largely bred for meat, with down as a byproduct. But any goose seems elegant on a pond and might be looked on as a discount swan.

Ducks for Fun

Most of the less common duck breeds fall into the ornamental category. These ducks are neither particularly prolific layers nor fast growers; rather, they are kept primarily for their attractive appearance. Ironically, among the ornamentals is also the duck from which most other domestic breeds have sprung. Since our domestic breeds came mainly from the wild mallard, this duck has been studied more than any other.

Mallard. Domestic Mallards are similar to their wild cousins, although they tend to lose their streamlined appearance and their desire to migrate. Hatchlings of both sexes start out looking alike, with black markings on downy yellow bodies, and their first feathers make them all look like females. At about 5 months of age, the males molt and take on their distinctive color — a shiny green head setting off a slate-gray body. Thereafter, for 2 months out of every year, during what's known as the eclipse molt, the males take on the brown hues of a female. People who are unfamiliar with waterfowl don't recognize male and female Mallards as being of the same breed. The female's coloring differs greatly from the male's; the females feature shades of brown and dark penciling, a pattern that offers ideal camouflage as she sits on the nest. Both sexes have an electric-blue band of feathers on each wing.

Mallards are fairly small as domestic ducks go, generally weighing 2½ to 3 pounds. They are seasonal layers that stop producing eggs and start nesting after laying 15 to 18 eggs. If a duck's first eggs fail to hatch (or are eaten by a passing 'possum) or the hatchlings don't survive (perhaps having been carried off by a hawk), she is likely to start laying and nesting again in an attempt to hatch a second, and sometimes even a third, batch.

Mallard

Call. Gray call ducks are nearly identical in color pattern to Mallards, but are somewhat smaller and more compact. These ducks make a persistent high-pitched quack once used by hunters to call wild ducks (hence the name), until a federal law of 1934 banned the use of live decoys. If you or your neighbors might be annoyed by persistent quacking, this breed would not be a good choice.

On the other hand, these adorable ducks are docile and friendly, and their size makes them good pets for someone with limited space or a tight budget. They require less living space than a larger breed, but given sufficient room for active foraging, they can be inexpensive to feed. Calls have a similar laying pattern to that of Mallards, but their hatchlings are much more delicate, especially during the first 3 or 4 days of life.

Calls are considered bantams, as they usually weigh 2 pounds or less. In addition to the original gray Call, with its Mallard-like plumage, this breed also comes in white and other colors developed by enthusiasts.

Call

Mandarin

WILD DUCKS

A few species of wild duck are popular with backyard fanciers. The most popular of these are two bantam breeds, the Mandarin and the related North American wood duck, both of which are valued for their stunningly colorful plumage. Wood ducks are native to the United States, and in this country, a permit is required from the Fish and Wildlife Service to keep them in confinement. Some states require local licenses as well.

Raising wild ducks is different from caring for domestic breeds. If you opt to keep wild ducks, consult suitable sources of information, as recommended by the person from whom you acquire your ducks.

Wood

Campbell

Runner

Ducks for Eggs

Some breeds of duck have been genetically selected for their outstanding laying habits, but unless this ability is maintained through continued selective breeding, the laying potential of a particular flock may decrease over time. For this reason, not all populations of a breed known for laying are equally up to the task. Some sources offer hybrid layers that are bred for efficient and consistent egg production, but their offspring will not retain the same characteristics and cannot be used to raise ducklings of your own. Since laying stops when nesting begins, the trade-off among laying breeds is that they do not have strong nesting instincts.

Although some ducks lay pale green eggs, most of the laying breeds produce white eggs.

Campbell. Originally bred in England at the turn of the 20th century, Campbells are fairly active foragers that can withstand cool climates. They weigh around 4 pounds, making them a fair size for eating. The original and most common color is khaki; the ducklings are dark brown and feather out to a seal-brown color, and the males' heads and wings turn darker brown. Good layers will produce 240 to 300 eggs per year.

Runner. Developed in Scotland from stock originating in the East Indies, where these ducks are put to work gleaning snails and waste grain from rice paddies, Runners have an upright stance that allows them to move around with greater agility than other breeds. They tend to be active and somewhat on the nervous side. Known also as Indian Runners, these ducks weigh around 4 pounds and can be expected to lay 140 to 180 eggs per year. They come in many colors, including white, fawn-and-white, chocolate, black, and blue.

Ducks and Egg Profile

Breed	Eggs per Year*	Weight per Dozen	Color
Campbell	250–325	30 ounces	White
Runner	150–300	32 ounces	White
Pekin	125–225	42 ounces	White
Muscovy	50–125	44 ounces	Waxy white
Rouen	35–125	40 ounces	Green
Mallard	25–75	26 ounces	Green
Call	25–50	18 ounces	White or green

**The precise number of eggs laid per year varies from strain to strain within each breed.*

Ducks for Meat

Any breed of duck may be used for meat, although some breeds have been developed to grow faster and larger than others while consuming less feed. You will find that the meat from the ducks you raise yourself is superior in quality to, and less expensive than, store-bought duck meat.

Pekin

Pekin. A Pekin is a big duck with snow-white feathers and an orange bill and feet. Pekins are not particularly outstanding layers and are only fair at nesting and caring for their young. At maturity, they weigh 8 to 10 pounds. Because they have snow-white feathers, they appear cleaner when plucked; their white pinfeathers don't show as much as the pinfeathers of colored waterfowl. If you order duck in a restaurant, you will most likely be served a Pekin.

Rouen

Rouen. The Rouen looks like an overstuffed Mallard. At maturity, it averages 9 to 11 pounds. Rouens are docile, tend to be relatively inactive, and are not particularly good layers. Because of their deep breast, they breed best on water, and their egg fertility tends to be low if they breed on land. Thanks to their dark feathers, Rouens are considered to be largely ornamental, although some people claim this breed has the most flavorful meat of any, and its size certainly makes it ideal for roasting.

Muscovy

Muscovy. The Muscovy doesn't look like any other domestic duck and, indeed, is only distantly related to the others. Muscovies are sometimes called quackless ducks because, in contrast to the loud quacking of other female ducks, the Muscovy female speaks in a musical whimper, although she can make a louder sound if startled or frightened. The male's sound is a soft hiss. Aging males take on a distinctive musky odor, giving this breed its other nickname of musk duck. Both the males and the females have a mask over the bridge of their beak that features lumpy red warts, called caruncles.

Muscovies are arboreal, preferring to roost in trees and nest in wide forks or hollow trunks. In confinement, they like to perch at the top of a fence, but they don't always come back down on the right side. To get a good grip while perching on high, these ducks have sharply clawed toes. If you have to carry one of these ducks, try to keep it from paddling its legs so those sharp claws don't slice your skin. Aging males can be aggressive, usually toward other male ducks but occasionally toward their keeper. With their powerful wings and sharp claws, they can engage in fierce battles over male dominance.

The male Muscovy matures to be the largest domestic duck, weighing up to 12 pounds. The drakes are thus twice the size of the average female, which tops out at around 7 pounds. Although white Muscovies are more suitable for meat, since they have a cleaner appearance when plucked, the original color is an iridescent greenish black with white patches on the wings. Throughout the years, Muscovy fanciers have developed about a dozen additional colors.

Muscovies are an entirely different species from other domestic ducks and, although they will interbreed with others, the resulting offspring will be mules — sterile hybrids that cannot reproduce. Commercial producers may deliberately cross Pekins with Muscovies to produce meat birds called Moulards, which have the large breast of the Muscovy with less fat than the Pekin. Muscovy meat differs in flavor and texture from that of other ducks, in part because it contains less fat.

Being native to Mexico and South America, Muscovies do best in warm climates, although they also do well in moderate zones. They adapt better than other breeds to an environment that lacks a steady source of water for bathing. They are intelligent and curious, and the females can be as friendly as puppies, making great pets. Both sexes have an enormous appetite for slugs, snails, and baby mice. With their massive bodies and large flat feet, though, they tend to be somewhat destructive to seedlings.

Duck Breeds: Average Mature Weight (in Pounds)

Breed	Male	Female
Call	2.0	1.5
Campbell	4.5	4.0
Mallard	3.0	2.5
Muscovy	12.0	7.0
Pekin	10.0	8.0
Rouen	11.0	9.0
Runner	4.5	4.0

Geese

Most breeds of domestic goose have been developed for meat, although some have been developed for their odd appearance. The Sebastopol, for instance, has curly feathers that look like a misguided perm. Nearly every breed has a tufted version, meaning that the geese have a puff of extra feathers on top of the head. No breed of goose lays as prolifically as a duck, and although a single goose egg makes a formidable omelette, the eggs are more often used for hatching or for creating craft items, such as decorative jewelry boxes.

African

Some backyard fanciers prefer geese that are not truly domesticated, such as the diminutive and dauntless Egyptian and the celebrated Canada. Being native to North America, Canada geese require a permit from the United States Fish and Wildlife Service to be kept in confinement.

African. The African is a graceful goose with a knob on top of its head and a dewlap under its chin. The brown variety, with its black knob and bill and a brown stripe down the back of its neck, is more common than the white variety with orange knob and bill. Being fairly calm, Africans are easy to confine and tend not to wander. They mature to weigh 18 to 20 pounds. If you are raising geese primarily for meat and are concerned about the fat content, select African or Chinese geese, both of which naturally have less fat than other breeds.

Chinese

Chinese. The Chinese (or China) goose is similar in appearance to the African but lacks the dewlap. It comes in both white and brown. In contrast to the typical goose honk, this breed emits a higher-pitched "doink" that can be piercing if the bird is upset or irritated. This is the breed most commonly used commercially as weeders. Because Chinese geese are both active and small, they do a good job of seeking out emerging weeds while inflicting little damage on established crops. Thanks to their light weight and strong wings, they can readily fly over an inadequate fence. Chinese geese are the best layers among geese and produce a high rate of fertile eggs even when breeding on land rather than on water. They grow relatively fast but are the smallest of domestic geese, reaching a mature weight of only 10 to 12 pounds.

Embden. The Embden most often matures to 20 or 25 pounds but can weigh up to 30 pounds. Because of its size and white feathers, the Embden is the most popular goose to raise for meat. In Europe, this breed was traditionally plucked throughout its life as a perpetual source of down for comforters and pillows, but the plucking of live geese is now considered inhumane. The yellow goslings have patches of gray down when they hatch, and some people claim they can distinguish female from male goslings by their higher ratio of gray to yellow.

Embden

Pilgrim. The Pilgrim goose is only slightly larger than the Chinese. It is the only domestic breed of goose in which the male and female mature to be different colors — the male is white like an Embden and the female is gray like a Toulouse. Because of this plumage pattern, Pilgrim hatchlings may be distinguished by sex on the basis of their down color: The males are yellow and the females are gray. In recent years, however, this distinction seems to be breaking down, as some flocks reputed to be Pilgrims produce white birds of both sexes. Pilgrims weigh 12 to 14 pounds at maturity and will fly over a fence when attracted to something on the other side.

Pilgrim

Toulouse. The Toulouse, named for a town in France, is the common barnyard goose. It has gray plumage set off with white feathers underneath. The Toulouse goose has been developed in two distinct populations. The common barnyard, or production, Toulouse matures to a weight of 18 to 20 pounds. The more massive giant, or dewlap, Toulouse matures to weigh 20 to 26 pounds.

Goose Breeds: Average Mature Weight (in Pounds) and Egg Production

Breed	Male	Female	Eggs per Year
African	20	18	20–40
Chinese	12	10	50–100
Embden	25	20	35–50
Pilgrim	14	12	20–40
Toulouse	20	18	35–50
Toulouse, dewlap	26	20	20–40

Toulouse

How Long Do Ducks and Geese Live?

When properly fed and protected from harm, a duck may live as long as 20 years, while a goose may survive to the ripe old age of 40.

Making the Purchase

Your first decision in selecting waterfowl is whether you wish to raise ducks, geese, or both. Ducks and geese are compatible and may be housed together in the same facilities, as long as space is sufficient. If you enjoy variety, select one breed of duck and one of goose rather than two breeds of the same species that will eventually interbreed and over time lose their original characteristics.

Your next decision is how many to get. Like most birds, ducks and geese are social animals that don't like to be alone. You will need at least two birds. If you want to raise young in the future, they must be of opposite sexes. If you don't expect to breed your birds, you can get by with two or more females, but they will lay eggs and may try to nest even if no male is around to fertilize the eggs.

The next decision is whether to start with hatching eggs, hatchlings, or mature birds. Experienced waterfowl keepers, and parents who are looking for a good project to educate their children, often start with hatching eggs, but that route carries certain risks. For a successful hatch, you need good-quality hatching eggs that have been properly stored and transported, an incubator that maintains consistent heat and humidity, and someone reliable to monitor the incubator.

Starting with mature birds has the advantage that you can see what you're getting, but they may be expensive and will not necessarily recognize you as a friend. If you cannot find the breed you want locally, having mature birds shipped is both expensive and a hassle.

For most people, starting with hatchlings is the way to go. They are relatively inexpensive, will grow up in familiar surroundings, and will quickly learn to recognize their keeper. Hatchlings are available from more sources than are hatching eggs or adult waterfowl.

One place to look is the local farm store. Either the farm store or your county Extension office can tell you if any hatcheries or waterfowl farms are in your vicinity. The larger operations might advertise in the Yellow Pages of the phone book, newspaper classified ads, or freebie newspapers. The farm store and Extension office can also tell you of any local waterfowl or poultry clubs in your area. Attending a poultry show at the county fair is another way to connect with sellers, as well as with like-minded folks who are willing to share their experiences and expertise.

If you have difficulty finding the breed you want locally, seek out a mail-order source, many of which advertise in farm and country magazines and on the Internet. If you desire one of the less common breeds, you might find a seller through the American Livestock Breeds Conservancy or the Society for the Preservation of Poultry Antiquities (see Resources, page 380). Hatchlings shipped on their first day of life usually survive the trip, although you'll be required to purchase a minimum number so they can keep one another warm. When the shipment arrives, open the box immediately and inspect the ducklings or goslings in front of the carrier, so you can make a claim in case they have died in transit.

When purchasing adult waterfowl, seek birds that are still young. Even though ducks and geese may live for many years, their fertility and laying ability decline with age. Ducks are at their prime age for breeding from 1 to 3 years of age, and geese from 2 to 5 years of age. Indications that a bird is still young are a pliable upper bill, a flexible breastbone, and a flexible windpipe. Another sign is blunt tail feathers, where the once-attached down has broken off. After a bird molts into adult plumage, it acquires the rounded tail feathers of an adult, and once it reaches full maturity, you'll have a hard time determining its age and will have to take the seller's word.

Before taking your birds home, make a final check for problems, such as crooked backs or breastbones. In geese, also check for nonsymmetrical pouches between the legs. Look for signs of good health: alertness, clear eyes, proper degree of plumpness, and normal activity level.

Naturally enough, the rarer the breed you have selected, the more expensive it will be. Young birds will generally be less expensive than hatchlings or old birds, with the exception of proven breeders.

Handling Ducks and Geese

The basic daily activities of waterfowl are eating, napping, and mating. If they have plenty to eat and a safe place to rest, they tend to be fairly content. After eating their fill, many breeds are happy to laze around the rest of the time. The active breeds tend to entertain themselves by marching around investigating things on the ground, or by flying up to get a better view from overhead.

Traits by Breed

Breed	Disposition	Noise Level	Tends to Fly
Ducks			
Call	Active	High	Yes
Campbell	Moderate	Moderate	No
Mallard	Active	High	Yes
Muscovy	Moderate	None	Somewhat
Pekin	Nervous	Moderate	No
Rouen	Calm	Moderate	No
Runner	Excitable	Moderate	No
Geese			
African	Moderate	High	Somewhat
Chinese	Active	High	Yes
Embden	Moderate	Moderate	No
Pilgrim	Moderate	Moderate	Somewhat
Toulouse	Calm	Moderate	No

Restricting Flight

Lighter geese will fly over a low fence when frightened (such as by a predator) or when attracted to something on the other side of the fence. If you can keep your young geese from flying, they probably won't attempt to fly over the fence after they mature. In this respect, Muscovies are more like geese than like ducks, although instead of flying over the fence, they are more likely to fly up and perch on a fence post or tree branch, and they may not come back down on their own side of the fence.

Ducks will fly for the same reasons as geese and, like geese, usually land on the other side of the fence. Mallards are the exception. They take great joy in flying high circles around their home, sometimes spying a nearby body of water and dropping in to investigate — which may not go over real well if that body of water happens to be a neighbor's swimming pool. When the ducks are ready to return home, they will circle around before landing inside their own yard, and if it contains a pond, they will land in the water with a splash. During breeding season, Mallards may seek nesting sites elsewhere, which can be frustrating if you've taken great pains to provide predator-proof digs. You can stop waterfowl from flying in two ways, one of which is permanent, the other temporary.

The permanent method is to clip a wing tip. The tip of the wing is called the pinion, and removing it at the first joint is called pinioning. Pinioning throws the bird's two wings out of balance, making it unable to fly. When hatchlings are pinioned, they seem to barely notice, and the wing heals rapidly. Pinioning gets more difficult as birds get older. Pinioning mature waterfowl should be done by a veterinarian or other waterfowl professional.

The temporary method of restricting flight is to clip the flight feathers of one wing, which serves the same purpose as pinioning. If you clip the feathers of young geese, you probably will not have to do it again once they mature. If you clip the feathers of flying ducks, you must clip again each time the feathers grow back.

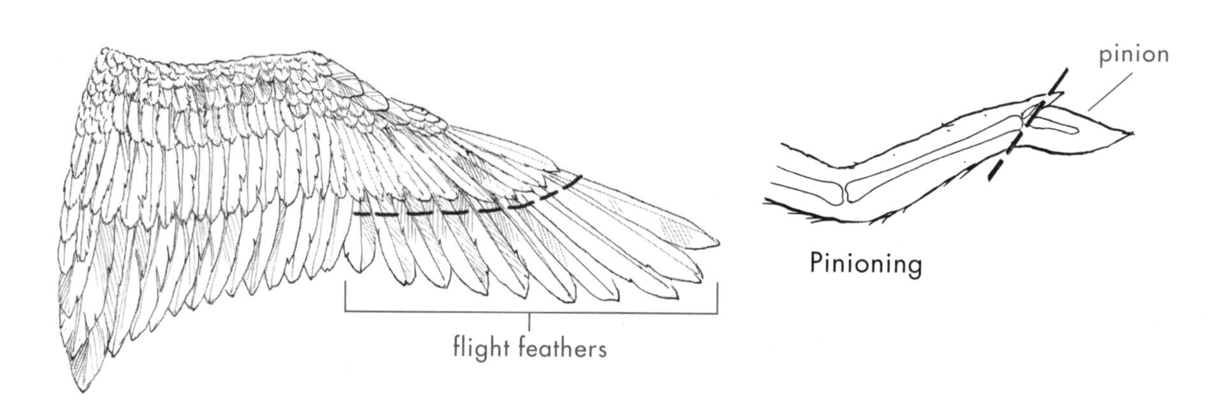

Clipping flight feathers

Molting

Over time, a bird's feathers get worn or broken. Its entire set of plumage is therefore periodically renewed in a process called molting. Young birds go through two consecutive molts. After maturity, they experience one complete molt each year, generally in late summer or early fall after breeding season but before cold weather sets in. They may also go into a molt at other times, as a result of some change in their environment or diet.

During the molt, Mallard drakes take on the camouflage pattern of the females, making it easier for them to hide when they drop their wing feathers and can't fly. As soon as the wing feathers have been renewed, the males go through a second, partial molt to resume their normal color pattern. All drakes of Mallard-derived breeds molt in a similar manner, although the color change may not be as obvious.

Your first clue that molting has started may be piles of feathers in the duck or goose yard, perhaps leaving the impression a predator has gotten one of your birds. You need only make a head count to be sure they are all there. The molt will last 6 to 8 weeks, during which ducks that are laying will decrease in production because they can't efficiently produce both feathers and eggs.

Waterfowl that are nutritionally stressed may experience a hard molt, meaning the entire process takes longer. Until their flight feathers fully return, molting birds are less able to get away from any predators that might sneak into their yard. If your birds receive proper nutrition with adequate protein, the molt goes more quickly. Increasing the dietary protein level by 10 percent — dry cat food is a good source of protein — will help the molt go smoothly.

Aggression

Flying is primarily a problem with ducks, whereas aggression is primarily a problem with geese and Muscovies. This behavior has a connection to species characteristics: Ducks are small and agile, can hide readily, and can fly to get away from predators, but geese and Muscovies are larger and clumsier and cannot readily hide, and so resort to aggression to ward off their enemies.

The best way to keep geese from getting mean in the first place is to take time to greet them whenever you meet. Notice how they greet one another; it's the polite thing to do. Geese get naturally aggressive during breeding season, when a gander will fiercely defend his mate while she is on the nest, and later with the same fierceness will defend his offspring.

When a gander is thinking about attacking, he will stretch his neck full length, throw his head back, and peer at you with one eye. This posture indicates that he isn't too sure about the wisdom of attacking. If you move toward him waving your arms and making a loud noise, he'll probably back off.

The gander to beware is the one that makes a beeline toward you with his neck stretched forward, head down, making hissing sounds. This gander is likely to bite. The worst thing you can do is run away. The best thing you can do is show him who's boss, but that doesn't mean giving him a swift kick with your boot or otherwise engaging in a tussle. Fighting with a gander will only make him meaner.

When a gander comes at you with the intent to bite, clap your hands and stamp your feet. If he keeps coming, swing your arms to make yourself look as big as possible, and run toward him. If that doesn't work, extend your arm with a pointed finger (creating the appearance of a long goose neck with a beak at the end), and move menacingly toward him. Tell him firmly, "Don't you dare!" If he gets close enough to bite, smack him on the bill. If the gander is young, it should take only one or two smacks on the bill to let him know which of you is the bigger, meaner gander.

If the smack doesn't work, step to one side as he rushes at you and grab him by the neck. Grasp the back of his neck with one hand and lift him off the ground, placing your other hand beneath his body. This action will cause the gander to suffer a great indignity he will not readily want to repeat.

Although you can keep your geese from getting mean enough to attack, you cannot stop their innate need to display aggressive behavior. It is part of the triumph ceremony, the purpose of which is to strengthen family bonds. A well-trained gander will pretend to attack you, running at you full tilt but turning back at the last moment. He'll then return to his mate with his wings outstretched (the gander's equivalent of thumping himself on the chest), honking triumphantly, and the two will discuss this brave incident between themselves. If they have goslings, the young will circle around and join in the discussion.

Two ganders or drakes may fight if the yard is too small for the gaggle, if there aren't enough mates to go around, if nesting sites are too few or too close together, or if a new bird is introduced into the group. Once the problem is resolved — a

larger yard is provided or some birds are removed, additional nesting sites are provided, and the newcomer finally makes friends — peace and calm usually returns.

Avoiding fights between, and with, Muscovy drakes is especially important, as their powerful wings and sharp claws make a formidable combination. When a Muscovy is excited or angry, its head feathers rise to create a sort of crest. The male, being larger to start with, raises a higher crest than the female. At the same time, the Muscovy may alternately jerk in and bob out his head; the rapidity of jerking and bobbing indicates the degree of anger or fear. Crest raising and head bobbing are also signs of sexual arousal, and slow head bobbing is often used as a form of social greeting. Muscovies wag their tails to indicate that some unpleasant encounter has been successfully resolved, and also for mysterious reasons as they go about their business.

Males versus Females

A good way to avoid aggression in your waterfowl population is to ensure that you have an appropriate proportion of males and females. Being able to distinguish the males from the females is therefore essential. Some color breeds, especially the Mallard, make it easy, because the male has a different color pattern from the female (except briefly during the molt, when males look like females).

Even when the male and female are of identical color, all duck breeds except the Muscovy give you two other clues. The fully feathered male has curly feathers, aptly called drake feathers, on top of its tail. And, whereas the male speaks in a barely audible, hoarse whisper, the female quacks raucously.

Muscovy females can also quack loudly, but they usually do so only when startled or frightened. Otherwise, they make a pleasant whimpering sound, while the male hisses somewhat like a goose. Male and female Muscovies of all varieties are nearly identical in color but are easy to distinguish from one another because the male is nearly twice as large as the female.

Distinguishing a goose from a gander is not easy, as you need to recognize subtle differences in voice and body posture. The gander's usual sound is a shrill alarm, while the female's is a repeated comfort sound at a much lower pitch. Both geese and ganders honk loudly when disturbed and hiss when angered. In stance, the gander walks more erect and in appearance has a slightly thicker neck and coarser head. The goose carries herself closer to the ground, and her neck is usually curved. When they walk together — and mated pairs don't often stay apart for long — the gander walks ahead and the goose follows.

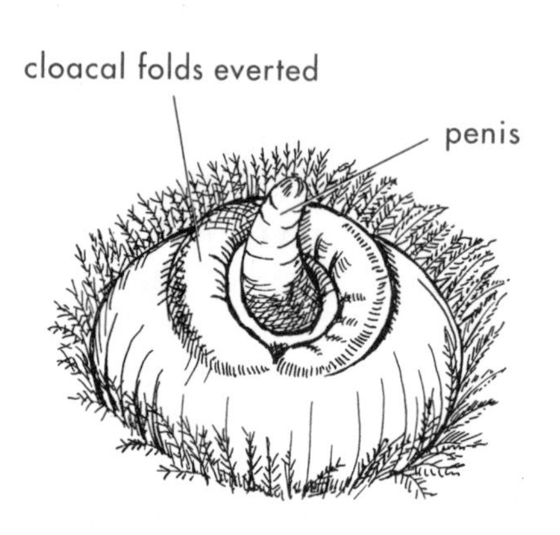

The male genital organ

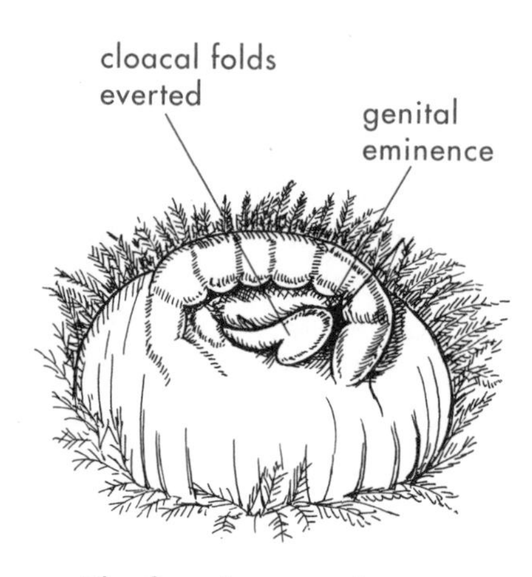

The female genital organ

In the African and Chinese breeds, the ganders have somewhat larger knobs on their heads than females. Traditionally, Pilgrim ganders are supposed to be white like an Embden, while the geese are gray and white like a Toulouse; these days, however, that distinction is becoming more theory than fact.

The only way to be 100 percent certain whether you have a goose or a gander is to vent-sex the bird, meaning you must examine its sexual parts. First, catch the bird and turn it on its back, taking great care not to get whacked by a wing, sliced by the claw of a paddling foot, or bitten on the back of your arm. Having someone help you hold the bird can be a great asset. Sit down, or hunker on the ground, with the bird's back over your knee, its head tucked under your arm, and its vent (the orifice under the tail through which the bird eliminates or lays eggs) facing away from you. With one hand on each side of the tail, use your fingers to push the tail downward while using your thumbs to press the sides of the vent (in a similar motion to extracting a blackhead, except here you must be gentle to avoid causing damage). If you are lucky, the bird will reveal its sex organ. If you are unlucky, it will first show its indignity by squirting a foul green mess in your direction. If the bird is tense, it will reveal nothing until you can get it to relax its vent muscles by inserting a finger and letting the muscles hold it tightly until they finally grow tired and relax. If you work things right, a penis will unwind from the vent of a gander, while a goose will reveal frilly folds of pink skin. To avoid getting hurt when you're ready to release the bird, turn it over and set it with its feet on the ground, holding your hand on its back for a moment before letting go. The bird will then stand up and move away from you.

Catch and Carry

Unless you have ducks that are tame pets, catching waterfowl can be a challenge. A poultry net (something like a butterfly net) comes in handy for the purpose, although if you have a sizable pond, the crafty creatures will swim out to the middle and wait for you to leave. If you have more than a few ducks or geese, you have to take care not to cause a stampede in which one could get seriously hurt, and you do not want to get hurt yourself if your waterfowl find themselves trapped in a corner and launch en masse into the air to get around you.

A safe way to carry a goose is with one arm underneath the goose for support and the other arm above the goose, securing its wings.

If your ducks can fly, move slowly and try not to frighten them into flight. If one crouches — a sign that it is about to spring into the air — you may be able to catch it just as it leaves the ground, especially if you have a net.

Herding your bevy or gaggle into a small area gives them less space in which to give you the run-around. The small area should be a place they're already familiar with; wary waterfowl do not herd easily into unknown territory. Muscovies are especially difficult to herd and do not readily go into enclosures, but they tend to be friendlier than other breeds and therefore easier to catch.

Ducks and geese don't move as agilely after dark, and they are readily confused by the beam of a flashlight. A duck will try to get away from the beam, letting you use the light to herd ducks from a distance into an enclosure. A goose tends to move toward the beam, giving you a chance to grab it when it gets near enough.

The safest way (for the bird, that is) to grab a duck or goose is by the neck. The neck is much stronger than a leg, which can easily snap if you pull in one direction while the bird tries to move in another direction. Once you have a bird by the neck, carry it by tucking its body under an arm, confining the wings so they can't flap. If you are carrying a Muscovy, watch out for those claws. A small person who has trouble carrying a large bird under one arm would be wise to carry a Muscovy with one hand on each wing joint, where the wing joins the body, keeping the bird's back toward you and its claws away from you. A goose may be held in a similar manner, or by hooking both wings under one arm while using the other arm to carry some of the weight.

When handling ducks and geese, take great care not to frighten or excite them. (You'll be able to tell that they're extremely frightened if they wobble instead of waddle or get around by using their wings as crutches.) Whenever you work around waterfowl, remain calm and move with deliberation. Handle your birds gently, and reassure them before releasing them.

Transporting

The ideal way to transport a duck or a goose is in a pet carrier, such as one designed for a cat or dog. Make sure the carrier is of adequate size for the bird. A sturdy cardboard box also works, provided it can be sealed so the bird can't break out and has airholes large enough so the bird can breathe but can't get its head through.

For short trips, you can transport a duck or goose in a feed sack. For a goose, cut off the tip of one corner of the sack, put the goose inside with its head through the hole, and tie off the sack's opening. For a duck, drop the duck into the sack and tie the hole with enough slack to admit air but not enough for the duck to escape. If you aren't going far and the temperature is not high, you may carry two or three ducks in one sack. Otherwise, put only one bird in each.

Any of these carriers may be loaded into the back of a pickup truck. If the bed has no camper cover, place the containers close to the cab, where the draft is less. If you're traveling in a car, avoid using the trunk, where exhaust fumes could kill your birds. Put the carrier on the back seat or on the floor.

Geese are more content when they can see where they're going and can peer out the window at passing scenery. When transporting a pair of mated geese, let them ride side by side, where they can see each other. If you transport them within earshot but not sight of each other, they will make you deaf with their lovesick calling before you get to your destination.

Housing Ducks and Geese

As picturesque as it may be to have ducks and geese wandering around your estate, giving them their own confined area is a good idea. Otherwise, they are likely to congregate on your doorstep, rat-a-tat for attention on your door, and leave their slippery "calling cards" on your stoop. Geese will try to chase away your visitors and otherwise amuse themselves by chewing the weatherstripping from around your doors, the wiring from beneath your family car, the telephone line going into your house, or anything else that catches their fancy.

Space

The larger the area in which waterfowl can roam, the less trouble they are likely to get into and the cleaner they will remain. An area that is too small for its population will soon become muddy, mucky, smelly, and fly-ridden. The more area available for grass to grow, the better chance the grass has against flat-footed waddling,

Minimum Living Space (Square Feet) per Bird

	Age (weeks)			
	1–3	4–8	9–17	17+
Duck	1.0	3.0	3.5	4.0
Goose	1.5	4.5	6.0	9.0

and the more widely duck and goose droppings will be spread, letting them dissipate naturally instead of building up to create a problem. Rather than considering minimum space requirements, the better plan is to provide as much roaming area for your bevy or gaggle as you feasibly can.

An ideal waterfowl yard has a slight slope and sandy soil for good drainage. The less ideal the area, the more space you should provide per bird. The yard should have both sunny and shady areas, letting the birds choose for themselves whether to rest in the warmth of the sun or away from its harsh rays. The yard should include a windbreak, which might consist of a dense hedge, the side of a building, or a portion of solid fence.

Fencing

Fencing is essential for keeping waterfowl safe from predators and from roaming too far afield. The fence must be sound and well maintained, as waterfowl are much more adept at finding ways to get out than they are at finding their way back in. How closely spaced the fencing material needs to be depends on whether or not you wish to protect vegetation on the other side. Geese especially, but ducks, too, will graze through the fence as far as they can reach.

A 3- or 4-foot fence will confine nonflying breeds. A 5- to 6-foot fence would be more appropriate for flying breeds of geese. If you can keep geese grounded until they reach maturity, they will have a hard time getting enough elevation to clear a fence. Flying ducks, on the other hand, can spring into the air from a standing start and must have a net ceiling over their yard (feasible only if you have a few ducks in a small yard) or have their wings clipped.

If you'll be confining hatchlings, secure a 12-inch strip of tight-mesh net along the bottom of the fence. Ducklings and goslings look a lot bigger in their downy coats than they really are, and they can slip through incredibly small spaces. The trouble is, after grazing on the wrong side of the fence, slipping back through with a full belly may not be quite as easy.

Electroplastic netting will keep out most predators and prevent across-the-fence grazing, but it won't keep in the downy babies. The biggest problem will come when the babies reach the same diameter as the netting's mesh, because a bird stuck in an electric fence will probably die. Netting that is 30 inches high will confine nonflying breeds; the 20-inch version is commonly used to confine weeder geese. Electroplastic netting deteriorates over time and must be replaced every few years. It will last a lot longer if, instead of being used as permanent perimeter fencing, it is used seasonally to provide additional grazing (or to restrict grazing) and is carefully stored during the off-season.

Perhaps the best fencing for waterfowl is narrow-mesh field fencing, stretched tight between posts spaced 8 feet apart and augmented with an electrified wire along the top. If the electric wire is no more than 2 inches above the fence, any predator that attempts to climb over the field fence will get a shock when it reaches for the electric wire, causing it to drop off the fence.

Predator Control

As much as you like your ducks and geese, predators abound that like them even more. Foxes, weasels, raccoons, and skunks have notorious appetites for waterfowl, their eggs, and their young. Dogs will chase and kill adult waterfowl for sport, and cats will certainly prey on downy hatchlings.

A strong fence goes a long way toward keeping out most predators. Burying the bottom of the fence belowground will discourage burrowing animals. A distance of 4 inches underground is usually a sufficient deterrent, but if you live in an outlying area where a predator might dig for hours undisturbed, going down 18 inches would be more appropriate. To keep the fence from rusting out, coat the part to be buried with tar or asphalt emulsion. If you have a bunker mentality and the yard isn't prohibitively large, a concrete footer along the fence line offers a permanent, albeit expensive, solution to the digging problem.

Keeping out flying predators is nearly impossible, unless you're ambitious enough to hang netting over the entire yard. Crows or bluejays will peck holes in your eggs, and hawks or eagles will swoop down and make off with your hatchlings. These problems may be largely prevented by providing a roofed shelter.

Shelter

Waterfowl aren't keen on staying indoors, but for their own protection, it's wise to provide them with a shelter and train them early to go in each night. Since most predators roam during the night, a well-built shelter will keep your birds safe. It will also provide a refuge from icy wind, scorching sun, and pelting rain.

The shelter need not be spacious, since the birds will be spending only their nights in it. About 6 square feet of floor space per duck and about 10 square feet per goose should be plenty. For only a pair or a trio, a doghouse-style shelter with a low roof is inexpensive and easy to clean. For a larger bevy or gaggle that requires more floor space, make the roof high enough for you to enter with ease for cleaning.

The best floor for the shelter is packed dirt. If your area has a serious predator problem, the expense of pouring a concrete foundation, if not an entire concrete floor, may be warranted. Small windows with small-mesh screening will let in air and allow the birds to see out, but take care to avoid making drafty conditions. A cold wind blowing through the feathers of a duck or goose will remove the body heat trapped by its protective undercoat of down.

Shavings on the floor will help control droppings and simplify cleaning. Fresh shavings will also keep eggs clean when they are laid in the shelter. Do not provide water for your ducks and geese while they are inside. The combination of the boredom of confinement and the attraction of water to play in will result in a soggy, smelly mess. Since they have no water, they should also have nothing to eat, but don't worry — they will get by just fine overnight.

Preventing Frostbite

The knobs on the heads of African and Chinese geese are subject to frostbite, making winter shelter for these breeds essential in cold climates. Suspect frostbite if a bird seems depressed and its knob first turns pale and then becomes bright red rimmed with white. Given time, knobs discolored by frostbite may return to normal.

The feet of all waterfowl are susceptible to frostbite, which may occur when ducks and geese congregate around a feeder surrounded by ice or packed snow. The signs of frostbite in feet are the same as those for knobs. Keep the area around feeders clear of ice and snow, or spread sand or straw over the frozen ground. Providing open water for swimming also helps prevent frostbitten feet. Try to keep at least a portion of the pond free of ice, perhaps by installing a recirculating pump.

Because waterfowl don't sleep at night but take catnaps, much as they do throughout the day, they will tend to be restless while confined. Lights, including moonbeams, shining through a window of the shelter and creating shadows on the walls can stir them up and cause a lot of quacking or honking. Where lights shining in from outside create shadow problems inside the shelter, hang some low-watt lightbulbs in the shelter, making sure to arrange them to avoid the birds themselves casting shadows as they move about. Protect the wiring in conduit and hang the bulbs high enough that geese can't dismantle your electrical efforts and get electrocuted in the bargain.

Waterfowl don't like being indoors and should not be confined any longer than is necessary. Even in the severest climate, as long as they can find a place away from the cold wind, they prefer to be outdoors. You may think you are doing them a big favor by breaking open a bale of straw to rest on and protect their feet from cold, but don't be surprised if they hunker down in the snow right next to the straw. Their double coat of feathers and down keeps them comfortable in weather that might send you scurrying for home and hearth. To keep their feet warm, waterfowl stand on one foot with the other tucked up, then switch and stand on the other foot while warming the first.

Because they don't like to be indoors, you'll need to train your bevy or gaggle from the start to go inside at night. Have a plan for getting them from their favorite resting place to the shelter door, perhaps by enlisting help or arranging doors and gates to funnel them in the desired direction. If the first time out they lead you around the barn and through the garden, be prepared to take that same circuitous route every evening from then on.

Even if your birds are well trained and go in without a hitch night after night, the time will come when they will balk. If you once let them get away with not going in, they will try it night after night, trying your patience as well. But, as obstinate as they can be about not wanting to go in at night, should you come home late one evening to discover no ducks or geese in the yard, don't panic. Chances are good they've gone in on their own and are impatiently waiting for you to come and shut the door.

A Pond

Waterfowl can get along just fine without water to swim in, but providing them with at least a small pond is a good idea for several reasons:

- It helps them keep themselves clean.
- It improves egg fertility.
- It lets them more comfortably endure hot and cold weather.
- It helps them evade predators.
- It gives them a place to use up energy in play.
- It gives you the pleasure of watching them float and frolic.

Providing water is especially important if you intend to raise the large, deep-breasted breeds, which have a hard time mating without water to make them buoyant. It also helps your birds evade predators; a goose will swim tantalizingly just offshore when tracked by a fox, and a duck will duck underwater when buzzed by a hawk.

Perhaps the biggest reason you should provide a pond is that waterfowl love water. Sure, it's hard to tell whether a duck is "happy" or a goose is "joyful," but who can doubt it after watching a duck or goose scoot across the water, disappear beneath the surface, pop up somewhere else, flap its wings, and start again? And even if we can't say for sure that the bird is happy, it will certainly make you happy to watch them. The pond needn't be expensive or elaborate. You can build your own, using a guide such as *Earth Ponds,* by Tim Matson.

Any waterfowl pond requires certain basic features to be safe for ducks and geese. It must have easy access in and out. The littlest swimmers may have problems if they can fall into the water but can't jump over a shallow edge to get back out. The pond must also be easy to clean or large enough to cleanse itself of waterfowl droppings. For these two reasons, a children's wading pool is not entirely suitable, unless you go to the trouble of installing a drain for ready cleaning and a hinged ramp that lets even hatchlings get in and out with ease, no matter what the water level.

Waterfowl are happier and healthier when they have access to a pond.

The water should be free of chemical runoff that can harm waterfowl. For this reason, and because waterfowl are not "pool-broken," it's a bad idea to let waterfowl swim in the family pool. If your baby ducklings can get through your fence and make a beeline for your neighbor's pool, it could end up being their last swim.

Fish and waterfowl may be compatible in a large pond or small lake. However, fish in a backyard pond won't take kindly to frequent draining and cleaning and certainly won't appreciate waterfowl deposits or the mud waterfowl stir up. In a larger body of water, bass and other fish may eat hatchlings, and snapping turtles can grab a grown duck by the foot and pull it underwater.

Introducing Home

If your first waterfowl are babies, confine them to an enclosed shelter until they are old enough to go out on their own. Baby waterfowl with no parents don't have the sense to come in out of the pond or out of the rain, and will get soaked through and take a chill. Until they feather out (at about 7 weeks for Mallard-derived breeds, 8 weeks for geese, and 14 weeks for Muscovies), confine babies indoors and provide heat as required by the prevailing weather.

Grown or partially grown waterfowl may be turned loose in a safe yard as soon as you bring them home. If they are of a breed that readily flies, make sure one wing of each bird is clipped so it doesn't take a notion to fly back to wherever it came from. After your ducks and geese get well acquainted with their digs and learn that they will always have plenty of good food to eat, fresh water to drink, and space to play, they will be content in their new home.

Feeding Ducks and Geese

The nutritional needs of waterfowl vary by species, age, purpose, and level of production. Ducks have slightly different needs from geese, growing birds have different needs from mature birds of the same species, and layers and meat birds have different needs from ornamental birds and from one another.

Under natural conditions, Mallard-derived ducks satisfy 90 percent of their nutritional needs by eating vegetable matter; the remaining 10 percent comes from animal matter, such as mosquitoes, flies, and tadpoles. The diet of a Muscovy tends to lean more toward meat, whereas geese are entirely vegetarian.

If you have only a few ducks or geese, you have no need to push them for egg or meat production, and you have plenty of space for them to forage, the birds can meet all of their nutritional needs on their own. During the winter and early in the breeding season, you may need to supplement their diet with grain or commercially formulated ration. It's a good idea to offer a little grain at least once a day, year-round, just to keep your waterfowl tame and used to your presence among them.

Commercial Feeds

Feed processors offer complete rations that are blended to meet the requirements of ducks and geese. Whether these feeds are available in your area depends largely on demand; if you are the rare person in your neighborhood who keeps waterfowl, you'll probably have a hard time finding commercial rations. In this case, look for a farm store that carries a national brand and is willing to order for you.

If you can't obtain duck or goose ration, you may have to make do with poultry feeds that are more widely available. Unless you expect your waterfowl to meet a high level of meat or egg production, most ducks and geese can get along nicely on commercial chicken rations that have sufficient protein. If you feed your waterfowl rations that were not originally formulated for ducks or geese, avoid formulas containing medications, some of which may be toxic or deadly to waterfowl.

In general, you'll have a choice of six basic formulations:

- **Starter ration** is formulated to get ducklings and goslings off to a healthy start during their first 2 weeks of life. The protein level of 18 to 20 percent is ideal for sturdy development.
- **Grower ration** has a protein level of 15 to 18 percent and is designed to promote growth. Ducks and geese raised for meat should be fed a higher level of protein for rapid growth; those to be kept for breeders should get less protein for more evenly balanced growth. Birds to be butchered stay on this ration until they are butchered (usually at 8 to 12 weeks of age). Waterfowl to be raised to maturity are switched over to developer ration at about 9 weeks of age.
- **Developer ration** is intended for growing birds that will be kept as layers or breeders. It has a protein level of 13 or 14 percent for more moderate growth. This ration is also suitable for mature waterfowl that are not breeding or laying.
- **Breeder-layer ration** has a protein level of 16 to 20 percent and includes vitamins and minerals that layers and breeders need to remain fit. Breeders and layers should be switched from developer to breeder-layer ration 2 or 3 weeks before they are expected to begin laying. Make the change gradually by mixing greater quantities of the breeder-layer ration into the developer ration until the switch is complete.
- **High-protein concentrate** is a commercial ration with a whopping protein content of 34 percent. Unlike the first four rations, which are complete feeds, this ration is intended to be combined with grains. By varying the ratio of concentrate to grains, as specified on the label, this ration may be used in place of any complete feed except starter.
- **Mixed grains,** sometimes called scratch grain or chicken scratch, usually consist of a mixture such as corn, milo, oats, and cracked wheat. Mixed grains should never be fed as the sole ration but may be used in combination with high-protein concentrate, fed in moderation to foraging waterfowl to keep them tame, and added to the ration in cold weather to provide a source of warmth. Whole grains are more nutritious than cracked grains, won't spoil as quickly, and, if dropped, will sprout and eventually be eaten.

Protein Requirements

Age	Stage	Protein Content
0–2 weeks	Starter	18–20%
3–8+ weeks	Grower	15–18%
9–20 weeks	Developer	13–14%
20+ weeks	Breeder-layer	16–20%
20+ weeks	Maintenance	14%

Storing Feed

To discourage raiding rodents, store feed in clean trash cans with tight-fitting lids. Secure the lid with a bungee cord if there's any chance of raccoons wandering by. Since commercial rations gradually deteriorate starting from the time they are mixed and since you have no idea how long the feed has been stored at the farm store, purchase no more than you will use in about 2 weeks. If you get home and find that the bottom of a bag has gotten wet and the feed inside is moldy, toss it. Moldy feed can be toxic to waterfowl.

Choosing the Right Form of Feed. Commercially prepared feeds come in three forms: mash (ground-up rations), pellets (compressed mash), and crumbles (crushed pellets).

Pellets are easiest for mature waterfowl to eat and result in the least amount of wasted feed (and, therefore, less wasted money). For birds that aren't yet old enough to handle pellets, crumbles are acceptable, but more feed will be wasted. Mash results in the greatest amount of wasted feed. This waste may be reduced by moistening the mash, but to avoid spoilage, mix only as much as the birds will eat before the next feeding.

Although not every feed is available in all three forms, you should have little difficulty obtaining the pellet form. The chief problem is getting a satisfactory pellet feed for the youngest birds, because they cannot eat pellets of the size that adult birds eat. Some companies make a small pellet about ⅛ inch in diameter, but it's relatively uncommon. You may have to settle for a crumble form for your ducklings.

How Much Feed Is Enough? The easiest way to feed waterfowl is to provide the rations free choice, meaning they have access to their feed anytime they wish to eat. Young waterfowl being fed starter or grower ration should definitely be fed free choice. However, if you have older waterfowl that spend most of their time foraging, they don't need any more commercial ration than they can eat within 15 minutes twice a day, morning and evening.

A growing duckling will eat 1 or 2 ounces of commercial ration per day, gradually increasing to about 8 ounces per day as it reaches full size. A mature duck of a laying breed will consume about 6 ounces of feed per day, whereas a duck of a nonlaying breed may eat 5 to 7 ounces per day when not laying and 12 to 15 ounces when laying. Geese of the heavier breeds may eat up to four times those amounts. Of course, these are only guidelines. The actual amounts your birds will eat depend on many factors, including the temperature, level of activity, and breed size. Birds naturally eat more when they are colder, more active, or of a larger breed. All waterfowl will eat less when forage is available.

Using Feeders. Feeders designed for other poultry may be used for ducks and geese. Hanging tube feeders may be adjusted for height as the ducklings or goslings grow, thereby minimizing waste. Rubber livestock feed pans with tapered sides are also suitable. Avoid any trough design that a duck or goose can easily tip by stepping

Foraging Ability

Breed	Ability
Call	Excellent
Campbell	Excellent
Mallard	Excellent
Muscovy	Excellent
Pekin	Fair
Rouen	Good
Runner	Excellent
All geese	Excellent

on the rim. A feeder used for free-choice rations should be protected from rain, which is easily done by building a little roof over it or by putting it in the entry to the waterfowl enclosure. If you put it in the enclosure, remove it at night when the ducks and geese are shut in. Fill the feeder only with as much as your waterfowl will consume in one day. Clean out any leftover feed at the next filling.

If the feed gets wet, clean the feeder within a few hours. Freshly dampened feed will not hurt ducks and geese. Damp feed left standing for more than a few hours may turn moldy and become harmful. In the course of eating, ducks and geese tend to get feed in their water and water in their feed. To keep both water and feed clean, space feeders and waterers 6 to 8 feet apart for mature birds. For ducklings and goslings, place feeders and waterers on opposite sides of the brooder.

Forage

In place of teeth, waterfowl have serrations called lamellae around the edges of their bills. Ducks use theirs to strain food out of water. Geese, like eternally teething toddlers, use theirs to chew on anything handy. Lamellae allow ducks to be foragers and geese to be grazers despite their lack of teeth. Given sufficient space with succulent vegetation, waterfowl can survive nicely on whatever they scrounge up. A grassy pasture is ideal; so is a lawn, provided it is not maintained with toxic chemical sprays. If you are a gardener, you would please your waterfowl by growing extra leafy greens, such as kale, cabbage, lettuce, alfalfa, or clover. Leftover fruits and vegetables make welcome additions to the waterfowl diet, provided they're fresh and not spoiled or moldy. Raw potato peels are not readily digestible and should not be offered.

Geese can get along almost exclusively on succulent greens and are often used as cheap labor for weeding crops. Although any breed may be used as weeders, Chinese are most often used, because they are energetic and active yet light enough not to inflict much damage on cash crops. A pair of mature geese in the average backyard garden would likely do more harm than good, but a few goslings may work perfectly as springtime weeders. Goslings begin to graze as young as 1 week of age, but they must be enclosed at night for warmth until they feather out at about 6 to 8 weeks.

Geese are especially adept at controlling troublesome grasses, such as crabgrass and Bermuda grass. They have been used to weed sugar beets, cane, potatoes, onions, tobacco, cotton, vine-grown berries, flowers, and ornamental plants. They are also kept in strawberry fields but must be removed before the berries ripen,

Range trough feeder

Grit

The digestive system of a bird starts with its crop, where swallowed food is stored. From there, it goes into a tough organ called the gizzard, where food particles are rubbed against pebbles and sand the bird has eaten. This grit acts as the bird's teeth to grind up food matter so it can be digested. Eventually the grit, too, gets ground up in the process and must be continually renewed.

Ducks and geese that spend time foraging pick up natural grit in the course of their daily activities. To make sure they get enough for proper digestion, insoluble grit should be available at all times. Granite grit may be purchased at a farm store. If you prefer, you may substitute clean coarse sand or fine gravel from a clean riverbed. Ocean sand may also be used, but it must be thoroughly leached of salt to avoid salt poisoning.

because they like strawberries almost as much as people do. Geese may be used in a vineyard or orchard to clean up windfall fruit, which otherwise attracts insect pests, but they should not be put in with young trees or vines, as they will strip away the flexible bark. Geese will keep an irrigation ditch free-flowing by clearing out grass and water weeds. They are also good at controlling weed growth along fence lines.

Weeder geese must have an abundant source of drinking water and a shady area where they can get out of the sun. Shade might be provided by shrubs and trees, or by a manmade device, such as a large beach umbrella. Weeders will concentrate their efforts near the source of water and shade, both of which may need to be moved daily to ensure that the entire area gets weeded.

Allow geese into the garden only after the crop has grown beyond the young succulent stage, when it becomes less appetizing than sprouting weed shoots. Should they get ahead of weed growth and not find enough to eat, the geese will start in on your crop. As succulent weeds become scarce, your geese may require supplemental feeding, but take care not to overfeed them, or they will tackle weeds with far less enthusiasm. To ensure that geese weed actively throughout the day, feed them only in the late afternoon or early evening.

Water

Having evolved in water environments, waterfowl cannot store water in their bodies and so, for their size, drink relatively large amounts of water. They may become ill if they are deprived of drinking water for more than a few hours. Laying breeds will stop laying and go into a molt when deprived of water. Some producers take advantage of this phenomenon by deliberately throwing a bevy into molt to reinvigorate egg production.

Unless your waterfowl yard has a free-flowing source of water, you will need to provide plenty of clean drinking water. The two most important features of a proper waterer are its ability to be easily cleaned and the ability of ducks and geese to dip in their entire heads. Waterfowl keep their bills free of caked mud and feed by squirting water through their nostrils. The water trough must be easy to clean to prevent an accumulation of sediment from all that face washing.

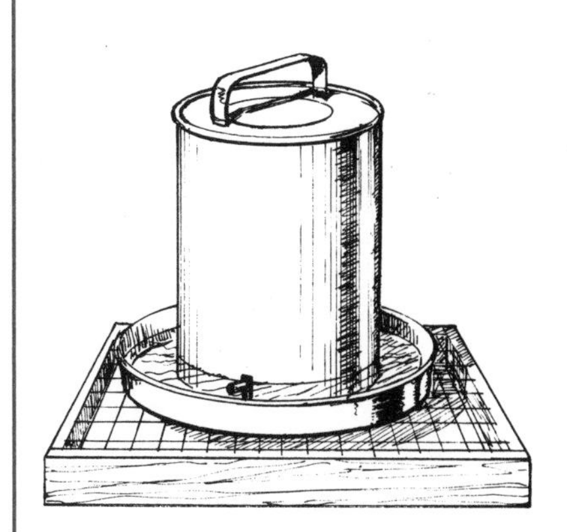

Raised wire waterer

Using an automatic watering device is an ideal way to ensure that your gaggle or bevy has a continuous flow of fresh drinking water. Hog waterers work well for geese, and chicken waterers are suitable for ducks.

In an area where temperatures dip below freezing, providing unfrozen drinking water can present a challenge. Water lines may be wrapped with heating tape, or a livestock heating device may be used in the trough, although don't be surprised if your geese work determinedly to dismantle any device they can get their bills on. A recirculating pump on a pond will help maintain an area of open water. If nothing else, place a rubber livestock feed pan beneath a faucet left open enough to maintain a steady drip.

Waterfowl like to bill water out of a trough, causing muddy puddles to form. The area around a waterer must therefore be able to drain quickly. Placing the waterer on a platform of wooden slats or narrowly spaced wire that is set on a bed of deep gravel will considerably improve drainage.

Automatic waterer

DUCK AND GOOSE EGGS

Most domestic ducks and geese start laying early in the spring of their first year. How well they lay and the length of their laying season depends on the following factors:

- Breed
- Strain (subgroup within the breed)
- Age
- Diet
- Overall health
- Weather
- Lighting conditions

Who's a Layer?

Campbells and Runners may begin to lay at 6 months of age and continue laying nearly year-round. In a warm climate, Muscovies can also be prolific layers. Among geese, Chinese are the best layers. Breeds, and strains within breeds, that have been developed for meat or ornamental purposes do not lay nearly as well as breeds and strains that have been selectively bred for their laying abilities. The best layers stop production only briefly during the fall molt.

Most geese hit their stride in their second year of life and peak at five, although they may lay for up to 10 years. Muscovies may lay for 6 or more years. Mallard-derived ducks normally lay well for 3 years, and some strains continue to lay efficiently for up to 5 years. If your purpose in keeping a bevy is egg production, replace older ducks before they lose efficiency.

Waterfowl tend to lay in the early morning hours. Ducks generally lay one egg every day, geese one every other day. The nonlayer breeds will lay a consecutive batch of eggs, called a clutch, and then take up to 2 weeks off before starting another clutch. If you leave a clutch in the nest, a duck or goose may start setting and stop laying for the season.

Egg size depends in part on the breed and the strain, but mostly on the size of the bird. Larger birds lay larger eggs. Smaller, yolkless eggs commonly signal the beginning and end of a bird's laying season, but they may also be laid during hot weather, especially if drinking water is in short supply.

Geese lay white eggs. Ducks may lay white or greenish eggs. Most consumers prefer white eggs, but the flavor of the egg is the same, no matter what color the shell is.

Feeding Layers

As the laying season approaches, ducks and geese require more nutrition. About 3 weeks before you expect them to start laying, gradually switch from maintenance ration to a layer ration that contains 16 to 20 percent protein. In warmer weather, fowl need more protein to keep eggs from getting smaller. If your farm store doesn't carry layer ration for ducks, layer ration formulated for chickens works well.

For sound, thick eggshells, provide calcium continuously throughout the laying season. Ducks and geese that have access to forage usually get plenty of calcium from the shells of bugs and snails they eat. Still, it's a good idea to provide them with a free-choice hopperful of ground oyster shell or limestone (but not dolomitic limestone, which can inhibit egg production).

Nests

Providing nests for your ducks and geese to lay their eggs in helps keep the eggs clean and protects them from being cooked by sun, washed by rain, or frozen in cold weather. Eggs laid in nests are easier for you to find than eggs hidden in the grass, but the latter are more difficult for predators to find. As a rule of thumb, furnish one nest per three to five layers.

The nest may be in the shape of a box, an A-frame, or a barrel on its side and braced to prevent rolling. A doghouse makes a good goose nest; for ducks, a larger doghouse with doors at opposite ends might be partitioned to create two nests. A good nest size is 12 inches square for a duck, 18 inches for a Muscovy, and 24 inches for a goose; the center of the top should also be 12, 18, or 24 inches high, respectively. The precise size is not critical, provided the nest is:

- Tall enough for a layer to enter and sit comfortably
- Wide enough for her to turn around, since waterfowl don't like to back out
- Small enough to offer a feeling of protective seclusion
- Separated physically or visually from the next nearest nest
- Situated in an area that is well protected from predators
- Open at only one end, so the layer can keep an eye on who might be coming
- Large enough to accommodate an abundance of soft nesting material

A thick layer of nesting material, such as shavings, dry leaves, or straw, will keep the eggs clean, reduce breakage, and prevent eggs from rolling out of the nest.

As laying season approaches, ducks and geese will start investigating available nesting sites. If they see an egg already in a nest, they will consider it a good safe place to lay and will be inclined to deposit their own eggs there. To encourage the use of the nests you provide, place an artificial egg in each one. Fake eggs were once commonly sold for this purpose. The toy eggs available around Easter work well and, for ducks, so do old golf balls. Sometimes fake eggs aren't enough to entice a duck or

A nesting box should offer the appropriate amount of space, privacy, and protection

goose to lay in the place of your choosing. If you keep track of the laying habits of your bevy or gaggle, you'll know when you are not getting your proper quota and must initiate an egg hunt.

Layer Problems

Ducks and geese encounter few problems associated with laying eggs. Still, for the health and safety of your layers, it's important to be vigilant.

Predators. A duck or goose on the nest is not as mobile as a bird that moves freely on the ground or in water. Freedom from predators in the laying yard is therefore essential. Predators may be attracted not only to the layer but also to her eggs in the nest. The predator may peck holes in the eggs and eat them right there (as crows do) or may consider them a take-out meal (as a skunk would). Guests that drop in from above are more difficult to control than those traveling along the ground. The only defense against aerial predators is well-hidden nests.

Egg Binding. Egg binding is a condition in which eggs become stuck and cannot be laid. It may happen when a duck or goose tries to lay an egg that is too big. The bird might, for example, be laying its first egg and not be quite ready. This problem is one of those associated with lighting layers before they reach maturity. Ducks or geese that tend to lay double-yolked eggs may also become bound. Double-yolked eggs occur when one yolk catches up with another, and both become encased in the same shell. Thin-shelled eggs resulting from a dietary inadequacy, such as low calcium intake, may cause egg binding, as may obesity.

Suspect egg binding if a duck or goose is listless, stands awkwardly with ruffled feathers, and has a distended abdomen. If you press gently against the abdomen of an egg-bound bird, you will be able to feel the egg's hard shell.

Sometimes you can lubricate your finger with a water-based lubricant or petroleum jelly, then work your finger around the stuck egg to help it pass. Breaking the egg is another option, although the duck or goose may be injured by sharp shards of shell.

Lighting Layers

To lay well, ducks need at least 14 hours of light per day. To keep the birds laying during the shorter days of late fall, winter, and early spring, you can use lightbulbs to augment natural daylight. Inside the shelter, use one 40-watt bulb for every 150 square feet; outside the shelter, use one 100-watt bulb for every 400 square feet. In both cases, use an inexpensive reflector (available at hardware stores) to direct the light appropriately, and hang the bulb about 6 feet off the ground to reduce shadows. Adjust your lighting times as needed to maintain 14 hours or more of light per day. Even a minor reduction in light hours can throw off the production of laying ducks.

To avoid risks associated with too-early laying, do not begin your lighting program until your ducks reach 20 weeks of age.

Prolapse of the Oviduct. Also known as eversion or blowout, prolapse of the oviduct is protrusion of the oviduct through the vent. It occurs when a duck or goose strains to pass an egg. It is caused by the same factors that cause egg binding or by weakened muscles due to prolonged egg production or excessive mating. Treatment may be successful if only a small portion of oviduct is everted and the treatment is begun right away. Gently wash the protruding part; coat it with a relaxant, such as a hemorrhoidal ointment; and push it back into place. Move the duck or goose to a warm, dry, secluded pen and reduce the protein in her diet to discourage laying.

Collecting Eggs for Eating

Just before an egg is laid, it is coated with a moist film called the bloom. If you ever find an egg just after it's been laid, it will still be wet with bloom. The bloom dries quickly to form a natural protective coating.

Duck eggs left in the nest for a long time tend to get dirty through the layers' comings and goings. But washing eggs removes the bloom and creates a risk that contamination will penetrate the porous shell. Lightly soiled eggs may be brushed clean. A mildly soiled egg may be washed in water that is warmer than the egg is, but it should be eaten soon, because it will not keep as well as an unwashed egg. Extremely soiled eggs should be discarded.

To avoid wasting a lot of eggs through soiling, keep conditions tidy in your waterfowl yard. Gather eggs each morning and discard any with cracked shells or holes pecked in them.

Egg Anatomy. If you break open an egg and examine its contents, you will find two stringy white spots on the sides of the yolk, one of which is usually more prominent than the other. These spots are called chalazae. As the egg was being formed, a layer of dense egg white (called the chalaziferous layer) was deposited around the yolk. At the two ends of the egg, the ends of the chalaziferous layer twisted together to form cords that anchored the yolk to the shell. When you break open an egg, the chalazae break away from the shell and recoil against the yolk. Misinformed people mistake the chalazae for the beginnings of a baby duck or goose.

When you crack open an egg, you will sometimes see a red or brown spot, called a blood spot, on the yolk. A blood spot is caused by minor hemorrhaging in the oviduct. Although it is harmless, it is not appetizing. The tendency to lay eggs with blood spots is inherited. If you can identify the duck laying such eggs, do not use her as a breeder.

Saving Eggs. Refrigerated duck eggs may be kept as long as 2 months but are best used within 2 weeks, while they're still at their peak of freshness. Excess eggs to be used for baking may be stored in the freezer during the laying season for use during the off-season. Since whole eggs burst their shells when frozen, break the eggs into a bowl, add a teaspoon of honey or a half-teaspoon of salt per cup (depending on your preference and the recipe you will use the eggs in), and scramble the eggs slightly, taking care not to whip in air bubbles. Pour the mixture into ice cube trays and freeze it solid, then remove the cubes from the trays and store them in well-sealed plastic bags. The frozen cubes take about 30 minutes to thaw and may be stored for up to 1 year at a temperature no higher than 0°F.

Parts of an egg

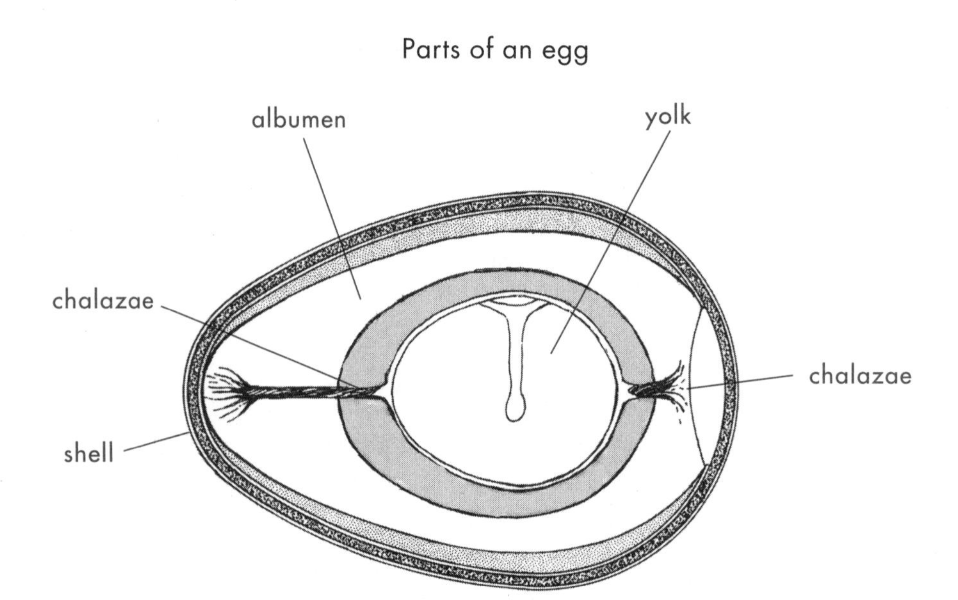

Duck Eggs versus Chicken Eggs

Duck eggs are not much different in flavor from chicken eggs and are more popular in many countries. When fried or boiled, duck eggs have a firmer texture that might be described as slightly chewy. Duck eggs have proportionally larger yolks compared with chicken eggs and therefore lend extra richness when added to batter for baked goods. The white of a duck egg, on the other hand, is difficult to separate from the yolk, making it unsuitable for meringues and similar recipes. Some people have a sensitivity to chicken eggs that can cause cramping but have no such reaction to duck eggs. The opposite is also true.

Breeding Ducks and Geese

Waterfowl naturally start laying eggs as yearlings, although under optimal conditions, production ducks may lay as early as 18 weeks of age. Ducks are at their prime age for breeding from 1 to 3 years of age; geese take a little longer to mature and are at their prime from 2 to 5 years of age. Yearling geese lay eggs, but those eggs won't hatch as consistently during the first year as in subsequent years. A drake is fertile by the age of 6 months and peaks after the third season. A gander reaches his prime of fertility at 3 years of age and peaks at 6 years. A duck or goose will lay eggs regardless of whether or not a male is present, but the eggs will not be fertile and cannot be used to produce ducklings or goslings.

Bonding

One of the most interesting things about waterfowl is the way they form attachments to the opposite sex.

Mallards and Mallard-Derived Breeds. Mallards usually bond in pairs, although if too few drakes are available to go around, some of the drakes will bond with two ducks. Even if you have an equal number of ducks and drakes, one drake may take on two ducks and leave another poor drake with none. This bonding lasts only for the current breeding season. As soon as the ducks start nesting, the drakes band together in a bachelor group. This ritual occurs each year, and one year's pairings may or may not be the same as the previous year's.

During the breeding season, each pair or trio of Mallards forms a close-knit unit. The duck will try to keep other ducks away from her drake, and the drake will run off other drakes. But while all this posturing is going on, actual mating occurs indiscriminately among and between the various pairs, and any time one drake mounts a duck, the others will hasten to join in.

Mallard-derived breeds show some of this fiery attachment to their mates, but the flame burns hottest for true Mallards.

Despite the bonding and apparent preferences for mates, you need only about one drake for each half-dozen ducks. Maintaining fewer drakes is favored by commercial breeders who wish to keep costs down by having fewer mouths to feed.

If a yard is home to more than one breed, the birds will not necessarily bond by breed and will almost certainly interbreed. To avoid having mixed-breed ducklings, the various breeds must be separated from one another at least 3 weeks before breeding season (or 3 weeks before nesting is allowed to begin) and must be kept apart until nesting starts.

Muscovies. Muscovies do not form similar allegiances; rather, the two sexes live separate lives throughout the year. The drakes seem to have insatiable sex drives, though, and won't discriminate between Muscovy ducks and others. Any offspring from the mating between a Muscovy and another breed will be a mule that is incapable of reproduction.

Muscovy drakes use their long, sharp claws and powerful wings in violent attacks on one another. This aggression may be minimized by supplying each drake with at least five ducks and by providing plenty of living space to a bevy containing more than one drake. If you have no need or desire for ducklings, peace and calm will reign in a Muscovy yard that contains only females.

Mating

The lighter breeds of both duck and goose can mate on land with good fertility, but the heavier breeds need water to give the males sufficient buoyancy for breeding. Since the drakes and ganders can get a better grip, breeding on water helps keep ducks and geese from having patches of feathers worn away during the breeding season.

Optimum Mating Ratios*

Duck Breed	Ducks per Drake (one male)	Ducks per Drake (multiple males)
Call	1–2	2–3
Campbell	4–6	5–7
Mallard	1–2	2–3
Muscovy	5–7	6–10
Pekin	2–5	3–6
Rouen	1–2	3–5
Runner	4–6	5–7

Goose Breed	Geese per Gander
African	2–6
Chinese	4–6
Embden	3–4
Pilgrim	3–5
Toulouse	3–4
Toulouse, dewlap	1–2

**During hot or cold weather, more males are needed per female to maintain the same degree of fertility.*

Don't let the idea of aggressive fowl scare you off. Aside from their breeding temper, Muscovy ducks are intelligent birds with sweet temperaments. A row of black-and-white ducks resting side by side in the shelter doorway looks like a pew full of habited nuns.

Geese. Geese generally mate in pairs or trios, although a gander of a light breed may take on as many as six geese. Once a goose and gander have bonded, they may chase away other females expressing an interest in joining them but may readily bond with daughters they raise themselves. Bonded geese will remain tightly bonded as long as they can see or hear each other. If one is removed or dies, the other will mourn. How deep the mourning and how long it continues varies by how long they have been together. The surviving older goose or gander in a pair may mope around, lose interest in eating, and eventually die. A surviving younger goose or gander may one day stop mourning and take on a new mate.

Nesting

The easiest way to get ducklings and goslings from your ducks and geese is to leave the hatching to your waterfowl. When a duck or goose starts getting reclusive, builds a nest to hide her eggs in, spends increasingly more time on her nest, and lines the nest with feathers pulled from her breast, she's getting ready to hatch her eggs. Gathering eggs in a nest and sitting on them until they hatch is called setting or brooding, and the resulting ducklings or goslings are the brood.

A duck or goose is likelier to get broody and stay broody if she can find a secluded place to make her nest. The best way to ensure that is to furnish your waterfowl yard with suitable nesting sites for laying. A marauder-safe nest has a closable front that may be latched at night. A nest with a latchable front must have air vents. Where temperatures get high during the spring and early summer, place nests in a shady area, such as beneath a tree or shrub, or pile brush on top. Since waterfowl like to brood in seclusion, cover nest openings with brush to give the birds a greater sense of security.

If your waterfowl shelter is large enough, you might partition off a section for nests, although waterfowl prefer not to nest in close proximity to one another. They prefer to be given a choice of nesting spots scattered throughout the yard. For successful hatching, provide one nest per female. If nests are in short supply, two females may lay in the same nest and may even set side by side to hatch the eggs, but the resulting hatchlings will probably get trampled in a squabble over whose they are.

A nest against the ground is better than one with a manmade bottom, since soil helps retain the moisture necessary for a successful hatch. Such a nest, however, may not be safe from predators that can tunnel through the soil to get at the eggs or the female bird. Sturdy fine-mesh wire fastened to the nest bottom, then topped with a layer of soil, will ensure both good humidity and protection from diggers. If you use buckets or barrels as nests, block both sides to prevent rolling and tamp clean soil into the bottom to form a level floor. Supply each nest with nesting material in the form of dry leaves or straw, and each female will shape her own nest.

A duck or goose will usually continue to lay in the same nest, provided it remains undisturbed. If one day she enters to find all the eggs gone, she is likely to look for a different place to lay. If you wish to hatch more ducklings or goslings than your waterfowl can hatch naturally, you'll need to collect some of their eggs to hatch in an incubator. Mark a couple of eggs and leave them in the nest to keep the duck or goose coming back to lay more. Later, you can let the duck or goose accumulate a nestful of eggs to hatch, or you might help her along by adding eggs from another bird to her nest. Be sure to remove the ones you marked, because they will be old and will rot rather than hatch in the heat under her breast.

During the brooding period, the mother duck or goose leaves the nest only occasionally.

When a goose has 6 to 12 eggs in the nest and a duck has 12 to 18, she will probably get the urge to set. If a duck or goose accumulates many more eggs than that number (as can easily happen when more than one female lays in the same nest), she may not be able to cover them all with her body; as a result, few or none will hatch. A duck or goose stops laying soon after she starts setting. During the brooding period, she will stay on the nest most of the time, leaving only occasionally to eat, drink, or swim. While she is off the nest, which may be as long as an hour, she will cover the eggs with feathers and nesting material to keep them hidden and warm.

When a duck or goose comes back from her swim and hunkers back down on her nest, moisture from her feathers will keep her eggs from drying out. As she settles back into the nest, she reorganizes the eggs by rolling them with her bill or paddling them with her feet. Sometimes during this process, an egg will roll out of the nest. A good broody will roll the egg back in with her bill, but a lazy mother will let it go. A duck or goose that is sloppy with her eggs may thus lose them one by one until none is left to hatch. To avoid this problem, add a lip at the front of the nest that is low enough for the female bird to step over but high enough to keep eggs from rolling out. A solid wall at the back of the nest will give the duck or goose a strong sense of security, since she'll then need to watch for predators approaching from only one direction.

A drake will stand guard while his duck lays an egg, but after she starts nesting, he loses interest and wanders off on his own. This attitude is probably a protective ploy to draw attention away from the nesting female. A gander, on the other hand, becomes fiercely protective, standing guard and warding off all intruders throughout the brooding period. This "meanness" is reserved for outsiders, for when the

Mothering Instinct and Brooding Time

Breed	Mothering Instinct	Days to Hatch
Ducks		
Call	Excellent	28
Campbell	Fair	28
Mallard	Excellent	28
Muscovy	Fair	35
Pekin	Good	28
Rouen	Poor	28
Runner	Fair	28
Geese		
African	Fair	30
Chinese	Poor	30
Embden	Fair	30
Pilgrim	Good	30
Toulouse	Fair	30
Toulouse, dewlap	Poor	30

goslings arrive, the gander shares equal responsibility for protecting them from harm. So strong is a gander's parental instinct that, on his own, he may adopt an orphaned brood of goslings or ducklings.

If all goes well, the hatchlings will appear in approximately 28 days if they are of a Mallard-derived breed, 30 if they are geese, or 35 days if they are Muscovies. The hatchlings will remain hidden under their mother for the first day or two before venturing forth into the big world. Even after leaving the nest, they will stay close to their mother and periodically crowd around her for warmth. On cold or rainy days, she may open her wings like an umbrella to provide additional protection from the elements. In early spring when the weather is still cold, or in rainy weather, confining the family to a shelter will help ensure the survival of the hatchlings.

Artificial Incubation

To develop into ducklings or goslings, waterfowl eggs must be subjected to a specific amount of heat and humidity for a specific length of time. Most mother ducks or geese know what to do. But even when they do everything right, you may wish to hatch the eggs in an incubator for one reason or another:

- Your breed may not be reliable setters.
- If you raise production layers, you may wish to keep the ducks laying and not setting.
- You may want more hatchlings than your females are able to hatch.
- Local wildlife may disturb the nests.
- Your waterfowl yard may otherwise not be conducive to encouraging broodiness.

Incubators

Incubators come in assorted sizes, shapes, and styles. The capacity of any incubator is measured in terms of how many eggs it will hold. Some incubators specify duck or goose eggs; others specify only chicken eggs. To estimate the equivalent number of waterfowl eggs, figure 100 percent capacity for Call duck eggs, 75 percent for Mallard eggs, 60 percent for Muscovy eggs, and 40 to 30 percent for goose eggs. An incubator may be either a still-air or a forced-air model, the difference being that the former has only a heat element, whereas the latter also has a fan to circulate the heat.

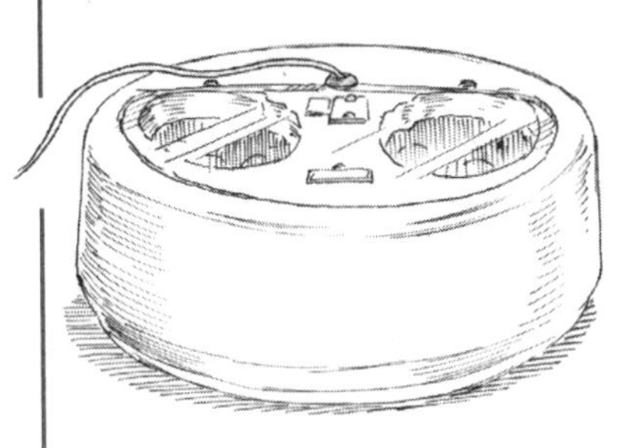

Still-Air Incubators. Still-air incubators are on the small side, starting at a capacity of about half a dozen eggs, and are popular primarily for educational projects. A still-air incubator may be used to hatch duck eggs, but only the largest models will successfully hatch goose eggs. Still-air incubators are more sensitive than forced-air incubators to environmental conditions; must be kept away from sunlight, drafts, and heaters; and must be operated where the room temperature remains steady (a temperature of 70°F is ideal). A still-air incubator is unlikely to have an automatic turning device. Without one, you will have to turn the eggs by hand three times a day.

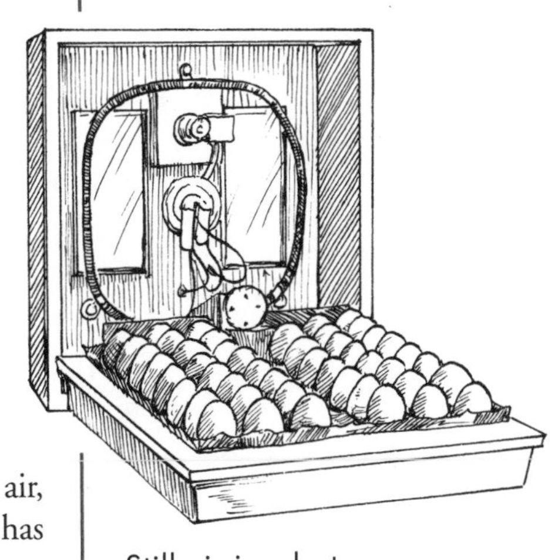

Still-air incubators

Forced-Air Incubators. A forced-air incubator has a fan to circulate the warm air, ensuring an even temperature throughout. The smallest forced-air incubator has about the same capacity as the largest still-air incubator. Big commercial models are

Location, Location, Location

The incubator should be set up where it is easy to monitor, so you can check often to see that the temperature is remaining steady. But it should not be in a high-traffic area where a person or pet is likely to trip over the cord and pull the plug.

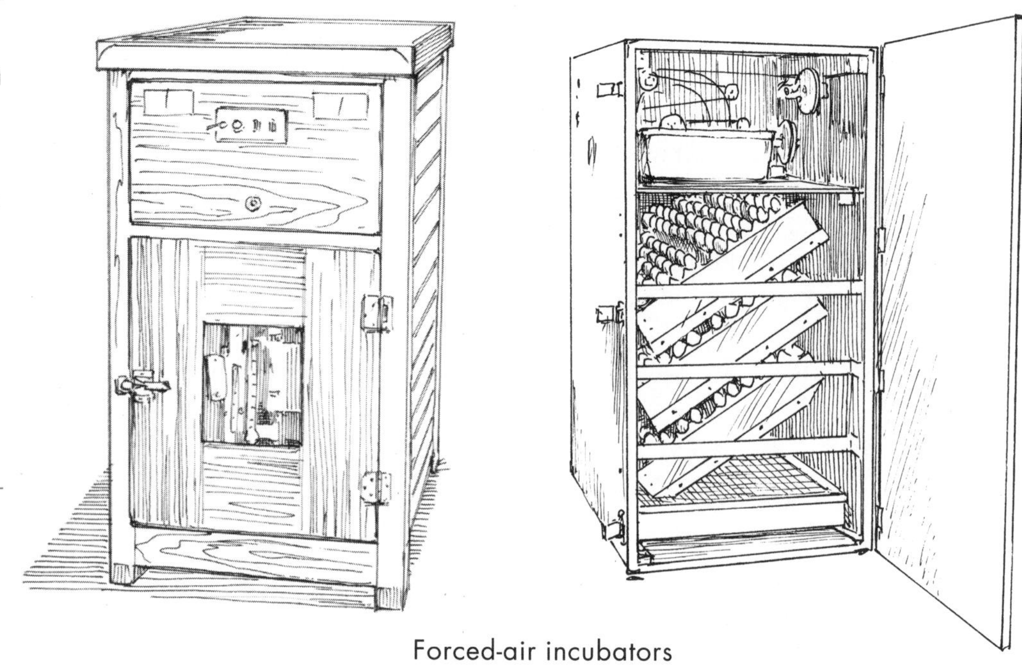

Forced-air incubators

room-sized and hold thousands of eggs. Forced-air incubators are less sensitive to ambient conditions, but they still operate best in a stable environment. A forced-air incubator will probably have a device that automatically turns the eggs, but that device must be correctly sized according to whether you are hatching duck or goose eggs.

Collecting Eggs for Hatching

Eggs collected for hatching must be handled carefully to preserve hatchability. Determining hatchability starts with an examination of the breeding bevy or gaggle. Collect eggs only from healthy, vigorous birds that are free of defects and deformities. Collect eggs daily and sort them by size, shape, and color. Select those with solid, thick, even-textured shells without cracks. Discard eggs that are damaged, deformed, or not typical for your breed.

Slightly soiled eggs should be wiped clean. Heavily soiled eggs should be discarded or rinsed in lukewarm water and immediately dried with a clean paper towel. A quick dip of all eggs in a solution of chlorine bleach and lukewarm water, mixed in the proportions listed on the label for household cleaning, will disinfect the shells and help keep bacteria out of your incubator. Be sure the rinse water or bleach dip is warmer than the eggs, or any bacteria on the shell may be drawn inside.

Mark each egg with the date on which it was laid. Use a grease pen, wax pencil, or crayon; do not use a sharp-pointed pencil or pen (which might pierce the shell) or a marker (which may contain toxins that could absorb through the shell). If you have more than one breed and their eggs look alike, mark each egg with a letter or symbol telling you what breed it is. Place these marks at the blunt end of the egg. If your incubator has no automatic turner, mark a prominent X on one side and a O on the opposite side to facilitate hand turning.

Before incubation, you can store the eggs where the temperature remains between 50 and 60°F and the relative humidity is around 75 percent. Such conditions are often found in a pantry or cellar. Store the eggs in an egg flat with the pointy end down. If you don't have an egg flat, you can make one by cutting the cover off an egg carton and using just the bottom. Elevate one side of the flat about 2 inches (a chunk of 2 x 4 lumber works great for this) and each day, switch the side that is elevated. Hatching eggs may be stored under these conditions for 7 days. After that, their hatchability gradually declines.

At the end of the week or when you have enough eggs to fill your incubator, whichever comes first, plug in the incubator and get it up to proper temperature and humidity. Load the eggs with the blunt end elevated and the pointed end downward. If your incubator has several trays and you don't have enough eggs to fill them all by the end of a week, you can fill the trays at weekly intervals until they are all full, or until the first tray hatches and is ready to be filled up again. Keep track of which week each tray will hatch.

Turning the Eggs

If your incubator is not equipped with a turner, you'll have to turn the eggs by hand three times a day, at regularly spaced intervals. Wash your hands first to keep body oil and grime from contaminating the shells. Turn all the eggs with the X upward at one turning, and those with the O upward at the next turning. Be sure to leave each egg with its pointed end lower than the blunt end.

Three days before the expected hatch, stop turning the eggs to give the fellas inside a chance to get oriented and plan their escape route. If your incubator has an automatic turner, remove the eggs from the turner, remove the turner from the incubator, and put the eggs back into the incubator. Don't be tempted just to unplug the turner with the eggs in it, because a delicate hatchling could get tangled in the turner and injure a leg.

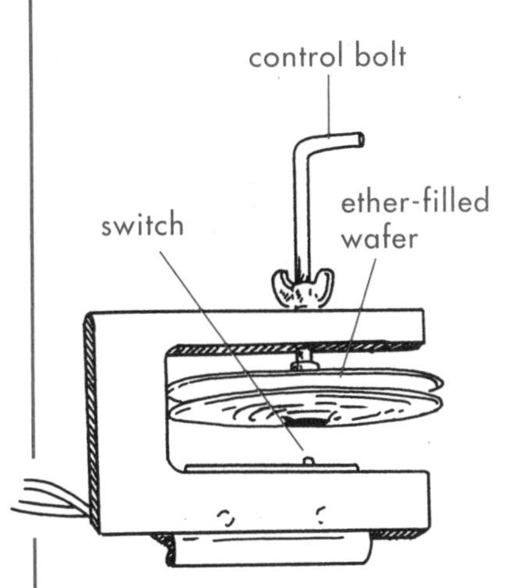

Incubator thermostat

Temperature

The instruction manuals that come with different incubators vary in their specified operating temperature, not because the actual temperature at which eggs hatch varies, but because of the location of the thermometer. Be sure that the thermometer is properly positioned according to the manufacturer's instructions. Adjust your incubator to operate as closely as possible to the temperature specified in the instructions for the eggs you are hatching. Standard temperatures for waterfowl are 99.5°F in a forced-air incubator and 102°F in a still-air incubator.

The thermometer is visible through a little window or sticks up through a hole in the incubator lid. A thermometer suitable for incubators registers temperature in increments of one-quarter degree. If a proper thermometer was not furnished with your incubator, you may purchase one at a farm store or through a mail-order poultry supplier.

The temperature is regulated by an ether-filled disk, called a wafer, that expands as it gets warm and contracts as it cools. As the wafer expands, it pushes against a thermal switch that causes the heat to go off. When the wafer cools and contracts, it releases the switch and causes the heat to go back on. The temperature is controlled by adjusting the distance between the wafer and the switch, usually by turning a screw on the outside of the incubator. Make sure the temperature is both correct and steady before putting eggs into the incubator. When you put in the eggs, the temperature may temporarily drop while the eggs warm up to the incubator's operating temperature. Do not adjust the thermostat during this time.

When hatching time approaches, the temperature in the incubator may rise because of all the body heat generated by hatching activity. Keep an eye on your thermometer and reduce the heat as necessary to keep it at the proper level.

In most incubators, the temperature doesn't remain perfectly steady; rather, it fluctuates as the heating coils heat up and cool down. In a properly designed incubator, the fluctuation remains within one-half degree on either side of the incubator's proper operating temperature. A poorly designed incubator that fluctuates more than that will produce poor hatches.

The temperature will vary more than usual when a storm moves through and the barometric pressure changes. Unless the storm front is predicted to hang overhead for an extended period, don't readjust the thermostat; otherwise, after the storm passes, your incubator's temperature will be out of whack. If you must adjust the wafer during an extended storm, keep a close eye on your incubator and readjust it when the storm passes.

Incubator Malfunctions. If the power goes out or the heater or thermostat in the incubator malfunctions, be prepared with an emergency plan. If you have a computer with an uninterruptable power source (UPS), shut down your computer and plug your incubator into the UPS. Without a UPS, you can keep an incubator warm for a short time by wrapping it in sleeping bags or down quilts. For extended periods, move the incubator near a heat source, such as a woodstove or a gas or oil heater, taking care not to let it get too hot. Until the power is restored, don't open the incubator, even to turn the eggs.

If the ether should leak out of the wafer, the heat will steadily increase until your eggs cook. Keep a spare wafer handy that is the right size for your model. Thermal switches sometimes get clogged with dust and down from the hatch and after a while will stop working. To keep the switch functioning properly, spray it with pressurized air after each hatch. Canned air is available from office supply outlets and computer stores.

An incubator malfunction may or may not affect your hatch, depending on how hot or cold the eggs get and for how long. Get your incubator back to normal operation as quickly as you can and proceed as if everything is okay until you find out otherwise. You might be surprised with a perfectly fine hatch. If not, rest assured that even a setting duck or goose occasionally blows it.

Humidity

To keep eggs from drying out during incubation, all incubators have a means of supplying humidity by evaporating water. The water pan may be designed with sloping sides or separate sections. Adding more water to a sloping pan, or filling more sections, creates a larger evaporating surface to increase humidity. Most models have vents that may be opened and closed for additional humidity control. Whenever you open the incubator, moisture will escape. Have a spray bottle ready, filled with clean lukewarm water, and use it to mist the eggs. Before closing the incubator, check the water pan and, if necessary, add warm water from your spray bottle.

Washing eggs before incubation removes the natural protective coating on the outside of the shell, hastening evaporation of the contents within. To hatch successfully, washed eggs therefore require a slightly higher humidity than unwashed eggs. Higher humidity is also required during the last 3 days of incubation, to ensure that the emerging ducklings or goslings don't dry out and get stuck in the shell.

Humidity is measured by using a hygrometer, which may or may not come with the incubator. Hygrometers are available from poultry supply outlets. Some hygrometers are calibrated in wet-bulb temperature, others in relative humidity. Either way, the hygrometer is essentially a thermometer with its bulb wrapped in a cotton

Incubation Humidity		
	Wet-Bulb	**Relative Percent**
Duck eggs		
Unwashed	82–84°F	55
Washed	84–86°F	65
Past 3 days	90–92°F	75
Goose eggs		
Unwashed	84–86°F	65
Washed	86–88°F	70
Past 3 days	92–94°F	80

wick. When the wick is immersed in water, the hygrometer measures the rate of evaporation from the wick. For an accurate humidity reading, the wick must remain wet at all times, be kept clean through periodic washing, and be replaced if hard water encrusts it with mineral deposits.

Placing a hygrometer in a small incubator would take up space that might best be used to hatch more eggs. Luckily, you have another way to determine the humidity, which is to monitor the rate at which the air cell at the blunt end of the egg increases in size during the hatch. If it enlarges at the rate shown in the accompanying sketch, your humidity is right on. If the air cell enlarges too rapidly, increase the humidity; if it is not enlarging rapidly enough, reduce the humidity. Air cells may easily be monitored by means of candling.

Candling

During incubation, an embryo develops from a spot on the yolk called the germinal disc, where fertilization takes place. You can find this spot if you break a fresh egg onto a plate. Not all eggs are fertile, even when a drake or gander is present, and not all fertilized eggs will hatch. During incubation, if you shine a light through an egg, you can see how it is developing. This is called candling, even though no one uses a candle anymore. Candling devices are available from poultry supply outlets, although a strong penlight flashlight works better than some bona fide candlers.

After the first week of incubation, candling will let you determine the size of the air cell and how the embryos are developing. In a darkened room, hold an egg against the light and rotate it gently until you see something or ascertain that nothing is there to see. An egg that is fertile and developing properly will reveal small blood vessels forming a small spiderweb pattern. Discard all other eggs. If you aren't sure, sacrifice a couple of eggs by cracking them open and examining the contents. You will soon be able to relate what you see through the shell to what's happening inside.

If you see nothing, the egg (called a clear) is either infertile or the embryo has died almost immediately, probably because of some problem related to breeder management. A thin dark ring around the circumference means the embryo started to develop but died. If you see murky clouds floating inside the egg, the contents have

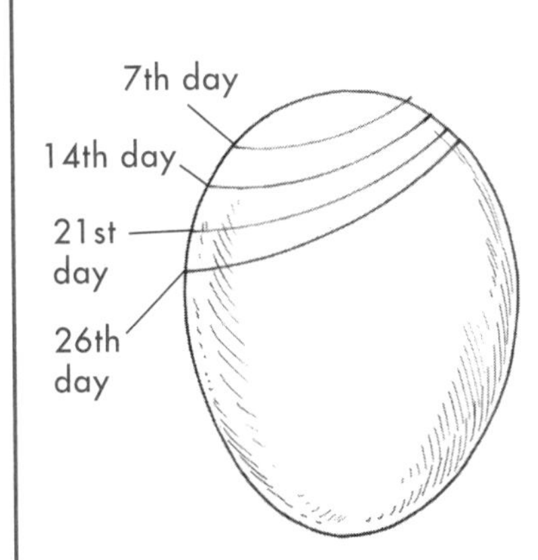

When candled, eggs that are dehydrating at the proper rate will show air cell sizes proportional to those here.

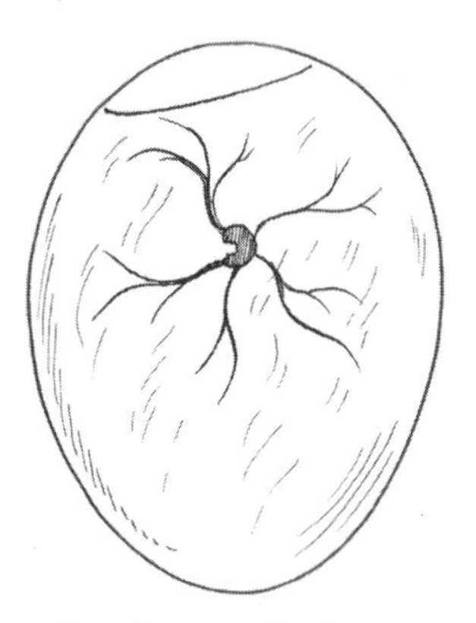
Fertile candled egg

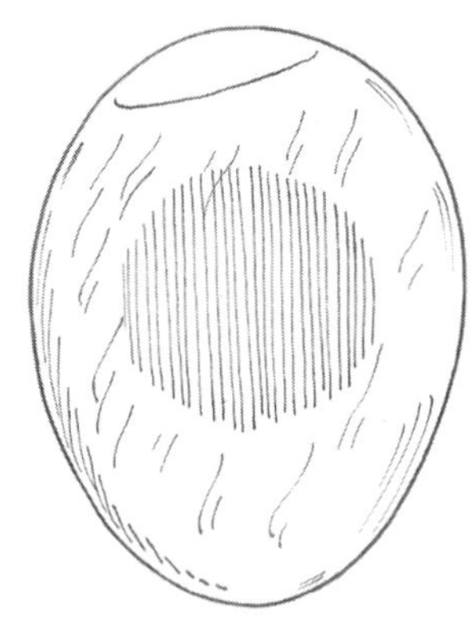
Nonfertile candled egg

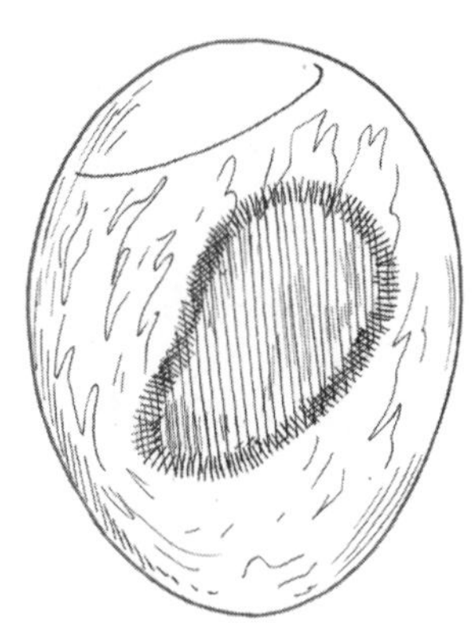
Dead candled egg

begun to rot. Rotting may be due to several causes, including partial development before incubation (perhaps because a duck or a goose started incubating the egg before you collected it), improper storage conditions, and improper operation of the incubator. Rotting eggs use up oxygen in the incubator that is needed by properly developing eggs, and a bad egg can pop at any time. You haven't lived until you've smelled the foul odor of an exploded rotten egg.

Candle the eggs again after 2 weeks of incubation to monitor air cell growth and make sure development is still on track. By this time, a properly developing embryo will fill much of the shell, revealing only a dark shadow. Watch closely, and you will see movement at the air cell end.

Hatching

Your eggs will hatch in about 28 days for Mallard-derived breeds, 30 days for geese, and 35 days for Muscovies. The hatch should take 24 to 48 hours for ducks of all kinds and 48 to 72 hours for geese. The exact incubation period and hatching time may vary depending on such factors as the vigor of your breeders and the precise temperature at which you operate your incubator.

Three days before you expect the hatch, stop turning the eggs and remove the turning device, if one is present. Closely monitor the temperature, especially in a small incubator, to make sure it doesn't increase because of heat generated by the hatchlings. Do not open the incubator again until the hatch starts, except as absolutely necessary to maintain or increase the humidity. Some incubators are fitted with a funnel-like device that lets you add water without opening the incubator. Some incubators, especially the smaller polystyrene (Styrofoam-like) models, may be rigged up with a hole in the lid and a short length of siphon hose with a funnel at the end. Keep the hole plugged until you need to add water, at which time remove the plug, insert the siphon or funnel, and pour in a measured amount of clean warm water. You may need a little practice to properly aim the siphon and get the water in the right place.

One of the most exciting moments of any hatch is when the eggs begin to peep — yes, you will hear peeping even before you see the first hatchling. Soon after the peeping starts, the first eggs will pip, meaning a piece of shell will be chipped away from the blunt end. After pipping, the hatchling turns itself around 360 degrees, cracking the shell all around with its egg tooth — a tiny sharp tool at the end of its bill that falls off soon after the bird hatches and no longer needs it.

Opening a Forced-Air Incubator

If you are operating a forced-air incubator with a built-in fan, turn off the power switch before opening the door to turn or candle the eggs or to remove hatchlings. If the fan is running when you open the door, cooler air will circulate in the incubator, making it more difficult to heat back up to proper temperature after you shut the door. Don't forget to turn the power back on when you shut the door. This style of incubator usually has a power indicator light, and a good habit to develop is to check the light before you walk away (or anytime you pass the incubator). Small forced-air incubators with the fan in the lid and still-air incubators that lack a fan need not be turned off.

When the egg is cracked all around, the hatchling gives a mighty shove and breaks free from the shell, exhausted and wet. If it is too weak to get out of the shell, avoid the temptation to help it out. A weak hatchling will probably not grow into a healthy, vigorous adult. If many hatchlings cannot get out, the cause is quite likely low humidity. Keep the water pan clean of surface fluff. In the future, try adding a second water pan.

The hatchlings will soon be rested up and ready to explore. By this time they will be dry, fluffy, and recognizable as downy ducklings or goslings. They may now be removed from the incubator and placed in a brooder. To avoid disrupting the hatch by frequently opening the incubator, remove the hatchlings in batches twice daily.

Hatchlings are tired and wet when they emerge from their shells but soon dry out and are ready to explore.

Cleaning Up

Hatching is a messy affair. It produces not only a lot of debris but also a lot of bacteria and other organisms that flourish in a warm, humid environment. As soon as the hatch is over and all the hatchlings, shells, and unhatched eggs have been removed, clean up the remaining mess. Remove all removable parts and wash all those that are washable. Vacuum away dry debris, spray the heater and thermostatic units with pressurized air, and sponge out the remaining debris. Scrub the incubator with warm, soapy water and follow up with a good disinfectant. Even the best disinfectant is ineffective against organic debris, which is why it's so important to first scrub the incubator clean.

Quaternary ammonia compounds are among the best disinfectants for incubators. They are sold in many places for all sorts of uses, from cleaning dog kennels to sterilizing surgeons' hands, and may be found under a variety of brand names. They leave no stain and have no odor, yet are strong and effective when used according to directions. A good alternative is chlorine bleach, diluted in hot water at a proportion of ¼ cup per gallon.

Thoroughly scrub and disinfect your incubator after each hatch, whether you plan to hatch another batch of eggs or store the incubator until next season. If your first hatch is successful but your subsequent hatch rate goes downhill, you can be pretty sure your incubator was not properly cleaned and disinfected.

Improving Success

Eggs fail to hatch for any number of reasons. If 60 to 75 percent of your fertile eggs hatch, you are keeping up with the average backyard incubator operator. You may achieve above-average hatching success in two ways: by improving your breeder management and by improving the way you operate your incubator.

If a significant percentage of your eggs are infertile, the problem lies with breeder management. Review your feeding program (are you increasing the protein level just before and during the breeding season?), availability of swimming water to breed on (especially for the larger breeds), age of your breeders (fertility is low in young and old birds), mating ratio, space available per bird, and general suitability of your facilities. Besides improving fertility, good breeder management also improves hatchability.

If most of your eggs are fertile but still fail to hatch, the problem lies in your incubator. A homemade incubator can be especially problematic. It may, for instance, not be well enough insulated to maintain a steady temperature, or it may have air pockets where the temperature is lower or higher than the thermometer reading. Even among commercial incubators, each has its own quirks, enhanced by the quirks of each unique environment in which the incubator is operated. Assuming that your incubator is properly built, take time to tweak your settings.

Start by keeping accurate records for each hatch: day set, day of hatch, how many hours between the appearance of the first hatchling and the last, temperature setting, spikes or dips in temperature (and what might have caused them), and humidity maintenance.

When the hatch is over, break open all the eggs that fail to hatch and note at what stage of development each died — first week, second week, fully formed but failed to pip, pipped but failed to hatch. Also note any problems among the hatchlings, such as weakness, crooked feet, and other deformities. Consult a comprehensive incubation troubleshooting chart, which may have come with your incubator, may be obtained from your county Extension office, or may be found in *The Chicken Health Handbook*. Incubation problems are similar regardless of whether you hatch waterfowl eggs or landfowl eggs.

You will probably find a pattern that points to a single problem, such as the temperature being too high or the humidity being too low. Even if your findings point to several different problems, select only one problem to correct and see if it improves your next hatch. If so, tackle the next single problem; if not, back off, reexamine your findings, and refine your tweaking. Hitting on the optimum combination of settings for your situation may take several hatches, but keep at it and you may improve your hatching success to 90 percent or better.

Ducklings and Goslings

In contrast to species that are born blind, naked, and helpless, waterfowl hatchlings are precocial, meaning that soon after the hatch they are out and about, exploring their surroundings and seeking out good things to eat. Their downy coats offer some protection from the elements, but if they have no mother duck or goose to shelter them from cold and rain (not to mention predators), they must be housed in a brooder until they are old enough to fend for themselves.

Spraddle Leg

A hatchling does not have strong legs. If it is forced to walk on a slick surface in the incubator, brooder, or shipping box, its feet can slide out from under and splay in two directions. To keep the feet together, bind the legs loosely with a piece of soft yarn tied in a figure eight or a rubber band knotted at the center. Provide a rougher surface, such as paper towels or hardware cloth, to ensure firm footing. Watch closely to make sure the hobble does not get too tight, and remove it after a couple of days as the legs toughen up enough to hold the little bird.

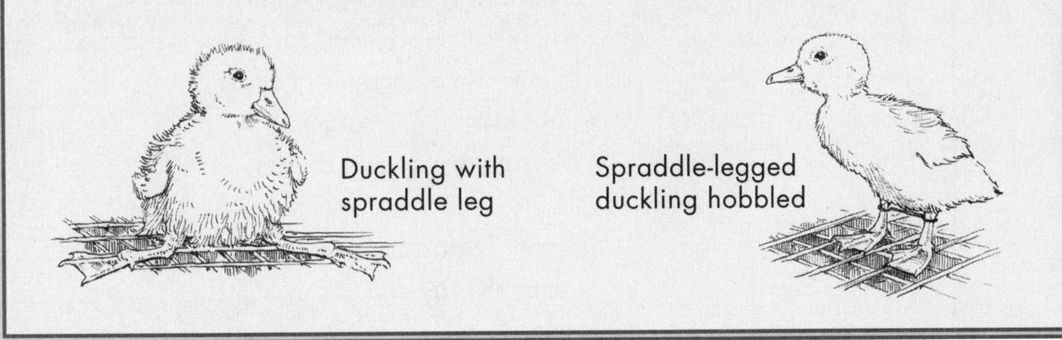

The Brooder

A brooder is a protected place that provides hatchlings and growing fowl with safety, warmth, food, and water. A home brooder can be made from a sturdy cardboard box. Line the box with newspapers over which is placed a layer of paper towels for the first few days to provide sure footing. A sheet of small-mesh wire fastened across the top of the box keeps out rodents, household cats, and other predators. An empty feed sack or a few sheets of newspaper covering the mesh guards against drafts.

For warmth, hang a light over the hatchlings, taking care that the bulb doesn't touch any cardboard or paper to avoid starting a fire. Using a reflector from the hardware store is an inexpensive way to deflect the heat downward and to keep the bulb from directly touching any paper parts. Hang the reflector by a chain that allows you to raise the bulb as your waterfowl grow. (See the illustration on page 48.) Be prepared to provide heat until the birds feather out. The rules of thumb are as follows:

Start the brooder temperature at 90°F. Decrease it five degrees per week until you get down to 70°F.

If the days are warm and nights are cool, you may need to provide heat only at night after the first couple of weeks.

To gauge brooder temperature, you don't need a thermometer. Just watch how your ducklings or goslings react to the source of heat. If they remain normally active, they are comfortable. If they huddle under the light and peep loudly, they are cold. If they spread as far as possible away from the heat and pant, they are hot. (See the illustrations on page 49 for details.)

As they grow, the hatchlings will naturally generate more body heat and rely less on artificial heat. The more rapidly you reduce the heat you provide (by raising the light or reducing the wattage), the more rapidly your waterfowl will feather out and become less dependent on artificial heat.

Identifying Hatchlings

Breed	Appearance
Ducks	
Call	Like Pekin or Mallard, only smaller
Campbell, khaki	Brown with dark brown bill
Mallard	Black and yellow, with yellow eye stripes
Muscovy, colored	Black and white, long toenails
Muscovy, white	Yellow with black head spot, long toenails
Pekin	Yellow with orange bill and feet
Rouen	Like Mallard, only larger
Runner	Vertical stance, various colors
Geese	
African	Greenish gray and yellow
Chinese, brown	Greenish gray and yellow
Chinese, white	Yellow
Embden	Yellow and pale gray
Pilgrim	Like Embden, only smaller
Toulouse	Like brown Chinese, only larger

At first, the hatchlings will require no more room than is needed for themselves plus containers for feed and water, but as soon as they get active they will need approximately ¼ square foot per duckling and twice that per gosling. Their space needs will double by the third or fourth week and double again from the fifth or sixth week until they are ready to live outdoors.

Good sanitation throughout the brooding period means dealing with all the moisture and fluid droppings your waterfowl will generate. For the first few days while the hatchlings are on paper towels, freshen the box by adding a new layer of towels, then periodically rolling them all up and starting over. As the young waterfowl get older, they will generate increasingly greater quantities of wet mess that can't be handled with paper. At that point, you must move them to absorbent litter, such as wood shavings.

Water

Baby ducks and geese love to play in water. This attraction leads to two serious problems: They can quickly make a mess of even the best-kept brooder, and they can just as quickly make an unsanitary mess of their drinking water. An open source of drinking water, such as a pan or bowl, is therefore unsuitable for brooding waterfowl. For the first 2 weeks, a suitable container is a circular chick waterer that screws onto a water-filled jar.

As the birds grow and require enough water to submerge their heads in, a deeper trough becomes more suitable. The trough must not be easy for the birds to tip by stepping onto the rim, should be placed over a drain that channels overflow and

No Swimming Allowed

Raised naturally, ducklings and goslings will happily swim with their parents. The older fowl will control how long the babies stay in the water and when they must come out. Left on their own, however, ducklings and goslings will play in the water long after they should come out, becoming soaked, chilled, and possibly ill (or worse). Artificially brooded waterfowl should therefore not have free access to water deep enough to swim in.

spills away from the brooding area, and should be covered with a wire grate to prevent swimming. If your ducklings and goslings can swim or walk in the water, they will leave behind droppings that can lead to disease. The grate or wire mesh must be big enough to allow them to get their heads through for a drink.

If, in a pinch, you need to use an open container, you can float a piece of clean untreated lumber on the surface. Cut the wood to fit the container's shape but make it slightly smaller, so your birds can drink from around the edges. Another ploy is to place a short cylinder of fine-mesh wire in the center of the container, leaving enough space between the cylinder and the rim to allow a satisfying drink.

This homemade waterer is made with a wire cylinder in an open pan.

Feeding

Hatchlings should have access to feed at all times. Feed them just enough twice a day to last until the next feeding. Since their appetites will grow at an alarming rate, you will need to constantly adjust the amount supplied at each feeding. Allowing waterfowl to go hungry will result in a frantic frenzy when they finally get something to eat, turning your cute downy ducklings and goslings into bug-eyed mini-monsters encrusted with caked-on feed.

Clean out the feeders frequently, taking care not to leave old, stale feed in the corners. For hatchlings, you may use feeders with little round holes designed for baby chicks. Watch carefully as your ducklings and goslings grow. When their heads get too big to fit through the holes, switch to troughs with wire guards that prevent the birds from walking in the feed.

Ducklings and goslings need water to wash down dry feed. They fill their mouths with feed, waddle to the waterer for a drink to wash it down, then waddle back for another mouthful. Before long, a trail of dribbled feed marks the path between the feeder and the waterer. If the waterer is close to the feeder, the feed will soon be a wet mess and the water will turn to sludge.

To avoid these problems, you can moisten the feed with water or skim milk, but you will have to offer less feed at a time and therefore provide feed more often. Left too long, moistened mash clumps together, discouraging young waterfowl from eating. Furthermore, in the warmth of the brooder, it may turn sour or moldy, causing illness if it is eaten.

Finding a commercial starter ration formulated for baby waterfowl can be difficult. A good substitute is mashed hard-boiled egg. Available grower ration is likely to be high in protein to promote fast growth in meat birds. If that is not your goal, reduce the protein level by supplementing the ration with high-fiber feed, such as chopped lettuce or succulent grass. The natural diet of goslings consists almost entirely of freshly sprouting grass. If you have access to a grassy area that has not been sprayed with chemicals, on sunny days you may move your ducklings and goslings outside to graze in a small portable enclosure that provides protection from predators and cold wind.

When warm weather allows ducklings and goslings to spend most of their time grazing, you can cut their feed back to the amount they can clean up in 15 minutes, twice a day. When they reach the age of 1 month, you may feed only once a day, preferably at night; that way, they'll be hungry in the morning and take greater advantage of the natural forage.

Excess protein in the diet and insufficient exercise may result in twisted wing.

Letting young waterfowl graze helps prevent twisted wing, a condition in which the flight feathers of one or both wings angle away from the body like an airplane's wings. This problem, also known as slipped wing, occurs because the flight feathers grow faster than the underlying wing structure. The heavy feathers pull the wing down, causing it to twist outward. When the bird matures, one or both of its wings remain awkwardly bent outward instead of gracefully folding against its body. Meat birds don't live long enough for this clumsy appearance to present a problem, but it is undesirable in mature waterfowl that are kept for breeding or for fun. You can easily prevent twisted wing by avoiding excess protein.

Hatchling Health

Ducklings and goslings are incredibly hardy and, when properly cared for, are not particularly susceptible to disease. Their health is easily maintained by providing a clean, dry environment, adequate heat and ventilation, proper nutrition, fresh feed, and clean water. Ill health among young waterfowl is uncommon and is likely to be one of the following six conditions.

Omphalitis. Omphalitis is a bacterial infection that occurs in any species of newly hatched fowl. The classic sign is red navels, followed by deaths in the brood for up to 2 weeks after the hatch. Omphalitis is caused by poor incubator sanitation. It is not contagious, but no cure is known. Prevention involves meticulously cleaning and disinfecting the incubator.

Sexing Ducklings and Goslings

If you must separate the ducks from the drakes or the geese from the ganders while they are young, the only way to do it is by vent sexing, which takes a considerable amount of practice in order to be accurate without injuring a delicate bird. If you don't have experience in vent sexing, find a professional to help you.

Baby ducks all sound the same, but as they grow, the drakes lose their voices. Soon, the drakes of Mallard-derived breeds develop conspicuous drake feathers that curl up and forward at the end of the tail. Among Muscovies, the drakes are about twice the size of the ducklings.

Among geese, vent sexing is the only accurate way to distinguish sex, even once they're mature. For details on vent sexing, see page 75.

Paratyphoid. Also known as keel, paratyphoid is caused by salmonella and may be transmitted by dirty eggs in the incubator or by infected droppings in the water, feed, or litter. Signs include trembling, gasping, breathing with a clicking sound, coughing, sneezing, poor growth, weakness, diarrhea, dehydration, and sometimes watery or pasty eyelids and wet, runny nostrils. A high rate of death occurs within a few weeks after the hatch; deaths increase due to such stresses as shipping, irregular feeding, chilling, or overheating. Since survivors are carriers, if an outbreak of paratyphoid occurs, the best approach is to destroy the whole group, completely disinfect the brooder and incubator, and start again. This disease may be avoided by observing strict sanitation in all phases of rearing waterfowl, especially in washing dirty eggs to eliminate bacteria on the shells before incubation.

Brooder Pneumonia. Brooder pneumonia, also known as aspergillosis, is an infectious disease caused by inhalation of a fungus that flourishes in damp feed and litter. Signs include gasping, inflamed eyes, bad breath, nervous activities, loss of appetite but increased thirst, and emaciation. Prevention involves removing wet feed and damp litter from the brooder and destroying it by burning. Keep the brooder warm, dry, and well ventilated without being drafty. Provide drainage under waterers to remove spillage.

Coccidiosis. Coccidiosis is transmitted on equipment and clothing and by insects and animals. Infected carriers shed disease organisms in their droppings and thus contaminate feed, water, litter, and soil. Signs include loss of appetite and weight, weakness and inability to stand, continual distressed peeping, and sometimes bloody droppings. Although this disease occurs much less frequently in waterfowl than in chickens and other landfowl, an outbreak may involve most or all of a brood, and the rate of death is high. Sulfa compounds provide effective treatment. Medicated feeds designed to prevent coccidioidal infections in chicks are of no use for waterfowl, because a different organism is involved. In fact, such feeds may be harmful to ducklings and goslings. Immunity develops through mild infections. Coccidial contamination may be avoided through good brooder sanitation.

New Duck Syndrome. New duck syndrome is the common name for anatipestifer, a contagious respiratory disease caused by pasteurella bacteria that spreads through water and the air. Signs include general depression, ruffled feathers, greenish diarrhea, inability to stand, bobbing or jerking of the head, and discharge through the eyes and nose. Signs in advanced cases include prostration and inability to lift the head, with mild coughing, incoordination, and head tremors. A bird may lie on its back and paddle the air with its feet. The death rate is highest between the ages of 5 and 10 weeks, and the cause of death is usually an inability to get to drinking water. Prevention involves good sanitation, as well as vaccination in commercial duck-raising areas where this disease is prevalent. Antibiotic therapy may save some infected birds.

Hepatitis. Hepatitis is a highly contagious virus that should be suspected if large numbers of sudden deaths occur. At first, baby waterfowl become sluggish, then squat motionless with their eyes shut or perhaps fall over onto their sides. Death follows within 1 to 2 hours of the first signs and is sometimes preceded by spasmodic kicking. Birds typically die with their heads drawn back. Treatment involves appropriate antibodies. Prevention requires good sanitation, meticulous cleaning of feeders and waterers, and disinfection of the brooder whenever the litter is replaced. A vaccine is available for areas where this disease is prevalent.

Probiotics

Probiotics are beneficial bacteria that are naturally present in the digestive tract of animals. These beneficial bacteria increase in numbers as the animal matures and, when the body is in proper balance, fight off disease-causing bacteria. By adding a probiotic formula designed for poultry to the drinking water of ducklings and goslings during their first week of life, you can give them a good start.

Signs of Trouble

Signs that something is wrong among waterfowl include:

- Weak and listless behavior
- Ruffled feathers
- Crusty or sticky eyes
- Unusual body movements
- Loss of appetite
- Increased thirst
- Change in the color or consistency of droppings
- Inexplicable deaths

Growing Up

Ducklings must be brooded until they are about 4 weeks old and goslings until about 6 weeks. If the weather is mild by then, they may be moved outdoors, but they still need shelter from wind, rain, and hot sun. In stormy weather, gather them up and bring them inside until they are fully feathered, which will be around 12 weeks for Mallard-derived ducklings and 16 weeks for Muscovies and goslings.

When you first turn them outdoors, keep an eye on them. Ducklings and goslings have an uncanny ability to find escape routes, get stuck in odd nooks, or get separated from their feed and water. If their yard is large or they are being turned out with mature ducks and geese, confine them to a small part of the yard for the first few days, enlarging their space only after they have had time to get their bearings.

Duck and Goose Health

Your ducks and geese are likely to remain healthy if you feed them a balanced diet, provide a continuing supply of clean, fresh water for drinking and bathing, and furnish adequate facilities for comfort and sanitation. Diseases become a threat largely where waterfowl are kept in crowded, unsanitary conditions.

It's a good thing ducks and geese are so naturally healthy, because it's nearly impossible to find a veterinarian who is knowledgeable about waterfowl problems. If a contagious disease should break out, your state's poultry pathology laboratory, located through your county's Extension office, can diagnose the problem. Most waterfowl conditions, however, must be dealt with by the owner of the gaggle or bevy — that's you.

Noncontagious Conditions

Health problems in a gaggle or bevy are more likely to involve a noncontagious condition than a serious disease. Following are some of the most common noncontagious conditions.

Drowning. As unbelievable as it may seem, ducks and geese can drown. Among domestic waterfowl, drowning most often occurs when they get into water they can't get out of. Ducklings or goslings may jump or fall into a kiddy wading pool, for instance, and by the time they've finished splashing around, the water level is too low for them to climb back out. Even older ducks can drown if, for example, they've gotten into a swimming pool and can't climb up the side to get out. Water that's contaminated with chemicals (including swimming pool chlorine), oil, sludge, muck, or mud can also lead to drowning. Prevention involves ensuring that waterfowl have no access to foul or chemically treated water and ponds or pools that do not provide easy exit.

Lameness. Lameness is the most common affliction of waterfowl, because they inherently have structurally weak legs. Their legs are made for flying and swimming, not walking. Lameness may result from grabbing a bird by the leg or pulling a leg free that's been caught in a fence. Lameness may also result from a glass sliver, a thorn, or a sharp stick lodged in the footpad. The feet of waterfowl kept on dry, hard-packed ground can develop abscesses that harden into calluses on the bottom of the pad. This condition, known as bumblefoot, may involve one or both feet and most often affects the heavier goose breeds. Treatment involves washing the affected foot, cleaning it with a bactericide, pressing any pus out of the abscess, and removing the hard core if one is present. Isolate the goose in a quiet, secluded place with clean litter or fresh grass and clean swimming water. To prevent this problem, keep feed and watering areas clean (or frequently move the feed and watering stations) and cover hard surfaces, such as concrete, gravel, or hard-packed soil, with clean litter. If the problem is caused by the geese stamping out the vegetation in their yard, it may be corrected by providing several paddocks and rotating the gaggle to periodically rest the vegetation in each paddock.

Crop Impaction. Crop impaction occurs in geese grazing on fibrous plants, including grass that has gotten tall and coarse. Geese are natural grazers, but tough, fibrous vegetation is difficult for them to digest and instead wads up in the crop. An affected goose loses interest in grazing, steadily loses weight, and may die unless the wad is surgically removed. You can check for impaction by feeling the crop. It should feel squishy. If, instead, you find a tight hard lump, the crop may be impacted, blocking food from passing through to the digestive tract. Prevention involves mowing the grazing area to keep it in the vegetative stage and providing other feed at times of year when vegetation stops growing.

Hardware Disease. Hardware disease results from the fact that waterfowl don't discriminate between tasty treats, such as bugs and slugs, and nasty things, such as shards of glass or bits of wire. Swallowing small, sharp objects may merely cause irritation or may poke a hole through some part of the digestive tract. The end result may be depression or death. Prevention involves meticulously patrolling the waterfowl yard to remove any small, sharp objects lying around and seeing that such things aren't tossed there in the first place.

Poisoning. Poisoning may result from the most common materials, such as salt used to deice winter paths, insecticides sprayed around your (or your neighbor's) yard or garden, rat poison, and bait set out for slugs, snails, or other garden pests. Perhaps the most common cause of poisoning is botulism from spoiled food scraps or contaminated water. If you feed kitchen scraps to your ducks and geese, first sort out anything that is spoiled or rotting. Any water your waterfowl have access to must be free of excessive droppings, decaying leaves, and dead animals.

Signs of botulism poisoning appear within 8 to 48 hours of consuming the toxin. In mild cases, the only sign may be a brief bout with weak legs before the birds return to normal. In severe cases, the birds may appear sleepy and unable to hold their heads erect, due to paralysis of the neck muscles. Their wings and legs may become paralyzed, too, causing them to lie on their sides. Waterfowl stricken while swimming may drown because of muscle paralysis, but most affected birds die of suffocation due to paralysis of the respiratory system.

If the poisoning is not so severe that the birds cannot drink, replace all sources of water with a solution of dissolved Epsom salts in the proportion of 1 pound per 15 gallons of water. Move them away from any possible source of contamination, including dead birds. By the time signs are first noticed, however, chances are good that the affected waterfowl are beyond help. The most prudent course of action is to avoid conditions that could lead to botulism poisoning.

Sticky Eyes

Crusty, sticky eyes may be the sign of a low-grade infection due to the duck or goose not being able to fully submerge its head in water and thereby flush out its eyes and nostrils. If swimming water is not available, make sure drinking water containers are deep enough that your birds can get in their entire heads.

Isolate an infected bird, clean its eyes with an eyewash recommended by your vet, and provide a nutritiously balanced diet to reduce susceptibility to infection. Sterilizing the drinking water by adding just a drop of chlorine bleach will help reduce the chance of infection.

Gout. Gout is a condition in which uric acid crystals, normally passed in urine, are deposited in joints or on internal organs. Gout can occur in any bird with kidney damage but can also occur if birds are overfed protein.

Penis Paralysis. Penis paralysis, also known as phallus prostration, is a sad result of the domestication of waterfowl. Under natural conditions, ducks and geese breed at maturity and have a short breeding season. At other times of the year, they are busy with more important things, such as migrating and finding enough to eat. Domesticated waterfowl, on the other hand, are required to do little to sustain themselves. They start breeding at a younger age, the males are often expected to breed more females than they would under natural conditions, and they breed for a longer period. As a result, sometimes a drake or gander overexerts his male organ to the point that he can't put it away anymore. It bounces around behind him in the dirt and mud, while the drake or gander wags his tail, hoping to get it to go back in. Eventually it becomes dirty and scabby, dries out, and perhaps gets infected, and the drake or gander may lose weight or die. This condition may be a genetically inherited weakness of the muscles, possibly aggravated by some dietary deficiency. Usually, by the time the condition is noticed, nothing can be done but to put the drake or gander out of his misery.

Contagious Diseases

If your bevy or gaggle exhibits several symptoms and sudden deaths occur, you can be pretty sure a disease is involved. Since the signs of various diseases can be confusingly alike, consult your state's poultry pathology laboratory for advice. The lab will probably want to examine several sick and recently dead birds to determine the cause of the illness. By finding out what the cause is, you can:

- Determine suitable treatment.
- Find out whether survivors will be carriers that continue to spread the disease.
- Learn how to clean your yard to make it safe for future waterfowl.
- Determine whether vaccination is possible to prevent future outbreaks.

Although they remain uncommon, the two most likely diseases of mature ducks and geese have the formidable names of cholera and plague.

Cholera. Fowl cholera affects not only ducks and geese but also many other species of domestic and wild birds. It is highly contagious, and different strains produce different degrees of severity, from mild to severe. The first sign is usually sudden deaths among apparently healthy birds, although sometimes a change in droppings to a greenish, watery diarrhea provides an early warning. Other signs, in addition to general signs of any disease, are nervous twitching and jerking, walking in small circles, watery encrustment around the eyes and nostrils, swollen legs and feet, and respiratory problems.

Fowl cholera is spread through contaminated feed, air, water, and soil. It is carried by rodents and other scavengers, and possibly by flies. An outbreak may follow cold, wet weather and is abetted by unsanitary conditions, overcrowding, poorly ventilated quarters, improper diet, or visits by wild waterfowl. Once a positive diagnosis has been made, antibiotic treatment is available. Prevention involves good sanitation, avoiding muddy conditions around watering stations, properly disposing of dead birds and animals, frequent paddock rotation, and vaccination.

Duck Plague. Duck plague, more formally known as duck virus enteritis, is a form of herpesvirus that affects all species of waterfowl. In addition to general signs of disease, specific signs include droopiness, watery or bloody diarrhea, swollen eyelids, and nasal discharge. Birds may be unable to walk and will lie with their wings spread out, then have convulsions and die. The virus spreads through contact with affected birds and their droppings and through contaminated stagnant or slow-moving water. The death rate is high, and deaths occur within 3 or 4 days after the first signs appear.

Viral enteritis has no cure, and it is a reportable disease, meaning that the law requires you to report an outbreak to local animal health authorities. They will quarantine your yard and kill all your birds to keep the disease from spreading to other domestic or wild birds. Happily, this disease is not common in domestic gaggles and bevies, and a vaccine is available for areas where outbreaks may be initiated by migrating wild waterfowl.

If you live in a flyway frequented by large numbers of migrating waterfowl or in an area with a concentration of commercial waterfowl farms, you may be advised to vaccinate your waterfowl against other diseases that are locally prevalent. Your county Extension agent should be able to tell you whether vaccination is necessary in your area and, if so, which diseases you should vaccinate against.

Ducks and Geese for Meat

Although some waterfowl breeds have been developed for efficient meat production, any breed is good to eat. If you purchase a duck or goose at the grocery store or butcher shop, the duck would most likely be a Pekin and the goose an Embden. These white-feathered breeds appear cleaner when plucked, because their white pinfeathers don't show as much as the pinfeathers of colored waterfowl.

Commercially raised waterfowl that are pushed for rapid growth are ready for butchering at an earlier age than most backyard waterfowl. Ducks of the Mallard-derived breeds may be ready as early as 7 weeks, geese at 8 weeks, and Muscovies at 14 weeks. Pasture-raised ducks and geese grow considerably more slowly, taking as much as three times longer to reach size, but are less expensive to raise and have less fat.

Feed Conversion

A typical feed conversion rate for meat ducks is 2½ to 3 pounds, meaning that each duck gains 1 pound of weight for every 2½ to 3 pounds of feed it eats. The feed conversion rate for geese is 2 to 3 pounds. The older the birds get, the higher the feed conversion rate, creating a trade-off between economical meat and more of it. To improve the conversion rate, commercially raised waterfowl are encouraged to eat continuously by being kept under lighting 24 hours and having their feed troughs topped off several times a day. At the same time, they are discouraged from burning off calories by being confined to a limited area.

Feathering Means Butchering Time

Weight gain is only one important factor in determining when a duck or goose is ready for butchering. The other is the stage of the molt. All those feathers that allow ducks and geese to swim comfortably in cold water take a long time to remove, and plucking is considerably more difficult if some of those feathers are only partially grown. Plucking is less time-consuming, and the result is more appealing, if a duck or goose is in full feather.

Soon after a duck or goose acquires its first full set of feathers, it begins molting into adult plumage and won't be back in full feather for another 2 months or so. You'll know the optimum time to pluck a duckling or gosling has passed when the feathers around its neck start falling out. From then on, the bird won't grow as rapidly as it has been and the feed conversion rate goes up. After this point, feeding waterfowl for meat purposes becomes more costly, the meat becomes tougher, and the meat of Muscovy males takes on an unpleasant musky flavor.

A duck or goose is in full feather when:

- Its flight feathers have grown to their full length and reach the tail
- Its plumage is bright and hard looking, and it feels smooth when stroked
- You see no pinfeathers when you ruffle its feathers against the grain
- It has no downy patches along the breastbone or around the vent

When all of these signs are right, a duck or goose is ready to be butchered. If you are doing your own butchering for the first time, have an experienced person guide you, or refer to a good book (see Recommended Reading, page 379). If you don't wish to do your own butchering, you might find a custom slaughterer willing to handle ducks and geese. Alternatively, a fellow backyard waterfowl keeper or a hunter might be willing to kill and pluck your ducks or geese for a small fee. If not having to kill your own waterfowl is important to you, determine before you start whether or not someone in your area can do it for you.

Using the Feathers

One of the advantages of cleaning your own waterfowl is getting the feathers and down as a by-product. The larger feathers may be used for various crafts, while the smaller, softer feathers may be used to make pillows or comforters. However, you'll need the feathers from a lot of birds to make a sizable pillow, let alone a comforter. One goose will give you about ⅓ pound of feathers, and a duck will yield about ⅙ pound. And remember, some of those feathers are too stiff for pillow making.

When saving feathers, discard any that are dirty or have the soft, gooey quill of an immature feather. After each round of butchering, wash your collection of feathers. Place the feathers in a tight-mesh laundry bag and submerge it in lukewarm water with a little washing soda and some detergent or borax. Gently slosh the feathers in the water, then rinse them in clean lukewarm water. Repeat until the rinse water runs clear. Gently squeeze the bagful of feathers and hang it outside to dry.

Waterfowl have an inner layer of feathers called down, which is extremely soft and light. Just as it keeps ducks and geese warm in cold water, it may be used to create vests and other articles of clothing to keep you warm in cold weather. Down gets its insulating ability from loft, which is its tendency to fluff up. The greater the loft, the greater the down's insulating ability. Goose down has more loft than duck down, which is why it's preferred by Arctic explorers. Because of loft, you may think you have a big bag full of down, but you'll soon find that it's mostly air. If you sneeze, or open a door or window, while you're gathering down, it will float out of the bag and waft through the air. Because down is so difficult to contain, it is often saved in smaller pouches rather than in larger bags. If you are careful while plucking, your down will remain clean and won't require washing. You can find suitable fabric for stuffing with down at many outlets that manufacture or renovate pillows, comforters, or sleeping bags.

Dressed Weight

A duck or goose loses 25 to 30 percent of its live weight after the feathers, feet, head, and entrails have been removed. Heavier breeds lose a smaller percentage than lighter breeds. The breast makes up about 20 percent of the total meat weight; skin and fat make up about 30 percent.

Storing the Meat

Freshly butchered duck or goose must be aged in the refrigerator for 12 to 24 hours before being cooked or frozen; otherwise, it will be tough. If you will not be serving the duck or goose within the next 3 days, freeze it after the aging period until you're ready to cook it. To avoid freezer burn, use freezer storage bags. Most ducks will fit in the 1-gallon size. A Muscovy female should fit in the 2-gallon size. Geese and Muscovy males should fit in the 5-gallon size. Remove as much air from the bag as you can by pressing it out with your hands or by using a homesteaders' vacuum device designed for that purpose. Properly sealed and stored at a temperature of 0°F or below, duck and goose meat may be kept frozen for 6 months with no loss of quality. To thaw a frozen goose or duck before roasting, keep it in the refrigerator for 2 hours per pound.

Cooking Methods

The most suitable method of cooking a duck or goose depends on its tenderness, which in turn depends on its age. Fast dry-heat methods, such as roasting, broiling, frying, and barbecuing, are suitable for young, tender birds; slow moist-heat cooking methods, such as pressure cooking or making a fricassee, soup, or stew, are required for older, tougher birds.

A whole duck or goose takes up a lot of freezer space. Unless you intend to stuff and roast the bird, you can save space by halving or quartering it, or filleting the breasts and cutting up the rest. Muscovy breast makes an exceptionally fine cut that is the most like red meat of any waterfowl. After removing the fillets, you might package the hindquarters for roasting or barbecuing and boil the rest for soup. A great use for excess Muscovy and goose meat is sausage. Small amounts may easily be made into sausage patties, whereas larger amounts might be stuffed into links.

Roasted to Perfection

Properly prepared, a homegrown duck or goose should not be greasy. Although ducks and geese have a lot of fat, the meat itself is pretty lean. All the fat is either just under the skin or near one of the two openings, where it may easily be pulled away by hand.

Proper roasting begins by putting the meat on a rack to keep it out of the pan drippings while the bird roasts. Remove any fat from the cavity and neck openings, and stuff the bird or rub salt inside. Rub the skin with a fresh lemon, then sprinkle with salt. Pierce the skin all over with a meat fork, knife tip, or skewer, taking care not to pierce into the meat. Your goal in piercing is to give the fat a way out through the skin as it melts during roasting. This melted fat will baste the bird as it drips off; no other basting is required. Do not cover the duck or goose with foil during roasting as you would a turkey.

Slow roasting keeps the meat moist. Roast a whole duck at 250°F for 3 hours with the breast side down, then for another 45 minutes with the breast up. Roast a goose at 325°F for 1½ hours with the breast side down, plus 1½ hours breast up, then increase the temperature to 400°F for another 15 minutes to crisp the skin.

Since not all birds are the same size and not all ovens work the same way, the first time around keep an eye on things to avoid overcooking your meat, which makes it tough and stringy. Once you settle on the correct time range for birds of the size you raise, you can roast by the clock in the future.

When the meat is done, it should be just cooked through and still juicy. You can tell it's done when the leg joints move freely, a knife stabbed into a joint releases juices that flow pink but not bloody, and the meat itself is just barely pinkish. To keep the meat nice and moist, before you carve the bird let it stand at room temperature for 10 to 15 minutes to lock in the juices. During this time, residual heat will cook away any remaining pink. Duck or goose does not have light and dark meat like a chicken or turkey; rather, it is all succulent dark meat.

3
Rabbits

INTRODUCING RABBITS

Rabbits belong to the *lagomorph,* or hare-like, group of the mammal class. Lagomorphs are subdivided into two families: pikas, and rabbits and hares. Pikas are small, short-eared animals that live in rocky, mountainous areas. Hares are closely related to rabbits; you can tell them apart by the following characteristics:

- Hares are usually larger than rabbits. Their legs are longer, their ears are longer, and they run with long, high leaps.
- Baby hares are born with fur, their eyes are open at birth, and they can move around shortly after they are born. Baby rabbits are born naked, their eyes are closed (they open at about 10 days), and they remain dependent on the mother for 5 to 8 weeks.
- Hares raise their young in simple nest areas. Because baby rabbits mature less quickly than hares, their nests tend to be more protected.

Rabbit

Where does the tame rabbit that we raise fit into the picture? The rabbit group is divided further into various species. Worldwide, there are 25 species of rabbits. Wild rabbits native to North America include the eastern cottontail, the desert cottontail, and the marsh rabbit. Tame rabbits that are raised in North America as pets and for fur and meat are descended from wild European rabbits.

Hare

Rabbit History

From fossils, scientists have determined that the rabbit has been around for about 30 to 40 million years. During that time, the rabbit has changed very little. When humans arrived on the scene, wild rabbits became part of their diet. Exactly when people decided to raise rabbits rather than hunt them in the wild is unknown. We do know that early sailors took rabbits along on voyages because they were easy to feed and care for and could be used as meat. These early sailors introduced rabbits to some of the new places they explored.

Early European settlers probably brought rabbits with them to the New World, but because wild rabbits and other game were plentiful here, settlers didn't need to raise many rabbits. However, that trend changed in the early 1900s with the Belgian Hare boom. The Belgian Hare is a breed of rabbit, not a hare, as the name might imply. In the beginning of the twentieth century, a legion of aggressive promoters led people to believe that raising Belgian Hares could be a highly profitable pursuit. Some of the newly converted breeders were successful, but many others lost money when they invested in these rabbits. Although the Belgian Hares didn't do all their promoters promised, they did greatly increase interest in rabbits in North America.

Since then, other breeds have been imported to North America and several American breeds have been developed. Forty-five breeds are currently recognized in North America by the American Rabbit Breeders Association (ARBA).

Why Raise Rabbits?

Rabbits have had an unusually broad range of uses:

- **Pets.** Rabbits make wonderful, easy-to-care-for pets, even for people with little space.
- **Fur.** Rabbit furs have been used to make warm clothing.

- **Show.** Rabbits come in many breeds and colors. This variety makes them an interesting and challenging animal to raise and show.
- **Meat.** Wild, and later tame, rabbits have provided meat for people throughout history.
- **Wool.** Angora rabbits produce wool rather than fur. Angora wool is used to make knitted garments that are softer and warmer than those made with sheep's wool.
- **Fertilizer.** Rabbit droppings are excellent fertilizer and can easily be used in the family flower or vegetable garden.

Choosing a Breed

Selecting a breed is probably the most important decision you will make. Your choice of breed affects many things, such as the size of the cage you will need and when you should first breed young rabbits. Research each of the breeds you're considering so you can choose the one that best suits you.

Crossbred rabbits, which are rabbits that have more than one breed in their family background, are appealing and readily available. However, it's best to begin with purebred rabbits. Purebred rabbits may cost more than crossbred rabbits, but they have many advantages that soon make up for the difference in price:

- A crossbred rabbit costs just as much to house and feed as a purebred.
- If you plan to breed and sell rabbits, purebred young bring a better sale price. You can therefore soon make up the difference in the initial cost of your parent stock (the mother and father).
- If you want to show your rabbits — and showing is fun! — purebred rabbits will qualify to exhibit at more shows.

A good way to begin learning about the different breeds is to become a member of ARBA (for contact information, see page 387). With your ARBA membership, you'll receive a guidebook that has pictures and a written description of every breed. The association also publishes *The Standard of Perfection,* which gives a detailed, up-to-date description of each breed.

Besides reading about the different breeds, find out where you can see rabbits in your area. If an agricultural fair is held in your community, be sure to check out the rabbit section, where you can meet local breeders. Local rabbit shows that are held under the auspices of ARBA usually include a great variety of breeds. These shows are called *sanctioned shows,* and they usually attract exhibitors from a fairly large area. If you become a member of ARBA, the magazine that comes to you with your membership will list upcoming sanctioned shows in your area. Your membership also provides a yearbook that lists all ARBA clubs and members. If you contact the club nearest you, it can provide a calendar of its scheduled events, including shows.

Whether you attend a fair or a sanctioned show, just look and learn the first few times you go — don't buy yet. This won't be easy, because you are sure to see several rabbits that you'd like to take home. Try to put off a purchase until you have found a breed that appeals to you *and* that will fill your needs as a breeder. Before buying, decide exactly *why* you are going to raise rabbits. Do you want to raise them as pets or because you would like to make a hobby of breeding or showing rabbits? Or do you wish to sell rabbit wool, fur, or meat? Some breeds are more suited to one or more purposes than to others.

Small Pet Breeds

Rabbits are ideal pets for busy families: Their initial cost is low, they are relatively inexpensive to keep, they can be kept indoors or out, they make no noise, they have few veterinary needs (including no vaccinations), and daily care is not demanding. Any breed may be kept as a pet, but if you want to raise rabbits primarily to keep or sell as pet rabbits, it's best to consider some of the smaller breeds.

Netherland Dwarf. The Netherland Dwarf is the smallest breed of rabbit. To qualify for showing, a mature Netherland Dwarf may weigh no more than 2½ pounds. This breed is known for its round body, broad head, short ears, and large, bold eyes; it comes in more than 30 recognized colors. Because the Netherland Dwarf is small and has such a bright appearance, it makes a very appealing pet. Netherland Dwarfs have small litters (two or three young in an average litter); if you want to breed them, keep in mind that you won't have as many to sell as you would if you were raising most other breeds.

Dutch. The distinctive markings of the Dutch make them attractive pets. They weigh only 3½ to 5½ pounds. Their small size and mild personality make them a good choice for young children. Dutch are available in six color varieties. Not all the young in a litter will have show-quality markings.

Mini Lop. The Mini Lop is popular because of the breed's outgoing personality and curious lop ears. Mini Lops can be seen in many colors. The two show categories are colored pattern and broken pattern (the latter includes any recognized color combined with white). The Mini Lop weighs, at the very most, 6½ pounds, making it a good choice even for families without a lot of space. The breed is also a good one for beginners, as Mini Lops are easy to raise.

Holland Lop. The Holland Lop weighs only 3 to 4 pounds when it is mature. You can tell by its broad, bold head and small, compact body that it is closely related to the Netherland Dwarf. Like other lops, the Holland comes in many colors. At shows, it will be placed in either a colored or a broken pattern category.

Mini Rex. This breed combines the beautiful short, plush fur of the Rex with small size to make it a perfect pet and exhibition rabbit. Recognized in many colors, the adult Mini Rex has an ideal weight of 4 to 4½ pounds. It is noted for its easygoing personality.

Angora Breeds

In recent years, natural-fiber products have become tremendously popular. This has brought about a new appreciation for the Angora rabbit breeds. Angora wool is obtained by pulling the loose hair from the mature coat. Because hand plucking helps the natural shedding process, it does not hurt the animal. The plucked wool may then be spun into yarn. Clothing made from Angora wool is soft and warm and brings a very high price. Although Angora rabbits need to be groomed often, raising them and selling their wool can be a profitable investment.

English Angora

English Angora. This is the fanciest of the Angora breeds. The English Angora is relatively small, weighing 5 to 7½ pounds. It's prized for its luxuriously soft

wool, which covers its entire body, including the ears, face, and feet. (The long, decorative wool on the head of the English Angora is called *furnishings.*) The breed includes two color varieties, white and colored, of which there are approximately 30 shades.

French Angora. The French Angora weighs in at 7½ to 10½ pounds. Wool from the French Angora is slightly coarser than that from the English Angora, and the French Angora has normal fur — not angora wool — on its ears, its face, and the ends of its feet.

Giant Angora. The Giant Angora is the largest of the Angora breeds. To qualify for showing, mature bucks must weigh at least 9½ pounds; mature does, 10 pounds. The Giant Angora comes only in white. Like the English, the Giant has wool on all parts of its body.

Satin Angora. Like the fur of the Satin breed (see page 121), the wool of the Satin Angora has a striking sheen, because its hair shafts are more transparent than those of normal fur. Like English and French Angoras, the Satin Angora comes in white and colored varieties.

Two other rabbit breeds have wool instead of normal fur: the Jersey Woolly and the American Fuzzy Lop. These small breeds produce less wool and are therefore more suitable for pets or show than for wool production.

Meat Breeds

If you are interested in raising rabbits for meat production, consider the large breeds, which weigh from 9 to 11 pounds when mature. Large breeds generally convert feed to meat at a profitable rate, and they yield an ideal fryer — 4 or more pounds — at 8 weeks of age, the preferred age for culling. Although not all large breeds are ideal for meat production, the following popular breeds are worth looking into.

New Zealand. This breed has long been a top-quality meat-producing breed. Although the breed comes in three color varieties — red, white, and black — the white variety has proven most popular for serious rabbit-meat-producing businesses. New Zealands are known for their full, well-muscled bodies and their ability to become market-ready fryers (4 to 5 pounds live weight) by 8 weeks of age. The New Zealand White is also an excellent fur breed.

Californian. The Californian is another outstanding meat breed. It is white with black coloring on the feet, tail, ears, and nose.

Champagne D'Argent. One of the oldest rabbit breeds, this silver-colored rabbit is born completely black and gradually turns silver as it matures. The breed is well regarded for both its meat and its fur.

Palomino. The Palomino combines an attractive color with a body type that is well suited for meat production. Developed in the United States, the breed is available in two colors: golden and lynx.

Florida White. Although the Florida White weighs only 4 to 6 pounds, it is a good meat rabbit. Like the Palomino, the Florida White was created in the United States.

Fur Breeds

Raising rabbits just for their fur is not practical, unless you want to tan the pelts at home or send them out to be processed. Fur markets are few and far between, and fur production is more suited to large rabbitries with many, many rabbits than to small home rabbitries. Most of the breeds that you might choose for fur production are also good for meat production, including the New Zealand and the Californian.

White-furred rabbits are desirable for fur production because their pelts may be easily dyed to a variety of colors. (Even though the Californian has black on its feet, tail, ears, and nose, it is considered a white rabbit to furriers because the usable portion of the pelt is white.) The fur breeds mentioned below are unique because of special features of their fur.

Rex. The Rex is the most valuable fur breed. It has very short, plush fur, making it the original "Velveteen Rabbit." This breed has its origins in two rabbits with unusual fur that showed up in a litter of rabbits with normal fur. These two rabbits with this mutation were mated, and their offspring inherited the plush fur. The breed is available in 15 color varieties, including white.

Satin. Like the Rex, Satins were bred from animals with a fur mutation. The hair shaft of the fur is transparent, which allows more light to pass through and gives the Satin breed its sheen. Satins come in 10 color varieties, including white.

Recognizing a Good Animal

After you have decided on the breed of rabbit you're going to raise, you must learn how to tell if a particular rabbit will make good stock. For the most part, that means making sure a rabbit is healthy and exhibits the proper characteristics of its breed.

Evaluating Health

Good health is the most important quality to consider when you select your stock. Examine all of the following:

- **Eyes.** The eyes should be bright, with no discharge and no spots or cloudiness.
- **Ears.** The inside of the ears should look clean. A brown, crusty appearance could indicate ear mites.
- **Nose.** The nose should be clean and dry. A discharge could indicate a cold.
- **Front feet.** The front feet should be clean. A crusty matting on the inside of the front paws indicates that the rabbit has been wiping a runny nose and may have a cold.
- **Hind feet.** The bottoms of the hind feet should be well furred. Bare or sore-looking spots may indicate the beginnings of sore hocks.
- **Teeth.** The front teeth should line up correctly, with the front top two teeth slightly overlapping the bottom ones.
- **General condition.** The rabbit's fur should be clean. Its body should feel smooth and firm, not bony.
- **Rear end.** The area at the base of the rabbit's tail should be clean, with no manure sticking to the fur.

Examining Breed Characteristics

Each rabbit breed has distinctive features that make it different from all other breeds. These features include colors, markings, body type (the proper size and shape for its breed), fur type, and weight. The more you know about the unique characteristics of your breed, the better job you will do of picking out a good one. Detailed descriptions of breeds are found in *The Standard of Perfection,* as well as in information produced by breed specialty clubs.

Rabbit Finances

Rabbit prices vary by breed, age, availability, and popularity; they range from $10 to more than $100 per animal. Because so many factors affect the price, it is difficult to give an average. However, if you are buying a purebred rabbit of one of the common breeds, you should be able to find a good-quality animal for $20 to $50. When you contact breeders, ask how much their rabbits cost, and choose a breeder whose prices fall within your budget. Make sure the purchase price includes the pedigree.

If you are purchasing purebred rabbits, they should have no conditions that would prevent them from winning prizes at a rabbit show or from being accepted for registration. Many of the health-related conditions mentioned above are disqualifications in all breeds. *The Standard of Perfection* contains a complete listing of disqualifications.

Making the Purchase

Before leaping into the rabbit-raising business, make a good evaluation of your own needs and preferences. Having the wrong type or number of rabbits can put a real damper on your enthusiasm. Once you know what you're looking for, it's time to find stock. If you've done your research, you'll probably already know where you can buy stock of the breed you're interested in.

How Many Rabbits?

Are you looking for one rabbit or several? If you are new to raising rabbits, you may decide to start with one. If you enjoy the experience, you can always get more.

If you want to start up with more than one rabbit, consider purchasing a *trio:* one buck and two does. Three rabbits is a reasonable number for a beginner to care for, and a trio can become the start of a breeding program. If you purchase a trio, buy animals that are not too closely related; you can determine their bloodlines by looking at their pedigrees. It's acceptable for the rabbits in your trio to have some relatives in common, but they should also have some differences in their family trees. They should *not* be brother and sisters; three rabbits from the same litter are too closely related to be used in a breeding program.

Choosing the Right Age Range

When you invest in rabbits, you want to enjoy them for as long as possible. Rabbits live an average of 6 to 8 years. For breeding purposes, rabbits are most productive between 6 months and 3 years of age.

2–3 months. If you're a beginner, consider buying a rabbit that is 2 to 3 months of age. A baby rabbit is cute, fun, and easy to handle, and you'll have the pleasure of watching it grow to adulthood.

4–5 months. Baby rabbits are adorable, but you can't always be sure exactly how they will look when they grow up. An advantage of purchasing a slightly older rabbit is that you can get a better idea of what it will look like when it matures. Rabbits that are 4 to 5 months old are still young, but their appearance will not change much further.

6 months or older. With a 6-month-old rabbit, you'll have a pretty accurate picture of what type of adult it will be. An older rabbit will also be ready for breeding sooner. However, you'll miss the fun of watching it grow up.

Age Limitations

Rabbits have a better start in life if they remain with their mother until they are 5 to 8 weeks of age. Be sure that your new rabbits have been weaned from their mother before you move them to a new home.

If you plan to breed your rabbits, don't buy animals that are more than 2 years of age.

Finding Stock

Rabbitries are scattered in great number across North America; you may be surprised to find one close to home. The following resources will help you find a rabbitry that carries the breed you're interested in raising.

Shows. Attending rabbit shows is a good way to find breeders. Shows also give you a chance to watch your breed being judged and to gather specific information about the characteristics of that breed, so you will recognize a good representative.

Breed Clubs. One way to locate breeders is to contact the specialty club that represents your breed. The ARBA yearbook lists each breed club and the address of the club's secretary, who can give you the names and addresses of breeders closest to you. Write, call, or e-mail those breeders to find out whether they have stock for sale.

Extension Service. Another network for locating breeders is your local Extension Service; every county has one. See Resources on page 380 to find your local office; the staff there should be able to refer you to a rabbit specialist in your area.

Inspecting the Rabbitry

When you've found a rabbitry, visit it, but don't rush into an immediate purchase. Take some time to look around. Ask the breeder to give you a tour. Ask as many questions as you need to. Along with information about the general care of rabbits, the breeder has in-depth knowledge about your breed.

Before you look at individual animals, you should feel comfortable with the rabbitry itself. If you do not like what you see, it's probably not a good place to purchase your rabbits. Remember, however, that a rabbitry does not have to be fancy and new to meet the needs of the rabbits. If the rabbits' cages are secure and clean and the rabbits are healthy and well fed, the breeder is doing a good job of providing for their welfare. Once the rabbitry passes your inspection, begin to look for a specific rabbit.

Things to Look for at the Rabbitry

- Is the rabbitry clean?
- Do the rabbits look healthy?
- Are clean water and food available to the animals?
- Are the rabbits properly housed?

Making Your Selection

The breeder's knowledge and experience can be very helpful to you *if* he or she knows what you're looking for. You should know the answers to the following questions before you visit the breeder:

- Do you want a buck or a doe?
- Are you planning to show the animal?
- What age rabbit are you looking for?
- About how much can you spend?

The breeder may have several rabbits that fit your criteria. Ask the breeder to remove the prospects from their cages so you can get a closer look. Placing the rabbits on a table will give you a chance to see them next to one another and to compare them in terms of size, markings, body type, and so on. Ask which rabbit the breeder thinks is the best, and why. Use the information you've learned as well. Many of the conditions you'll want to know about require handling the rabbit. If you are not an experienced handler, ask the breeder to help you check the rabbit's teeth, sex, and toenails.

If you're interested in a young rabbit, ask to see its mother and father. This will give you an idea of what the rabbit should look like at maturity. Ask to see written records for the parents that indicate how well they have performed. These include the parents' pedigrees and records that provide information such as how many young the mother has raised, how many young the buck has sired, and perhaps information on how well the parents have placed at rabbit shows.

Don't make a final decision until you are sure that the animal is healthy; has no eliminations or disqualifications; meets the size, color, and other characteristics required of its breed; and meets your own criteria, including price. If you have examined the animal carefully and it has passed your inspection, the parent stock looks good, the rabbit meets your criteria, and you and the breeder agree on price, you have found your rabbit.

Final Purchase Details

Once you've found your rabbit, there are some details to attend to before taking it home.

Pedigree Papers. If you are purchasing a purebred rabbit, the price should include pedigree papers. An organized breeder will have pedigree papers ready. (See pages 164–165 for more about pedigree and registration.)

Feed. Be sure to ask the breeder what kind of food the rabbit is currently receiving, how much it receives, and how it's fed. This information will help you smooth the rabbit's transition to its new home. If you plan to use a different feed from that used by the breeder, make the change gradually. Ask the breeder to supply you with about 1 pound of the rabbit's current feed. Give this feed to the rabbit for the first day or two. Then begin to mix the breeder's feed with yours, gradually making a complete changeover to the new feed.

Guarantee. Ask about the breeder's policy concerning problems that may arise with your new rabbit. Most breeders will guarantee the health of their rabbits for at least 2 weeks. After 2 weeks, it becomes difficult to know whether a problem started at the breeder's rabbitry or was acquired after the rabbit was sold.

Handling Your Rabbit

Repeated gentle handling makes your rabbit easier to handle when you need to, but you must observe certain points of rabbit etiquette. With practice, your skill will increase.

Being picked up can be scary for a rabbit. When you lift a rabbit, you're taking away its most effective method of defense — the ability to run away from danger. If your rabbit is frightened, it will try to run away. Its toenails help it grip surfaces, enabling it to run faster. When you lift the rabbit, you become the only surface it has to grip. This desire to grip often results in scratches for the handler and a panic situation for both the handler and the rabbit. Many new owners become discouraged when their rabbit scratches. Remember, your rabbit is scared — it's not mad at you or scratching you on purpose. You can make things easier on yourself by wearing a long-sleeved shirt when you handle a rabbit.

A "football hold" helps your rabbit feel safe while you're carrying it. For extra security, place your free hand on the rabbit's back.

Rabbits have powerful hind leg and back muscles. If not properly supported, a rabbit can kick out and break its own back, causing paralysis.

Some breeds handle more easily and respond better than others. Most rabbits enjoy being stroked, especially when they are young. Remember always to talk calmly and quietly. Loud noises easily startle rabbits.

The time you spend learning proper handling skills will pay off in many ways. If you show your rabbits, animals that are used to being handled generally perform better on the show table. They will also sell better, since you will be able to show them off to their best advantage and prospective buyers will be able to get a better look at them. And rabbitry chores are easier if your animals cooperate during cage cleaning, grooming, and breeding.

Getting Acquainted

A new rabbit has many adjustments to make. Give it a few days to get used to you and its new surroundings before you handle it a lot.

Get to know your rabbit in a setting where both of you are comfortable. A good place to get acquainted is at a picnic table covered with a rug, towel, or other covering that will give the rabbit secure, not slippery, footing. Your rabbit will be able to move safely around on the table, and you can safely pet it without having to lift it. Rabbits will usually not jump off a table; nonetheless, *never leave the rabbit unattended when it is out of its cage.*

Once your rabbit seems comfortable on the table, start to practice picking it up. The table offers a handy surface on which to set the rabbit safely back down.

Lifting Your Rabbit

If the rabbit starts to struggle while you're holding it, drop to one knee. If the rabbit escapes from your arms, it won't have far to fall.

The best way to lift a rabbit is to place one hand under it, just behind its front legs. Place your other hand under the animal's rump. Lift with the hand that is by the front legs and support the animal's weight with your other hand. Place the animal next to your body, with its head directed toward the corner formed by your elbow. Your lifting arm and your body now support the rabbit; it's like tucking a football against you for a long run. Place the animal gently back on the table and repeat this lift.

Practice handling skills often, but for short periods of time — about 10 to 15 minutes daily is all it takes. Have a short practice session each day until you and your rabbit are comfortable with each other.

Once you feel at ease, begin to move around while holding your rabbit. You may need to steady and secure your animal with your free hand until it gets used to being carried. At first, just walk around the table; if your rabbit becomes frightened, you have a safe place nearby to set it down promptly.

If you need extra control, you can pick up your rabbit with one hand by grasping the loose skin at the nape of its neck, providing support under its rump with your other hand. This method can damage the fur and flesh over the rabbit's back, however, and is especially harmful to the more delicate coats of Rex and Satin rabbits. Once you have lifted the rabbit, move it close to your body where it will feel more secure. After it's tucked in "football style," you may use a one- or two-hand carry. The rabbit's behavior will help you decide whether to use one or both hands.

Sometimes an overactive bunny will struggle and get out of control. When this happens, drop to one knee as you work to quiet the animal. Lowering yourself lessens the distance the rabbit has to fall and provides a position from which you can easily set the animal on the ground, if necessary. After a short rest on the ground, carefully and securely lift the rabbit again. Even the most mild-mannered rabbit can have a bad day, so be prepared to handle any situation in a calm, controlled manner.

Turning Your Rabbit Over

After you master lifting and carrying your rabbit, you should learn how to turn it over. In the future, as part of your health maintenance program, you'll need to get your rabbit on its back in order to examine its sex, teeth, and toenails, so it's worth your time to help your rabbit become accustomed to being held on its back.

Turning a rabbit over puts the animal into an unnatural position. First, its feet are off the ground, so it cannot run away, and second, its underside is now exposed. Your rabbit has good reasons to resist this type of handling — a wild rabbit in this position is probably about to be eaten by a predator — so be especially careful and patient. Again, a rug-covered picnic table is a good place to practice.

Use your upper hand to control the rabbit's head; when the head is restrained, the rabbit is unable to move.

To turn the rabbit over, use one hand to control the head and the other hand to control and support the hindquarters. Place the hand that holds the rabbit's head so you are holding its ears down against its back while you reach around the base of the head. If you prefer, place your index finger between the base of the rabbit's ears and then wrap your other fingers around toward its jaw. With your other hand, cradle the rump. With your hands in place, lift with the hand that is on the head and at the same time roll the animal's hindquarters toward you. Try to do this movement in a smooth, unhurried manner.

If your rabbit has cooperated, you will now be able to let the table support its hindquarters so the hand that was holding the rump is free to check the rabbit's teeth and toenails. If your rabbit fights against this procedure, keep trying, but be sure to do so in a place where it hasn't far to fall, and try to support it securely.

Part of the trick to successfully turning your rabbit over is being able to grasp its head so it cannot wiggle away. A way to gain some extra control is to grasp the lower portion of the loin instead of cradling the rump. Another approach is to pick up your rabbit and allow it to rest against the front of you instead of tucking it against your side. The animal will face upward with its feet against you. When it is in this upright position, place one hand on its head, as suggested above, and keep your other hand supporting its hindquarters. Now bend forward slowly at the waist and lower the animal to the table, where it should arrive in the proper turned-over position.

An easy trick to getting the rabbit on its back is to cradle it against your chest and then slowly lean forward over a table, until the rabbit's back is on the table's surface.

If you need to turn your rabbit over for a closer look and no table is handy, let the animal rest on your forearm. Having the rabbit in this position gives you better control over the hindquarters, because they are tucked between your elbow and your body. This maneuver is easier if you sit in a chair. You can use your legs instead of the table to support your rabbit. In fact, you may find that when you are seated, you can hold your rabbit more securely for such procedures as trimming toenails.

Rabbit Housing

Rabbit housing, called a hutch, can be the largest single expense in rabbit keeping. Whether you buy or build a hutch, you must be sure it meets the needs of your animals. Proper housing will contribute greatly to the health and happiness of your rabbits.

An all-wire hutch is the first choice of an experienced rabbit raiser. Wire hutches come in two basic styles. The first type is intended to be suspended so the manure falls to the ground. The second type includes a pull-out tray that the manure falls into. These cages, too, may be hung, but they may also be stacked on top of one another. Stacking can make it a little harder to reach each cage, but it saves space if you have a very small indoor rabbitry. You must have excellent ventilation if you plan to keep several rabbits indoors.

When picking out a cage for your rabbits, evaluate the following factors.

Choosing the Right Size

Your rabbit will spend most of its time in its hutch and will need space to move around, as well as space for feeding equipment. A general rule is to allow ¾ square foot of space for each pound your rabbit weighs. For example, a cage 2 feet wide and 2 feet long is 4 square feet and would be perfect for a 5- to 6-pound Mini Lop. Find out how much a mature animal of your rabbit's breed weighs, and use this weight as a guide to plan the size of the hutch. Young rabbits and bucks may do fine with slightly less space, but does that will be having litters will need the whole amount.

After you have figured the width and length of the cage, decide how high it should be. Most cages are 18 inches high, although small breeds, such as the Netherland Dwarf, need a height of only 12 to 14 inches.

Protection from Weather and Predators

With proper care, rabbits have fewer health problems when raised outdoors. However, you must be able to adjust the housing to changes in weather. Where you live determines what weather conditions you will have to deal with. Rabbits are most comfortable at temperatures of 50 to 69°F. In hot climates, keep the cage in the shade and make sure it has plenty of air circulation around it.

Rabbits do well in cold weather and can survive temperatures well below zero. However, they need to be protected from wind, rain, and snow. Indoor cages provide this protection by default. If your cages are outside, you'll need to add protection as the temperature drops. Most outdoor cages can be enclosed by stapling plastic sheeting around three sides. If the weather gets cold in your region, cover the front with a flap of plastic as well. Do not enclose the whole cage so tightly that it is difficult to feed and care for your animal or that the ventilation is poor. Empty plastic feed bags may be used as covering material. Outdoor cages should also have a secure

Wrap plastic sheeting around three sides of the cage to protect the rabbit from cold winds and precipitation.

Baby-Saver Wire

Some manufacturers offer cages made from *baby-saver wire,* a special wire mesh that may be used for cage sides. The upper portion of baby-saver wire has 1" x 2" mesh, and the bottom 4 inches has ½" x 1" mesh. Having this smaller mesh next to the cage floor keeps babies safely inside the hutch. Baby-saver wire is more expensive than 1" x 2" wire mesh, but if your cage will house a producing doe, it's a wise investment.

roof to keep out rain and snow. Try to take advantage of the winter sun. If your outside cages are movable, place them in a location that receives a lot of direct sun. Make sure the cage is not exposed to the wind.

A sturdy, secure hutch will contribute to the safety of your rabbits. Most cages are built on legs or hung above the ground, which protects them from dogs, cats, rodents, raccoons, and opossums. Keeping your cages in a fenced area or indoors is even better but may be impossible when you're just getting started.

Ease of Cleaning

Clean cages are important to the health of your rabbits. If your cages are easy to clean, you will be able to do a better job of caring for your rabbits. Cages should have wire floors that allow droppings to fall through to a tray or to the ground. Choose all-wire cages or wood-framed cages that do not have areas where manure can pile up. Plan your cages so you can easily reach into them and clean all parts.

Wire Choices

As you look at cages, you will discover many types of wire. The most common wire used for the sides and tops of hutches is 14-gauge wire woven in a 1" x 2" mesh. Smaller wire mesh also works well, but it will probably be more expensive. Sometimes you may have a choice of 14- or 16-gauge wire. The 14-gauge wire is heavier and stronger. While 16-gauge wire will work for the smaller breeds, always use 14-gauge if you are raising medium or large breeds. A cage made from 14-gauge wire may cost a little more than one made from 16-gauge wire, but it will last longer and is usually worth the extra money.

You may also need to choose between wire that is welded before being galvanized and wire that is welded after being galvanized. Wire that is galvanized *after* welding is stronger and smoother for the rabbit's feet.

The floor of the cage should be made of 14-gauge welded wire woven in a ½" x 1" mesh. This smaller mesh gives more support to the rabbit's feet but still allows manure to pass through. Never use larger wire for the floor of the hutch. Rabbits can get their feet caught in the openings and break or dislocate their rear legs.

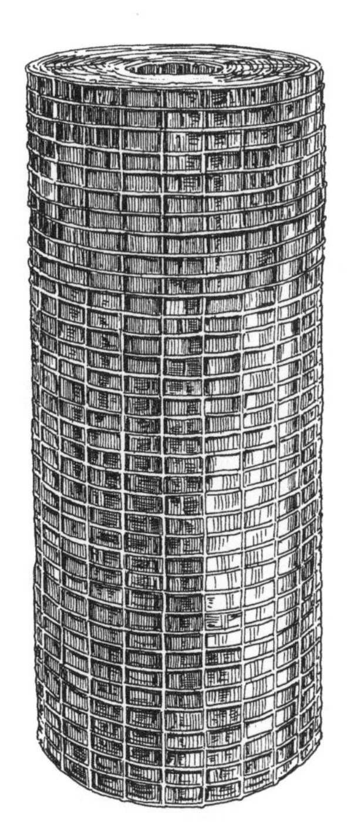

Welded galvanized wire is sold in big rolls; you can ask the staff at the feed store or garden center to cut it to size for you.

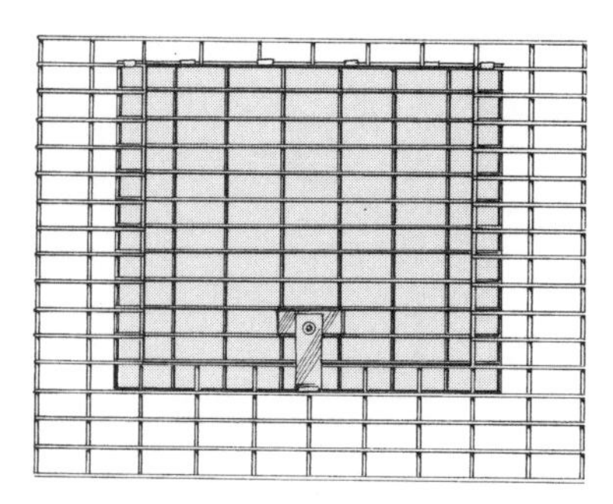

The door should overlap the cage on three sides. If it opens out, attach it to the cage on one side; if it opens in, attach it at the top.

Types of Doors

The door opening is cut into the front of the cage. The door itself should be larger than the opening so it overlaps the opening on all sides. Doors should be placed toward one side of the front of the cage, so there is enough space left for the feeder and waterer.

A door that is attached at the top and swings into the cage when opened will fall back to a closed position even if you forget to latch it, and the rabbit cannot push it open. Cages with doors that open toward the outside make it easier for you to reach into the cage, but if you forget to latch the door, your rabbit may push it open and tumble out of the cage.

Check the door and door opening for sharp edges. You don't want a cage that will scratch you or your rabbit. Some cage doors are lined with metal or plastic to protect you from sharp edges.

Feeders

Most rabbit cages are displayed with an attached feeder. However, the list price of the cage usually does not include the feeder. If the feeder doesn't come with the cage, you'll need to buy one. (See pages 139–140 for information about feeding equipment.)

Urine Guards

A urine guard is a common extra feature on rabbit cages. Four-inch-high metal strips attached around the inside base of the cage help direct the rabbit's urine so it falls directly below the cage floor. A urine guard also helps prevent baby bunnies from becoming stuck in or falling through the sides of the cage. Urine guards are *not,* however, a substitute for baby-saver wire. Cages with urine guards are more expensive than those without them.

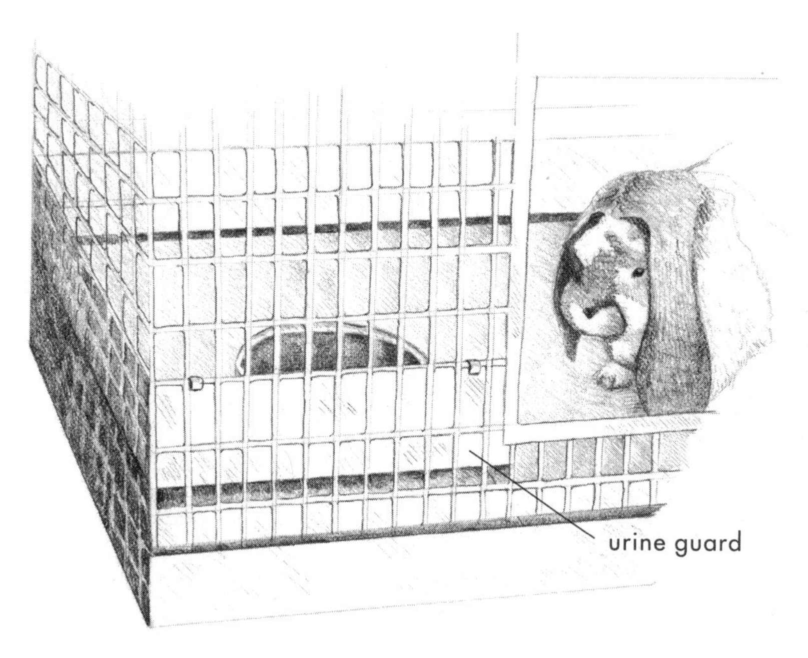

A urine guard helps keep the area around the cage clean.

Buying a Cage

Purchase the best possible hutch you can afford; your rabbit will, after all, be spending most of its time there. Rabbit cages are available from several sources:

- **Mail-Order Rabbitry Supply Companies.** Rabbitry supply companies specialize in cages and equipment for rabbit breeders. They generally offer the best quality, selection, and price. Most rabbitry supply companies have on-line catalogs or will send you a free catalog if you request one. A list of several suppliers is included in the appendix.
- **Farm Supply Stores.** The same store where you buy your feed may also sell cages. Prices will probably be fairly reasonable and the quality acceptable.
- **Local Rabbitry Suppliers.** Many small local suppliers also sell cages. These folks are often rabbit enthusiasts who build and sell cages and other supplies as a part-time business. However, it may be a challenge to find one of these suppliers near you. Talk with other rabbit raisers to find a supplier that serves your area.
- **Pet Shops.** Pet shops often carry rabbit cages, but unless they specialize in rabbits, their supply is often limited. Ask a salesperson to assist you in selecting a cage, and if the selection seems limited, ask him or her if you can browse through the stock catalogs and have a cage special-ordered for you.

Preassembled versus Do-It-Yourself

Suppliers often give you a choice of buying the cage assembled or unassembled. Assembled cages are more expensive and, if you order them from a catalog, cost more to ship. If you choose to save money and buy an unassembled cage, you'll need a tool called *J-clip pliers.* If you have one or just a few cages to assemble, you may be able to borrow J-clip pliers from a local rabbit raiser. If you are going to assemble lots of cages, purchase one of these tools.

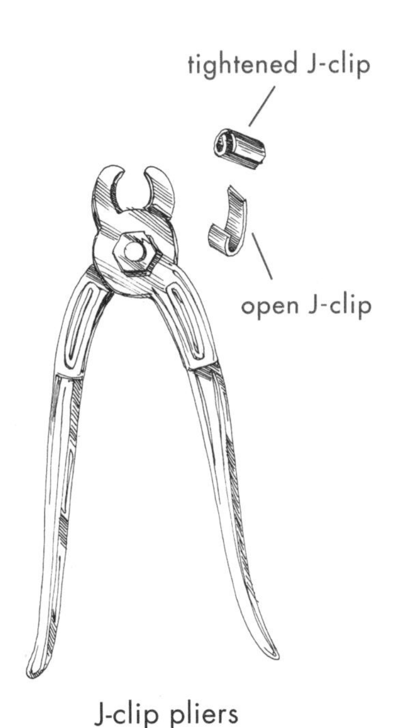

J-clip pliers

Buying a Secondhand Cage

Many rabbitkeepers begin with used cages. This option is usually a less expensive way to get started, but be sure to do the following before you use secondhand equipment:

- Check carefully for holes, rusted wire, chewed wooden areas, and sharp edges. Repair problems before you introduce your rabbits to their new home.
- Carefully clean and disinfect the cage before you use it. First, scrape off old manure and hair. Then, wash the entire cage with a solution of bleach and water (mix 1 part household chlorine bleach with 5 parts water). Set the cage in a sunny spot and allow it to dry thoroughly before you put a rabbit into it.

Building Your Own Cage

If you're planning a small rabbitry, buying cages may be less expensive than building them. The prices suppliers charge for cages are based on the cost of the cage materials. By purchasing materials in large quantities, suppliers get a discount and can charge less per cage. When you purchase materials to build just one or two cages, you may find that your cost per cage is as much as or more than that of a ready-built cage. So, why build your own cages?

- You can design your own cages to best fit your space and best meet the needs of your rabbits.
- If raising rabbits is a family "business," building a hutch can become a family project.
- Once you learn how to build hutches, you can become a hutch supplier for other rabbitkeepers in your area.

Build-your-own rabbit hutches come in many types. The two described here are fairly simple and can be adapted to fit into a wide variety of small rabbitries.

Materials and Tools for the Single Hutch

Materials

- 12½ feet of 18-inch-high 14-gauge baby-saver wire, or 14-gauge 1" x 2" welded wire mesh (for the sides and door)
- 3 feet of 30-inch-wide 14-gauge ½" x 1" welded wire mesh (for the floor)
- 3 feet of 30-inch-wide 14-gauge 1" x 2" welded wire mesh (for the top)
- J-clips (about ¼ pound)
- Latch for the cage door

Tools

- Tape measure
- Wire cutters
- Pliers
- Hammer
- J-clip pliers
- Section of 2x4 lumber, about 2 feet long

The Single Hutch

This hutch will be 3 feet long, 2½ feet wide, and 18 inches high, an adequate size for all except the giant breeds.

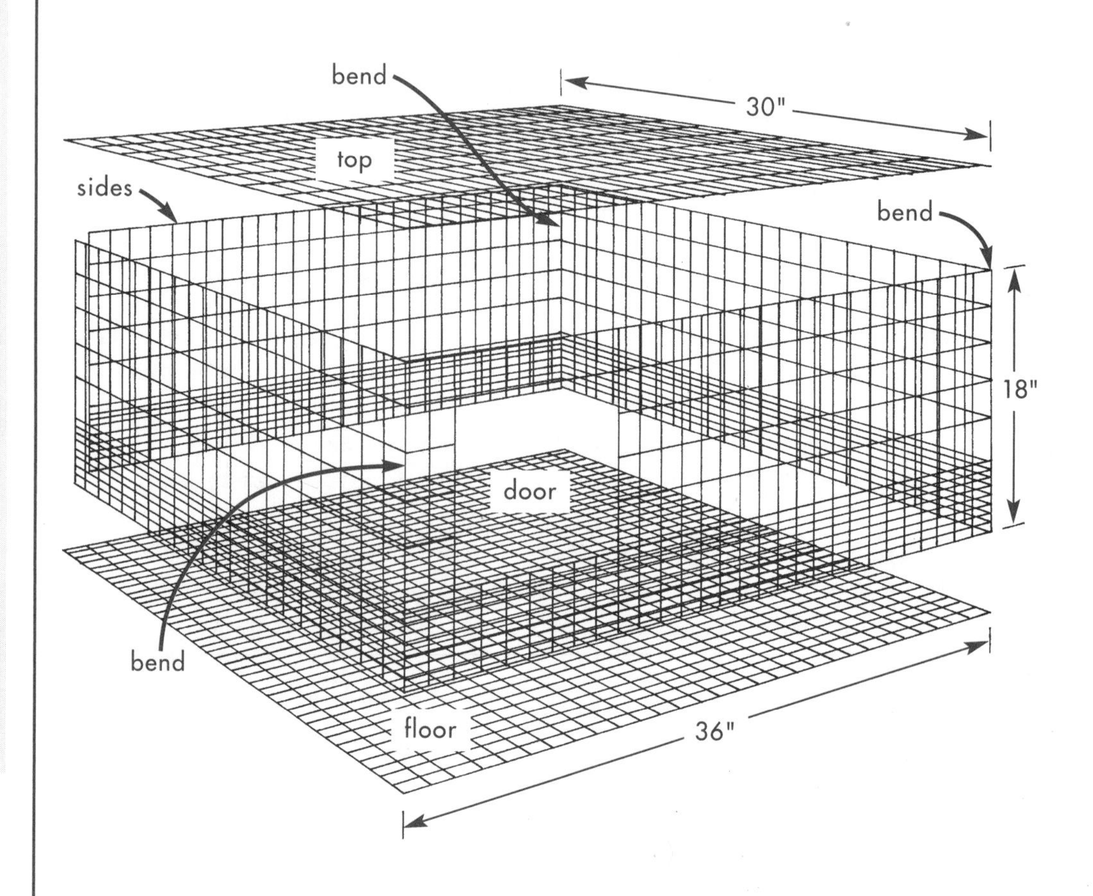

The Sides of the Cage

1. Cut an 11-foot length of the 18-inch baby-saver wire.
2. Measure a distance 36 inches from one end. At this spot, use the section of 2x4 lumber and bend the wire over it as shown at right to form the first corner of the cage.
3. Measure 30 inches from the first corner and form a second corner, using the same method.
4. Measure 36 inches from the second corner and make the third and final corner.
5. You will now have a 30-by-36-inch rectangular shape. Close the rectangle by joining the ends together with the J-clips. Clips should be placed at the top, at the bottom, and every 2 to 3 inches in between. The four sides of your cage are now complete.

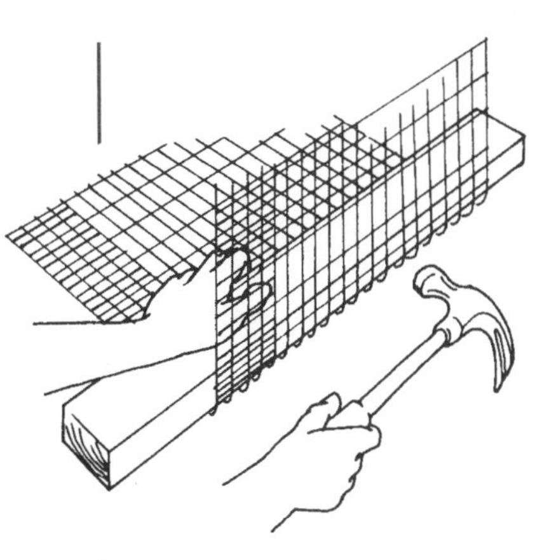

Step 2

The Floor and the Top

6. Create the floor by attaching the ½" x 1" wire mesh to the sides of the cage with J-clips. (Remember that the smaller mesh of the baby-saver wire forms the lower part of the cage, near the floor, and the larger 1" x 2" mesh forms the upper part, near the top.) The 1-inch side of the mesh should face the ground; the ½-inch side provides a smoother surface for your rabbit to stand on and helps prevent sore hocks.
7. Use J-clips to attach the 1" x 2" wire mesh section to form the cage top.

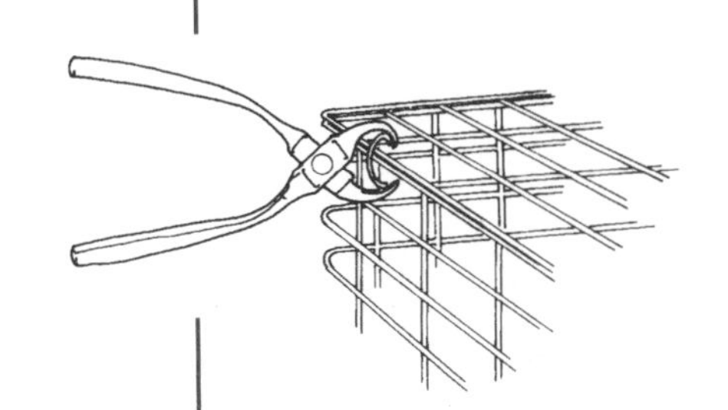

Step 7

The Door Opening

8. Decide how large the door will be and where to position it on the front of the cage. The door must be big enough to allow you to put a nest box (see page 153) into the cage. For most breeds, a 12-inch-square door opening is about right. If you plan to attach an outside metal feeder, place the door to one side of the front of the cage. The bottom edge of the door opening should be 4 inches above the floor of the cage, to prevent the rabbits from falling out when the cage door is open.
9. Measure the location and size of the opening, then use wire cutters to make the opening. As you cut, leave an end of wire about ¾ inch long sticking out.
10. After the door opening is cut out, use regular pliers to bend these protruding ends over. This process takes some time but creates a smooth finish that will not scratch you or the rabbit. An alternative would be to cut the door opening as close as possible — don't leave any ends protruding — then purchase plastic or metal edge protectors to cover the edges.

The Door

11. Make the cage door from the wire remaining from the piece you cut for the sides. The door should be larger than the door opening by at least 2 inches. Thus, if the opening is 12 inches square, make the door at least 14 inches square.
12. Instead of attaching the door to the edge of the door opening, attach it at least one mesh block beyond the edge. This position makes it sturdier. If you've decided to have the door open into the cage, attach it with J-clips along the top of the opening. Doors that open out are usually attached along one side.
13. Attach the door latch.

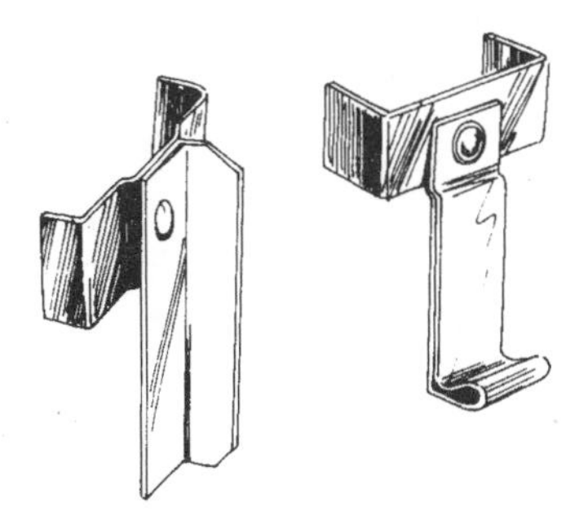

Galvanized metal door latches come in a variety of styles.

The Multiple-Unit Hutch

Multiple-unit hutches for rabbits are like apartment buildings for people. Because adjoining units share walls, less wire is need, and the construction cost per unit is less than the cost to construct individual units. If you plan to keep several rabbits, a multiple-unit hutch may be right for you. You will use the same tools, types of wire, and techniques that are required to build a single unit.

The all-wire hutch described here will be 7 feet long, 30 inches wide, and 18 inches high. It may be subdivided into four units for small breeds, three units for medium-sized breeds, or two units for large breeds.

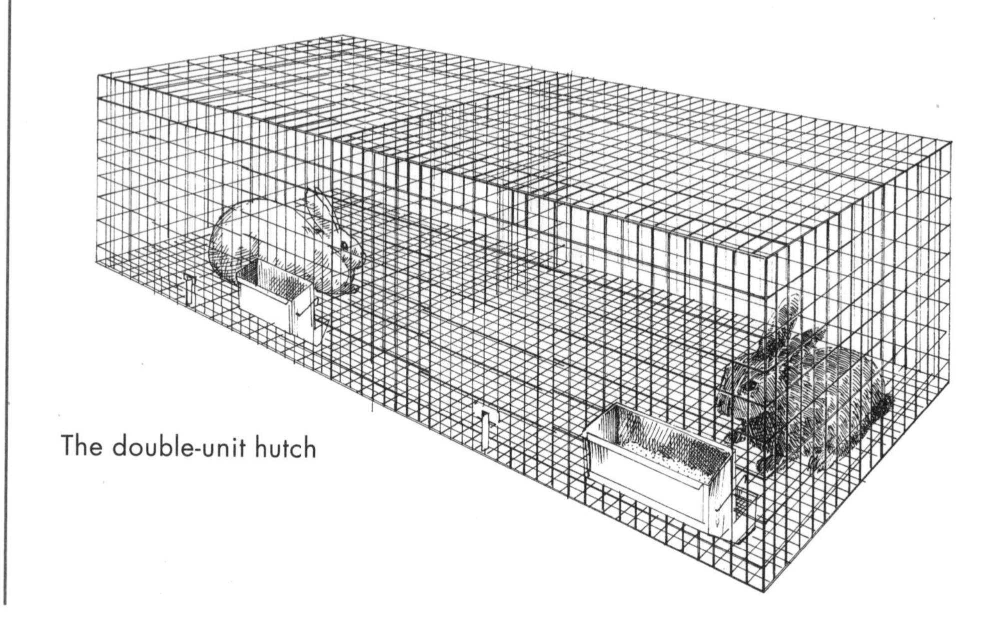

The double-unit hutch

Materials and Tools for the Multiple-Unit Hutch

Materials for 2 Units

- 19 feet of 18-inch-high baby-saver wire (for the sides)
- 8½ feet of 30-inch-wide 1" x ½" 14-gauge wire (for the floor and partitions)
- 8½ feet of 30-inch-wide 1" x 2" 14-gauge wire (for the top and doors)
- 2 door latches

Materials for 3 Units

- 19 feet of 18-inch-high baby-saver wire (for the sides)
- 10½ feet of 30-inch-wide 1" x ½" 14-gauge wire (for the floor and partitions)
- 10 feet of 30-inch-wide 1" x 2" 14-gauge wire (for the top and doors)
- 3 door latches

Materials for 4 Units

- 19 feet of 18-inch-high baby-saver wire (for the sides)
- 12 feet of 30-inch-wide 1" x ½" 14-gauge wire (for the floor and partitions)
- 10 feet of 30-inch-wide 1" x 2" 14-gauge wire (for the top and doors)
- 4 door latches

Tools

- Tape measure
- Wire cutters
- Pliers
- Hammer
- J-clip pliers
- Section of 2x4 lumber, about 2 feet long

The Sides

1. Measure a distance of 7 feet from one end of the baby-saver wire. At this spot, use the section of 2x4 lumber and a hammer, as shown on page 133, to form the first corner of the cage.
2. From this bend, measure 30 inches. Make a second corner.
3. Measure 7 feet and form the last corner.
4. Join the two ends with J-clips to form a large rectangle (30 inches by 7 feet).

The Floor

5. Cut a 7-foot length of ½" x 1" mesh. Use J-clips to connect this floor piece to the four walls. Attach the flooring wire as described in step 6 on page 133.

The Interior Partitions

6. Cut the remaining ½" x 1" mesh into 18-by-30-inch pieces for use in subdividing the cage. You will need one piece to partition for a two-unit cage, two pieces for a three-unit cage, and three pieces for a four-unit cage. (You may also use solid metal or plastic dividers, available from rabbitry suppliers.)
7. Use J-clips to attach the partitions securely to the floor, front, and back of the cage. Space the partitions 21 inches apart for a four-unit cage, 28 inches apart for a three-unit cage, and 42 inches apart for a two-unit cage.

The Top

8. Cut a 7-foot section of 30-inch-wide 1" x 2" mesh for the top of the cage. Attach the cage top to the front, back, ends, and interior partitions by using J-clips.

The Door Openings and Doors

9. Review steps 8–12 on door openings and doors for a single hutch (page 133). Cut a door opening in the front of each unit. Use the remaining 1" x 2" mesh to cut the doors. Attach the doors, and attach a latch to each door.

Frame for a Single Hutch

If your all-wire hutch will be used outdoors, you will need to adapt it so that your rabbit is protected from various weather conditions. The most common way to protect a hutch is to build a wooden framework into which you set the cage. Constructing this framework will be a bit more of a challenge than cage building. The plans for a single-hutch frame shown on page 136 can be modified for a multiple-unit hutch.

Cut the Legs

1. Cut two 60-inch lengths from one of the 10-foot 2x4s to form two front legs.
2. Cut two 56-inch lengths from a second 10-foot 2x4 to form two (shorter) back legs.

Cut the Support Pieces

3. From each of the two remaining 10-foot 2x4s, cut:

- One 35¼-inch piece (for the side of the cage-support frame)
- One 38-inch piece (for the front and back of the cage-support frame)
- One 44⅛-inch piece (for the front and back roof support)

4. Cut the 8-foot 2x4 in half. These pieces will be the side roof supports. You will later recut them to length.

Materials and Tools for a Single Hutch

Materials

- Four 10-foot lengths of 2x4 lumber
- One 8-foot length of 2x4 lumber
- One 4-by-5-foot sheet of exterior-grade plywood (¾-inch thickness)
- 12d common galvanized nails
- 6d common galvanized nails
- 8 L-brackets
- Wood screws

Tools

- Saw
- Tape measure
- Hammer
- Screwdriver
- Pencil

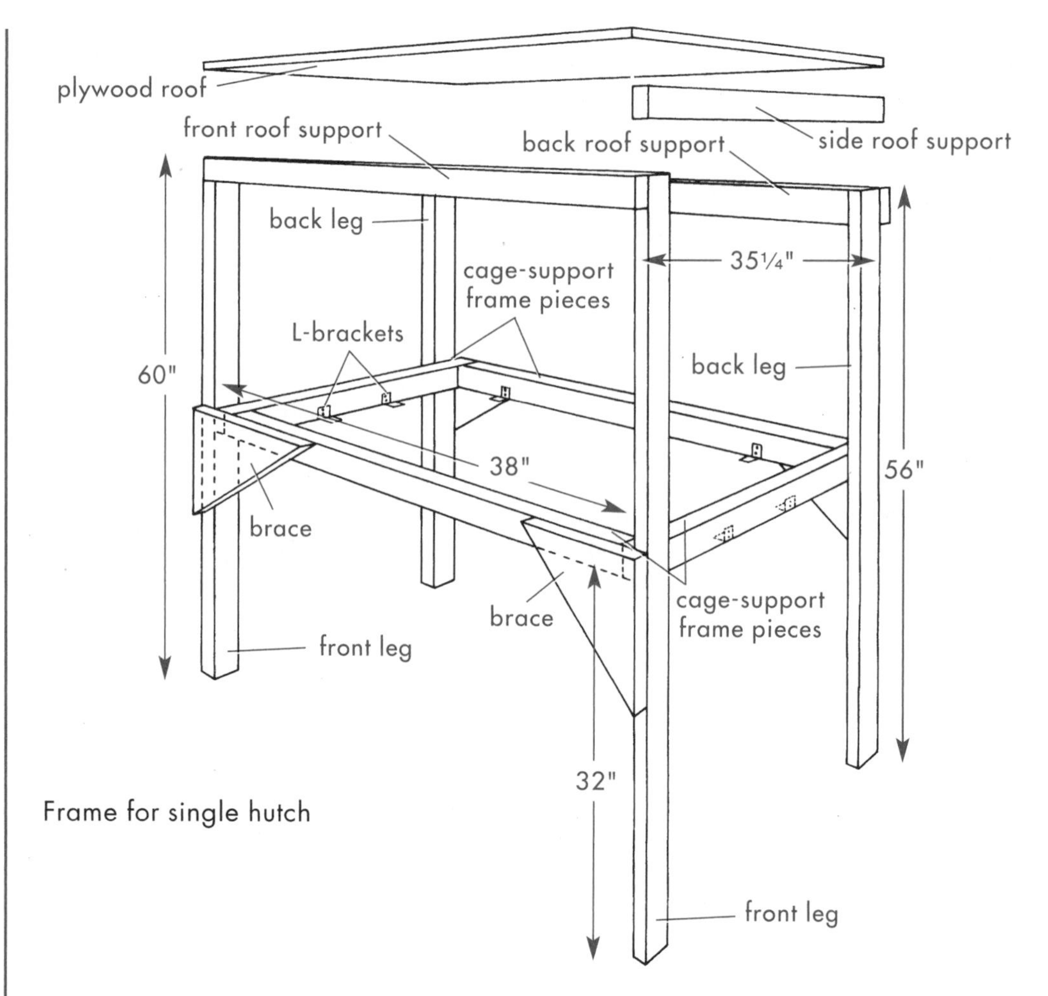

Frame for single hutch

Put It All Together

5. Lay the four cage-support frame pieces on edge on a flat surface. They should form a large rectangle, with the 38-inch front and back pieces between the 35¼-inch side pieces. Drive two 12d nails through the outside face of a side piece into the end of the back piece to make the corner. Nail the three other corners together in the same fashion to complete the cage-support frame.

6. Screw the L-brackets to the inside face of the cage-support frame, as shown in the illustration above. Make sure the bottom of each bracket is flush with the bottom edge of the frame.

7. Mark one end of each of the legs with a pencil to indicate the bottom and to avoid confusion during assembly. Measure up from the bottom of each leg 32 inches, and draw a straight line across the inside face of each leg at this point. This mark represents the height of the cage-support frame.

8. Use 12d nails to attach the legs to the ends of the cage-support frame (as shown above), making sure that the bottom of the frame is even with the lines marked on the legs.

9. Use 12d nails to attach the front and back roof supports to the legs. The top edge of the front roof support should be slightly higher than the top of the front legs to prevent the front legs from being in the way when it comes time to put on the plywood roof.

10. Take one of the side roof supports and hold it in place against the legs on the right side of the hutch. With a pencil, mark the angled cut you'll need to make at the side roof support's ends so it fits snugly between the tall front legs and the short back legs. Cut the piece to size, and attach it with 12d nails to the front and back roof supports. For added strength, nail it to the tops of the legs as well. Repeat this procedure for the left side.

11. Use scraps of plywood to cut four isosceles triangles, with the equal sides being 12 inches long. Nail these braces to the hutch with 6d nails, as shown.

12. Use 6d nails to attach the 4-by-4-foot plywood roof. The roof should overhang all four sides, to help keep rain or snow out of the hutch. A large overhang along the front will give extra protection to the attached feeders.

Advantages of a Wooden Frame

A wooden frame has features that make it adaptable to a variety of situations:

- Sides may be left open for good air circulation.
- Removable, solid wooden sides or plastic sheeting may be attached for protection against cold weather.
- The cage may be removed from the frame, making cleaning easy.
- Since the cage may be removed from the frame, it is easy to move an expectant doe to a warmer location for a winter kindling.

Home Away from Home: Rabbit Carriers

If you plan to show your rabbit or will need to transport it from one location to another, you'll need a rabbit carrier. Like cages, the best rabbit carriers are made of wire mesh. You can purchase a good rabbit carrier, or you can build one yourself. If you have built a cage, building your own carrier will be easy and economical. Many small pieces left from cage construction can be used to build a carrier. If you don't plan to build a cage and would have to purchase wire cutters, J-clips, J-clippers, and wire just to make a carrier, you're probably better off buying a carrier.

Bottom Tray

One of the largest costs involved in building your own carrier is the metal pan that encloses the bottom. You can cut this cost by using a substitute, such as a plastic cat litter tray or a large roasting pan.

Tips on Building a Carrier

The sides of the carrier may be made from one long piece or four smaller pieces of wire mesh. The length of the four sides will vary based on the size of the bottom tray. The height of the sides will vary according to the size of the breed for which you are building the carrier. Small breeds require a height of 8 inches, medium breeds require 10 inches, large breeds require 12 inches, and giant breeds require 14 to 16 inches. Add an extra 2 inches to the height. This height allows you to attach the floor 2 inches above the tray so droppings can pass through.

The floor of your carrier should be made from ½" x 1" welded wire. Be sure to place the ½-inch side toward the top to protect the rabbit's feet.

Rabbit carriers are designed so the top opens, making it easy to reach in and safely remove the rabbit. Allow an extra 1 to 2 inches on each end and on one side of your top piece. This extra wire may be bent so that it overlaps three of the sides. The straight edge is attached with J-clips to form a hinge on the fourth side, allowing the top to open and close. Attach a door latch or hook so you will be able to securely close the carrier. The clip from an old pet leash also makes a secure latch.

Use S-hooks or springs to attach the cage to the tray. These items are fairly inexpensive and make it easy to remove the tray for cleaning.

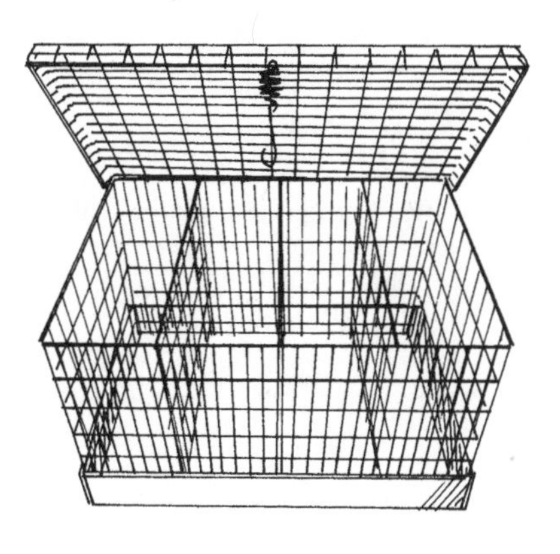

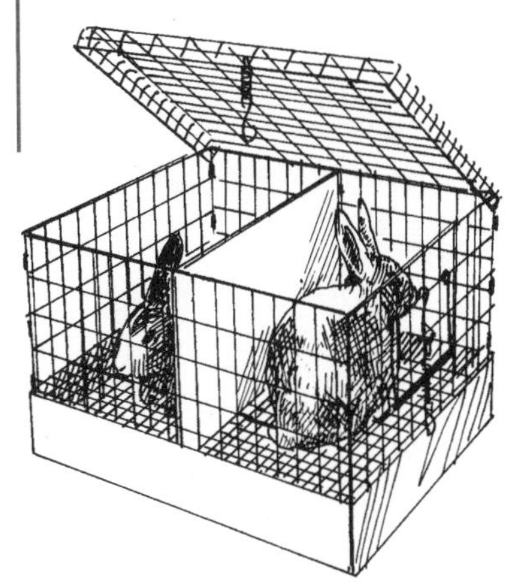

Rabbit carrying cages

Feeding Your Rabbit

Providing a healthy, balanced diet is key to successful rabbit raising. In the past, rabbit owners mixed different kinds of feeds to get the proper nutritional balance for their animals. Today, stores carry feed that is already balanced. However, a good feeding program requires much more than simply filling up the feed dish.

Commercial Rabbit Feed

Feeding commercial rabbit pellets is the best and easiest way to provide proper nutrition for rabbits. A great deal of research goes into producing a balanced ration, and feed companies are well qualified to formulate rations. Commercial feeds vary in quality and price, and different feeds are available in different locations. Talk with other breeders and your feed store clerk to learn about brands of feed that are available in your area.

Like cat food and dog food, rabbit food is available in several types, each intended to meet the needs of a rabbit at a different stage in its life. The ingredient that usually varies among these types of feed is the protein content. Mature animals usually need feed that provides 15 to 16 percent protein. Active breeding does usually need a feed that provides 17 to 18 percent protein.

The label on a feed package will tell you not only how much protein but also how much fiber and fat the feed contains. For optimal health in your rabbits, follow the feed guidelines in the chart below.

Salt

Salt is important to a balanced feed. You may see advertisements for inexpensive salt spools that can be hung in your rabbit's cage, but these spools are not necessary (and they'll cause your cage to rust). The commercial pellets you use as feed include enough salt to meet your rabbit's needs.

Rabbit Feed Guidelines

Rabbit	Optimal Feed Content
Mature rabbits and developing young	12 to 15 percent protein 2 to 3.5 percent fat 20 to 27 percent fiber
Pregnant does and does with litters	16 to 20 percent protein 3 to 5.5 percent fat 15 to 20 percent fiber

Hay

Rabbits enjoy good-quality hay, and it adds extra fiber to their diet. If you feed hay, be sure it smells good, is not dusty, and is not moldy. And remember that feeding hay will add more time to your feeding and cleaning chores.

If you choose to feed hay on a regular basis, place a hay manger in each cage. Hay mangers are easy to construct from scraps of 1" x 2" wire mesh (if you make your own cage, you can use the leftover wire). If your cages are inside and are of all-wire construction, you may just place a handful of hay on the top; the rabbits will reach up and pull it through the wire. Do not place hay on the floor of the cage, because it will soon become soiled and droppings will collect on it, promoting disease and parasites.

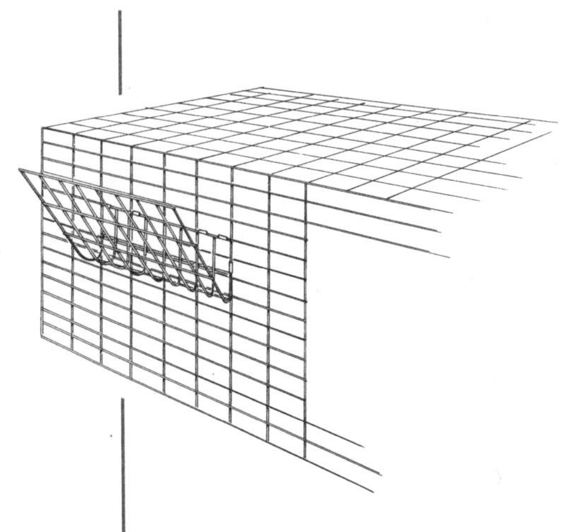

To make a simple hay manger, bend a scrap of wire mesh and attach it to the sides of your cage with J-clips. Make sure there are no sharp wire edges protruding into the cage.

Special Feeds and Treats?

Experienced rabbit raisers, especially those trying to prepare their animals for shows, give their animals special feeds, such as oats, sweet feeds, corn, and sunflower seeds, which they feel give their animals an edge. However, any additions to the diet can upset a rabbit's digestion. It is best to keep to a simple diet of pellets and water, perhaps supplementing with some hay. Avoid additional feeds until you have several years' experience in successful rabbit raising.

Many new rabbit raisers like to pamper their animals by giving them fresh greens and vegetables as "treats." In fact, many of these treats can actually be harmful to rabbits. Because young rabbits have especially sensitive digestive systems, it's best not to give them any treats. (Absolutely do not give young rabbits any greens, including lettuce; they can cause diarrhea.) As the rabbit matures, its digestive system becomes hardier, but treats should never become a large part of your rabbit's diet.

What might you give as an *occasional* treat to an adult rabbit? They'll enjoy the following:

- Alfalfa
- Apples
- Beets
- Carrots
- Lettuce
- Pumpkin seeds
- Soybeans
- Sunflower seeds
- Turnips

Feeding Equipment

Pellets are usually fed from crocks (heavy plastic or pottery dishes) that sit on the cage floor or from self-feeders that attach to the side of the cage. If you choose to use a crock, select one that is heavy enough that your rabbit cannot tip it over and waste its feed. To make feeding easier, place the crock close to the door of the cage. Never place the crock in the area that your rabbit has chosen as its toilet area. Because rabbits can soil the food in crocks, feed only as much as will be eaten between each feeding time.

Self-feeders attach to the wall of the cage and are filled from outside the cage. If you use a self-feeder, choose the type that has a screened, not solid, bottom. The screen allows *fines* (small, dusty pieces that break off from the pellets) to pass through. Otherwise, fines tend to build up, since rabbits will not eat dust from feed. An advantage of self-feeders is that young rabbits are less likely to climb into them and soil the feed.

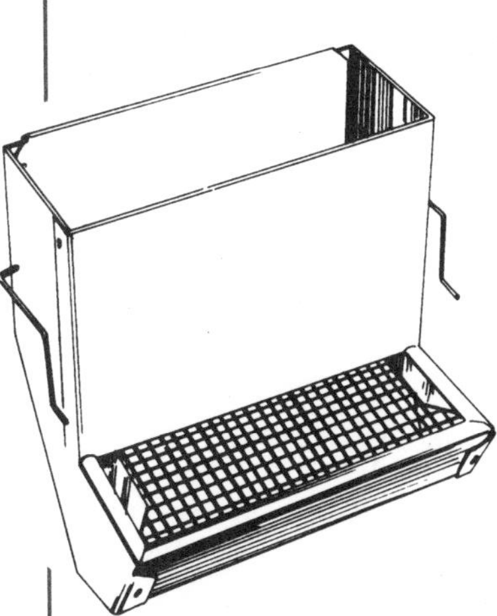

This self-feeder clips to the cage wire and may be filled from outside the cage.

If you need to keep costs low as you start out (and most everyone does), make a simple feeder from a 12-ounce tuna fish can. Tack the can to a 10-inch square of ½-inch plywood. The plywood base will make your feeder hard to tip, but you will still be able to remove it from the cage for regular cleaning.

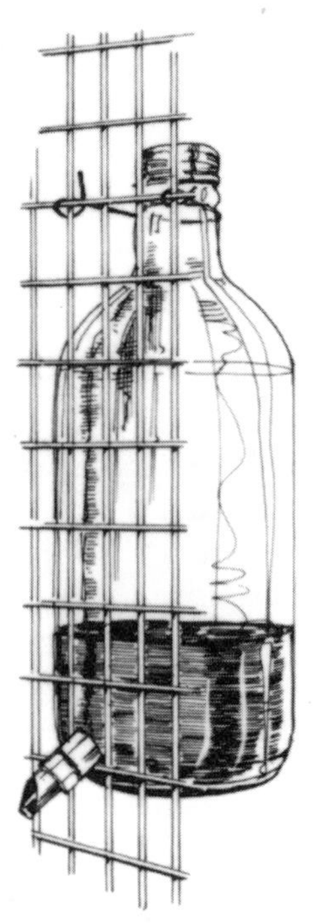

A soda bottle may be fitted with a drinking valve to make an inexpensive waterer.

Watering Equipment

Your rabbits will not thrive unless you provide them with a constant supply of clean, fresh water. As with feed, there are several ways to provide water to your rabbits. The ideal type of waterer is a plastic water bottle that hangs on the outside of the cage. It has a metal tube that delivers water to the rabbit. This type of waterer keeps water cleaner than crocks do and cannot be tipped over, but water bottles freeze easily and therefore cannot be used outdoors in cold weather.

You can also make a water container from a plastic bleach or soda bottle and a drinking valve, such as that used in automatic watering systems. If your local farm store does not have these valves, you can order one from a rabbitry supply company (see Suppliers, page 390). Wash the bottle well, then cut or drill a hole into the side of the bottle near the bottom. Spread epoxy cement on the rear part of the valve, and insert it into the hole. Use wire to hang the finished waterer on the outside of the cage.

If you live in an area where freezing temperatures are common and you keep your rabbits outside, you might try heavy plastic crocks. You can remove frozen water from these crocks without breaking them or waiting for them to defrost. Hit the plastic crocks against a fence post or other solid object to dislodge the frozen water, then refill them. This procedure can save time during winter feedings.

You can make a watering dish from two empty bleach bottles. For each waterer, you'll need a 1-gallon bottle and a ½-gallon bottle. Cut the base from each bottle, about 4 inches up from the bottom. Wash each piece well. Place the ½-gallon piece in the center of the gallon piece. Fill the space between the two with concrete. When the concrete hardens, you will have a low-cost, hard-to-tip waterer.

When to Feed

The feeding habits of domestic rabbits are similar to those of wild rabbits. Wild rabbits are nocturnal — you don't see them much during the day, but they often appear in the early morning or late afternoon to have their big meal of the day.

If our tame rabbits could pick a mealtime, they would probably agree with their wild relatives and choose late afternoon. Using this information, most rabbit breeders feed the largest portion of their rabbits' daily ration in the late afternoon or early evening. Although different breeders develop schedules that work for them, two feedings a day — one in the morning and one in the late afternoon — seem to be most common. If you choose to feed twice a day, serve a lighter feeding in the morning and a larger portion in the afternoon.

Whatever feeding schedule you choose, it's important to follow it every day. Your rabbits will settle into the schedule that you set, and variations in that schedule may upset them.

Feeding Time Is Checkup Time

Feeding time is important for more than just feeding; this time spent with your rabbits is an opportunity for observing many important details:

- Has a youngster fallen out of a nest box?
- Is an animal sick or not eating well?
- Does a cage need a simple repair before it becomes a major repair?
- Does your rabbit need extra water because of extreme heat or freezing conditions? Remember, your rabbit needs clean, fresh water at all times.

Because a lot can happen to your rabbits in 24 hours, feed, water, and observe them at least twice a day. The serious rabbit breeder sees this expenditure of time as a worthwhile investment.

How Much to Feed

Knowing how much to feed is probably the most complicated part of feeding rabbits. No set amount is right for all animals, but the chart on page 142 gives some average amounts.

Homemade feed measure

To use the guidelines in the chart, you'll need to measure how much you feed to each animal. You can make your own measuring container from an empty can. You'll need a scale that measures weight in ounces. To adjust for the weight of the can, place it on the scale and set the weight dial back to zero. If this is not possible with your scale, remember to figure the weight of the can into your measurements. Add pellets to the empty can until the scale reads the number of ounces you want to feed, then mark a line at that level. Repeat this procedure as many times as necessary to ensure that you give each of your rabbits the correct amount of feed at each feeding time.

Feed guidelines are exactly that: guidelines only. Some rabbits need a little more feed, and some need less. Adjust each rabbit's ration according to its physical condition. Some reasons a rabbit's feeding may need to be adjusted are:

- A young, growing rabbit needs more feed than a mature rabbit of the same size.
- Rabbits need more feed in cold weather than in hot.
- A doe that is producing milk for her litter needs more feed than a doe without a litter.

By stroking each rabbit from the base of its head and following along the backbone, you can tell who is too fat and who is too thin. When you run your hand over the rabbit, you will feel the backbone below the fur and muscles. The bumps of the individual bones that make up the backbone should feel rounded. If the bumps feel sharp or pointed, your rabbit could use an increase in feed. (However, extreme thinness may indicate health problems.) If you don't feel the individual bumps, it probably means too much fat is covering them, and the rabbit should receive less feed.

If you run your hand over each rabbit at feeding time, you'll know whether you should increase or decrease the ration.

Feeding Chart

	Amounts of Pellets Needed Each Day		
	small breeds	medium breeds	large breeds
Buck	2 oz.	3–6 oz.	4–9 oz.
Doe	3 oz.	6 oz.	9 oz.
Doe, bred 1–15 days	3 oz.	6 oz.	9 oz.
Doe, bred 16–30 days	3½–4 oz.	7–8 oz.	10–11 oz.
Doe, plus litter (6–8 young), 1 week old	8 oz.	10 oz.	12 oz.
Doe, plus litter (6–8 young), 1 month old	14 oz.	18 oz.	24 oz.
Doe, plus litter (6–8 young), 6–8 weeks old	22 oz.	28 oz.	36 oz.
Young rabbit (weaned), 2 months old	2–3 oz.	3–6 oz.	6–9 oz.

If you need to adjust the amount you feed, make the change gradually. Unless the rabbit is very thin or very fat, increase or decrease by about 1 ounce each day. You should see and feel a difference in 1 to 2 weeks. It's good to form the habit of giving each rabbit a weekly check. It takes only a few seconds to feel down the back of an animal. If you check regularly, you can adjust the ration before your rabbit becomes too fat or too thin.

In general, overfeeding is a more common problem than underfeeding. Overfeeding is troublesome for several reasons. For one thing, it costs you more to care for your rabbits if you feed them too much. More important, if you are breeding your rabbits, overweight rabbits are generally less productive than animals of the proper weight. In fact, fat rabbits often do not breed, and an overweight doe that does become pregnant often develops health problems.

Rabbit Health

Four major factors contribute to your rabbits' health: housing, environment, observation, and nutrition.

- **Housing.** Hutches should offer protection from extremes in temperature, from wind, and from rain and snow. Hutches must be cleaned often to prevent disease-causing germs from flourishing.
- **Environment.** Prevent situations that stress or upset your rabbit, such as exposure to loud noises, encounters with wild or frightening animals, and summertime heat.
- **Observation.** One of the most important things you can do each day for your rabbit is to observe it. Through observation you'll learn what is normal

for your rabbit, and you'll be better prepared to identify abnormal behavior. Changes in the way your rabbit acts or looks often indicate that it is not feeling well. The sooner a problem is spotted, the better are your chances of solving it before it becomes too serious.

- **Nutrition.** Feed the proper amounts of the right foods on a regular schedule. Be familiar with the amount of food your rabbit normally eats. A change in eating habits is often one of the first signs that a rabbit is not feeling well. And don't forget that your rabbit needs fresh water at all times.

Abscesses

An abscess helps the body rid itself of infection that was usually acquired through a cut or a scratch. If an abscess forms on your rabbit, you may feel it as a bump on the rabbit's skin. If you part the fur and see a raised, reddish area that feels warm, it is probably an abscess.

Over time, an abscess grows, and the skin covering the area opens to allow pus to escape. Although abscesses will drain by themselves, the pus carries germs that can infect other areas of the body. To prevent the infection from spreading, it's better for you or your veterinarian to start the draining process. If you do it yourself, you will need a helper.

1. One person should hold the rabbit while the other uses medicinal disinfectant to clean the area around the abscess. Cleaning may be easier if you clip away some of the hair from the area.

2. Use a sharp, clean instrument, such as a new razor blade, to make a small opening in the abscess. Because the pus in the abscess has been pushing hard against the rabbit's skin, this small cut will probably not hurt your rabbit. In fact, because the cut releases the pressure of the pus, it will actually relieve some of the pain.

3. Use clean cotton balls to help push out the pus.

4. Use medicinal disinfectant to wash out the wound and surrounding area.

5. Apply an antibiotic ointment to the wound. You may use an over-the-counter medication sold for human use. Do not put a bandage on the wound, since your rabbit would only chew it off.

6. Discard the cotton balls and thoroughly wash your hands with soap and warm water.

7. Check the wound daily and apply more antibiotic cream until it is healed.

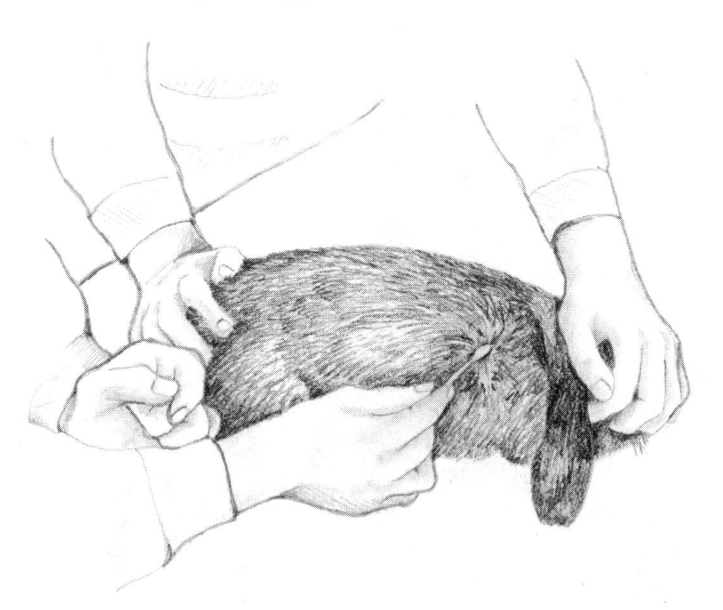

Use a cotton ball or swab to apply antibiotic cream to the wound.

A Health Checklist

Occasionally, an animal becomes sick. If you suspect that a rabbit is not well, the following checklist will help you pinpoint what might be wrong.

❑ Yes ❑ No The rabbit is eating less than its normal amount of feed.

❑ Yes ❑ No The rabbit's eyes look dull, not bright.

❑ Yes ❑ No Discharge is coming from its eyes or nose.

❑ Yes ❑ No The rabbit's droppings are softer than usual.

❑ Yes ❑ No The rabbit's coat looks rough.

❑ Yes ❑ No The rabbit is dirty and has manure stuck to its underside.

❑ Yes ❑ No The rabbit is having trouble breathing or is breathing rapidly.

If you checked "yes" for one or more of the above items, the rabbit may be ill. Begin a more thorough evaluation, or consult with your veterinarian to identify the problem.

Cuterebra Infestation

A parasite uses a living animal as its food source. If you notice a large bump that feels like an abscess (you may even see pus) but has a hole near the top, you're dealing with a parasite. In this case, a Cuterebra fly larva has penetrated or migrated to the skin of your rabbit and is now growing under its skin. If left alone, this parasite will continue to eat away at the rabbit until it is mature and crawls out of the hole to develop into a flying insect.

Removal of Cuterebra larvae is usually no more difficult than treating an abscess, but because the larvae are so harmful to a rabbit if they are ruptured in the removal process, seek the help of your veterinarian.

Coccidiosis

A common disease that causes diarrhea is coccidiosis. The disease can kill rabbits if it goes untreated. Coccidiosis may cause any of the following:

- Soft droppings (diarrhea)
- Rough-looking fur
- Animal not growing as well as you think it should
- A potbelly, although the rabbit is rough-feeling over the backbone

Coccidiosis often occurs in young litters of rabbits. The protozoans that cause this disease are present in droppings and in soiled feed and bedding. A cage that was a clean home for one doe gets soiled much more easily when she has to share that space with eight growing youngsters. But older rabbits living alone may also contract coccidiosis. No matter what their age, rabbits kept in clean living conditions are much less likely to suffer from this disease.

Treatment of coccidiosis has two parts: thorough cleansing of the hutch and medical therapy for the animals.

Coccidiosis Treatment: Housing

- Remove all soiled bedding and food.
- Scrape all built-up manure from the cage at least every 3 days.
- Clean cage with a bleach–water solution (1 part household chlorine bleach to 5 parts water). Let dry thoroughly.
- If litter is old enough (5 to 6 weeks), separate them into smaller groups. Of course, additional cages are necessary for this.

Coccidiosis Treatment: Medical. Sulfaquinoxaline is used to treat coccidiosis. It should be available in most farm supply stores, probably as a treatment for coccidiosis in calves, pigs, or chickens. If the product label doesn't give directions for treating rabbits, use the dosage suggested for poultry or mink.

Sulfaquinoxaline is mixed in water when given to animals. Mix up a batch of medicated water, following the instructions on the product label. Provide the medicated water for 5 days, or as recommended on the label. If your rabbit doesn't like the taste of the treated water, add a bit of gelatin powder to it (about 1 teaspoon to a gallon of water). The treatment plan is generally as follows:

- Give sulfa-treated water for 5 days.
- Give plain water for 10 days.
- Give sulfa-treated water again for 5 days.
- Repeat this sequence two to four times a year.

Why is repeated treatment necessary? During the first few days of sulfa treatment, the original organisms that make a rabbit sick are killed, and your rabbit will seem to be much better. The organisms that cause coccidiosis produce eggs, however, and a second treatment period is needed to kill organisms from the eggs that later hatch.

If the animal is being raised for meat, it must not be given medication within a certain number of days before it is sold. Follow the instructions on the label.

Other Causes of Diarrhea. Coccidiosis is not the only trigger for diarrhea in rabbits. A rabbit may also have diarrhea because it lacks fiber in its diet, has experienced a sudden change of feed, is suffering from a disease of the digestive system, or has a bacterial infection. Administering Tetracycline (by mixing it in with the rabbit's water) may cure some cases of bacterial diarrhea. Consult your veterinarian about this treatment and others.

Colds and Snuffles

When a rabbit has a cold, it exhibits some of the same symptoms as humans do, such as sneezing and a runny nose. Unfortunately, 90 percent of the rabbits that appear to have colds actually have a much more serious disease called *snuffles,* which is caused by *Pasteurella multocida.* Snuffles can be treated, but it's difficult to cure. You're better off putting your time into preventing this illness, and the best prevention is to keep your rabbitry clean and well ventilated.

Sneezing is usually the first sign of a cold or snuffles, but many things other than disease can make a rabbit sneeze. If you hear sneezing, be sure to look for other signs of illness. A white discharge from its nose or matted, crusty fur on the front paws (caused by a rabbit wiping a runny nose) indicates that the rabbit probably has a cold or snuffles.

Rabbits and humans don't communicate colds to one another, but rabbits communicate colds to other rabbits. If you suspect that a rabbit has a cold or snuffles, isolate it from your other rabbits. Be sure to thoroughly disinfect its cage before moving another rabbit in.

If you want to try to treat the sick rabbit with antibiotics, contact your veterinarian for advice. Many rabbitry owners, however, would recommend that you simply have the rabbit put down. (Your veterinarian can assist you in this if you don't want to do the slaughtering yourself.) The risk of contagion is high, and snuffles can infect an entire rabbitry in no time at all.

Ear Canker

Ear canker is caused by tiny mites that burrow inside the rabbit's ear. A rabbit with ear mites will shake its head and scratch at its ears a lot, and the inside of its ears will look dark and crusty.

Mites can easily be treated with common household oils. Mites breathe through pores in their skin. If they come into contact with oil, it blocks these breathing pores and they soon die.

1. Use an eye dropper to put several drops of miticide-containing oil or olive, cooking, or mineral oil into the rabbit's ear. Gently massage the base of the ear to help spread the oil around.

2. Use cotton swabs to remove some of the loose, crusty material.

3. Repeat this treatment daily for 3 days, wait 10 days, and repeat the treatment for 3 more days. Wait another 10 days and repeat the treatment again for 3 days.

Ivermectin, either injected under the skin or swabbed into affected ears, is another very effective treatment for ear mites. Consult your veterinarian about the proper formulation and dose. Be aware that there will be a withholding period if you are treating rabbits raised for meat.

If you see mites on one rabbit, check all your rabbits. Mites are more likely to appear in rabbits that live in dirty cages. To prevent ear canker, keep your cages clean.

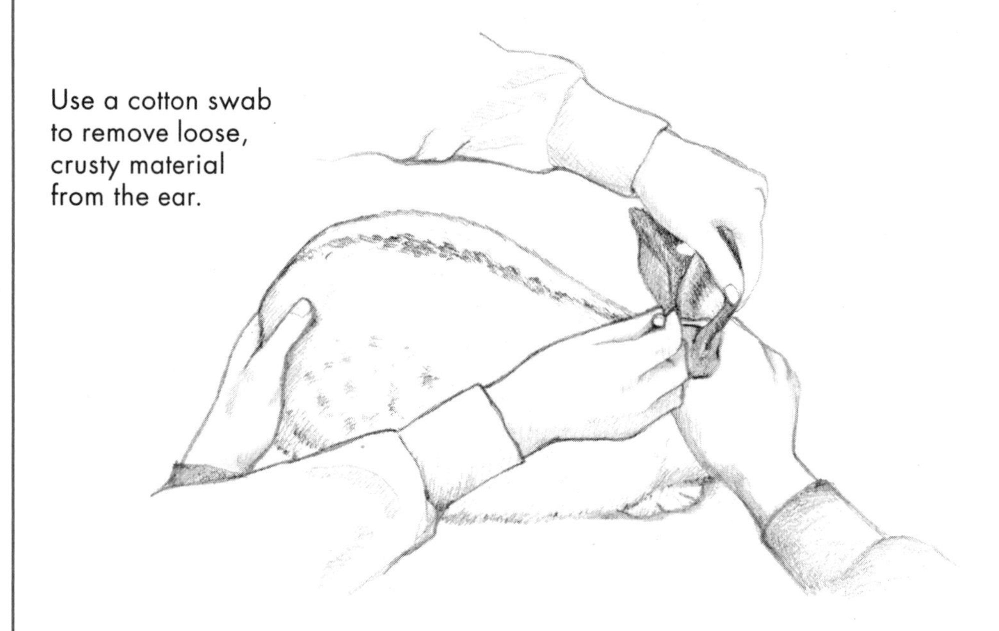

Use a cotton swab to remove loose, crusty material from the ear.

Malocclusion

A rabbit's teeth grow continuously, about ½ to ¾ inch per month. In a normal rabbit, the top front teeth slightly overlap the two bottom teeth. Normal chewing keeps a rabbit's teeth at a proper length. If the teeth do not meet properly, this normal wearing down doesn't occur and the teeth can grow overly long. This condition is called *malocclusion, buck teeth,* or *wolf teeth.* The top teeth may curl around and grow toward the back of the rabbit's mouth, like a ram's horns. The bottom teeth may grow out in front of, instead of in back of, the top teeth. These bottom teeth may grow so long that they stick out of the rabbit's mouth and may even grow into its nostrils. The rabbit may have a hard time closing its mouth. When it tries to chew, these long teeth can stick into its mouth and cause it pain; they also make it hard for the rabbit to pick up its food.

You should watch for two warning signs of malocclusion:

- Your rabbit is losing weight but does not seem sick.
- Your rabbit drops its food or seems to have difficulty chewing when it eats.

If you suspect malocclusion, remove your rabbit from its cage to check its teeth carefully. Good handling skills will really pay off, because you must turn your rabbit over to examine its teeth. To help your rabbit comfortably eat its food, you will have to trim these overgrown teeth. It's easiest to have a helper for this job — one person to hold the rabbit securely and another to do the trimming.

1. Use fingernail clippers to trim the teeth as close to normal length as possible. Be careful not to cut shorter than normal, or you may cause bleeding and discomfort. You may have to trim each tooth more than once to get it down close to normal size. You may need a larger tool for some large breeds. Your rabbit may not like having you stick the clippers in its mouth, but the actual clipping does not hurt it, because the teeth have no nerves.

2. Check the rabbit every 2 or 3 days to see how its teeth are coming along. Trimming should enable your rabbit to begin to eat normally again.

Most cases of malocclusion are inherited. To avoid perpetuating this trait, animals with malocclusion should never be used as breeders. Malocclusion is also a disqualification for showing.

Some rabbits develop malocclusion from an injury, as when a rabbit chews on its wire cage. Whatever the cause, once the teeth are out of position, they rarely return to the proper position. In most cases, the teeth must be clipped every 2 to 3 weeks for the rest of the rabbit's life.

Wool Block

Rabbits are very clean animals. During their normal grooming process, they naturally swallow some loose fur. Sometimes they swallow so much that a ball of fur forms in the digestive tract, and the rabbit's digestive system cannot dissolve it. This condition is called *wool block* or *hair balls.* It is seen more often in Angoras than in other breeds but can also occur in rabbits with other types of fur.

If your rabbit is eating very little and is losing weight, wool block may be the problem. Check the rabbit. Are its teeth overgrown? Does it have diarrhea or a runny nose? If not, wool block is a good possibility.

An animal with wool block has a big lump of fur in its stomach, so it thinks it is full and will not eat, causing it to lose weight. It will continue to drink, so treatment is usually effected through its drinking water.

Pineapple and papaya contain a substance that can break down the fur balls. Add pineapple juice or crushed papaya tablets (available at most health food stores) to the rabbit's water. To ensure that the pineapple or papaya mixture gets into your rabbit's system, use an eyedropper to hand-deliver a dose. Severe cases may require surgery.

As a preventive measure, many Angora breeders offer chunks of fresh pineapple or papaya tablets on a regular basis.

Sore Hocks

The hock is the joint in a rabbit's hind leg between the upper and lower leg bones. Because the rabbit carries most of its weight on its hind feet, the area between the hock and the foot suffers a lot of wear and tear. Although the rabbit has fur on this part of the leg, it is not always thick enough to protect it. Sometimes the fur wears away and the unprotected skin develops sores. A sore hock can cause much discomfort.

Suspect sore hocks if your rabbit shows signs of being uncomfortable when it moves. Your rabbit will put its foot down and then quickly reposition it, as would a person trying to walk barefooted over sharp stones.

If you suspect that a rabbit has sore hocks, remove the animal from its cage and examine its hock area. Are there areas where the fur has been worn off? Are these areas bleeding or infected? Also examine the front feet for sores.

If you spot sores, clean the hock area and apply a medicated cream, such as Bag Balm or Preparation H.

You can help your rabbit recover more quickly by checking its cage for things that may have caused the sore hock, such as sharp spots on the cage floor or wire that has been used upside down. Place a board or piece of carpeting in the cage to keep your rabbit's feet off the wire. Be sure the cage is clean to reduce the chance of infection.

Sore hocks take a long time to heal and, if untreated, can become infected. The sooner you start treatment, the better your chance of success will be.

Causes of Sore Hocks

Some rabbits seem more likely than others to get sore hocks:

- Nervous rabbits that stamp their feet a lot may develop sore hocks more often than their calmer relatives.
- A Rex's soft, short fur makes it more likely to develop sore hocks.
- Thinly furred foot pads may result in sore hocks.
- Some of the large breeds seem to get sore hocks more often. Because these animals carry more weight, they put more stress on their hock area.
- Animals whose feet are small compared with their body size are likely to get sore hocks.

Hot-Weather Care

Special care is more important during hot weather than during cold. The fur coats that keep rabbits cozy in the winter sometimes provide too much warmth during the summer months. To keep your rabbits comfortable during a heat wave:

- Place outside hutches in shady locations.
- Provide good ventilation. Remove plastic sheeting or the hutch cover. If your hutches are inside, open windows or provide a fan to increase air circulation.
- Provide lots of cool, fresh water.
- Be prepared for extreme heat. Use empty plastic soda bottles to make rabbit coolers. Fill the bottles two-thirds to three-quarters full with water, and keep them in your freezer. In periods of extreme heat, lay a frozen bottle in each cage. The rabbits will stretch out alongside their cooler. It's also a good idea to place a cooler in the carrier if you are traveling with a rabbit in hot weather.

Sunstroke and Heatstroke

Hot weather can be hard on rabbits, and high temperatures can kill them. During a hot spell, check your animals during the day. It is normal for a rabbit to stretch out in the coolest part of its cage. During hot weather, supply a wet towel for the rabbit to lie on during the hottest part of the day to help keep it cool. If your rabbit is breathing heavily and its muzzle is dripping wet, it may have heatstroke or sunstroke. You must act immediately to save the rabbit's life.

1. Move the rabbit to a cool location.

2. Wipe the inside and outside of the rabbit's ears with ice cubes. Because the blood vessels in a rabbit's ears are closer to the surface than those in other parts of its body, ice applied to the ears will cool it faster than applying ice to any other spot. Wrap the ice in a cloth so that you can hold it more easily.

In advanced cases of sunstroke, the rabbit may be almost lifeless. In this situation, every second is important. To cool the rabbit as quickly as possible, gently place it in a container of water. Use room-temperature water, not cold water, so you don't give the rabbit a sudden shock. Do not let the rabbit's head go under the water; just wet it up to its neck. Once it is wet, place it in a quiet, shaded area to recover.

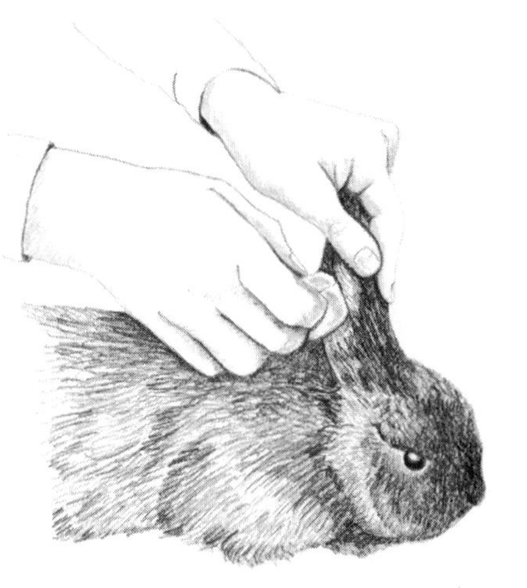

Wipe the inside and outside of your rabbit's ears with an ice cube to counteract heatstroke.

Trimming Toenails

Properly trimmed toenails are important because they decrease the chances that your rabbits will be injured. Long toenails can get caught in the cage wire and cause broken toes and missing toenails. The time spent trimming toenails will also decrease your likelihood of being scratched while handling your rabbits.

Trimming toenails is not hard to do, but many rabbit owners put off this task because they are afraid of hurting their rabbit. Learning to trim toenails is easier if

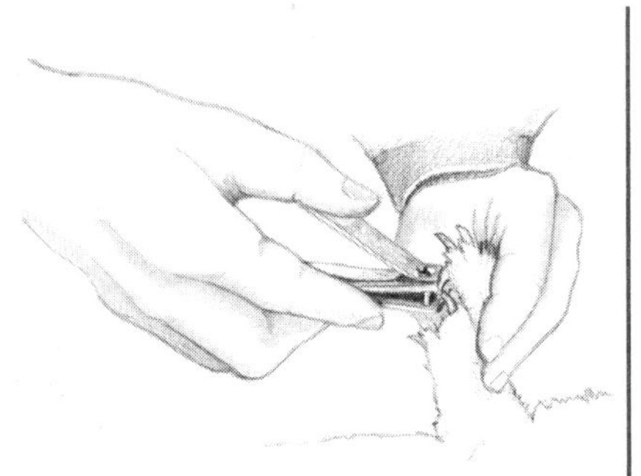

Use toenail clippers to trim your rabbit's nails.

you have a helper — one person can hold the rabbit while the other does the trimming. For most breeds, human fingernail or toenail clippers work well; larger breeds sometimes need dog toenail clippers.

The person holding the rabbit should sit with the rabbit supported on his or her legs and turned over for trimming. The person doing the trimming can now use both hands — one to push the fur back to see the nail, the other to do the trimming.

If you can, choose a white rabbit for your first trimming job. When you look at a white toenail, you will be able to see a pinkish line extending from the toe. This line is the blood vessel that brings nourishment to the growing nail. Toenails that need trimming have a clear white section after the pink. This white area can be trimmed without hurting the rabbit; trimming the pink area will hurt the rabbit and cause bleeding. Leave about ¼ inch of nail beyond the end of the blood vessel. On colored toenails, seeing where the blood supply ends is more difficult. If you are not sure, trim only a little at a time. If you cut too closely and the rabbit begins to bleed, use a clean cloth to apply direct pressure and stop the bleeding.

A Rabbit First-Aid Kit

If you start with healthy rabbits and take good care of them, they should have few health problems. If a problem arises, early treatment will increase your rabbit's chances for a full recovery. Having the necessary things on hand will help you start treatment as soon as you notice a problem. Assembling a rabbit first-aid kit is easy; you may already have some of the supplies in your home. Every couple of months, go through the kit and check the condition of your supplies. Discard and replace expired medications.

For items you have to purchase, try to find other rabbit owners who would also like to make a first-aid kit for their rabbits. Most medicines come in fairly large amounts, so it helps if you can share the supplies and the cost with several others.

Your rabbit first-aid kit should contain the following:

- **A large container** to hold your first-aid materials. A school lunch box is about the right size.
- **Several small bottles or jars.** Empty containers from prescription drugs work well.
- **Labels** for marking each container with the following information:
  Name of the medicine
  What the medicine is used for
  How the medicine is given
  How much medicine is given
  How often the medicine is given
- **Cotton balls** to clean wounds or apply medicines. Store cotton balls in a plastic bag.
- **Cotton swabs** to clean wounds or remove ear mite crust. Store swabs in a plastic bag.
- **Hydrogen peroxide** to clean wounds.
- **Disposable rubber gloves.**
- **Nail clippers** for keeping nails trimmed. Clippers may also be used to trim teeth. Use ordinary nail clippers intended for humans or, if you have a large breed, use dog nail clippers.
- **Small pair of sharp scissors** to trim hair from around wounds.
- **Plastic eyedropper** to give liquid medicines or put oil into ear to treat ear mites. Wash well between uses. Store in a small plastic bag.
- **Miticide or cooking oil** to treat ear mites.

The following products can be purchased at local farm or drug stores:

- **Antibiotic cream** for wounds.
- **Bag Balm or Preparation H** for sore hocks.
- **Papaya tablets** for prevention and treatment of wool block.
- **Sulfaquinoxaline** to treat coccidiosis.
- **Eyedrops** for nest-box eye infections (page 157).
- **Your veterinarian's phone number** so you can quickly call for help when you need it.

BREEDING RABBITS

Producing your own stock is an exciting part of raising rabbits. Before you make plans to breed rabbits, you'll want to be sure you have some potential markets for the young that are born. Once you decide to breed, you should decide when to begin your breeding program, make sure your rabbits are in good health, and decide which rabbits to breed.

When to Breed

Animals that are to be used as breeders must be mature and healthy. The larger the breed, the more time it takes for the rabbits to reach maturity. Small breeds should be at least 4½ months old before their first breeding. Medium- to large-sized breeds — those that mature at 6 to 11 pounds — are generally ready to breed at 6 months of age. Giant breeds are not ready for breeding until they are 9 to 12 months of age.

Selecting Breeding Pairs

Producing young is strenuous for the rabbit, so take the time to check each animal's health before you consider breeding it. Breeders should be in good physical condition, not too fat and not too thin, and free from diseases, especially coccidiosis.

If your rabbits are the proper age and are in good health, your next step will be to decide which rabbits to breed. If you have only one buck and one doe, the decision is made for you. But if you have several bucks and does, choose the two rabbits that have the greatest potential for producing good-quality babies. A good-quality buck and a good-quality doe are more likely to pass on desirable characteristics than are poorer-quality breeders. If one of your rabbits has a characteristic that you would like to improve on, try to select a mate that is especially strong in that area. For instance, if your doe would have better type (shape and proportions) if her shoulders were broader, mate her with a buck that has broad shoulders.

Once these decisions are made, you are ready to begin your breeding program.

Signs That Your Doe Is Ready for Breeding

- The doe rubs her chin on her food dish or on the edge of her cage. Rubbing is how rabbits leave their scent to mark territory and to let other rabbits know that they're around.
- The doe's genital opening looks reddish. A pale or light pink genital opening indicates that the doe is probably not ready to mate.

Putting the Doe and the Buck Together

Once you have selected the two rabbits you wish to breed, the next step is to put the doe and the buck together for mating. Bucks are almost always willing to breed; does sometimes resist.

When you are ready to put the male and the female together for breeding, always take the doe to the buck's cage. Rabbits, especially does, are territorial. If the

Be There for Mating

Do not leave your rabbits unattended during mating. That way, you can be sure that mating has taken place, and you can separate the animals if problems arise.

buck is put into the doe's cage, the doe may respond by defending her territory from the buck rather than mating with him. By taking the doe to the buck's cage, you avoid arousing her territorial instinct, and she is more likely to mate.

When a doe and a buck are placed together, the buck is usually ready to mate and chases after the doe. If the doe is ready to mate with the buck, she will not run away and will raise her hindquarters after the buck starts the mating motions. To hold his position on the doe, the buck grabs a mouthful of her fur. Sometimes he pulls out some of the doe's fur, but he usually does not hurt her. You will be able to tell when mating has occurred because the buck will make a noise (usually a groan or a squeal) and then fall off to the side of the doe.

Sometimes the doe does not seem to want to mate with the buck. She will not raise her hindquarters and will run around the cage to try to get away from him. She may change her mind and be ready to mate after a few minutes, so leave them together for a short while. However, if the doe does not show interest within 10 minutes, take her back to her own cage and try again in another day or two.

Occasionally, a doe is not ready to mate and will fight with the buck. If the buck and the doe fight, separate them and try again another day. If the doe is upset, she may bite. Be prepared: Have a pair of heavy gloves nearby in case you need to remove the doe quickly.

After Mating

Once mating has taken place, return the doe to her own hutch. Next, write down when the doe will kindle (give birth). You can expect the litter to be born in 28 to 35 days; the average length of gestation is 31 days. You might want to keep this information on a hutch card attached to the doe's cage. Hutch cards have space to record information about the doe and her babies. Preprinted hutch cards are often available from feed companies, or you can make your own.

Sample Hutch Card

Doe's name: ____________________

Doe's number: ____________________

Buck's name: ____________________

Buck's number: ____________________

Date litter is due: ____________________

Number born: ____________________

Number weaned: ____________________

Comments: ____________________

____

____

____

Feeding Pregnant Does

Fatness causes problems, such as pregnancy toxemia, in pregnant does. Resist the urge to give extra feed to the doe during the early stages of pregnancy. Most breeders keep the doe on her normal ration for the first half of the pregnancy (15 days). During the second half of the pregnancy, increase the doe's feed gradually. (See the chart on page 142 for guidelines.) Just before a doe kindles, she will eat less, and you can decrease her feed during the last 2 days.

Nest Box Guidelines

- Small breeds: 14" long, 8" wide, 7" high
- Medium breeds: 18" long, 10" wide, 8" high
- Large breeds: 20" long, 12" wide, 10" high

The Nest Box

Your doe will need a nest box in which to have her litter. Nest boxes come in many sizes and types. The larger the breed, the larger the box required. Nest boxes may be constructed of wood, wire, or metal.

Wooden Nest Box

Homemade nest boxes are commonly made of wood. Use ⅜-inch plywood. Many does love to chew on the edges of the nest box, and plywood holds up well. Use wood glue and nails to assemble the nest box. Drill a few holes in the bottom of the nest box to permit drainage, which helps keep the box from becoming damp. Damp nest boxes can contribute to diseases in young rabbits.

Some breeders in cold climates place a cover over part of the nest box to help retain heat. A disadvantage of a partial cover is that it traps moisture. Use your best judgment to decide whether your nest box needs a cover.

Wooden nest boxes need to be thoroughly disinfected between litters. Use a bleach–water solution (1 part household chlorine bleach to 5 parts water) to clean the box and let it dry thoroughly in the sunshine. Store nest boxes where they won't be contaminated by other animals, such as dogs, cats, or rodents.

The wire nest box has a corrugated cardboard liner.

Wire Nest Box

You may also build a nest box from ½" x 1" wire mesh. Line the box with cardboard in cooler weather to prevent drafts. To help keep the nest box clean, discard cardboard liners after each litter.

Metal Nest Box

Metal nest boxes may be purchased in a variety of sizes. An advantage of metal nest boxes is that they are very easy to clean. During cold weather, many breeders use cardboard liners as insulation.

Nesting Materials

A doe needs clean, dry, soft materials in the nesting box so that she can make a nest. Hay and straw are the most common nesting materials. Place about 2 inches of wood shavings in the bottom of the nest box and add lots of clean hay on top. Use less bedding in the summer. In winter, place layers of cardboard — for insulation — in the bottom of the box, followed by wood shavings, with hay packed in tightly on top.

Putting the Nest Box in Place

Place the prepared nest box in the doe's cage on the 27th or 28th day after mating. If you put the box in too early, the doe may soil it with droppings or urine. Any later may be too late for a doe that kindles early. Do not place the nest box in the part of the cage the doe has chosen as her toilet area.

Cold-Weather Kindling

Although healthy adult rabbits don't suffer when it's cold, newborn rabbits are vulnerable to cold temperatures. If you're expecting a litter during cold weather, be sure the doe has a well-bedded nest box. You may want to move the doe into a warmer location, such as a cellar or a garage. After the litter is about 10 days old, the cage, with the doe and the nest box of babies, can be moved back outside. Some breeders keep the box of newborn rabbits in their warm homes. At feeding times, they take the nest box out to the doe or bring her inside to nurse her litter.

Birth

As the big day approaches, the doe will begin to prepare a nest. Some does jump into the nest box as soon as it is placed in the cage. They will busily dig and rearrange the bedding to suit themselves. You may see the doe carrying a mouthful of hay around the cage as she decides where to place it. When you see this activity, kindling time is near. She also uses fur that she pulls from her own chest and belly for nest material. This behavior not only provides soft nest material but also exposes her nipples, making nursing easier for her babies after they're born.

Some does, like some people, leave everything to the last minute. They show little interest in the nest box when you first put it in. You may feed your doe one evening and see an untouched nest box only to come back the following morning and see that it's all over: The nest is made, fur is pulled, and babies have already arrived.

The doe will eat less the day or two before kindling. This change in appetite is normal. She also may act more nervous than usual.

When the doe starts to carry straw around the hutch, it's almost kindling time.

Checking on the New Litter

It's important to check on the well-being of the young and the doe after kindling. Some does do a neat job of kindling, and you may not even notice that the litter has arrived. If you look closely, you will see that the fur in the nest appears all fluffed up. On closer observation, the fur seems to move.

Be sensitive to the doe's "new mother" anxieties. Be sure to keep the doe's area quiet. If she appears nervous when you approach to check on the litter, distract her

Caring for Orphans

Sometimes a doe dies after giving birth. In that case, you may wish to try to feed and care for the babies until they are mature enough to take care of themselves. To feed the young, make up the following mixture and store it in the refrigerator.

- 1 pint skim milk
- 2 egg yolks
- 2 tablespoons Karo syrup
- 1 tablespoon bonemeal (available in garden supply centers)

Use an eyedropper or drinking straw to feed this mixture to the babies twice a day. Feed them until they stop taking the milk (usually about 5 to 7 mL).

In addition to keeping the young warm and fed, you must be sure they urinate and defecate regularly. Stimulate elimination by stroking their genitals with a cottonball after you feed them. Follow these procedures until the young are 14 days old.

first. A nervous doe will often jump into the nest to protect her babies, and she may step on and injure them in the process. Take a handful of fresh hay, a slice of apple, or another of the doe's favorite treats if she needs a distraction. Once she is occupied, you can check things out. Gently place your hand into the fur nest. You should feel warmth and movement. If the doe is still calm, part the fur and take a peek. You should see several babies cuddled together for warmth. Newborn rabbits have no fur, and their eyes will be closed. Take a quick count. Counting may be a challenge, as the bunnies will be piled on top of one another.

Check the new litter for babies that did not survive. If you find any dead babies, remove them, as well as any soiled bedding.

Fostering

Medium- and large-sized does usually have eight nipples from which their young will nurse; smaller breeds may have fewer nipples. If a doe has more babies than nipples, some of the babies are not going to get as much to eat as the others. To prepare for this situation, many rabbitry owners plan the breeding so more than one doe is kindling at a time. Then, if one doe has an unusually large litter, some of the babies may be moved into the nest box of a doe that has had a small litter. Most does will accept additional babies if they are about the same size as their own and smell like their own.

Some people go to great lengths to trick the doe into thinking these additional babies are her own. One method is to put a strong-smelling substance on the doe's nose. The idea is that a little dab of perfume or vanilla makes everything smell the same to the doe, and therefore she will not notice the extra babies. But you can be successful in fostering babies without using special scents. Here's how: Rabbits nurse their young only at night. If you need to move babies to another nest box, move them in the morning. The babies cuddle in with their new littermates, and by feeding time, all smell the same to the doe.

Feeding the Doe

Keeping the new mother in good health is the most important thing that you can do for the babies. Give the doe a limited amount of feed for the first few days after kindling. She will continue to eat and drink, but you will notice that she eats a little less than her normal amount. After a few days, her appetite will increase. She will then require an increased amount of feed so that her body can meet its own nutritional needs *and* produce milk for the growing litter. The doe's need for water will also increase.

Caring for the New Litter

Check the litter every 2 or 3 days to make sure the bunnies are healthy and remain in the nest with their siblings. It's normal for the young to react to a human hand invading their nest, so don't be surprised if the babies jerk away from your touch. Just be careful that they do not injure themselves. Do not let individuals squiggle away from the warmth of the group.

Occasionally, a baby will fall out of the nest box. Mother rabbits do not carry their babies like cats carry their kittens. If you find a young bunny out of the nest box, place it back with its littermates. In addition to needing the warmth of the group, babies that stray from the nest are likely to miss mealtime. The mother will nurse only one group of babies. Young that are separated won't get fed and will soon die.

In cold weather, young that leave the nest too early are at particular risk for dying of exposure. Rabbits are born without fur, and if they leave the warmth of the nest, they will soon become chilled. On cold days, begin your morning chores with a quick check for straying babies. If you find a baby outside the nest and it still feels warm to the touch, place it back with its littermates. If the baby feels cold to the touch, it may need some extra help regaining its body temperature (see the box below).

Baby rabbits begin to grow fur within a few days of birth and by 2 weeks will be completely furred.

Methods of Warming a Chilled Bunny

- If you have a heating pad, turn it on, cover it with a towel, place the baby rabbit on it, and place the pad and rabbit in a small box. Make sure the box provides protection if you are taking the rabbit into an area where there are pets, such as dogs or cats.
- Share your body heat. Tuck the baby inside your shirt so it can rest against you and gain warmth. It's a wonderful experience to feel a cold, almost lifeless bunny slowly regain consciousness.

Nest Box Maintenance

Most does will enter the nest box only at feeding time. Once in a while, however, a doe spends more time than necessary in the box. (She is more likely to spend excessive time in a nest box that is too big.) As a result, the nest box becomes dirty and damp — a perfect setup for disease. If you see droppings in the nest box or the bedding materials feel damp, clean the nest box.

1. Remove the nest box from the doe's cage.

2. Remove and save any clean and dry fur the doe has pulled. Place this fur in a container — a cardboard box or a clean bucket with some clean bedding in the bottom — and gently place the litter in this temporary home.

3. Remove the soiled bedding and replace it with clean bedding.

4. Use your hand to form a hole in the clean bedding. Line this area with some of the saved fur, place the babies in it, and cover them with the remaining fur.

5. Return the clean nest box and litter to its location in the doe's hutch. Your bunnies now have a clean and healthy nest to grow up in.

Eye-Opening Time

The babies should open their eyes at 10 days of age. It's important to watch for this event. Sometimes a baby cannot open its eyes even though it is old enough to do so. This situation usually indicates an eye infection. An infection may affect one or both eyes. Eye infections may occur in one or several littermates. Tending to this condition as soon as possible is your best means of avoiding blindness.

A baby that cannot open its eyes usually has dry, crusty material sealing the lids together. Use a soft cloth, cotton ball, or facial tissue soaked in warm water to wash and soften the crusted eyelids. Then use your fingers to gently separate the eyelids. Gently wash away any remaining crusty material. If the edges of the eyelid look reddish, place a drop of over-the-counter eyedrops for humans (such as Visine or Murine) in the eye.

Check the babies for several days and repeat this procedure as necessary. Once the eye is cleaned and opened, most rabbits recover quickly.

If you see pus when you open an eye as directed above, it indicates a more serious condition. After gently washing away the crusty material and pus, use medicated eyedrops containing the antibiotic neomycin; these eyedrops are available from your veterinarian. You will need to clean the eye and apply medication for several days, after which the infection should clear up.

Some rabbits have a normal-looking eye after recovering from an infection of this type. However, some may develop a cloudy area on part or all of the eye. Cloudiness usually means the rabbit will be partially or totally blind in that eye.

Beginning to Handle the Babies

As the babies approach 3 weeks of age, they will begin to come out of the nest box. At this age, they also can find their way back into the box, so you don't have to worry about constantly watching for strays. The bunnies will begin to eat pellets and drink water. You'll need to start providing more food and water, even though the babies are still getting milk from their mother. Baby rabbits continue to nurse until they are about 8 weeks old.

Baby rabbits are at their cutest from 3 to 8 weeks of age. This age is an excellent time at which to begin to handle them. Be especially careful when you pick them up for the first time. Young rabbits are easily frightened, but they adapt quickly. Try to spend a few minutes each day handling the babies. Rabbits that are handled properly as youngsters will be more easily handled throughout their lives.

Sexing the Litter

It takes some practice to sex young rabbits accurately. Try to sex the litter at around 3 weeks of age, when you are beginning to handle the youngsters on a regular basis. Part of the young rabbits' handling experience should include getting used to being turned over for examination, which is necessary for sexing.

The small sexual openings on young rabbits make it difficult to be absolutely sure which sex an animal is. To help you gain confidence in sexing your litters, start to check them at an early age and reexamine them weekly. Use a black marking pen to write a "B" in an ear of those you believe are bucks and a "D" in an ear of those you believe are does. Each time you reexamine the litter, check to see if you agree with what you thought when the rabbits were a week younger. You may find that you will change the letter in the ears of several animals. If you continue to check through weaning, you will find that sexing young rabbits becomes easier as the rabbits get older. Even experienced rabbit breeders can make mistakes in sexing young rabbits, so don't become discouraged.

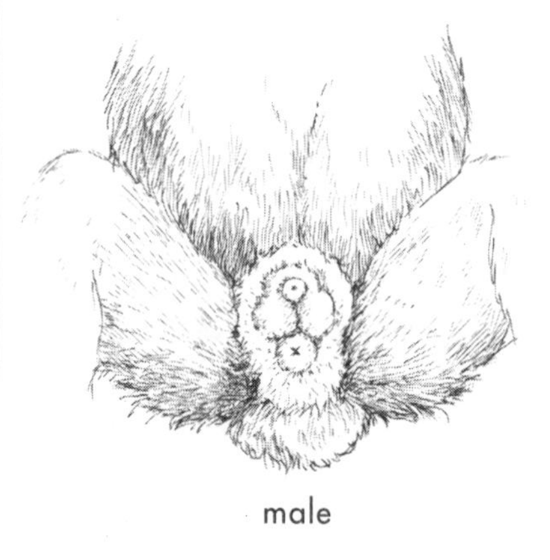

male

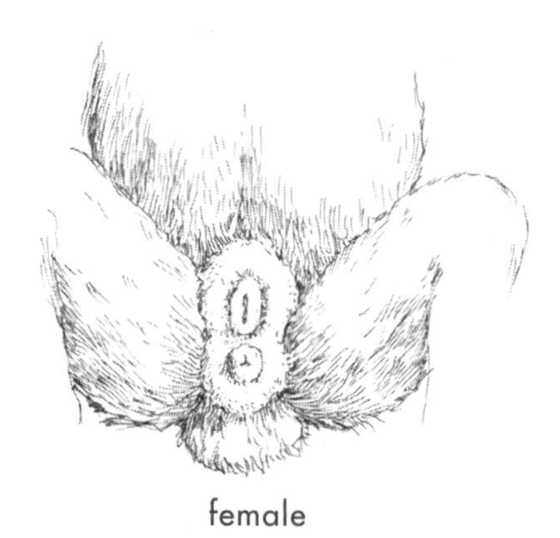

female

When the litter has reached 3 weeks of age, examine each young rabbit to determine its sex.

Hutch Maintenance

As the litter grows, it will become more of a challenge for you to keep the hutch clean. The crocks used to feed and water the litter will become soiled frequently, as young bunnies climb into the crocks to reach the feed and leave their droppings behind. Feeding smaller amounts more often will save more feed than feeding a large amount once a day. Giving several small feedings will also give you more opportunities to clean out the feed dishes. The cleaner the litter's environment, the healthier it will be.

THE DOE'S REFUGE

You may wish to allow the nest box to remain a little longer than 3 weeks, more for the doe's sake than for the litter's. As the babies explore their hutch, they pester their mother for extra attention. A nest box with a partial top provides a place for the doe to get away from the demands of the litter. Once the babies are big enough to climb on top of the box, the doe's hideaway no longer serves its purpose, and it's time to remove the nest box.

Removing the Nest Box

When the babies are out and about at 3 weeks of age or so, you can remove their nest box, unless the weather is very cold. Once the babies start eating pellets, they also produce droppings that dirty the box. If you choose to leave the nest box in the cage longer, be sure to clean it and replace the bedding whenever it becomes soiled.

WEANING AND SEPARATING THE LITTER

As the litter reaches 5 to 8 weeks of age, the young rabbits are ready to be weaned, or removed from their mother. By 8 weeks, young rabbits are used to eating pellets and drinking water, and the doe's body needs a rest before she is ready to raise another litter.

In addition, by 8 weeks of age, the youngsters begin to exhibit adult behaviors. They express instincts about territory and breeding. Young bucks will begin to chase after their sisters. Rabbits can mate and produce litters before they are full grown. Because having a litter at a young age is very stressful for a doe, and because it's not desirable for close relatives to breed, part of the weaning process includes separating the litter by sex.

Many breeders feel that the weaning process is easier on the mother if not all the young rabbits are removed at once. Since most litters are about half bucks and half does, I suggest that the young bucks be separated first. A few days later, the young does can be moved. This two-part weaning process allows time for the doe's body to adapt gradually to the need for reduced milk production.

It's also a good idea to attach a card to each cage that identifies the rabbit housed there. The card should include when the animal was born, what sex it is, and who the parents are. Having this information handy will make it easier to show your rabbits to prospective buyers.

Weaning is a good time to attend to three other tasks that need to be accomplished with each litter: culling, writing pedigrees, and tattooing.

Culling

Culling is the process of deciding which animals in the litter are best suited to be show animals, breeders, pets, or meat animals. In culling, apply the same qualifications that you would use in evaluating an animal to purchase:

- ◆ Check the health of each animal.
- ◆ Check for any disqualifications.
- ◆ Evaluate each animal for how closely it meets the standards set for its breed.

The youngsters that do well in all three areas become your show prospects, future breeding stock, or stock to sell as purebreds. Rabbits that are healthy but not outstanding as examples of their breed most generally become pets, or, if they are of the medium or larger breeds, they may be sold as meat rabbits. Animals with health problems should be caged away from the others and should receive proper treatment.

Once you have selected the best animals from the litter, house these rabbits in individual cages if possible. As rabbits mature, they do best if they have their own private space. Rabbits that are housed together often fight to decide who will be the boss of the hutch. Although rabbit fighting seldom leads to life-threatening injuries, torn ears, scratches, and bite injuries are common. Young rabbits being raised for pets or meat may be kept more than one to a cage, but the cage should provide adequate space, and each cage should house just one sex. Rabbits may be kept more than one to a cage until they are sold or up to 4 months of age. Be alert to any fighting, and separate those involved.

Pedigrees

Write pedigrees for all stock that will be kept for breeding and that you hope to sell as purebreds. You need not write a pedigree for pets and meat animals. See page 164 for information about how to fill out a pedigree.

Tattooing

Tattooing is generally done at weaning. Once a litter is split up into several cages throughout the rabbitry, remembering who is who becomes difficult. Rabbits you plan to sell as purebreds should be tattooed. You do not have to tattoo pet stock or those destined to become meat.

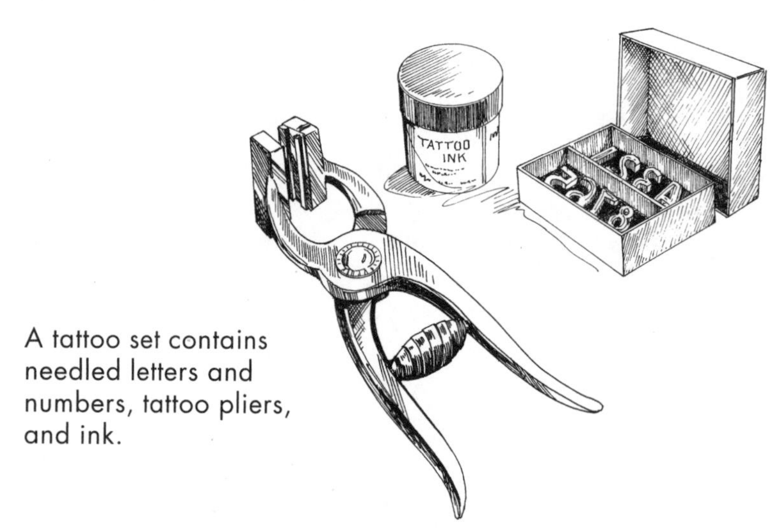

A tattoo set contains needled letters and numbers, tattoo pliers, and ink.

If you have a small rabbitry, it may be impractical to spend money on tattooing supplies. In that case, ask other breeders if they will tattoo your animals for a small fee.

The tattoo may be a number, a word, or a combination of letters and numbers. Some breeders use short names, such as "Joe" or "Pat," so the tattoo names the animal as well as identifies it. Others develop systems that provide more information about each animal. For example, in one system, a letter stands for a certain buck, another letter stands for a certain doe, an odd number means the animal is a buck and an even number means it is a doe. Thus, if we had an animal with the tattoo "TB3," we would know that it was a buck whose father was named Thumper and whose mother was named Bambi. The possibilities are endless. Develop a tattoo system that works best for you. Most tattoo sets are limited to five numbers or letters per tattoo, so be sure that your system takes this limitation into account.

The letters and numbers in a tattoo set are made up of a series of small needles. When the tattoo pliers are closed on the rabbit's ear, they make a series of small punctures. Tattoo ink is then rubbed into the area. The tattoo ink that goes into the punctures becomes permanent, while the ink on the ear's surface soon wears away. Tattooing is a bit painful for the rabbit, but only for a short time.

To learn how to tattoo your own animals, observe others when they tattoo. Choose rabbits that are not show quality for your first tattoos. Tattooing has no one right way. Here is one successful technique:

1. Practice placing the numbers and letters in the tattoo pliers. Use a piece of paper to check what you have in the pliers. It's very easy to get a 21 when you really wanted a 12. By checking on paper first, you'll avoid an incorrect tattoo.

2. Restrain the rabbit. The rabbit will pull away when the pliers are clamped, and it could be injured. If you have someone to help you, one person can restrain the animal and the other can do the tattooing. The holder can wrap the rabbit in a towel so only its head and ears stick out. The towel will keep the rabbit from moving and from scratching the holder. If you have to tattoo alone, you might place the rabbit in a narrow show carrier. The carrier keeps the rabbit confined and allows you to use both hands for the tattooing. One hand secures the left ear and the other holds the tattoo pliers.

3. Use a little rubbing alcohol on a cotton ball to wipe out the ear before tattooing. Cleansing the ear surface allows the ink to adhere better.

4. Look at the inside of the rabbit's *left* ear. You'll be able to see some of the blood vessels that are just below the surface of the skin. Find a space where no large vessels cross, and plan to place your tattoo there. Even with careful planning, you may hit a blood vessel. Hitting a vessel will cause some bleeding that can be easily stopped by applying some gentle pressure to the area. Bleeding can wash the tattoo ink out of the punctures, so avoiding blood vessels usually makes for a more successful tattoo.

5. Clamp the pliers down on the rabbit's ear in the spot you've chosen. This part is the hardest, because no one wants to hurt an animal. Take comfort in the fact that the pain is brief. Release your grip on the pliers immediately after you close them, and the worst is over. Just how hard to close the pliers to ensure a good tattoo is difficult to say. The thin ears of young rabbits need only the pressure used to pretest the tattoo on paper. Older animals have thicker ears that require a slightly firmer grip.

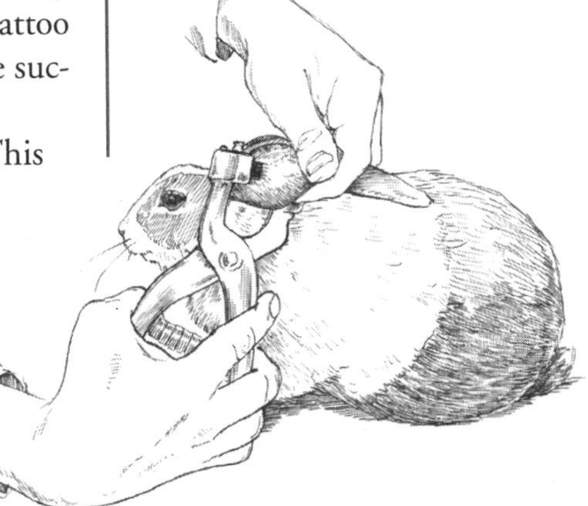

Step 5

Step 6

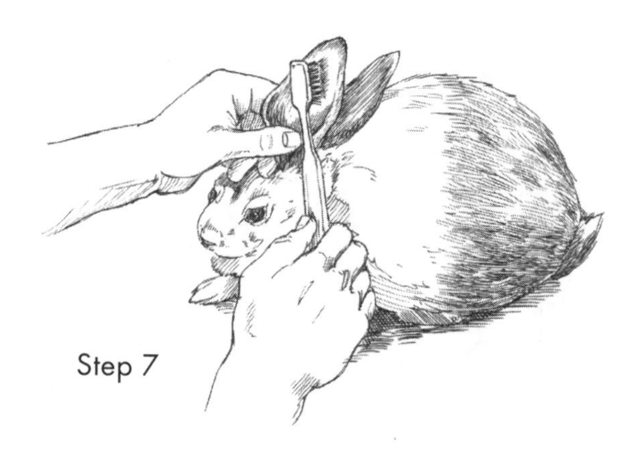

Step 7

6. Give the rabbit a few seconds to calm down, and then look inside the ear. Your tattoo should appear as a series of dots.

7. Now apply the tattoo ink. An old toothbrush works well for this purpose; the bristles help work the ink into the small punctures. Most rabbits do not seem to mind this part of the procedure.

8. After the ink is applied, some breeders apply a thin layer of petroleum jelly. The jelly forms a seal that helps keep the ink in the punctures. The excess ink will wear off normally, or the mother may clean it off while grooming her baby. If the ink has not worn away by the time your rabbit is entered in a show, you will need to remove it to make the tattoo readable. Use a tissue and a little petroleum jelly to wipe the excess ink away.

Meat Rabbits

You can sell live meat rabbits to a processor, who will, in turn, sell them to a slaughterhouse. You can usually locate processors by talking to other meat rabbit breeders in your area. Processors often attend rabbit club meetings, so you can also find them by attending the meetings of your local club. A processor will tell you what size rabbits he or she wants.

You can also sell live rabbits directly to a butcher or other meat market. The butcher will take care of slaughtering and processing the meat.

You cannot sell dressed rabbit meat to any market, whether a commercial enterprise or your next-door neighbor, without some sort of license from your local health department. You may, however, slaughter and butcher rabbits for your own family's consumption. For guidance on home butchering, consult a good reference book (see Recommended Reading, page 379), and solicit the advice of other meat rabbit raisers in your area.

Managing Your Rabbitry

All the skills and knowledge you have gained will benefit your rabbits only if you apply them through a sound rabbitry management program. As the manager of your rabbitry, you are responsible for its success. A good management program will allow you to use your time effectively to accomplish the many tasks that are part of the job.

Daily Tasks

- **Feed and water regularly.** Establish a daily schedule and stick with it, every day of the week.
- **Observe your rabbits and their environment.** Daily observation helps you catch small problems before they become large problems.
- **Keep the hutch clean.** Attend to small cleaning needs so that they don't grow into large cleaning chores.
- **Handle your rabbits.** Regular handling will make your animals more gentle, and you will become more aware of their individual condition. This may be impractical in a large rabbitry, but you should strive to handle some of your animals each day and each one of your animals every week.

Weekly Tasks

- **Clean cages.** Solid-bottom cages and cages with pullout trays must be cleaned and rebedded weekly. On wire-bottom cages, use a wire brush to remove build-up of manure or fur.
- **Clean feeders.** Rinse crocks with a bleach-water solution (1 part household chlorine bleach to 5 parts water). Check self-feeders for clogs or spoiled feed.
- **Check rabbits' health.** If you did not have time to handle and check some of your animals earlier in the week, do so now.
- **Check supplies.** Make sure that you have enough feed and bedding for the coming week.
- **Make necessary repairs.** Have you noticed a loose door latch or a small hole in the wire flooring? Take the time to make small repairs before they lead to larger problems.
- **Prepare for coming events.** Is a doe due to kindle in the next week? Is a show entry due soon? Check your rabbitry calendar, where these things should be noted, and prepare yourself.
- **Check growing litters.** Is the nest box clean? Is it time to remove the nest box? Do any babies show evidence of eye infections?

Monthly Tasks

- **Check toenails.** Check each animal and trim the nails of those that need it.
- **Update written records.** Catch up on writing pedigrees. Record feed costs and other rabbitry expenses.
- **Provide preventive medicines.** Most preventive medicines (such as those for coccidiosis) are offered on a monthly schedule. Your preventive program will be more effective if you administer it regularly.
- **Tend to the needs of developing litters.** Young rabbits grow a lot in a month. Litters should be weaned by 8 weeks of age. This age is also the time to tattoo and to separate littermates by sex.
- **Check fans and air vents if your rabbitry is indoors.** Good ventilation is important to the health of your rabbits.
- **Restock your first-aid kit.** If you've used any supplies from your first-aid kit, restock. Check the expiration dates on the medications and replace those items whose expiration date has passed.

Keeping Records

If you own one or two rabbits, it will be easy to remember all the important details about them. As your rabbits begin to produce litters and the size of your rabbitry grows, however, it will be difficult to recall everything about each animal. Keeping accurate written information about your rabbits is therefore important.

- You'll have information to help you decide which doe gets bred to which buck.
- Your rabbitry will be more productive because your records will show which animals are your top producers. You can use this information to help you decide which animals to keep in your herd.
- You'll know how much your rabbits cost you and how much money they make for you.
- You'll have the information you need to make a written pedigree for each animal. A rabbit with a pedigree usually brings a higher price.

Pedigree Papers

A rabbit's pedigree papers are a written record of its family tree — who its parents, grandparents, and great-grandparents were. Usually, pedigrees also include such information as the color, tattoo number, and weight of each animal. You can use the information on a pedigree to help you make decisions that will affect your breeding program. Are you trying to increase the weight of your rabbits? If you breed your doe to a buck whose pedigree shows relatives with high weights, you will have a better chance of producing heavier rabbits than if you breed to a buck whose relatives are on the light side. Are you trying to produce a certain color? Even if your buck and doe are not of that color, if the color appears in their pedigrees, it's more likely to appear in their offspring.

As a breeder, it's your responsibility to keep accurate pedigrees for the rabbits that you raise. Most breeders fill out pedigrees for animals that are shown and bred. Pedigree papers are not necessary for animals sold for pets or meat.

Filling out pedigrees can be confusing at first. You have to take information from the buck's pedigree and combine it with information from the doe's pedigree. Pedigree forms use the term *sire* for the father or buck and *dam* for the mother or doe. Write up pedigrees *before* you sell a purebred rabbit; they take time to fill out, and if you have it written up ahead of time, you won't have to mail it to the buyer later.

Sources of Blank Pedigrees

- The American Rabbit Breeders Association (ARBA) sells pedigrees in books of 50.
- Most breed associations have pedigrees that feature an illustration of their breed.

Registration Papers

Many people mistakenly think pedigreed animals are the same as registered animals. A registered rabbit must have a pedigree, but it has to meet additional requirements as well.

- ◆ The rabbit must be examined by a registrar who is licensed by ARBA. The registrar will come to your home and check to see that your rabbit has no disqualifications for its breed. The registrar will need a copy of the rabbit's pedigree.
- ◆ You will have to show that you are a member of ARBA.
- ◆ You will have to pay a registration fee.

If your rabbit passes its examination, is at least 6 months old, and has a permanent tattoo number in its left ear, it may be registered. The registrar will fill out papers to send to ARBA and will tattoo a registration number in your rabbit's right ear. The American Rabbit Breeders Association will mail you your rabbit's official registration papers.

If you decide to register your purebred rabbits, plan ahead and have several done at once. You may want to get together with other rabbit owners and plan a time for a registrar to come. You may also have your rabbits registered at ARBA shows, which have a registrar on duty. Just remember to take each rabbit's pedigree.

Hutch Cards

A pedigree tells who your rabbit is. A hutch card, on the other hand, tells what he or she can do. Hutch cards, as their name suggests, are attached to the rabbits' cages. They are kept so that you can write down important information along the way. For a doe, a hutch card records such information as which buck she is bred to, when her litter is due, and how many young she kindles. For a buck, the card lists which does he is bred to, when they kindle, and how many young he has fathered. Hutch cards help you see how productive your breeders are.

Hutch cards are often available free from feed companies. You may be able to pick them up at your feed store, or you may have to write to the feed company and request a supply. You can also make your own (see page 152).

The information recorded on the hutch cards is important. Since they can be easily soiled or lost, record the information in a second location. On a regular basis — perhaps once a month — transfer the information from your hutch card to a permanent management record.

Management Records

Management records usually cover a full year. They bring lots of information together and give you a good overall picture of your rabbitry. Using your management records, you can keep track of the cost of feed, the total number of young born, the total income from rabbits sold, and your show results. If you keep good management records, you will know whether your rabbits are productive, and you will know how much you spend keeping them. Your management records will help you decide whether your hobby can become a profitable small business.

Monthly Management Charts

Post in a location that will be convenient to record the facts about your rabbitry. Monthly sheets can be totaled to determine an annual report.

Expenses

Date	Animals Purchased No.	Animals Purchased Cost	Pounds of Feed Used	Cost of Feed	Other Costs (litter, supplies, etc.) Item	Other Costs Cost
TOTAL						

Income
(list of income from sale of equipment, breeding fees, etc.)

Date	Item	Amount Received
	TOTAL	

Breeding Record

Date Bred	Name or Numbers Doe	Name or Numbers Buck	Date Due to Kindle	Date Kindled	Number of Live Young Born
				TOTAL	

Show Participation

Date	Name and Place of Show	Placing or Award Received	Entry Fees	Value of Premiums Won
		TOTAL		

4

Goats

Introducing Goats

Goats serve many purposes worldwide. They produce delicious milk, healthful low-fat meat, and fiber for spinning. They are excellent at brush control, and they may be used to carry camping supplies on hiking trips or hitched up to help with light chores around the yard. They are inexpensive to maintain, require simple housing, do not take up a lot of space, and are easy to handle and transport.

Scientifically, goats belong to the suborder Ruminantia — that is, they are ruminants, like cows, deer, elk, caribou, moose, giraffe, and antelope. Ruminants are hoofed animals with four-part stomachs. Within the suborder Ruminantia, goats belong to the family Bovidae, which includes cattle, buffalo, and sheep. Of the six species of goat, one, *Capra hircus,* is domesticated.

One nice thing about goats is that they do not require elaborate housing. All they need is a shelter that is well ventilated but not drafty and provides protection from sun, wind, rain, and snow. You can easily convert an unused shed into a goat house. Each goat requires at least 15 square feet of space under shelter and 200 square feet outdoors. A miniature goat needs at least 10 square feet under shelter and 130 square feet outdoors. You'll also need a sturdy fence — don't underestimate the ability of your goats to escape over, under, or through an inadequate fence.

Goats are social animals that like the company of other goats, so you'll need at least two. If you will be breeding your goats, the herd will probably grow larger than you initially expect. Plan ahead by providing plenty of space.

Goats are opportunistic eaters, meaning they both graze pasture and browse woodland. Those that harvest at least some of their own food by grazing or browsing will cost less to maintain in hay and commercial goat ration. Each year the average dairy goat eats about 1,500 pounds of hay and 400 pounds of goat ration. Nondairy goats do well on hay and browse, with little or no ration.

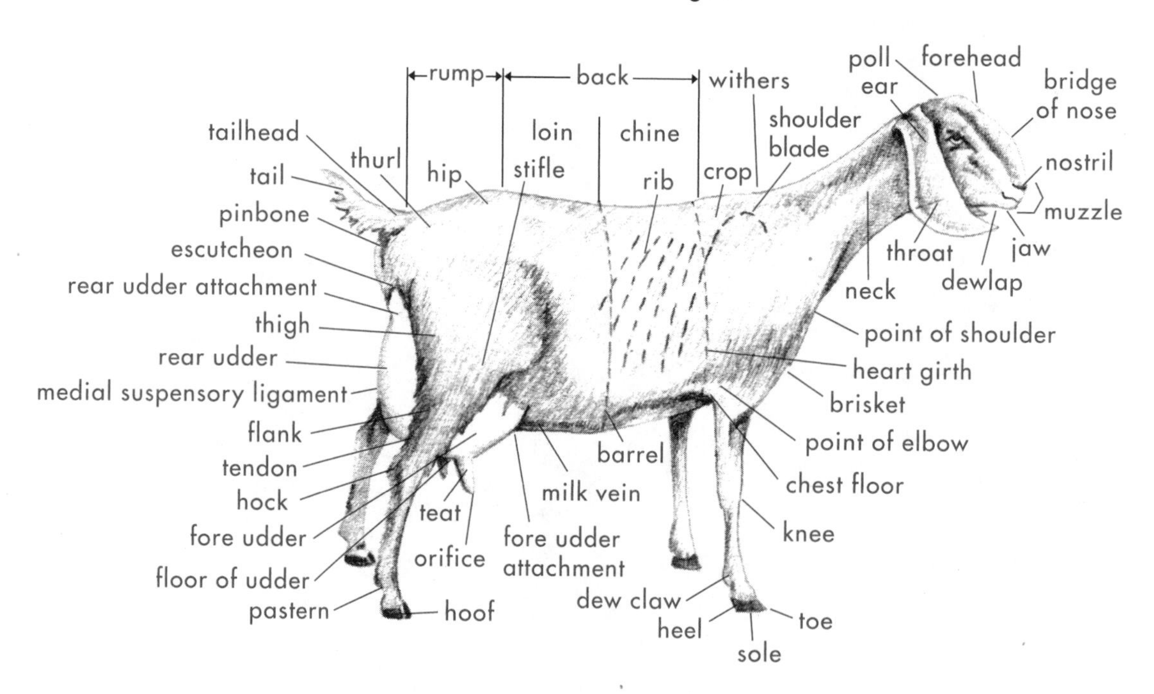

Parts of a goat

Despite what you may have heard, being opportunistic eaters does not mean goats eat things like tin cans. A goat learns about new things by tasting them with its lips. Young goats like to carry things around in their mouths, as puppies do. If you see a goat with an empty can, it could be playing with it or eating the label, which, after all, is only paper made from wood. Although the goat may look cute carrying a can, it's a bad idea to leave such things where a goat can find them; the goat may cut its lips or tongue on the sharp rim.

Another myth is that goats are smelly. A goat is no more smelly than a dog, unless you keep a breeding buck, which will smell pretty strongly during the breeding season. But unless you plan to breed your does, you don't need a buck. And even if you do plan to breed, you may find it more convenient and economical to use someone else's buck if you have only a few does.

So what's your reward for keeping goats? If you raise dairy goats, each doe will give you about 90 quarts of delicious fresh milk every month for 10 months of the year. You and your family might drink the milk or use it to make yogurt, cheese, or ice cream. Surplus milk may be fed to puppies, chickens, pigs, calves, or orphaned livestock and wildlife.

From each meat wether (castrated buck), you will get 25 to 40 pounds of tasty, lean meat, which may be baked, fried, broiled, stewed, or barbecued. If you raise fiber goats, from each adult Angora, you will get 5 to 7 pounds of mohair twice a year. From each cashmere goat, you will get just less than 1 pound of down per year.

Each doe you breed will produce one kid or more annually; some does kid twins year after year. Every day, each goat will drop a little more than 1 pound of manure, which makes good fertilizer for the garden.

The ultimate reward, of course, is the fun of raising healthy, contented goats.

Buying Goats

More than 200 breeds of goat may be found worldwide. Each breed has characteristics that are useful to humans in different ways. Some are efficient at turning feed into milk or meat, others at turning feed into hair for spinning. Some breeds are small and produce less milk or meat than larger breeds, but are easier to keep in small spaces. Your purpose in keeping goats will determine which breed is right for you.

Dairy Goats

A dairy goat, sometimes called a milk goat, is one that produces more milk than it needs to nurse its kids. In the United States, there are six main dairy breeds.

Alpine. An Alpine has a long neck and a two-tone coat, with the front end a different color from the back. A mature doe weighs at least 135 pounds, and a mature buck weighs at least 170 pounds.

LaMancha. LaManchas come in many colors and are the calmest of the dairy breeds. A LaMancha is easy to recognize because it has only small ears or no visible ears at all. A mature doe weighs 130 pounds or more. A mature buck weighs 160 pounds or more.

Dairy Goats

Alpine

LaMancha

Nubian

Oberhasli

Saanen

Toggenburg

Nubian. Nubians come in many colors and are the most energetic and active of the dairy breeds. You can tell a Nubian from any other goat by its rounded face (called a Roman nose) and long floppy ears. A mature doe weighs 135 pounds or more. A mature buck weighs 170 pounds or more.

Oberhasli. The Oberhasli looks something like a refined deer. Its coat is bay (reddish brown) with black markings. A mature doe weighs at least 120 pounds, and a mature buck weighs at least 150 pounds.

Saanen. A Saanen is all white or cream colored. A goat of this breed in any other color is called a *Sable.* A mature doe weighs 135 pounds or more. A mature buck weighs 170 pounds or more.

Toggenburg. A Toggenburg has white ears, white face strips, and white legs setting off a coat that may range in color from soft brown to deep chocolate. A mature doe weighs 120 pounds or more. A mature buck weighs 150 pounds or more.

Alpines, Oberhaslis, Saanens, and Toggenburgs are closely related and are similar in shape. They all originated in the Swiss Alps and are therefore referred to as the Swiss breeds or European breeds. These goats have upright ears and straight or slightly dished faces. They may or may not have wattles consisting of two long flaps of hair-covered skin dangling beneath their chins. These breeds thrive in cool climates.

LaManchas and Nubians, on the other hand, originated in warmer climates and are therefore grouped together as tropical or desert breeds. The Nubian originated in Africa, and the LaMancha comes from the West Coast of the United States. As a general rule, both breeds are better suited to warm climates than the Swiss breeds.

If you buy a young female, or doeling, you can't tell for sure how much milk she will give when she matures, but you can get a good idea by looking at her dam's milk records. An average doe yields about 1,800 pounds, or 900 quarts, of milk per year. A doe's dairy character gives you a fair idea of whether she will be a good milker. Characteristics of does that prove to be good milkers include:

- A soft, wide, round udder
- Teats that are the same size, hang evenly, and are high enough not to drag on the ground or get tangled in the doe's legs when she walks
- A well-rounded rib cage, indicating that the doe has plenty of room for feed to fuel milk production
- A strong jaw that closes properly, so the doe has no trouble eating
- Strong, sturdy legs
- Soft skin with a smooth coat

A dairy goat may be born with horn buds that will eventually grow into horns. Kids with buds are usually disbudded, because mature dairy goats without horns are easier to manage and are less likely to injure their herdmates or their human handlers. If they are to be registered or shown, they are not allowed to have horns. Goats born without horns are called polled.

If your dairy herd includes polled does, make certain your buck is disbudded rather than polled. The polled trait is linked to a gene for infertility; if you breed a polled buck to your polled does, half of their offspring will be incapable of reproducing.

Meat Goats

In many countries, more goats are kept for meat than for any other purpose, and many people prefer goat meat to any other. Since slightly more than half of all goat kids are male and only a few mature bucks are needed for breeding, most young bucks are raised for meat. Surplus goats of any breed may be used for meat, but a breed developed specifically for meat puts on more muscle, and does so more rapidly, than other breeds. In the United States, three types of goat are kept primarily for meat.

Boer. The main meat breed today is the Boer goat. Boers originated in South Africa, where they were developed for their rapid growth, large size, high-quality meat, and uniformity of size, meat quality, and color. The Boer has a white coat, a brown or dark red head with a white blaze, and horns that curve backward and downward. A mature doe weighs 150 to 225 pounds. A mature buck weighs 175 to 325 pounds.

Spanish. Before Boer goats became popular in the United States during the latter part of the 20th century, most meat goats were essentially those that were left to roam over brushy range or forest land in the South and Southwest to keep the land cleared of brush and undergrowth. These goats are often called Spanish goats because the first feral herds were brought to this country by Spanish explorers and sometimes left behind to furnish meat for future expeditions. Because these goats vary widely in shape and color, the term *Spanish* doesn't really refer to a specific breed. Mature does weigh 80 to 100 pounds; bucks weigh 150 to 175 pounds.

San Clemente. During the 1500s, Spanish goats were left on San Clemente Island, off the California coast near San Diego. A few descendants still survive as a kind of living history, showing us what goats must have looked like 500 years ago. At one time, so many goats populated San Clemente that they nearly destroyed the island's vegetation. However, because of a successful eradication effort, the goats are now in danger of disappearing. San Clemente goats are smaller and more fine-boned than other Spanish goats, and their horns grow more upright. They come in all colors, the most common of which is tan or red with black markings. A mature doe weighs 30 to 70 pounds. A mature buck weighs 40 to 80 pounds.

Myotonic. A rare goat formerly raised for meat, but that is today more of a curiosity, is the myotonic goat. This animal is also called the Tennessee fainting goat, the Texas nervous goat, or the wooden leg goat. Myotonic goats are not a specific breed, but they share a genetic disorder called myotonia. When a goat with myotonia is frightened by a loud noise, its muscles contract and its legs go stiff. If the animal is caught off balance, it falls to the ground and can't get up again until its muscles relax. Frequent tensing and relaxing of the muscles gives myotonic goats heavy thighs, making them suitable as meat animals. Myotonia also keeps these goats from becoming aggressive, making them good pets. Because they cannot climb or jump like other goats, they are more easily confined, but they also make easy prey for dogs and coyotes.

The origin of myotonic goats has been traced back to four goats brought to Tennessee in 1880 by a man from Nova Scotia who later disappeared, leaving the goats behind. When those goats were bred, their odd genetic trait was inherited by their offspring and passed on through other generations. Myotonic goats come in a variety of colors. Mature does weigh about 75 pounds; mature bucks weigh up to 140 pounds.

Miniature Goats

Miniature goats are smaller than full-size goats and therefore produce less milk or meat. Minis eat less, require less space, and have scaled-down housing needs that make them ideal for cold climates, where they spend a lot of time indoors. The two miniature breeds are African Pygmy and Nigerian Dwarf.

African Pygmy. Pygmies are blocky, deep, and wide, and their faces are dished. The most common color is agouti, meaning they have two-tone hairs that give the coat a salt-and-pepper look. The Pygmy has the muscular build of a meat breed. Mature does weigh 35 to 60 pounds, and mature bucks weigh 45 to 70 pounds.

Nigerian Dwarf. The Nigerian Dwarf is a miniature dairy breed. It is smaller and finer-boned than a Pygmy and has longer legs, a longer neck, and shorter, finer hair. Nigerian Dwarfs are lean and angular, with faces that are flat to slightly dished. Dwarfs come in all colors. Mature does weigh 30 to 50 pounds, and mature bucks weigh 35 to 60 pounds.

A Dwarf yields about 300 quarts, or 600 pounds, of milk per year, which is one-third the amount you would get from a regular-sized goat. Despite its stockier build, a Pygmy doe produces about the same amount of milk as a Dwarf. The milk from miniature goats tastes sweeter than other goat milk because it is higher in fat.

Fiber Goats

Some goats have long hair that may be spun into yarn and woven or knit into fabric to make clothing, drapes, and upholstery. Two kinds of goat are known for their fine hair or fiber.

Angora. The Angora goat originated in the Himalayas and came to the rest of the world via Turkey. The name *Angora* is derived from Ankara, the capital of Turkey. Angoras are raised for their long, silky, wavy hair, called mohair. Like sheep, Angoras are sheared twice a year, in spring and fall. The average amount of mohair sheared from a doe per year is 10 to 14 pounds; a wether averages slightly more.

When selecting an Angora, spread the hair with your hands and notice how much pink skin you see. The less skin you see, the better. The best Angoras have hair that is neither light and fluffy nor dark and greasy. Avoid a goat with a chalky white face and ears; it is likely to have lots of straight, brittle, chalky white hairs, called kemp, that are undesirable because they do not produce quality yarn.

Pure mohair is creamy white. Colored hair results from crossing an Angora with some other breed. Naturally colored mohair is popular among hand spinners, even though the hair of a crossbred goat is usually lower in quality and quantity than the hair of a pure Angora.

Angoras have floppy ears and short faces that may be straight or slightly rounded. A mature doe may weigh 75 pounds or more. A buck usually weighs about 150 pounds.

Cashmere. The cashmere goat is prized for its fine undercoat, called cashmere. The word *cashmere* derives from the eastern Himalayan state of Kashmir. Goats originating in this area and in other cool climates grow downy coats for winter warmth.

Cashmere is not a breed but a kind of downy hair that is softer and finer than mohair. Cashmere is found on more than 60 breeds worldwide. In the United States, it most often occurs on Spanish and myotonic goats. Cashmere is usually white but may be gray, tan, brown, or black.

The best way to determine whether a young goat will produce cashmere is to ascertain that both of its parents are good producers. Cashmere is valuable because of its rarity; the average cashmere goat produces only about one-third pound of down per year. You may be able to find a good cashmere goat at a reasonable price, but top-quality mature animals cost thousands of dollars.

Making the Purchase

Once you have selected a breed, you must decide whether you will raise does, wethers, or bucks, and whether you will register them.

If you are keeping goats for milk, you must, of course, have does. Even if the milk from one doe is plenty for your needs, your doe will need a companion, which may be another doe or a wether. If you want milk year-round, you must have a second doe; a doe must be bred to produce milk, but during the 2 months just before she gives birth, she will not produce any milk. A wether is a good choice if you want to engage in goat packing or driving, since it can handle more weight than a doe. If you raise fiber goats, a wether produces more hair per shearing than does a doe, and the quality of the hair is more consistent for a longer part of its life. If you are raising a goat to butcher for meat, a wether is cheaper and grows bigger than a doe.

Getting a buck (uncastrated male) as your first goat is not a good idea. A buck must be housed separately so he won't fight with other goats or breed does that are too young. During breeding season, a buck becomes aggressive and hard to handle. In addition, a buck develops a strong odor that gets on your skin and clothing when he rubs against you. Unless you have a lot of does to breed, keeping a buck is an unnecessary expense. You'd be better off finding a buck owner nearby who is willing to breed your does.

If you choose to keep a buck and you are raising dairy goats, you'll need to provide separate facilities so the buck smell won't affect the taste of the milk. Bucks are often relegated to a back shed, which may have poorer living conditions than those enjoyed by does or wethers. This practice is not fair to the animal; bucks require the same amount of shelter and grazing land as the others. A buck kept confined to a dark shed or stall will become bored and difficult to handle. Except when bucks get excited during breeding season, they are generally just as gentle as does or wethers.

Registration Papers. When purchasing a doe or a buck, find out whether the animal is registered. A registered goat has official papers issued by an organization that keeps track of production records, show records, and pedigrees for that breed. A good goat need not be registered, although you may want registered animals if you wish to compete at shows or you might someday sell your goats. Insist on receiving the registration papers when you pay for the goat. A registered goat will cost more than a goat without papers. Exactly how much more depends on how easy it is to find the breed you want in your area. The more common the breed, the less your goat should cost, regardless of whether it is registered.

Before You Bring Home a Goat. Once you have selected the goat you wish to purchase, the final matter is to make sure the animal is healthy and sound. A healthy goat has a clean coat and bright, alert eyes. It should be as curious about you as you are about it. A good goat has a strong, wide back, straight legs, sound feet, and a wide, deep chest. Avoid a goat with a swayback, a narrow chest, a potbelly, bad feet, lame legs, or a defective mouth.

How Long Do Goats Live?

The normal life span of a goat is 10 to 12 years, but some goats live as long as 30 years. The productive life of a dairy goat or fiber goat, during which you can expect it to produce a reasonable amount of milk or fiber, is about 7 years.

Ask the seller to make a list of medications or vaccinations the goat has had, and the date of each. Ask for recommendations regarding future vaccinations. If you live in the same general area, find out who the seller's veterinarian is — good goat veterinarians are hard to find.

Ask what the goat has been eating and obtain enough of the same feed to last at least 1 week. If you plan to alter the feeding program, make the change gradually to ensure that your new goat remains healthy.

Before you take the goat home, ask the seller to trim hooves, remove horn buds, vaccinate, and perform any other necessary procedures. Even the most routine procedures cause some degree of stress, which is considerably reduced when the animal undergoes them in familiar surroundings.

Handling Goats

Working with goats can be frustrating or rewarding: frustrating if you try to work against their nature, rewarding if you put their nature to work for you. By understanding why goats act as they do, you will better know how to treat them, and the experience will be more rewarding.

Goats are like cats in that they are curious and independent and do pretty much as they please, whether or not their behavior pleases you. If you know what pleases them, you can get them to do what pleases you.

Goats are social animals. A goat should have at least one companion, which may be another goat or some other type of animal. No goat should be housed alone.

As soon as you put two or more goats together, one of them takes over. You can easily tell which goat is the herd boss — it's the one in the lead. The herd boss is usually the oldest doe, called the herd queen. The other goats won't move until the herd queen leads the way. If anything happens to the boss, the herd falls into confusion until a new leader takes control. When you visit a herd, if you don't give your attention first to the queen, she'll display jealous misbehavior.

Goats protect themselves by butting enemies with their hard heads. They also butt heads with each other in play and to determine the pecking order. Baby goats play by pushing each other with their heads and will try to push against your leg or hand. Don't let them, because as a young goat grows up, pushing turns into butting. If you teach your goats while they are young not to push or butt you, they will be easy to handle as they mature.

Goats and Other Animals

Goats get along well with other animals. They are often kept in the same pasture with cows, sheep, horses, or donkeys. Goats do especially well with cows because they eat some plants that cows won't eat, whereas cows eat inferior hay the goats turn up their noses at. Keeping cows and goats together is an economical way to manage available feed.

Goats are sometimes kept with sheep because they are generally calm, while sheep are easily frightened. Sheep tend to remain calmer with goats around.

Donkeys are sometimes housed with goats to fend off predators, especially coyotes and dogs, which donkeys will kick at or chase away. Goats also get along great with horses. In the nineteenth century, a common practice was to house a goat with a racehorse. Sometimes before a race, a competitor would sneak in and steal a goat.

The horse, missing his buddy, could become upset enough to lose the race. The horse's owner then became angry because someone "got his goat" — a familiar expression to this day.

Goats get along well with dogs and cats, too. Certain breeds of guardian dogs are used to protect goat herds. Cats are often kept as mousers in dairy barns, an arrangement that works out well. Treat your barn cats to a daily saucer of warm milk fresh from the goat, and they'll stick around.

Housing chickens with goats may be picturesque but is not a great idea. The chickens will nest in the hay and roost over the manger. Much hay will be wasted, because goats will not eat it once it's been soiled by the chickens.

Goats and Stress

Any unusual, painful, or unpleasant experience causes goats stress. Such experiences include being chased by dogs, teased by insensitive people, or handled roughly. Many ordinary events in a goat's life are inherently stressful, including being weaned, castrated, disbudded, transported, isolated, or artificially bred. How a goat reacts to stress depends somewhat on its genetic background. Some breeds, especially Nubians, are more excitable than others. Reaction to stress also depends on individual temperament, past experiences, and familiarity with surroundings.

Developing a routine for managing your goats helps reduce stress. Goats like to be fed by the same person at the same time every day. If you are late, your goats will misbehave. The same goes for milking. If you fail to milk on time or send someone unfamiliar to do the milking, your goats may act up.

Oddly enough, although regular routine reduces stress, an overly rigid routine can also cause stress. The trick is to make change part of your routine. Instead of always taking care of your goats by yourself, occasionally ask a friend or family member to come along and help. Then, if one of them takes over while you are away, your goats won't be upset by the presence of a stranger. Similarly, if you don't feed or milk at the exact same time every day, your goats won't be upset if you arrive early or late once in a while.

Because goats are naturally curious, not all new situations are stressful. Forcing a goat to confront a new situation, however, always causes stress. When a goat balks, give it a little time to check things out, and it will probably soon proceed on its own.

Preconditioning goats to new procedures goes a long way toward reducing stress. If you regularly take a doeling to the milk stand for a brushing and an udder massage, she will be comfortable with the idea of getting on the milk stand long before she starts giving milk. Handle a kid's feet frequently while it's young, and it won't balk when it needs its first hoof trim. Run the electric clippers when you handle a young goat — bringing the clippers gradually nearer until they eventually touch the goat's body — and when it's time to clip its coat, the goat won't be frightened by the noise and vibration.

Another stress reduction measure is to reassure each animal by repeating its name. Talk or sing to your goats while you milk, feed, groom, shear, medicate, or perform other chores. After a goat has had an unpleasant experience, talk calmly or sing quietly until the animal has calmed down.

Training your goats, especially the herd queen, to be cooperative and well-mannered reduces stress for the entire herd. To minimize squabbles among the herd, start with the queen whenever you perform any procedure, such as feeding, grooming, milking, hoof trimming, and shearing.

Well-Behaved Goats

Goats that aren't handled often tend to become shy. You will have a hard time getting them to come for milking, hoof trimming, or weighing, and if you plan to show them, they will behave poorly in the show ring.

Handling goats to keep them friendly takes little time. Whenever you enter your goat house, greet each animal by name, starting with your herd queen. Scratch each goat's ears and face. Your goats will crowd around, happy to see you. If you always handle them in the same order, they will learn to come to you each in turn.

As soon as your goats are big enough, give each one a collar. A plastic chain makes a good collar. It is sturdy enough to lead a goat by but will break if the goat gets hung up somewhere and pulls away, preventing the goat from being choked by its own collar. Collars work for all goats except Angoras. Since a collar will tangle in the Angora's long hair, teach the goat to be led by its horns or chin hairs, or guide it with one hand under its chin and the other hand on its rump.

A well-behaved goat will learn to follow you when you talk gently, use its name, and put your hand on its collar or chin. A stubborn goat will plant all fours on the ground and refuse to budge. If the goat balks, grab one ear and pull firmly. A goat doesn't like to have its ears pulled and will usually come just to make you stop pulling.

A frightened goat may rear up on its hind legs. If a goat rears, let go and move out of the way to avoid being hurt. Talk gently until the goat calms down, then try again.

Packing and Driving

A benefit of having well-behaved goats is that you can use them for driving and packing. Goats make wonderful draft and pack animals because they are active, friendly, and enjoy being with people. They eat less than larger animals used for these purposes, their smaller hooves do less damage to the environment on pack trips in the wilderness, and their droppings are indistinguishable from a deer's.

A *draft goat* pulls a cart, small wagon, sled, or other load. A draft goat may be hitched up for fun or to help with light chores, such as hauling hay, bedding, or firewood. Goat-sized cultivators are available that let you use your goat to turn garden soil.

A *pack goat* wears a saddle that may be loaded with supplies. The most common use for pack goats is to haul camping gear on wilderness hikes, but they have also been used to carry sensitive research equipment and other fragile goods into mountainous areas.

Draft goat halter, harness, and reins

The same goat may be trained to do both driving and packing. Wethers of a European breed are most often trained for draft or pack use because they are calmer than bucks but stronger than does. A doe, on the other hand, tends to move more lightly (which may be important if she's carrying fragile items) and will supply fresh milk on camping trips.

A goat is trained to pack or drive with lots of patience and gentle handling. Use the same methods as for training a donkey, mule, or horse (for which plenty of good reference material is available). Like any animal trained for packing or driving, a goat learns through repetition, so work with it for at least a few minutes every day.

Transporting Your Goats

Because goats are small, they are easy to transport. A baby goat can easily be transported in a pet carrier. Many goats ride in the back seat of the family car. On a long trip, though, you'll have to stop once in a while and take the goat for a walk.

For long distances, the back of a pickup truck works better for both the goat and the human passengers in the cab. The goat must not be able to jump out and must be protected from wind. The pickup bed should be covered with a camper shell or a sturdy stock rack wrapped in a tarp. Add a little bedding to help the goat keep from slipping during curves or sudden stops.

HOUSING GOATS

A goat's housing needs are simple: fresh air; a place to get out of hot sun, blowing wind, and cold rain and snow; a clean place to sleep; and safety from predators.

Goats are incredibly curious creatures. They constantly check things out. If the fence develops a hole, in no time they will find the hole and wiggle through. If you accidentally leave a gate open, before you know it they'll be down the road checking out the neighborhood. Goats investigate not only with their eyes but also with their lips, which they use to test new foods and objects, including gate latches. If a latch moves, they'll keep working it until it falls open, and out they go.

Goats also chew on things. If you use a rope to tie a gate shut, a goat will chew through the rope, open the gate, and go exploring. Electrical connections are especially dangerous for a goat to chew. Since a goat will stand on its hind legs and stretch to investigate anything that looks interesting, make sure all electrical wiring and fixtures are well out of reach.

Goats are famous for their ability to jump. Kids love to leap against a wall and push off with all fours. If the wall has a glass window in it, the glass could shatter and the kid could be seriously cut. Goats also love to climb. Make sure your goats can't climb onto the roof of their house. Their sharp hooves could cause the roof to leak, and even though goats are surefooted animals, one could fall off the roof and break a leg.

Every time you visit your goats, check around for things that could hurt them. A nail sticking out of the wall can rip open a goat's lip. A loose piece of wire can get wrapped around a goat's neck or leg. A rake or pitchfork lying on the ground can pierce a goat's foot.

The Goat House

Goats require a dry, clean shelter, which need not be fancy. Any sturdy structure will do as long as it provides shade and protection from rain, snow, and wind. Goats will keep one another warm down to temperatures as low as 0°F, provided they can get out of wet and drafty weather. To find out whether your goat house is too drafty, go out on a cold or windy day and squat down to goat height. If you feel an uncomfortable draft, your goats will feel it, too. Seal the gaps where the wind comes through. On the other hand, take care not to make your goat house too tight, since good ventilation promotes good herd health.

A goat house should be sturdy and provide protection from the elements.

Goats suffer more in warm weather than in cold weather. Swiss breeds, because they originated in cool climates, suffer more in warm climates than desert breeds do. Using electric animal clippers to give them a trim will keep these longhaired goats cooler.

When temperatures get above 80°F, make sure your goats have shade and cool water. Stir up a breeze indoors by opening the doors and windows. To keep your goats from escaping through the wrong door, create a screen door of sorts by using a stock panel or other sturdy open wire structure that lets in the breeze but keeps the goats confined.

In warm climates, give your goat house a south-facing wall that may be removed in summer to increase air movement. In climates that remain mild, the house will need only three walls. Some large dairy herds are sheltered by only a roof and one or more half walls mounted with hayracks. Sometimes meat herds, and often brush herds, have no housing other than a wooded area for protection against the weather.

A small family herd, however, should have a shelter, no matter how rudimentary. Allow at least 15 square feet of housing per goat, 10 square feet per miniature. If you're building from scratch, plan now for future herd expansion.

In addition to the main area, you will need at least one smaller stall to hold one goat. An extra stall will come in handy for housing a sick or injured goat, a pregnant goat that is about to kid, or kids you wish to wean.

You will need space to store feed, supplies, and dairy equipment away from the goats' living area. Separate the storage area from the goat area with a wall at least 4 feet high; the half wall lets you watch your goats and your goats can watch you while you work in the storage area.

Goats should be able to reach their heads into, but not enter, the feeding area.

Feeding Arrangements

A well-designed goat house lets you feed and water your goats from the storage area without entering their living space and getting mobbed. Head-sized openings to the feeding area let a goat reach through to munch on hay in a manger. A manger is nothing more than a trough designed to keep the hay off the ground, where it won't get trampled or dirty. Clean hay is important, because goats like to snack all day long.

Through additional openings, the goats can drink water from a bucket. Keep the bucket outside their living area so the goats can't easily fill it with droppings or kick it over.

Layout for a typical goat house

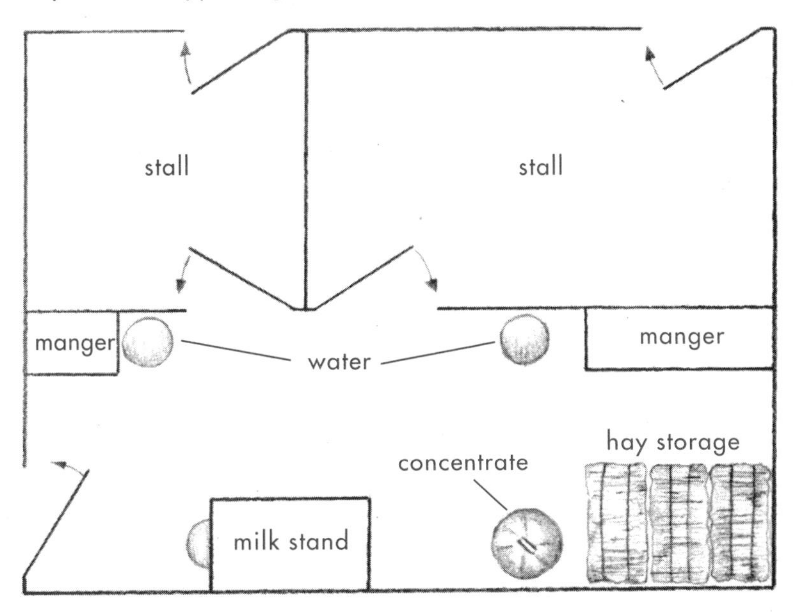

Allow one opening per goat for hay so all your goats can eat at the same time. Because they are herd animals, goats tend to all eat and sleep at the same time. However, they won't all drink at once; you'll therefore need fewer additional openings for water. One opening per six to eight goats should be adequate, depending on how rapidly they empty the bucket and how often you refill it. If you have a lot of goats and you install watering devices that refill automatically, you'll need fewer water openings.

Make the feed and water openings just big enough for a goat to push its head through but not big enough for its whole body to get through. An opening big enough for a doe's head is big enough for a kid to pop through — which is another good reason for housing weaned kids in a stall separate from adult goats.

To keep Angoras from getting bits of hay in the hair of their heads and necks, build their manger of stiff wire or slats of wood spaced 2 inches apart.

Bedding

Packed dirt makes a good floor for a goat house because it lets urine and other moisture drain away. Better yet is to cover the dirt with a slat floor made from 2x4 lumber set on edge with ¼-inch gaps between the boards. Design the floor in sections that may be lifted for cleaning.

A plain dirt floor should be covered with a thick layer of bedding. Shavings make excellent bedding. Straw also works fine, if you clean it out often. Otherwise, it gets packed down so tight that you'll hurt your back removing it, unless you have a tractor with a front loader and tines. Waste hay that isn't moldy is the most common bedding for goats. Hay is the primary food of goats, but they tend to eat only the best parts and leave the rest. Using waste hay as bedding saves you money by making use of leftovers. The goats will help you out by pulling hay into their living area and spreading it around to make themselves comfortable.

Goats that spend their days on wet or filthy bedding may develop a bacterial infection in their udders or hooves. Keep the bedding clean and dry by periodically removing and replacing it. Between cleanings, spread a fresh layer on top as often as necessary to provide your goats a clean place to sleep. Some people replace the bedding weekly; others replace it only in the spring and fall. The more often you clean out the bedding, the easier the job.

Each day, each goat produces about 1¼ pounds of manure and 1¼ pounds of urine. Some of this manure and urine gets mixed into the bedding. Used bedding from a goat house makes good fertilizer for flower beds, vegetable gardens, or fruit and nut trees. To fertilize crops and gardens, spread the used bedding in the fall and work it into the soil so it is well rotted by spring planting time. Before mulching trees and shrubs, compost the bedding to avoid damaging plants with fresh manure.

The Goat Yard

Goats need space outside to wander around and get fresh air and sunshine. Allow at least 200 square feet of outdoor space per regular-sized goat, 130 square feet per miniature.

A yard on the south side of the house gets more sun and stays drier than a yard on the north side, an important factor in avoiding damp ground that can lead to bacterial infections of the hoof. The yard should slope away from the house for good drainage. If your land is level or drains poorly, erect a wooden platform or a concrete pad in the yard to give your goats a dry place to stand. A concrete pad also gives goats a place to scrape their feet, which helps wear down excess hoof growth.

An empty cable spool is a popular toy.

Goats of all ages love to jump and play. An outcropping of rocks makes an ideal play area. A popular and inexpensive toy is an empty cable spool with a board securely nailed over the holes in both ends so that the goats can't slip during play and break a leg.

If you provide your goats something to climb on, take a good look to see what they can climb onto from there. Make sure the climbing object is far enough from the house that they can't jump onto the roof, and far enough from the fence that they can't jump over and escape.

A Goat Fence

Goat owners love to say that "a fence that won't hold water won't hold a goat." Of course, that's a slight exaggeration — but only slight. Most goat troubles occur because of inadequate fencing. Goats are curious, agile, and persistent. If there's a way to escape, they'll find it. They can flatten their bodies and crawl under a fence or spring off the ground and sail over it. If they can't get under or over a fence, they'll lean on it until they crush it down.

Goats seem to believe the grass is greener on the other side of the fence, and they'll stretch their necks to eat that grass and whatever else is growing out there. They will nibble on trees and shrubs growing within 2½ feet of the fence. In doing so, they push against the fence until they bend it out of shape.

A goat loves to scratch its back by leaning sideways against a fence and walking along in one direction. It will then turn and go the other way to scratch the other side. Sooner or later, all that pushing and rubbing will knock down a flimsy fence. A properly built fence not only keeps goats where they belong but also protects them from predators, such as stray dogs, coyotes, wolves, and bears.

A goat-tight fence may be made of woven wire or electric strands. A woven-wire fence should be 4 feet high for calm breeds such as myotonic goats and Angoras and 5 feet high for active breeds such as miniature goats and Nubians. Woven wire comes with 6-inch or 12-inch openings. Wire with 6-inch spacings is best because kids can't slip through it. Attach the wire to 8-foot posts driven at least 2½ feet into the ground every 8 feet. Place corner post and gatepost bracing on the outside of the fence; if you brace on the inside, your goats will use the braces to climb up and out.

Electric fence wire may be used in conjunction with a nonelectric fence or to build an all-electric goat-tight fence. One strand of electrified wire placed 12 inches off the ground on the inside of a fence will keep goats from pushing against the fence. Another strand placed about nose high will keep goats from leaning on the fence or jumping over.

Electrified wire is the way to go if you wish to goat-proof an existing non-electric fence. If you are building a new fence, it's cheaper to make it all electric from the start. Use high-tension smooth wire and a high-energy, low-impedance energizer. Goats will stay away from it because they don't want to get shocked; however, the fence must always be on. Your goats will test the fence constantly, and the moment the juice goes out, so will they.

An electric fence need not be as high as a woven wire fence; up to 40 inches will suffice. String the bottom wire 5 inches from the ground. String the second wire 5 inches from the first, the next wire 6 inches up, the next wire 7 inches up, the next wire 8 inches up, and the top wire 9 inches up. Connect every other wire to the energizer, and the alternating wires to the ground. For safety reasons, you must get all the connections right,; if you aren't sure how to build an electric fence, consult an expert or a good book (see Recommended Reading, page 379).

The Goatproof Gate

Goats are expert gate-crashers, so take special care in designing your gate. Make the gate as high as the rest of the fence. Use a latch that goats have trouble opening. A latch that requires two different motions, such as lifting and pulling, is more difficult for a goat to open than a latch that simply flips up.

No matter what kind of latch you use, secure it with a bolt snap, available from any hardware or farm store. Attach the snap to the gate with a short chain so you can't drop it, put it into your pocket, or otherwise misplace it

Install the latch partway down the gate, on the side away from the goats. They won't be able to reach over the gate to work the latch, but be sure *you* can reach the latch from the inside — if a tall person installs the latch, a shorter person may have trouble reaching it.

Hinge the gate to open into the yard, toward the goats. Even if the goats manage to open the latch, they'll be pushing against the gate and keep it shut. Make the gate strong enough to support the weight of a goat standing on its hind legs to peer at the world outside the goat yard.

Feeding Goats

Goats, like humans, need a balanced diet to remain active and healthy and to produce kids, milk, meat, and fiber. A goat's digestive system fills up one third of its body. Like cows and sheep, goats are ruminants. (See pages 313–314 for a detailed description of the ruminant digestive system.)

Even among ruminants, goats eat the widest range of plants. Their ability to use plants other animals can't digest makes them popular as livestock worldwide.

Left to their own devices, goats browse on shrubs and trees, much like deer.

Grazing and Browsing

Humans and other animals with simpler digestive systems need dietary fiber to stimulate digestion, even though their stomachs cannot digest the fiber itself. A goat's digestive system breaks down fiber into nutrients the animal needs for survival. Fiber is a goat's main food, and the goat consumes it in the form of grass, hay, twigs, bark, leaves, cornstalks, and various other plant parts.

Whereas some animals are grazers, reaching down to munch on grass and other low-growing plants, and others are browsers, reaching up to snack on leaves and bark, goats are opportunistic feeders — they eat whatever is available. Given a choice, they wander from one food source to another, grazing and browsing as necessary.

Because they can eat such a varied diet, goat herds can live in diverse circumstances. Some goats roam entirely in wooded areas, others are kept on pastures, and still others are confined to barns and have all their food brought to them. Each arrangement has advantages and disadvantages.

Letting your goats browse or graze reduces feeding time, labor, and expense. Since feed is about 70 percent of the cost of keeping goats, letting them forage adds up to big savings. Foraging works well for meat and fiber breeds. Allowing dairy breeds to browse forested areas, however, may not be a good idea for two reasons. First, brambles and low branches can scratch udders. Second, some plants give milk an unpleasant flavor. Dairy goats are more often allowed to graze on improved pasture (pasture that is maintained by the owner), where their udders are safe from harm and weed control eliminates wild onion, garlic, mint, and other plants that give milk a bad flavor.

Goats kept on pasture produce more milk, but the milk has a higher water content than the milk of goats not allowed to graze. For cheese makers, the higher percentage of water is problematic because the milk produces less cheese than does milk with a lower water content. Goats in commercial dairies are therefore confined to loafing sheds, where all their feed is brought to them. The confinement system is also suitable for goats living on a small lot with insufficient land for browsing or grazing.

Feeding Hay

Whether goats browse, graze, or are kept in confinement, they need hay. Hay is nothing more than pasture plants that have been cut, dried, and stored loose or compressed into square or round bales.

On average, a goat eats 3 percent of its body weight in hay each day, which adds up to about 4 pounds for a large goat and 2 pounds for a miniature. If a square bale weighs 40 pounds, two full-grown goats will consume approximately 73 bales per year. Since a goat won't eat more hay than it needs, feed hay "free choice" — that is, always keep the manger full, so that the goat can eat hay whenever it wants to.

Fresh green pasture plants contain a high percentage of water. If pasture is a goat's sole source of feed, the animal will have a hard time satisfying its hunger. Hay takes the edge off, making the goat less likely to scarf down poisonous plants or overeat proper pasture plants. In addition, fermentation in a rumen full of nothing but fresh pasture produces excess gas that cannot escape fast enough, causing the rumen to bloat dangerously. A goat that eats plenty of hay before going out to pasture will not graze frantically to curb its hunger and will be much less likely to bloat. A goat that can come and go as it pleases, munching on free-choice hay and wandering out for a mouthful of pasture, has little chance of bloating.

The quality of hay varies, as does a goat's nutritional needs. A growing goat, a pregnant doe, and a lactating doe all have higher nutritional needs than a mature wether or an open dry doe (one that is neither pregnant nor giving milk). During breeding season, a buck's nutritional needs are higher than at other times of year.

Legume hays, such as alfalfa, clover, soybean, vetch, and lespedeza, provide excellent nutrition for kids, pregnant does, and lactating does. Grass hays, such as timothy, red top, sudan, bromegrass, and fescue, are less nutritious. A good all-purpose hay is a 50-50 grass–legume mix.

Look for early-cut hay that is fine-stemmed, green, and leafy. Buy hay sold for goats or horses; hay sold for cows is often too stemmy. Goats, especially milking does, can be pretty persnickety about eating hay with coarse stems; they'll nibble down the tender parts and leave the rest. Bucks and wethers will eat stemmy hay when forage is sparse, but if they have a choice they, too, will waste most of it. Stemmy hay does not even make good bedding.

Hay sellers advertise through the classified section in the newspaper, especially from late spring through early fall. Various county agriculture offices keep lists of hay growers. The clerk at your feed store may know someone who sells hay, or the feed store may stock it.

Most goat owners prefer to handle small square bales, even though many growers have switched to large round or square bales that are easy to transport by tractor. Small square bales that are easy to move by hand are ideal for the small goat herd. For a large herd, the large square or round bales may be more convenient, but they require a feeding system suitable for doling them out.

You can buy hay by the load delivered to your door. To save money, you can sometimes buy hay in the field, right after it has been baled. The grower will expect you to pick it up and probably load it yourself onto your truck. If you do not have room to store enough hay to last a year, find a grower who will store it for you and let you pick it up as you need it. Expect to pay more for stored bales than for hay purchased in the field.

Keep your hay under cover and off the ground on pallets. Properly stored hay retains its nutrients for a long time, but one good rainstorm can ruin baled hay. Never feed your goats moldy or musty-smelling hay.

Unless the hay is exceptionally good or your goats are exceptionally hungry, about one third of the hay will go to waste. Remove leftover hay from the manger every morning and use it as bedding or toss it onto a compost pile. Don't put the hay directly onto your garden soil, or you will introduce unwanted weed seeds.

Concentrate

A young growing goat and a mature animal that produces kids, milk, or fiber needs more nutrients than even the most nutritious hay can provide. Such a goat requires a ration that contains grains and other nutrient-rich feeds combined into a dietary supplement variously called goat feed, goat chow, goat ration, or concentrate (because it is a concentrated source of nutrients).

Not all goats in a single herd require the same amount of concentrate at the same time. One doe may be dry while another is lactating. One may be still maturing while another is about to give birth. Even two does of the same size and age, both dry or both lactating, may require different amounts of concentrate to maintain the same body weight or to produce the same amount of milk.

Feeding guidelines are therefore nothing more than estimates. Always use your own best judgment. If you raise miniature goats, feed them approximately one half the amount required by large goats. Keep written records to remember who gets what. Adjust concentrate levels according to the following four factors:

The quality of the roughage the goats eat. Goats that eat fresh browse, green pasture, or good hay need less concentrate than goats that get little or no browse or pasture and poor-quality hay.

Each goat's physical condition. A well-conditioned goat is fleshed out but not too fat. A dairy goat is too fat when you can't feel her ribs. A fiber goat or meat goat is too fat when you can grab a handful of flesh behind the elbow. If a goat is too thin, feed more grain. If a goat is too fat, feed less grain.

The goat's age. Let kids nibble on concentrate as soon as they are interested. At first, they will just mouth the ration with their lips, but as they grow they will learn to relish their little taste of grown-up feed. After the kids are weaned, gradually work up to 1 pound of concentrate per day. When feeding any goat 1 pound or more per day, divide the concentrate into two feedings, morning and evening. Feeding too much concentrate at once upsets the rumen's balance.

The goat's level of production. Mature goats that are not pregnant or lactating require a maintenance ration that provides just enough nutrients to maintain the animal's health and body weight. A maintenance ration for wethers and open dry does on good browse or pasture need not include concentrate. A supplemental feeding of ¼ to ½ pound (1 to 2 cups) of concentrate per day, however, increases the growth rate of a meat goat, improves the hair growth of a fiber goat, and keeps all goats easier to handle because they look forward to your regular visits with the feed can.

Open dry does or wethers raised in confinement may be fed up to 1 pound of concentrate per day. The same applies to open dry does and wethers that normally browse or graze but whose feeding pattern is disrupted by bad weather or whose forage supply has been curtailed by drought. Since open dry does and wethers have low nutritional requirements, you may save money by feeding them shelled corn, barley, oats, wheat, sorghum, or milo instead of commercial concentrate. Whole grains that are dry and hard do not digest as well as grains that have been rolled, crimped, cracked, or flaked.

A pregnant dry doe that is not a dairy breed should be kept on a maintenance ration until 6 weeks before she gives birth. Then feed her a little concentrate, gradually increasing the amount to 1 pound. After she gives birth, continue feeding 1 pound a day (1¼ pounds if she has twins) until her kids are 6 weeks old, then begin gradually decreasing the concentrate. By the time her kids are 3 months old, the doe should be back on a maintenance ration.

Dietary Changes

Anytime you change a goat's diet, you run the risk of disrupting the rumen's digestion activity. To keep your goat from getting sick, make all dietary changes gradually. Whenever you increase or decrease the amount of concentrate a goat gets, do it gradually over several days. If you change from one kind of hay or concentrate to another, mix the new feed with the old in increasing quantities until the switch is complete.

Concentrate Feeding Guidelines*

Kid	Nursing Weaned	Nibble 1–2 lbs.
Maintenance ration	Fresh forage available	¼–½ lb.
	No fresh forage available	1 lb.
Wether or open dry doe		Maintenance ration
Nondairy goat	Pregnant dry	Maintenance ration
	6 weeks before kidding	Increase to 1 lb.
	Nursing	1–1¼ lbs.
	3 months after kidding	Maintenance ration
Dairy goat	Pregnant dry	1 lb.
	2 weeks before kidding	Increase to 3 lbs.
	Lactating	½ lb. per 1 lb. milk (1 lb. minimum)

**Reduce amounts by half for miniature goats. Adjust all amounts to each goat's condition.*

If your pregnant dry doe is a dairy breed, feed her about 1 pound of concentrate per day. During the last 2 weeks of pregnancy, gradually increase the concentrate to about 3 pounds a day by the time she gives birth. During early lactation, when her milk production is increasing, feed her a minimum of 1 pound of concentrate per day plus an additional ½ pound for each pound of milk she produces over 2 pounds. During late lactation, when her production has leveled off, feed her ½ pound of concentrate per pound of milk. After the doe has been bred, gradually decrease her concentrate to 1 pound per day, and start the feeding cycle again.

Above all, remember that concentrate is a supplemental ration and not a goat's main diet. No matter how little or how much concentrate a goat gets, it should have access at all times to as much hay as it wants.

Concentrate Feeders and Storage

Some herd owners feed concentrate from a communal trough, but doing so can mean that timid goats don't get their fair share. Feeding concentrate to each animal separately means each goat gets an appropriate portion and the portion is tailored to the individual goat's needs.

Placing feeders in a manger the goats can access through the usual feed hole ensures that each animal eats its own ration. To keep fast eaters and bullies from stealing feed from others, you may find it necessary to run a chain across their feed holes so that they can't get out until everyone is finished eating.

Lactating does are often fed on the milk stand to keep them from getting restless. However, if you train your does to stand calmly without being fed, they won't get restless when they finish eating before you finish milking.

Concentrate comes in 50-pound bags. Store unopened bags away from moisture, out of the sun, and off the ground. Your feed store may let you have a wooden or plastic pallet to store sacks on. Otherwise, raise sacks off the ground by using bricks, concrete blocks, or pieces of lumber.

After opening a bag, pour its contents into a clean plastic trash can with a tight-fitting lid. A 10-gallon can holds 50 pounds. Storing concentrate in a can keeps it from getting stale or absorbing moisture from the air. Moisture causes concentrate to become moldy; never feed moldy grain to goats. Empty the can completely before pouring in another bag, so that stale concentrate won't build up at the bottom of the can.

Store concentrate where your goats can't get into it. Otherwise, they will jump on unopened sacks until they tear a hole. They will work on the lid of a storage can until they get it open. The resulting feast will disrupt rumen fermentation, and the goats will become ill, like kids who eat too much candy the day after Halloween. In the case of goats, however, all that fun could prove fatal. As a safety measure, in case a goat should get out, secure the lid of the storage can with a bungee cord.

Besides preserving the feed, a storage can keeps out munching mice, which consume an astonishing amount of concentrate while fouling what they leave behind, until the goats may turn up their noses at it. When you transfer feed from the sack to the can, take care to avoid spillage, since mice are attracted by spilled grain. Sweep the floor of your storage area regularly. In spring and fall, when barn mice are most active, reduce the population by using traps baited with peanut butter. Better yet, keep a cat in your barn and reward it with a daily saucer of warm goat milk.

Soda and Salt

The rumen ferments its contents best within a narrow range of acidity. Feeds that ferment rapidly increase the rumen's acidity. If the acidity goes up too fast, the microorganisms that cause fermentation multiply too fast. As a result, the rumen's balance is upset, and the goat becomes sick. Your goat's health therefore depends in part on proper rumen acidity.

Alkaline substances, such as sodium bicarbonate (common baking soda), reduce acidity. A goat eats soda to keep its rumen acidity within the proper range. The goat knows when it needs soda, and how much. All you have to do is make sure your goats can get soda when they need it. Feed-grade baking soda may be purchased at any feed store and is less expensive than baking soda from the grocery store.

Each day, a goat will lap up an average of 2 tablespoons of soda. Lactating does, and all goats on summer forage, eat more soda than at other times. The choice should be theirs.

Sodium chloride, or common salt, helps control rumen acidity, aids digestion in other ways, and helps keep a goat's body tissues healthy. Besides regular salt, goats require many other minerals in minute amounts. They obtain some of these trace minerals from good hay, fresh forage, and concentrate. To make sure your goats get all the minerals they need, give them free-choice trace mineral salt, which is a combination of trace minerals and salt and is sold at any feed store. It comes in loose form, like table salt, or compressed into a block. Loose salt is easier for many people to handle than a heavy block and is easier for goats to lick, especially at times when their salt need is high.

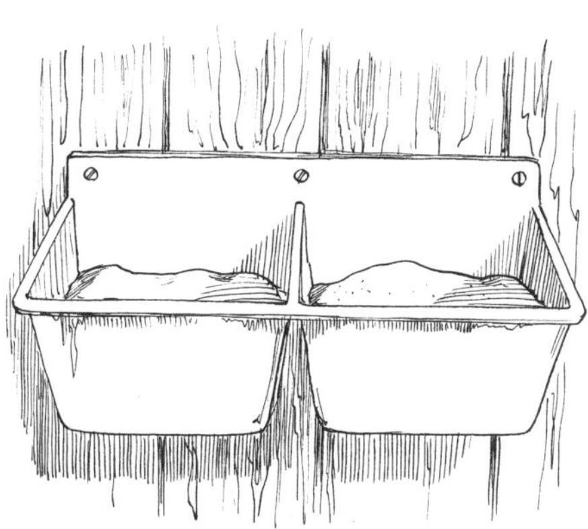
Soda and salt feeder

Your feed store may sell a trace mineral mix formulated specifically for goats. If not, get the mix for horses or cows. The mix must contain copper, iodine, and selenium. Copper is essential for dairy goats. If your goats browse on trees and other deep-rooted plants, or if their drinking water flows through copper pipes, they already have some copper in their diet and need less from the trace mineral mix. Iodine and selenium are both essential for a doe to give birth to healthy kids. Iodine and selenium do not occur in plants grown in regions where these minerals are lacking in the soil.

Be careful not to overfeed trace minerals. Excess iodine in a doe's diet makes her milk taste unpleasant. And because excess selenium in the diet can be toxic, do not give your goats trace mineral mix containing selenium if the soil in your area is high in selenium. The soil tends to be high in selenium in Colorado, Kansas, Montana, Nebraska, North Dakota, South Dakota, Utah, and Wyoming. The soil is deficient in selenium in some areas of the Northeast, the southern Atlantic seaboard, and the Pacific Northwest in the United States and in the Maritimes and parts of British Columbia in Canada. Your local veterinarian or county Extension agent can tell you if the soil in your area is high or low in selenium.

An easy-to-clean plastic salt feeder is available for a few dollars from nearly any farm store. Clean and refill the feeder often. Salt attracts moisture that causes the surface to crust over. Soda and salt both turn lumpy from the water dripping off the chin of a goat that has just had a drink. And your goats will periodically delight in backing up to the feeder to fill it with droppings.

Water

The most important and least expensive item in a goat's diet is water. Goats should have access to clean, fresh water at all times. Water aids digestion, controls body heat, and regulates milk production. The more water a doe drinks, the more milk she gives. Lactating does drink more water than dry does. All goats drink more water in warm weather. They drink less when they graze on spring pasture, because fresh grass contains water.

An ideal way to make drinking water available at all times is to have waterers that refill automatically, which requires installation of some plumbing. In winter, wrap the lines with heater tape to keep the water from freezing, or install a heater device in the water fount.

A 5-gallon plastic bucket works well as a water container for a small herd. Place the bucket outside the stall, where it may be accessed through a head hole. Keeping the water bucket outside the stall means your goats can't knock it over and wet the bedding, fill it with droppings, or accidentally drop a kid into it while giving birth. A goat won't be able to get its head through the handle and spend the day wandering around with a bucket hanging around its neck.

Goats will not drink water that has been contaminated. Many items may contaminate the drinking water, including hay, hair, manure, insects, and a drowned mouse. Empty and refill the bucket at least once a day, and scrub it with a plastic brush and bleach at least once a week.

To encourage your goats to drink, fill the bucket with cool water in warm weather and warm water in cool weather. In cold weather, you must keep the water from freezing. If the goat house has electricity, keep the water from freezing by setting the water bucket on an electric pan heater or by using a plug-in bucket with a self-contained heater. Both devices are available through farm supply sources (see Resources, page 380). Otherwise, check the water at least twice daily and break ice as necessary.

Breeding Goats

Part of a doe's normal annual cycle is giving birth. Whether you keep goats for milk, meat, or fiber, selling kids is one way to earn income that pays for their upkeep. You may choose to raise the kids yourself as meat for your family. If you keep a doe as a pet or for fiber, you may choose not to breed her at all. A doe used for dairy purposes, on the other hand, must be bred each year to renew milk production.

Most dairy goats produce more milk than their kids need. They give more milk over a longer period than meat or fiber goats, but over time, their milk production gradually lessens. Does that give milk steadily for more than a year are exceptions.

When to Breed

In deciding when to breed each doe, consider three factors:

Her age and size. Do not breed a young doe until she reaches at least 75 percent of her mature weight, which usually happens when she is about 8 to 10 months old. A doe may think she is ready to breed when she is 6 months old, but breeding her before she is large enough may stunt her growth. A doe that is bred early will produce fewer kids that are smaller than normal.

When she was last bred. Renewing a doe's milk production by breeding is called freshening. Most dairy goats are milked for 10 months of the year, then are given 2 months off before they freshen again. Meat and fiber goats are often bred every 8 months. Once-a-year breeding, however, will give your doe time to rest and produce more kids per breeding over her lifetime.

The season. A doe must come into heat, or estrus, before she may be bred. Spanish goats may come into estrus every 3 weeks year-round. For most other goats, the breeding season is August, September, and October. Most dairy goats are bred during September and October so they will give birth when spring's green pastures provide the extra nutrition a freshened doe needs. Angoras are usually bred from August through November, after the fall shearing, so they will kid after the spring shearing. Cashmere does are bred no later than mid-November so their kids will be weaned by the time down starts growing in late June; otherwise, lactation may decrease fiber production.

Estrus

Throughout the breeding season, a doe comes into estrus every few weeks. Estrus lasts for 2 to 3 days. The time between the start of one estrus and the start of the next is called the estrous cycle. Different does have different estrous cycles, ranging from 17 to 23 days. The average doe has a 19-day cycle. Keep accurate records on each doe to track her exact cycle.

Some does show little or no signs of estrus, a phenomenon known as silent heat. Most does show some signs, but each has different signs or different combinations of signs. As you record each doe's estrous cycle, note the signs she displays so you'll know what to look for next time.

Signs of estrus are as follows:

- The doe may "talk" more than usual. She may bleat so loudly you think she is in pain. Don't worry, she isn't.
- She may urinate more often than usual.

- The area under the doe's tail may become dark or swollen and wet. You may see sticky mucus that will be clear early in estrus and white toward the end of estrus.
- The doe may be restless. She may flag, or wag her tail. She may let you handle her tail, or she may move away if you try to touch it.
- The doe may give more milk than usual just before coming into heat, then give less milk than usual for a day or two.
- She may mount another doe as if she were a buck, or let other does mount her.
- If a buck lives nearby, you'll have no doubt when a doe is in heat. She'll get as close to the buck's yard as she can. The buck will wag his tongue, slap a front hoof against the ground, urinate on his own face, and otherwise act the fool.
- If no buck is around, you might trick the doe into displaying signs of estrus by using a buck rag, which is a piece of cloth rubbed on the forehead of a mature buck and placed in a sealed container. When you open the container near a doe in estrus, she will show clear signs of interest.

Flushing

A doe gets pregnant more easily and has more kids if she gains weight starting 1 month before she is bred until 1 month afterward. When a doe gains weight, more eggs "flush" from her ovaries during estrus. Getting a doe to gain weight at breeding time is therefore called flushing. Flushing is more important for a thin doe than for a well-conditioned doe. To get the maximum benefit from flushing, worm your does before breeding season.

You may flush does in one of three ways:

- Move them to fresh pasture.
- Feed them an extra 1 to 1½ pounds of alfalfa cubes, pellets, or hay each day.
- Feed them an extra ½ to 1 pound of grain or concentrate each day.

The Buck

For a small herd of does, keeping a buck may not be economical, but it's convenient if no suitable breeding buck is nearby. Otherwise, by the time you recognize the signs of estrus, make an appointment to have the doe bred, and transport her to the buck's location, it may be too late. If you'd rather not own a buck, you might arrange to lease one during breeding season.

If you find a suitable buck nearby, make breeding arrangements with the owner in advance. Find out whether the owner will be reachable on short notice and what the breeding fee will be. Some stud owners like to keep the doe overnight to make sure she is bred. Others expect you to wait while the doe is bred so you can take her back home with you.

No matter which breeding arrangement you prefer, seek a strong, healthy buck that's been well cared for. If you plan to do some serious breeding, sell the kids for breeding, or show the offspring, the buck should be the same breed as your doe. If you plan to raise or sell the kids for meat, the buck need not be the same breed as the doe. Mating your doe to a buck of a different breed, in fact, often produces larger, faster-growing kids.

Breed a fiber doe to a buck with good fiber quality. Breed a dairy doe to a buck whose dam had a good record in milk production. To improve the udders of your doe's offspring, select a buck whose dam has a good udder. If the buck has been bred previously, try to look over his offspring as well. If he has produced quality kids in the past, he will probably continue to do so.

Breed your does only to a buck with sound, strong, well-trimmed feet. A buck with poor feet may have a hard time mounting the does.

Like does, the buck should be wormed and his concentrate ration increased in preparation for breeding season. These measures help produce healthy, motile sperm.

If you cannot find a suitable buck within a reasonable distance, you may wish to breed your does by artificial insemination (AI). To find an AI practitioner, contact the nearest goat club, check the ads in one of the goat magazines (listed in the appendix), or ask your veterinarian. The AI practitioner will help you select a buck through descriptions in a catalog and arrange to purchase that buck's semen for local storage until your does come into heat. When your doe is in estrus, the AI practitioner may come to your place or may ask you to take your doe elsewhere for insemination. You will be charged for both the semen and its placement. Artificial insemination can be costly; be sure to find out in advance how much it will cost.

Mating

A doe is ready to breed when she is in standing heat. A doe that is not in standing heat will move away from a buck that tries to mount her, whereas a doe in standing heat waits patiently to be mated. Some bucks have to go through their whole clownish routine before getting down to business, but the act of mating itself takes just a few seconds. You can distinguish a trial run from the real thing by the way the buck arches his back before uncoupling from the doe.

By remaining on hand for the mating, you can be sure the act has been completed and you can make a close estimate of the kidding date. If you are away a lot, you may prefer to run a buck with your does during breeding season. This is a lot less work for you, but you never know for sure when each doe has been bred and, therefore, when she is likely to kid. In these circumstances, chances are greater you won't be on hand to help out if something goes wrong at kidding time, such as the doe having trouble during labor, the newborn kids suffering frostbite in freezing weather, or kids getting trampled by anxious herdmates or butted to death by a suspicious first-time mother.

As each doe is bred, note the date and the names of the doe and the buck. You might jot this information on a calendar in the goat barn and transfer it to a notebook or computer database at the end of breeding season. If you have your doe bred by a registered buck and you plan to register the kids, you will need a formal service memo signed by the buck's owner that shows the date, the names of the doe and the buck, their registration numbers, and the owners' names. The organization that registers your breed supplies pads of preprinted service memos to buck owners. Since the owner may be busy, be sure to ask for this record at the time your doe is bred.

When a doe has been successfully bred and becomes pregnant, she is said to have settled. A doe that does not settle usually comes back into heat on her next cycle. If she settles, she will not come back into typical heat. She may show some signs of estrus when her next cycle is due, but they won't be as strong as usual. If you put her together with a buck at that time, she will display little interest in him. For the next 150 days, the doe will spend much of her time sleeping.

Gestation and Kidding

The gestation period for goats is approximately 150 days, give or take 2 weeks. First-time mothers often kid early, especially if they are carrying more than one kid. A doe that kids late may give birth to extra-large kids and may therefore have trouble in labor. Keep a record of the gestation period for each doe; you will find that each follows her own pattern, which will help you predict pretty precisely when she is likely to kid next time. Note the time of day as well as the date, since each doe tends to be fairly consistent whether she gives birth in broad daylight or in the wee hours of the morning.

Managing the Expectant Doe

Except for dairy goats, most does will no longer be giving milk by the time they are bred; they will have dried off naturally after their previous kids were weaned. A dairy doe may be milked up to 2 months before she is due to kid, then should be dried off so that her body can rest. Drying off consists of discouraging milk production by putting the doe on a maintenance diet and by no longer milking her. Some does dry off on their own. Others will stop producing milk within a few days after you stop milking. A really good milker may continue producing milk. Relieve the doe of the excess milk about 1 week after you stop the regular milking. But do not continue milking her regularly, because that encourages further milk production.

As kidding time draws near, you will need to attend to a few management chores:

Adjust the doe's diet. Using the guidelines in the section on feeding goats (page 187), adjust the doe's diet to ensure that she is getting hay of the proper quality and the right amount of concentrate for her age, breed, and condition.

Give vitamins and minerals. Unless you live in an area where the soil is high in selenium, give each doe a selenium injection 1 month before she is due to kid. If your herd does not have access to fresh forage, also give each doe an injection containing vitamins A, D, and E. Obtain the doses from your veterinarian or the owner of a large goat herd. If you have never given an injection, ask a veterinarian or experienced goat handler to show you the proper procedure.

Crotch the doe. Before the due date, use electric clippers to trim the hair from the doe's udder and tail and beneath her tail. This procedure, called crotching, makes the doe more comfortable and makes kidding cleaner. Crotching lets you more easily watch the doe's udder development and helps newborn kids find her teats. A kid can starve by sucking on a doe's long hair by mistake. Crotching also makes it easier to clean postkidding fluids from the doe's hindquarters and udder.

Gestation Checklist

Date of breeding ________

Signs of return of estrus (17–23 days) ________

Dry off (90 days) ________

Start increasing feed (100–120 days) ________

Inject selenium and vitamins A and D (135 days) ________

Kidding day (150 days, more or less) ________

Clip your own nails. As kidding time gets close, keep your fingernails cut short. If you have to help out, long fingernails may scratch and injure the doe.

Prepare a kidding stall. Have a clean stall ready with fresh bedding where the doe can kid in a relatively sanitary environment. Straw or waste hay makes better bedding than shavings, which tend to stick to a wet newborn kid. Make sure the water bucket is positioned where a kid can't be accidentally dropped in and drowned. The best place for the water bucket is outside the stall, so that the doe has to drink through a head hole.

Arrange for help. If this kidding will be your first, try to find an experienced goat or sheep keeper who is willing to be on hand at kidding time. That person can at least reassure you that everything is going fine (which it usually is). If you can't find someone to be on hand in person, find one who is willing to offer advice over the phone at a moment's notice (and at any time of day or night) and come right away should you feel you need help.

Assemble kidding supplies. Before kidding starts, gather the following supplies and keep them in a handy place. After each doe kids, replenish your supplies in preparation for the next kidding. You don't want to be running around looking for things when a doe starts to kid; you need to be there to watch. If you keep your supplies in a clean container with a lid, you can leave them at the goat barn. Store medical items that might freeze in a separate smaller container, such as a lunch bucket, and leave it where you can grab it as you hurry out the door.

- Paper and pencil to write down the date and time of kidding, order of birth, and any unusual events or problems
- A pair of overalls and a washable jacket (kidding can be messy)
- A snack for yourself, in case you're at the goat barn a long while and can't or don't want to leave to get something eat
- A large box to put the newborn kids in
- Plenty of old towels or large absorbent rags to line the bottom of the box and dry the newborn kids
- A hair dryer to dry the kid fast in cold weather
- A heat lamp in case a kid is weak or sickly (needed only in freezing weather)
- Tamed iodine (such as Betadine) to coat the kid's navel
- Soap and a container to hold warm water to wash your hands
- Long surgical gloves (available in drugstores) in case you have to reach inside the doe
- Water-based lubricant (such as K-Y jelly) to lubricate your hands in case you have to reach inside
- Two uterine boluses of antibiotic (available from a veterinarian or farm store) per doe to prevent infection in case you have to reach inside
- A stack of old newspapers, so the second kid won't land in the mess left by the first kid and to wrap the afterbirth for disposal
- A scale so you can weigh each kid as soon after birth as possible
- A washcloth to clean the doe's udder after she kids
- A bottle and rubber nipple in case you need to feed a kid with a weak sucking instinct
- Popsicle sticks and surgical tape to splint a kid's legs if it can't stand properly.
- Ear splints and surgical tape to straighten out folded ears (needed only if your doe is a Nubian; see page 199)
- Molasses to mix with warm water to feed the doe after she gives birth
- A bowl or bucket to hold the warm molasses water
- A handful of raisins or roasted peanuts to reward your doe

It's Almost Time

When a doe is due to kid, you should separate her from the other goats. The stall where the herd lives is probably not clean enough for a newborn kid. The other goats may panic at the sight of kids and butt or trample them. And the other goats may become curious when birthing starts, crowding around and causing the doe distress. The new mother needs space to bond with her new kids without being disturbed.

Signs that kidding is near may be distinct or undetectable. A few days before a doe kids, her udder may swell and look shiny; however, some does won't swell with milk until after the kids are born. The hairless area beneath the doe's tail may redden and swell. The doe may show little interest in eating.

A doe carries kids on her right side. When her time for kidding draws near, you can put your hand on her side and feel the kids moving. As long as you can still feel them, they probably won't be born for at least 12 hours.

Just before birth, the kids will move back toward the birth canal. This shift causes the areas around the doe's tail and hips to look bony. Her muscles may loosen so much that she can't hold her tail down, and it will remain in an upward position. The doe will probably become extremely affectionate toward you, perhaps nickering when you approach her and trying to lick your face. She will discharge thick mucus. She will get restless. She will paw the ground and repeatedly lie down, then stand up.

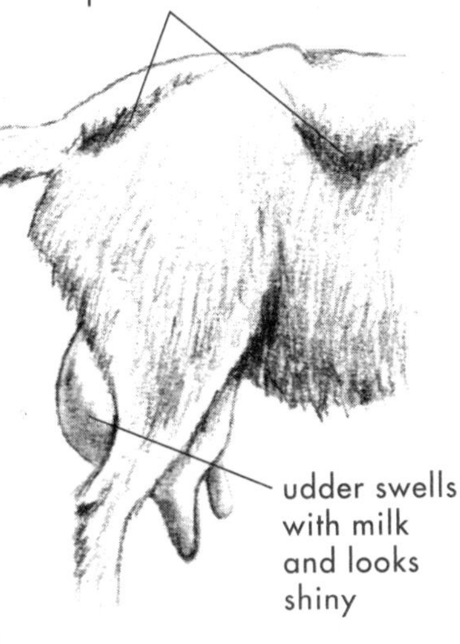

Signs that kidding is near

Pregnancy Toxemia

If a doe stops eating late in pregnancy, she may have pregnancy toxemia, also known as ketosis. This condition occurs when a doe draws energy from her own body to feed her developing kids. It is seen most often in first-time young mothers, in does carrying multiple kids, in extremely fat does, and in does that have not received sufficient nutrition throughout pregnancy.

As a result of pregnancy toxemia, the doe becomes progressively weaker, may wobble when she walks, has trouble getting up, and may appear lame. A good preventive is to keep a bag of feed-grade dry molasses on hand to sprinkle on top of the concentrate of any pregnant doe that shows signs of going off feed. If a doe becomes weak before you notice she has gone off her feed (as may happen if you have more than a few does), stronger measures are needed to keep the doe alive. Give her 2 to 3 ounces of propylene glycol twice a day until she kids. Pregnancy toxemia may be prevented through proper nutrition.

It's Time!

Most kids are born without human help. Chances are good, in fact, that you will go out to check on a doe only to find she has already given birth. It's a good idea to try to be there during the event, in case the doe needs help. However, unless a problem occurs, let your doe kid on her own. The most she will likely need from you is moral support. Some does seem to wait for the comfort of your arrival before they start giving birth. Others seem to deliberately wait until you leave.

When a doe lies down and starts straining and groaning, she is in labor. Soon the water sac will emerge; do not break it. When it breaks on its own and fluid spills out, a kid will soon follow. Usually, the first thing you see is a pair of tiny white hooves. Then you will see a little nose resting on the hooves. If you see two hooves and no nose, don't worry. The kid could be coming out back legs first, which is perfectly normal — the birth will just take a little longer.

Once you see the hooves, the doe will strain a few more times, and out will come a newborn kid. The doe should lick her kid, which stimulates it to breathe and creates a bond between mother and offspring.

How many kids a doe has depends on her age and breed. An older doe usually has more kids at a time than a doe giving birth for the first time. Angora and Spanish goats have either one or two kids. A Pygmy usually has twins. Most other breeds have either twins or triplets. Nubians and myotonic goats may have four or five kids at a time. A doe that is herself a twin, triplet, or quadruplet is more likely to have twins, triplets, or quads. Flushing a doe before breeding increases her chances of producing multiple kids.

If the doe starts straining again after the first kid emerges, dry the first kid and put it into a clean box out of harm's way. You don't want it to be trampled or rolled on while the second one is being born. Scatter fresh bedding over the mess left by the first birth, or spread newspapers behind the doe to give the next kid a clean place to fall.

Stay with your doe until you are certain she has had all her kids. You don't want to come back later to find that a kid has died because the doe was so busy taking care of the first one that she didn't clean off the second one's face to help it breathe.

The birth of kids is exciting but can also be scary the first time around. Talking to your doe and repeating her name in a reassuring tone will help keep both of you calm. No matter how nervous you are, stay as calm as possible so the doe will remain calm as well.

When a Doe Needs Help

If the doe is straining and you see two hooves but no nose, the kid's head may be folded back. You might see a nose and only one hoof, or no hooves, because one or both legs are folded back. In these cases, the doe needs your help. Here's what to do:

1. Wash the area under the doe's tail with soap and warm water.

2. Scrub your hands and arms with soap and water and put on surgical gloves, if you have them. Lubricate your hands with water-based lubricant (K-Y jelly) or liquid soap.

3. Reach inside the doe and feel for the kid's feet and nose.

- If the kid's head is turned back, push the feet back until you have room to move the head forward.
- If one or both legs are folded back, push the kid back and try to bring both legs forward, one at a time. Cup your hand around each hoof so it doesn't tear the doe.

Take care to ascertain exactly what it is you are feeling. If the doe is carrying more than one kid, they may be tangled together. Follow each leg to the body so you know that both legs belong to the same kid. If two kids are coming through at once, push one back to give the other more room.

4. When you have one kid in proper position, take hold of both of the kid's legs and pull gently, but pull *only* when the doe strains. *Do not attempt to pull out the kid when the doe is not straining.*

5. To prevent infection due to your intrusion, after all the kids are born, scrub your hands, take two uterine boluses, reach inside the doe again, and deposit the boluses.

If you fail to get things sorted out or your doe strains for more than 45 minutes without giving birth, get help immediately. A doe that has trouble giving birth may die, her kids may die, or both may die.

Normal Kidding Positions

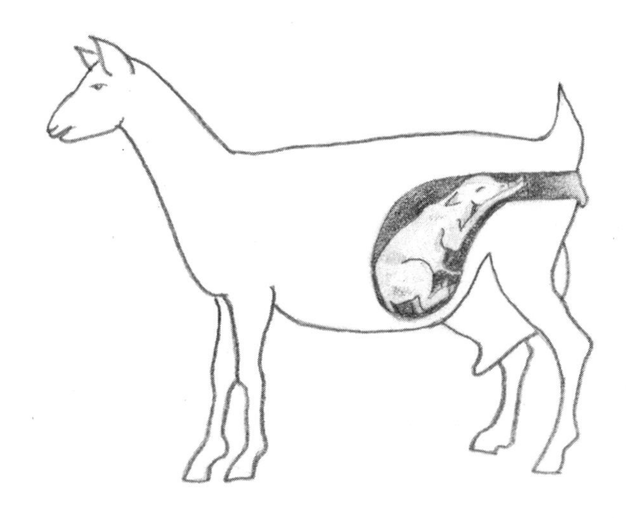

Problem Kidding Positions

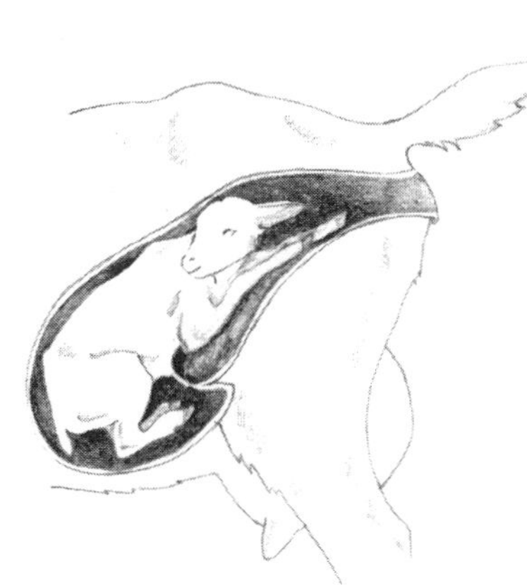

head back

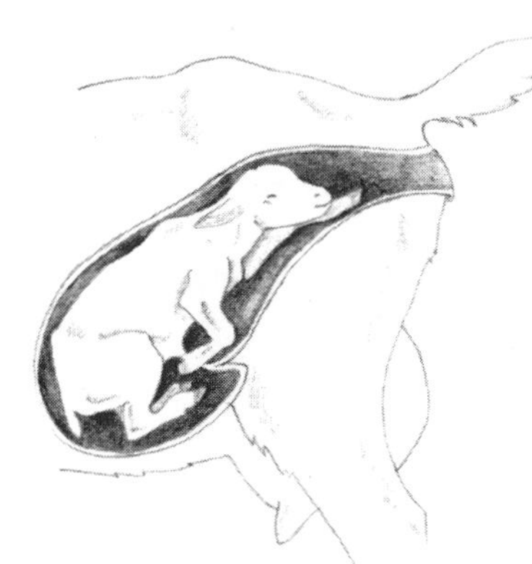

one leg back

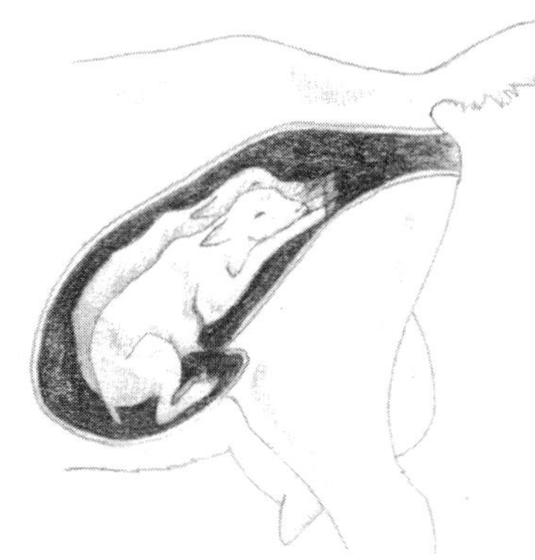

two at once

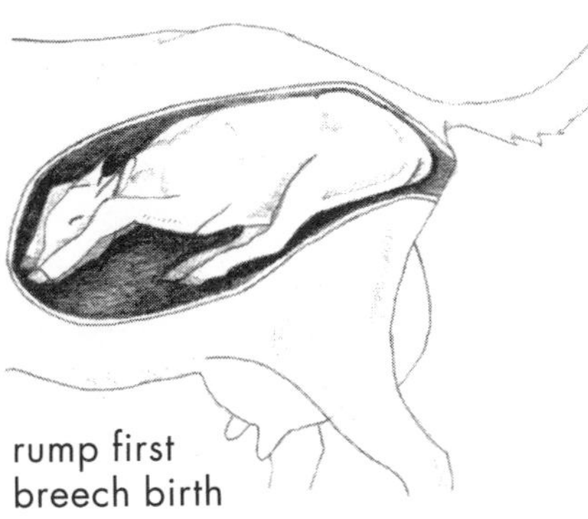

rump first
breech birth

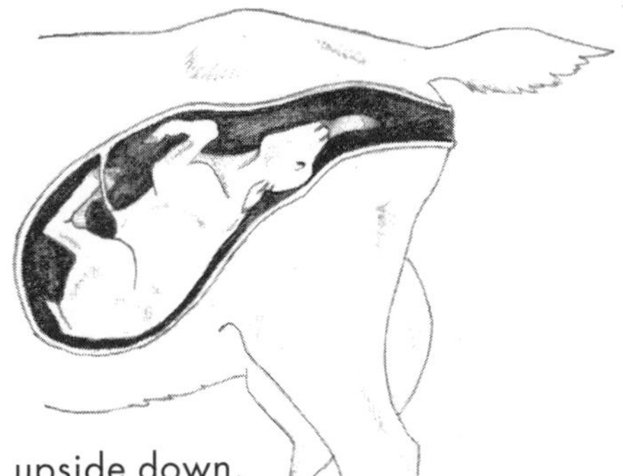

upside down,
one leg back

After the Birth

Thirty minutes to 12 hours after a doe gives birth, she will pass the afterbirth — a mass of bloody tissue. Most does produce only one afterbirth, but occasionally a doe will pass two. The doe may try to get rid of the afterbirth by eating it. Even though goats are vegetarians, eating the afterbirth is a normal and instinctive act designed to avoid attracting predators to the birth site. If you find the afterbirth before it is eaten, wrap it in newspapers and dispose of it.

Never pull on afterbirth while it is hanging from the doe. It may still be attached, and you could cause the doe to bleed to death. If the afterbirth does not fall out on its own within 24 hours, call the vet.

Giving birth causes a doe to lose a lot of fluids, and she will probably be thirsty afterward. Give her a bucket of warm water with a little molasses mixed in. Besides replacing lost fluid, warm water will help her relax and pass the afterbirth, and the molasses will provide energy.

As a reward for a job well done, offer the doe a handful of raisins or roasted peanuts. Feed her as usual, but don't worry if she doesn't eat right away. Depending on how hard she labored and how many kids she produced, she may be tired and prefer to rest. Some goats, however, have their heads in the hay manger the moment kidding is over.

Breeding and Birth Record

No.______

On (date) __________ Buck (name of buck) __________

Registration No. ______ Registry __________

Was Bred to Doe (name of doe) __________

Registration No. ______ Registry __________

Owned by __________

Address __________

____

Signed (buck owner) __________

Date of Birth __________

Name of 1st Kid __________ Sex ______

Description __________

__________ Tattoo: L ______ R ______

Name of 2nd Kid __________ Sex ______

Description __________

__________ Tattoo: L ______ R ______

Name of 3rd Kid __________ Sex ______

Description __________

__________ Tattoo: L ______ R ______

The new mother will discharge bloody fluid for up to 2 weeks after kidding. Keep her comfortable by cleaning her tail and udder twice a day. Use warm soapy water, pat her dry with a paper towel, and coat the hairless parts with dairy balm (such as Udder Balm or Bag Balm) if they appear dry and chafed.

Newborn Kids

As each kid is born, note the date, time of day, and duration of labor. Weigh the kid and jot down its birth weight and sex. When a doe has two or more kids, note any identifying marks that will help you remember their order of birth. If two kids look exactly alike, put temporary marks inside their ears with a marking pen until you can identify some other distinguishing feature. You may choose to name them at this time, or you may prefer to wait until some feature of appearance or personality hints at an appropriate name.

First Things First

A kid is born covered with slimy mucus. The doe licks the kid's face to remove the mucus so it can breathe. If the doe is distracted by another kid on the way, help her out by cleaning the first kid's nose and mouth. If the kid does not start breathing immediately, tickle the inside of its nose with straw. If you hear rattling sounds when the kid breathes, lift it by its back legs so mucus can easily drain from its airways. You will know the kid is okay when it bawls, an indication that it has gotten enough air into its lungs.

In cold weather, dry the kid rapidly with old towels or clean rags so that it won't become chilled. If the weather is freezing, complete the job with a hair dryer or towels warmed in a clothes dryer. If you find a newborn kid that is not moving or feels cold, warm it up in a hurry. A kid that is shivering but otherwise seems fine may be warmed by holding it inside your jacket. A kid that is not breathing or moving may be warmed by soaking it in warm (not hot) water for 15 minutes, then drying it thoroughly and placing it on a heating pad or under a heat lamp while it revives. Even a kid that appears to be dead may be brought around, so don't give up too easily. A kid that has been cold or wet for too long, on the other hand, may not revive.

Do not put a kid under a heat lamp unless it is premature, sickly, or suffering from extreme cold. As soon as a newborn kid is dry, its body should adapt to normal temperatures. Using a heat lamp prevents healthy kids from adapting and therefore does more harm than good. As long as the kids have a draft-free place to curl up and sleep, they should be fine. In the coldest weather, a small doghouse provides the kids a place to sleep where their own body heat will keep them warm.

A newborn kid has a bloody umbilical cord hanging from its navel. If the cord is long enough for the kid to step on, reduce its size by cutting it with a clean, sharp knife or scissors a few inches from the body, where the cord starts to thin. Do not cut through the cord close to the body, where it is thick and pink, and do not pull the cord off. To prevent bacteria from invading through the navel area, coat the cord and navel with a tamed iodine solution, such as Betadine, or chlorhexidine (Nolvasan) solution.

Folded Nubian Ears

Nubian kids are sometimes born with folded ears that don't straighten out right away. If you don't flatten them, the ears will remain permanently folded. Keep a supply of ear splints cut from stiff cardboard in this pattern. You will need two splints for each folded ear. As soon as the kid is dry, sandwich the folded ear between two splints, keeping their narrow ends upward. Tape the splints firmly together. Remove the splints in 3 or 4 days.

Ear splint pattern

If a kid tries for several minutes to stand up but can't, its legs may be weak. Strengthen a weak leg by sandwiching it between two Popsicle sticks and wrapping tape firmly (but not tightly) around the splint. Remove the splints after 2 or 3 days. Replace them only if the kid has trouble walking without them.

Keep the doe and her kids apart from the herd for a few days while they learn to respond to one another's calls. The kids need time to learn where the udder is and to grow strong enough to scamper away when they get butted for trying to steal milk from the wrong doe.

Colostrum

The first milk a doe produces after kidding is called colostrum. It is thicker and yellower than regular milk, because it contains extra nutrients. It also contains antibodies that help protect a newborn kid against disease.

Some kids are ready to nurse as soon as they can stand. Others want to rest first. A newborn kid can absorb antibodies from colostrum for only about 24 hours; it should have taken in several helpings by then. Don't let more than 2 hours go by without ensuring that a kid gets its first colostrum.

The easiest way to feed colostrum is to let the kid nurse. Clean the doe's udder with a clean cloth and warm soapy water, and dry it with a paper towel. Make sure the teats are functioning by gently milking a stream from each side. Getting the milk started may also help remove the waxy plug that seals each teat, making it easier for the kid to start nursing. If the doe is a heavy milker, the pressure from her milk-swollen udder may already have caused the plugs to pop out.

After you squirt a bit of colostrum into a kid's mouth, the kid should start sucking right away. Some kids, however, don't initially have a strong sucking instinct. If a kid won't suck, milk the doe and put the colostrum into a small flexible bottle, such as a clean single-serving soft drink bottle. Fit the bottle with a nipple, such as the Prichard Teat designed for this purpose; it is available at some farm stores and all mail-order goat supply outlets. Warm the filled bottle in hot water for a few minutes until a squirt of colostrum placed on your wrist feels neither hot nor cold.

Put your finger into the kid's mouth and rub gently back and forth along the length of its tongue. As soon as you feel the kid attempt to suck, squeeze a little warm colostrum onto the back of its tongue and stroke the throat until it swallows. Keep at it until the kid drinks ½ cup of colostrum. After a nap, the kid should be stronger and ready to nurse on its own.

Bottle Feeding

Bottle feeding can be a nuisance, since kids need to eat every few hours around the clock. The milk must be warmed before each feeding, and the bottles and nipples must be scrupulously cleaned afterward. These conditions may not sound intimidating until it's 2 A.M. on a cold winter morning.

Sometimes you have to bottlefeed kids because the doe won't let them nurse. She may try to get away from the kids or butt them, injuring or killing her own offspring. A doe that will not accept her kids was probably herself raised on a bottle. Think about the long-term ramifications before starting your kids down the bottle-feeding trail.

Despite the drawbacks, some goat owners choose to bottlefeed their babies. One reason is to keep their goats friendly and people-oriented. Kids raised by hand

Treating Colostrum and Milk to Prevent CAE

Colostrum is sensitive to heat. If it gets too hot, it turns into a thick pudding and its antibodies are destroyed.

To properly heat colostrum, put it into a double boiler (to prevent scorching), clip on a candy thermometer, and stir steadily. Have a thermos bottle filled with hot water ready. While heating colostrum, do not remove the thermometer or stirring spoon and then return it to the pot, or you may reinfect the colostrum. When the colostrum reaches 135°F, pour the water out of the thermos and pour in the colostrum. Screw on the lid and wrap the thermos in towels.

After 1 hour, open the thermos and test the colostrum with a clean thermometer. The temperature must be no less than 130°F. If the temperature dips below 130°F before 1 hour is up, you'll have to start over.

Milk is far easier to treat than is colostrum, since pasteurization destroys the CAE virus. Directions for pasteurizing milk are given on page 223.

Cool the colostrum or milk to 100°F before feeding it to the kids. The temperature is just right when you can't feel a drop placed on the inside of your wrist. Some pasteurizers have a setting for colostrum, making heat treatment easy and foolproof.

are like puppies — always happy to see you. Kids raised by a doe, on the other hand, tend to be shy unless you take time every day to work at socializing them.

Another reason to bottlefeed kids is to keep some of the doe's milk for yourself. An alternative to bottle feeding for this purpose is to house the doe at night separately from the kids when they reach 2 weeks of age. Milk the doe in the morning for yourself, then put the kids in with her so they can nurse the rest of the day.

Yet another reason to bottlefeed kids is to avoid spreading caprine arthritis encephalitis (CAE), a virus for which no cure and no preventive measures are known except to break the disease cycle. (The disease does not affect humans.) The best way to avoid this virus is to purchase only certified CAE-free goats. A doe that is not certified may be infected with the virus and pass it on to her kids. Kids usually get CAE by nursing from an infected doe (although they may also be infected before they are born or through contact with infected goats after birth).

Bottlefeeding kids colostrum and milk known to be CAE-free helps break the disease cycle. Colostrum and milk from cows, as well as milk replacer available from the farm store, are virus free and may be fed to kids instead of goat milk. Kids raised on cow milk or milk replacer, however, don't grow as well as kids raised on goat milk. CAE is more likely to be transmitted through colostrum than through milk. Goat colostrum and milk must be heated to destroy the CAE virus.

Each kid should get milk or milk replacer amounting to not less than 15 percent nor more than 25 percent of its body weight each day. Kids kept at the maximum of 25 percent of their body weight will grow fast and sleek. Kids kept closer to the 15 percent minimum will start nibbling on hay and other forage earlier and be easier to wean.

To make sure kids are getting the right amount of milk, weigh them at least every other day and increase their ration accordingly. Divide the total daily amount into four or five evenly spaced feedings for the first couple of days. By the time the kids are 1 week old, they may be fed three or four times daily. Gradually work down to two feedings per day at evenly spaced intervals.

Kids may be fed from individual bottles or from a community bucket with several nipples around the bottom. Using separate bottles ensures that each kid gets the right amount. If you have more kids than you can feed with a bottle in each hand, you can make or purchase a rack that holds multiple bottles, allowing all kids to nurse at the same time. A bucket is easier to handle than multiple bottles and reduces the amount of time needed to clean the bottles, but slow eaters may not get their full share.

To keep track of feeding times and amounts, make a chart showing what time the kids are scheduled to be fed and how much each should get. Update the chart each time you weigh the kids. Place a checkmark on the chart after each feeding — an important procedure if you are busy and forgetful, or if more than one person helps feed the kids.

Weaning

When a kid is born, only one of its four stomach chambers — the abomasum — functions. A newborn kid therefore digests milk like a puppy, kitten, or human baby with a single-chamber stomach. By the time the kid is 1 week old, it begins nibbling on hay, grain, and grass. The more solid feed it eats, the more quickly the three other chambers develop. Keeping bottlefed kids on the hungry side encourages early eating of solid foods, provided free-choice hay or browse is available. As soon as your kids start eating solid foods, make sure they have access to clean drinking water at all times. They may not drink any at first, but they should have the option to do so.

How much concentrate kids need is influenced by the quality of their hay; the lesser the quality, the more concentrate they need for proper nutritional balance and growth. The proper amount for each kid also depends on the animal's size and growth rate. Cut back on the concentrate if a kid is getting too fat — if you can't feel the ribs of a large-breed kid or you can grab a handful of flesh from behind the elbow of a miniature breed. Like milk, concentrate should be divided into two evenly spaced feedings. As a general guideline, by weaning time, regular-sized kids may be gradually worked up to 1 pound of concentrate per day, miniatures to ½ pound.

Most kids no longer need milk by the time they reach three times their birth weight or 8 weeks of age, whichever comes first. Angoras grow more slowly than other breeds and are usually not weaned until they reach 4 months of age.

Weaning a bottlefed kid encourages early rumen development and frees up your time for other things. To start weaning bottlefed kids, substitute water for a small portion of the milk. Gradually decrease the amount of milk and increase the amount of water. The kids will be weaned without even noticing.

If your kids nurse, weaning can be upsetting for the kids, the doe, and you. When you separate a kid from its mother, both parties will carry on dramatically. They won't get so upset if you put them in side-by-side stalls where they can see each other. After a few weeks, the kids will be weaned and may be put back with their mother.

If you raise meat goats or fiber goats by nursing, don't worry about weaning. Your doe will produce only as much milk as her kids need. When her milk is no longer needed, she will dry off naturally and her kids will be automatically weaned.

Raising Kids for Meat

When raising kids for meat, your goal is to get the fastest weight gain at the least cost. The meat that is the least expensive to produce and has the mildest flavor comes from milk-fed kids that are 6 to 8 weeks old and weigh less than 35 pounds.

Raising grain-fed kids is more costly, but their meat is more flavorful. In addition to their milk ration, these kids get a small amount of grain from the age of 6 weeks to about 12 weeks, when they weigh 50 pounds or more and are ready for market.

The least expensive way to get full-flavored meat is to wether your bucklings and let them nurse as long as they like (and the doe will allow), and then put them on pasture. By the time a wether is 1 year old, it should weigh 80 pounds or more.

If you are raising Boer goats for meat, the best marketable live weight is about 85 to 95 pounds, yielding a carcass weighing 45 to 50 pounds.

Although goat meat is delicious, it has never caught on in mainstream America, perhaps in part because the English language has no one-word name for it. "Goat meat" is a more direct term than such words as "beef" and "pork," and people in the goat meat industry are trying to come up with a suitable name for their product. Meanwhile, those who market goat meat borrow words from Spanish or French. The meat of a 2- or 3-month-old kid weighing less than 50 pounds is called cabrito or chevrette. The meat of a 1-year-old wether weighing about 80 pounds is called chevon. Meat from older goats is called chivo or mutton.

For information on butchering goats, cuts of meat, and meat storage, consult a comprehensive book (see Recommended Reading, page 379).

Tracking Weight

The average newborn kid weighs about 7 pounds. Doelings may weigh less, bucklings more. Triplets and quads weigh less than twins or singles. Miniature kids weigh about half as much as full-size kids. Weigh each kid at birth, as soon as it is dry but before it has its first meal. Track each kid's growth by weighing it every other day for the first 4 weeks, then once a week until it reaches maturity. Record the dates and weights on a chart. (For instructions on how to weigh a kid, see page 209.)

After the first week, a kid should gain ¼ to ½ pound per day. Some grow faster than average; some grow slower. Except briefly during weaning, at no time should a kid lose weight.

If a kid fails to gain weight or loses weight, look for a reason. Perhaps it is not getting enough milk. If the kid is nursing, make sure that the mother is producing sufficient milk. Check her udder: Perhaps a teat is plugged, or the udder is infected and sore, causing the doe to push the kid away when it tries to nurse.

If the doe is nursing more than two kids, she may not have enough milk for them all. A strong kid may push a weaker kid aside at nursing time. In that case, you may have to bottlefeed the slow-growing kid.

If a kid you are bottlefeeding grows slowly, you may not be feeding it enough. Charts apply only to the average goat. In real life, no average goats exist. Make adjustments to suit the needs of your individual animals.

Take care not to go overboard by feeding your kids too much milk at once, or they will get diarrhea. Diarrhea, also known as scours, is a dangerous cause of weight loss.

Scours

A kid's first bowel movement is black and sticky. Then, for the next few days, the kid passes yellow, pasty material that may stick to the hair around the hind legs and rear end and need to be cleaned off. When a kid is about 1 week old, it starts dropping small brown pellets that resemble the droppings of mature goats, only smaller. Droppings that are loose and white, light yellow, or light brown indicate diarrhea.

Diarrhea, or scours, is a fairly common problem, especially during a kid's first few days of life. It more often strikes bottlefed kids than kids that nurse naturally. Scours may be caused by chilling, erratic feeding, dirty bedding, dirty milk bottles, overeating, and milk that is too rich. If the problem is not corrected immediately, the kid will die.

Avoid scours by washing bottles and nipples after each feeding and rinsing them in warm water mixed with a splash of household chlorine bleach. Feed kids small amounts at a time, and space feedings evenly throughout the day.

If a kid gets diarrhea, stop feeding milk. Substitute an electrolyte fluid in an amount equal to the amount of milk the kid would otherwise drink. Some goat supply outlets sell electrolytes mixed specifically for goats. In a pinch, you can use Gatorade or mix up a homemade electrolyte drink from the following ingredients:

- 2 tablespoons table salt (sodium chloride)
- 2 tablespoons Lite salt (potassium chloride)
- 2½ gallons water

The scours should clear up within 2 days of substituting electrolyte fluid for milk. If not, consult your veterinarian. The kid may have a disease.

Scouring in a 3- or 4-week-old kid is probably due to coccidiosis, caused by microscopic parasites called coccidia that are always present in the soil. Properly managed kids are exposed to coccidia gradually and develop immunity. Kids that live in filthy conditions or that must drink water with manure in it are exposed to too many parasites at once. The main symptom of coccidiosis is diarrhea, which is sometimes tinged with blood. Treatment is with a coccidiostat; this medication is available from farm stores, goat supply outlets, and veterinarians. Even if a kid recovers from coccidiosis, it may never grow to be fully productive. Since coccidia flourish in warm, humid weather, careful scheduling of breeding may help prevent infection. Manage your does to give birth during the cold days of winter, which gives the kids time to grow and develop immunity before warm weather comes.

Identification

Even if your goats are not registered, each animal should have a unique means of identification to help you sort out the look-alikes and help your veterinarian keep track of health tests. Federal regulations require the identification of goats in certain categories (see the box on the facing page). If you plan to do any pedigree breeding or to register your goats, you need a way to positively identify each animal. Different registries recommend different systems of identification, so before you mark your goats, check with the appropriate breed organization. Some registries recommend using two methods to ensure that at least one endures for the animal's life. The most common forms of identification are ear tags, tattoos, and ear notching.

Ear Tags are commonly used for meat and Angora goats. Tags are sold at farm supply stores; they come in plastic or metal and in a variety of shapes and sizes. Goats

Identification Regulation

The United States Department of Agriculture requires certain categories of goat to bear some means of identification, especially if they are involved in interstate travel. The purpose of this program is to eradicate a relatively rare disease called scrapie.

Ear tags are the preferred method of identification, although some states allow tattoos. For details, contact your state's U.S. Department of Agriculture (USDA) Animal and Plant Health Inspection Service office, which you may locate by calling the federal office toll free at 1-866-873-2824. Information is also available online at www.aphis.usda.gov/vs/nahps/scrapie.

are so active that some styles are easily torn off, injuring the goat's ear. The safest tags for goats are round button tags that attach to the center of the ear, where they are less likely to be torn off than tags that attach to the edge. Another option that is easier to read than a button tag is a small rectangular tag that clips to the front of the ear.

Ear tags come in various colors and may be preprinted or blank. Depending on the manufacturer, preprinted tags may be numbered sequentially (you assign the next available number to the next kid born) or may include other information, such as your name or initials, your herd or farm name, or your city and state. If you use blank tags, you must mark them yourself with an indelible marker. Over time, the information wears off; unless you check often, the original information may be lost before you get around to renewing the markings.

For your goats to comply with federal and state scrapie regulations, you must use USDA-approved ear tags. Along with any other information you wish to include, each tag must bear your premises number (which you will be assigned) and the animal's unique identification number.

Tattooing. Proper tattooing as a means of permanent identification is used for dairy goats, and sometimes in combination with ear tags for meat and Angora goats. Tattoo kits with goat-sized numbers and letters are available from any goat supply catalog. Instructions come with the kit. Additional instructions may be obtained from registry associations. All dairy breeds, except LaMancha, are tattooed in the ear. Since LaManchas have no external ears, they are tattooed in the tail web, the hairless area under the tail.

A common system for dairy goats is to tattoo the right ear with three letters designating the herd name. If you join the American Dairy Goat Association, three letters will be assigned to you. If your herd designation is XYZ, for example, all your kids will have "XYZ" tattooed in their right ears.

The left ear has a letter indicating the year of birth and a number indicating the kid's birth sequence. All kids born in 2004, for example, would have the letter T. The first kid born in your herd in that year would be "T1," the second kid "T2," and so forth.

Not all breeds have such well-organized tattooing systems, nor do they all tattoo both ears. Angora herd owners, for instance, make up their own herd's numbering system. Whereas no two dairy goats could have the same tattoo combination, Angoras belonging to neighbors may conceivably be tattooed with the same identification

Tattoo Year Code

R = 2002
S = 2003
T = 2004
V = 2005
W = 2006
X = 2007
Y = 2008
Z = 2009
A = 2010
B = 2011
C = 2012
D = 2013
E = 2014
F = 2015
H = 2016
J = 2017
K = 2018
L = 2019
M = 2020

(To avoid confusion, G, I, O, Q, and U are not used.)

number. Furthermore, some owners tattoo the left ear and some tattoo the right, so you have to look in both ears before you can be sure the goat is tattooed at all.

If your goats must comply with federal and state scrapie regulations, you may be allowed to identify them with tattoos. Regulations vary from state to state; find out what regulations pertain in your state.

If you expect few kids each year, you may not wish to purchase your own tattoo kit or ear tag applicator. Goat clubs sometimes hold demonstrations where you can take your kids to be tagged or tattooed. A neighboring goat keeper may be willing to tag or tattoo your kids for a small fee.

Ear notching. Angora goats may have their ears notched as the sole means of identification or in combination with being tagged or tattooed. Notches are made using a hog ear notcher, available from farm stores. Each notch location is associated with a numeric code, as recommended by the American Angora Goat Breeder's Association and shown in the illustration above. The number of notches appearing in any one location on the goat's ear is multiplied by the numeric code for that location. The sum obtained by adding the various notched locations together equals the animal's identification number, which may or may not be tattooed in the goat's ear.

Referring to the drawing, let's say a goat has two notches in the middle of its left ear (2 x 1,000), two notches on the inside of its left ear (2 x 10), and three notches on the outside of its right ear (3 x 3). These notches would correspond to an identification number of 2,029. The numeric code allows you to be economical with your notching; note, for instance, that one notch on the inside of the right ear (30) has the same numeric value as would three notches on the inside of the left ear (3 x 10).

No system of identifying the goats in your herd makes sense unless you keep careful records for each animal. Your records should include information and identification details not only for the goats currently in your herd but also for those you no longer have, including how they were disposed of, such as to whom they were sold and when.

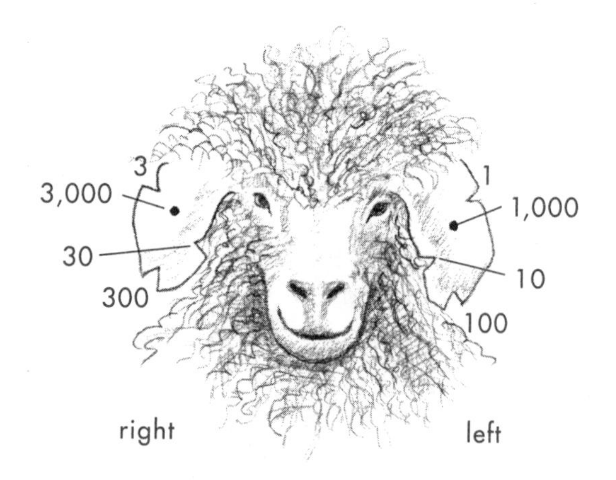

Angora ear notches

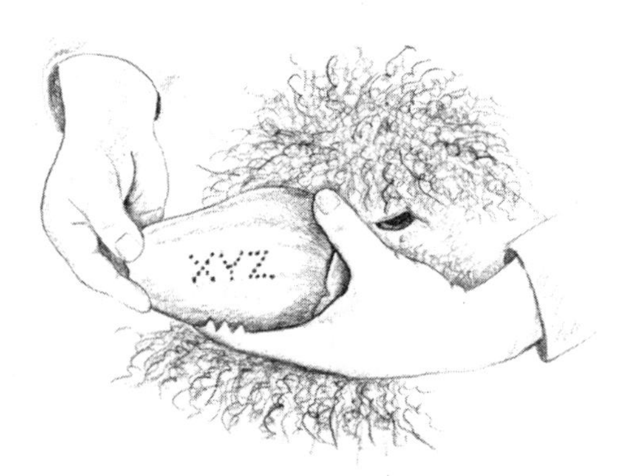

Angora, notched and tattooed

Disbudding

All kids are born without visible horns. Some kids have horn buds that develop into horns. Other kids are polled, meaning they have no horn buds. You can tell the difference between a horned kid and a polled kid from birth. The wet hair on the head of a polled kid lies smooth, whereas the hair on a horned kid is twisted at the two spots on its head where horns will grow.

Wild goats use their horns for protection. When goats are raised in a barn, their horns can become dangerous weapons. Without meaning to, a goat with horns can injure herdmates, or you, simply by lifting or turning its head at the wrong time. For that reason, a dairy goat kid born with horn buds should be disbudded. Calmer breeds such as Angora and African Pygmy, and goats that roam outdoors in herds, usually have their horns left on.

Disbudding should be done as soon as you see little horns, since the procedure becomes more difficult as the horns grow. Different breeds grow horns at different rates. Most dairy goat kids are ready for disbudding between the ages of 1 and 2 weeks. Older goats may have their horns surgically removed by a veterinarian, but the procedure is painful and the resulting wound takes a considerable amount of time to heal. Disbudding is painful in young kids as well, but the kids forget so quickly afterward that within a few minutes they may start butting heads in play.

Disbudding requires a disbudding iron and a box to hold the kid. For just a few kids each year, you may wish to invest in these items jointly with another goat keeper, or you might seek a nearby goat keeper who is willing to disbud your kids for a small fee. Often the same person who tattoos kids will also disbud them.

Do not attempt to disbud a kid without first having an experienced person show you how. Use the following tips to help the kids through the ordeal:

- Give each kid a baby aspirin, or half an adult aspirin, to dull the pain.
- Place a small bag of crushed ice on the kid's head immediately after both buds have been burned.
- If you are disbudding bottlefed babies, give each one a small amount of warm milk as soon as you remove it from the disbudding box.
- To prevent infection, give each disbudded kid an injection of tetanus antitoxin — 500 IU under the skin. One bottle holds 1,500 IU, or enough for three kids. Tetanus antitoxin is available from any farm store or veterinarian.

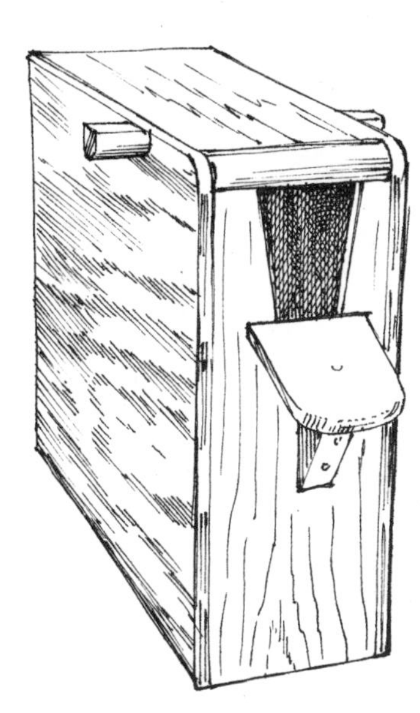

A disbudding box may be purchased or made and used to confine a kid for tattooing and disbudding.

Castrating Bucklings

Unless you have a prime buckling you plan to keep or sell for breeding, any buckling kept past weaning age should be castrated. Because a castrated buck is called a wether, castrating is sometimes also called wethering. A wether makes a better pet than an intact buck; is easier to train for pack or draft; and, if raised for meat, grows faster and tastes better. Castrating your bucklings simplifies management, since wethers need not be separated from does or doelings to avoid unwanted mating.

Castrate a buckling as soon as his testicles descend into the scrotum, usually between the ages of 1 and 3 weeks. To do the job, you will need a 9-inch lamb Elastrator. Before trying it for the first time, have an experienced person show you how to use it. If you raise only a few kids each year, you might seek a nearby goat or sheep owner willing to castrate your bucklings for you.

Saying Goodbye

One of the hardest things about raising goats is giving up the kids. It's easy to get attached to the cute little guys. But unless you have a plan for disposing of excess kids, you will soon find yourself with more goats than you need, want, or can afford to keep in terms of management time and the cost of expanding your goat facilities. The longer you keep kids and get to know their individual personalities, the more difficult becomes the decision about which ones to sell.

Your plan for disposing of excess kids might include contacting a local meat goat buyer about selling your kids when you can expect the greatest return — usually around the Christmas and Easter holidays. If you are selling goats to other breeders, they may wish to arrange to purchase a buckling or doeling from a certain mating before the kids are even born. By having a plan, you will find it easier to give the kids up, because you'll know from the start they aren't yours.

Of course, if you plan to expand your herd or replace older or less productive animals, you have the happy situation of getting to keep at least some of the kids your does produce.

Keeping Goats Healthy

Each goat herd is unique in its combination of breed, age, local climate, prevalent pathogens, and other elements that influence health. Your health care plan must therefore fit your particular situation and cannot be developed by blindly following a formula established for another herd elsewhere.

The most important health care measure is disease prevention, which must start when you acquire your first goats. Purchase only healthy animals. If you feel unsure about how to tell a healthy goat from a sick one, take along an experienced person to coach you. You'll avoid a lot of future expense and heartbreak.

Sickness in goats is usually caused by poor management. Good management includes protecting your goats from illness and watching for early signs of problems. Take care not to allow your goats to become crowded, because crowding causes stress and stress decreases resistance to disease. Transporting and showing goats are stressful and may bring your goats into contact with potentially unhealthy animals. If your goats regularly come into contact with others outside your herd, you will need a stronger routine health care program and greater attention to keeping vaccinations current than someone whose herd is isolated from other goats.

All herds require a health care routine that includes weighing, hoof trimming, and regular worming, as well as vaccinations and booster shots as determined by the diseases that are prevalent in your area. Consult your veterinarian and other goat owners to develop a suitable health care program for your herd.

Measuring heart girth

Weighing

Each time you have a goat confined for hoof trimming, weigh it and record the date and weight on a chart. Your weight records will tell you such things as that a young goat is growing properly, a young doe is big enough to breed, or a pregnant doe is getting enough to eat. Loss of weight is often the first sign that something is wrong.

Estimating Your Goat's Weight

Heart Girth (in inches)	Weight (in pounds)	Heart Girth (in inches)	Weight (in pounds)	Heart Girth (in inches)	Weight (in pounds)
10.75	5	21.75	37	32.75	105
11.25	5.5	22.25	39	33.25	110
11.75	6	22.75	42	33.75	115
12.25	6.5	23.25	45	34.25	120
12.75	7	23.75	48	34.75	125
13.25	8	24.25	51	35.25	130
13.75	9	24.75	54	35.75	135
14.25	10	25.25	57	36.25	140
14.75	11	25.75	60	36.75	145
15.25	12	26.25	63	37.25	150
15.75	13	26.75	66	37.75	155
16.25	14	27.25	69	38.25	160
16.75	15	27.75	72	38.75	165
17.25	17	28.25	75	39.25	170
17.75	19	28.75	78	39.75	175
18.25	21	29.25	81	40.25	180
18.75	23	29.75	84	40.75	185
19.25	25	30.25	87	41.25	190
19.75	27	30.75	90	41.75	195
20.25	29	31.25	93	42.25	200
20.75	31	31.75	97		
21.25	35	32.25	101		

Small kids are easy to weigh. Pick up the kid and hold it while you stand on a bathroom scale. Then weigh yourself without the kid. Subtract the second number from the first; the difference is the kid's weight.

If you raise dairy goats, use a dairy scale to weigh kids. Place a kid in a grocery bag with handles and hang the bag from the scale. Be sure to keep the kid's head outside the bag so it can breathe.

When a goat gets too heavy to lift, the best you can do is estimate its weight. Even though the estimate is not 100 percent accurate, it will tell you whether the animal is gaining or losing weight. To estimate a weight, measure the goat's heart girth — the distance around the goat's middle, just behind its front legs, over its heart. Measure an Angora after shearing, or its long hair will make it seem heavier than it really is. Have the goat stand on a level surface with its legs solidly beneath it.

Use a dressmaker's tape measure or a weigh tape from a goat supply catalog. The weigh tape automatically converts heart girth to estimated weight. If you use a dressmaker's tape, the chart above will help you convert inches to pounds.

Hoof Trimming

Hooves are made of keratin, the same material your fingernails are made of. Like fingernails, hooves grow uncomfortably long if they aren't trimmed. Wild goats generally live in rocky areas, where abrasion wears down their hooves as they move about seeking fresh forage. When a goat spends its days in a barn or on pasture, however, its hooves keep growing. Unless they are trimmed, eventually the goat will be unable to walk properly. Hooves left untrimmed for too long fold under, trapping dirt and moisture that create an ideal environment for bacteria that can cause foot rot. Excessive growth also alters an animal's stance, eventually crippling the animal permanently.

To learn what a properly trimmed hoof looks like, study the feet of a newborn kid. Note how its hooves are flat on the bottom and look boxy. When a hoof has been properly trimmed, the bottom is parallel to the growth rings and the two toes are of equal length.

How often you need to trim hooves depends on how fast they grow. Some hooves need trimming every 2 weeks; others may not need to be trimmed more often than every 2 months. The rate of growth of a particular goat's hooves is influenced partly by genetics and partly by environment. Like wild goats, domestic goats can wear off some hoof growth if they have access to an abrasive surface, such as a rock outcrop or a concrete pad on which to play.

A goat that is not used to having its hooves trimmed may struggle and kick. Avoid this problem by training kids to the procedure early and by continuing to check and trim hooves often throughout each animal's life. Frequently trimming small amounts at a time is easier than infrequent major trimming.

You will need a pair of sharp shears, such as good garden pruning shears or hoof trimmers designed specifically for the purpose; the latter are available from any livestock supply store or catalog. Keep the shears sharp by periodically rasping the blades with a file.

You'll need to confine the goat so it doesn't attempt to wander away while you're trimming its hooves. If you have dairy goats, use the milk stand to hold each goat during the trimming. Otherwise, fasten the goat's collar to a wall or a fence and crowd the animal with your body so it can't move around. Face the goat in one direction to trim the two hooves away from the wall, then turn it around to trim the other two hooves.

While a kid is still small, you can trim its hooves by turning it on its back and holding it firmly on your lap. A kid's hooves shouldn't need much trimming, but starting early is a good idea to get the animal accustomed to the procedure.

Angoras, too, may be set on their backs for hoof trimming. Place the Angora on a clean floor with its head between your legs. An Angora will lie quietly when set on its rump — a position used to shear mohair as well as to trim hooves. However, don't try this with a mature dairy goat or meat goat, or you're likely to get kicked in the face.

Trim when hooves have been softened by rain or dewy grass. To prevent injury to the goat, work in good light and trim away a small amount at a time. Here's the procedure:

1. Grasp one leg by the ankle and bend it back to see the bottom of the foot.

2. Use the point of the shears to scrape away dirt (A).

3. Cut off long growth at the front of the hoof (B).

4. Snip off flaps that fold under the hoof.

5. Trim the bottom, one tiny slice at a time, cutting toward the toe. Stop trimming when the hoof looks pink, indicating that you are getting close to the blood supply. Do not keep cutting, or the foot will bleed. If that happens, pour hydrogen peroxide over it.

6. If you reach the blood supply before an overgrown hoof is properly trimmed, avoid causing damage by continuing the trimming a few days later.

7. Optional: Smooth the hoof with a woodworker's rasp (C).

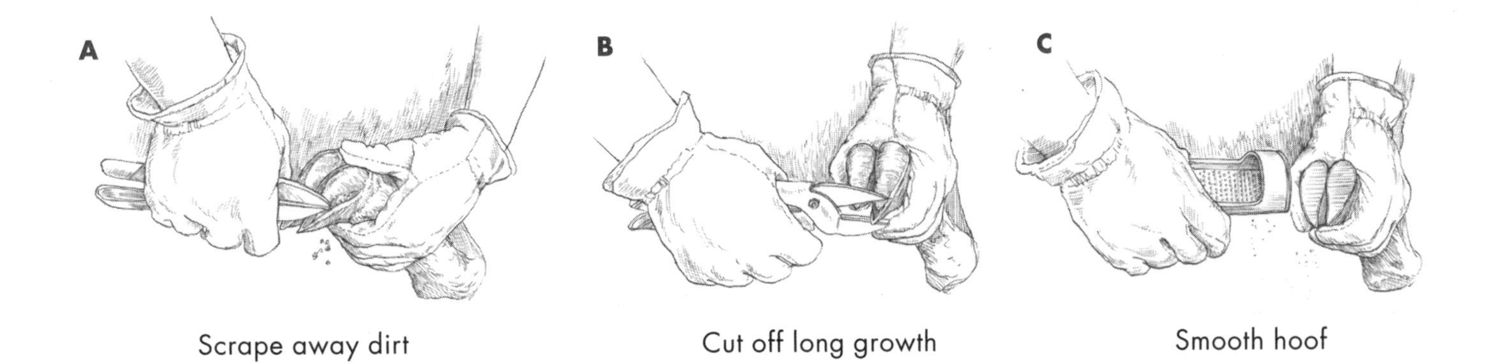

Scrape away dirt — Cut off long growth — Smooth hoof

Signs of Illness

Take time to study your goats while they are healthy so you will readily notice any changes in the way they look, eat, or move. The sooner you realize a goat is getting sick, the quicker you can do something about it, and the greater the chance it will recover.

Notice the size, shape, firmness, color, and smell of your goats' droppings. Any change may indicate a dietary imbalance, the beginning of a disease, or an infestation of parasites. Parasites may be internal, such as worms and coccidia, or external, such as lice and ticks.

Listen for teeth grinding, a sign the goat is in pain. Look for changes in the color of the gums and the lining around the eyes. These areas should be bright pink. If a goat is in shock or has lost blood, these areas may turn pale. A purple or blue color may indicate damaged airways or other breathing problems. If the color is pale gray or blue and the goat has a hard time breathing, call your vet immediately.

Before calling a veterinarian about any goat's health, first take the animal's temperature and be prepared to relay the information to the vet. Inexpensive digital thermometers, complete with instructions for their use, are available through veterinary supply catalogs. Restrain the goat as you would for hoof trimming and insert the thermometer into its rectum. If you know the animal's normal temperature at rest, you will have a better idea when its temperature is abnormal. A good idea is to take each goat's temperature whenever you weigh the animal, and record the temperature along with the weight. Any time you note a change from normal, look for the reason.

Before calling the vet, you might also take the goat's pulse by using one of the following two methods:

- Place your fingertips on both sides of the lower rib cage and count the beats for 1 minute.
- Place your finger on the big artery on the upper inside part of one of the rear legs and count the beats for 1 minute.

Signs of Health in Adult Goats*

Pulse rate: 70 to 80 beats per minute

Breathing rate: 12 to 20 breaths per minute

Rumen movements: 2 or 3 every 2 minutes

Rectal temperature: 101.5°F to 105°F

*Kids have higher pulse and breathing rates than adult goats

Signs of Illness in an Adult Goat

Behavior	Inactive Grinds teeth Makes complaining sounds Coughs frequently Takes quick, shallow breaths Any change in normal behavior
Digestion	Eats less or not at all Urinates more or less than usual Manure changed in color or consistency
Milk	Inexplicable drop in production Change in color, odor, or consistency
Coat	Becomes rough or dull Hair falls out or scabs appear Goat scratches or bites itself
Body temperature	Above or below normal

Common Goat Problems

If one of your goats becomes sick, don't try to treat it yourself unless you know for sure what the problem is. Giving medications incorrectly can do more harm than good. The first time you give your goat a shot or a drench (liquid medication given through the goat's mouth), have an experienced person show you how.

The following list describes the most common problems in goats. Alternative names are included to help you discuss the problem with others or look up the condition in a comprehensive veterinary manual, such as *The Goat Health Handbook*.

Abscesses

Also called: Casseous lymphadenitis, pseudotuberculosis.
Symptoms: Firm swellings under the skin.
Cause: Bacterial infection of lymph nodes by *Corynebacterium pseudotuberculosis*. Bacteria most likely enter through skin wounds.
Prevention: Provide a safe environment to reduce the risk of skin wounds (see page 179). Cull affected animals.
Treatment: Antibiotics are not effective. Surgical removal can be attempted for valuable animals.

Bloat

Also called: Acidosis or ruminal tympany.
Symptoms/effects: Swelling on left side, kicking at stomach, grunting, slobbering, lying down and getting up; can lead to death.
Cause: Excess gas in rumen.
Prevention: Feed balanced rations. Make any dietary changes gradually. Prevent overeating of concentrate or lush pasture.

Treatment: Keep the goat on its feet (propped between hay bales, if necessary). Rub its stomach to eliminate gas. Drench with 2 cups of mineral oil followed by ¼ cup of baking soda dissolved in 1 cup of water. Call veterinarian immediately.

Caprine Arthritis Encephalitis
Also called: CAE or CAEV.
Symptoms: Weak rear legs in kids; stiff, swollen knee joints in mature goats.
Cause: Infection by caprine arthritis encephalitis virus.
Prevention: Purchase certified CAE-free goats. Do not feed kids raw colostrum or milk from infected does.
Treatment: No effective cure.

Chlamydiosis
Also called: Chlamydial abortion, enzootic abortion.
Symptoms: Abortion during last 2 months of pregnancy, or kids stillborn or very weak.
Cause: Infection by *Chlamydia psittaci.*
Prevention: Vaccinate. Avoid contact with infected goats. Burn or deeply bury dead newborn kids and afterbirth (or submit to a pathology lab for diagnostic testing); isolate aborting does until birth discharge stops. Antibiotic as prescribed by a veterinarian may control the spread of this disease to uninfected does.
Treatment: None.
Human health risk: Contagious to humans. Wear plastic gloves when attending a birth or disposing of birthing materials.

Coccidiosis
Also called: Cocci, coccidia.
Symptoms/effects: Loss of appetite, loss of energy, loss of weight, diarrhea (sometimes bloody); can cause death.
Cause: Infection by Eimeria species of protozoal parasites.
Prevention: Keep bedding, feeders, and water pails scrupulously clean.
Treatment: Coccidiostat (sulfa drug) used as directed on the label.

Enterotoxemia
Also called: Clostridium perfringens infection, overeating disease.
Symptoms/effects: Twitching, swollen stomach, teeth grinding, fever; can lead to death.
Cause: Infection by *Clostridium perfringens* bacteria.
Prevention: Avoid abrupt changes in diet. Vaccinate.
Treatment: No effective cure.

Foot Rot
Also called: Hoof rot.
Symptoms/effects: Lameness, ragged hoof, grayish discharge, smelly feet, foot deformity, loss of weight, tetanus; can cause death.
Cause: Infection by *Fusobacterium nodosum* bacteria.
Prevention: Keep bedding clean and dry. Trim hooves regularly.
Treatment: Trim hoof to healthy tissue. Soak foot in copper sulfate solution (½ pound per gallon of water) for 2 minutes. Severe cases may require antibiotics, as recommended by your veterinarian.

Ketosis

Also called: Pregnancy disease, pregnancy toxemia (the disease occurs in does just before or after kidding).
Symptoms/effects: Sweet-smelling breath, urine, or milk; loss of appetite; doe lies down and can't get up; can cause death.
Cause: Metabolic disorder triggered by a sudden change in diet, overfeeding during early pregnancy, or underfeeding during late pregnancy.
Prevention: Proper feeding.
Treatment: Drench with 1 tablespoon of baking soda dissolved in ¼ cup of water, then drench with 2 ounces of propylene glycol (or, in a pinch, 1 cup of honey or corn syrup) twice daily. If the doe is weak or can't swallow, call vet immediately.

Lice

Also called: Pediculosis, body lice, external parasites.
Symptoms: Scratching, loss of hair, loss of weight, reduced milk production.
Cause: Contact with infested animals, living in damp housing.
Prevention: Avoid contact with infested animals. Spray or dust fiber goats 6 weeks after every shearing.
Treatment: Powder, dip, spray, injectable, or pour-on insecticide approved for livestock. (For milk goats, use an insecticide approved for dairies.) Repeat in 2 weeks to kill newly hatching lice eggs. If infestation is severe, repeat again in 2 weeks.

Mastitis

Also called: Infected udder.
Symptoms: Doe stops eating; may have a fever; udder is unusually hot or cold, hard, swollen, or painful; milk smells bad or is thick, clotted, or bloody (slightly pink milk at freshening is normal).
Cause: Various bacteria, often following injury or insect sting to the udder.
Prevention: Keep bedding clean. Remove objects from facilities that could damage the udder. At each milking, check udder and milk for symptoms. Use proper milking techniques, and apply a teat dip after milking. At least once a month, check the first stripping from each side with a mastitis test. The California Mastitis Test kit is sold by most farm stores and dairy suppliers. The test is designed for cows, which have four teats, so you will use only two of the four cups in the kit.
Treatment: Isolate the doe. Apply hot packs four or five times a day. Milk three times a day. Milk the infected doe last. Contact your veterinarian about antibiotic treatment.
Human health risk: Do not drink infected milk or feed it to goat kids. Do not drink milk from treated does until the time span specified on the drug label or by your veterinarian has passed.

Orf

Also called: Contagious ecthyma, sore mouth.
Symptoms/effects: Crusty sores on the lips and muzzle of young animals can cause difficulty nursing and eating and loss of condition.
Cause: Viral. The virus can survive for years in the environment.
Prevention: A vaccine is available but should be used only in infected herds under a veterinarian's supervision.
Treatment: Sores will heal by themselves in 1 to 4 weeks. Give extra milk or feed to kids that aren't eating well.
Human health risk: This virus will also cause skin infection in people. Use gloves when handling affected kids and sterilize nipples, bottles, and feed tubs after use.

Pinkeye

Also called: Conjunctivitis, infectious keratoconjunctivitis.
Symptoms/effects: Red-rimmed, watery eyes; squinting; may cause blindness.
Cause: Various bacteria; less commonly viruses, or other microorganisms.
Prevention: Avoid dust, eye injuries, and contact with infected goats.
Treatment: Antibiotic drops or ointment under the eyelids. Treat all goats, even those without symptoms.

Plant Poisoning

Also called: Toxic reaction.
Symptoms/effects: Frothing at the mouth, vomiting, staggering, trembling, crying, rapid or labored breathing, altered pulse rate, convulsions; can lead to death.
Cause: Ingestion of plants sprayed with pesticides or toxic plants such as black nightshade, bracken fern, death camas, hemlock, horse nettle, laurel, Japanese yew, milkweed, oleander, and rhododendron, as well as the wilted leaves of trees that produce stone fruit — cherry, peach, and plum. Limp leaves that are still green or are partially yellow are the most dangerous; fully dried leaves are no longer toxic. Ask your county Extension agent for an illustrated list of poisonous plants in your area.
Prevention: Feed a balanced diet, including free-choice hay. In autumn, keep goats away from stone-fruit trees that are dropping their leaves.
Treatment: Try to figure out what the goat ate. Place 2 tablespoons of salt on the back of the goat's tongue to induce vomiting. Call the veterinarian.

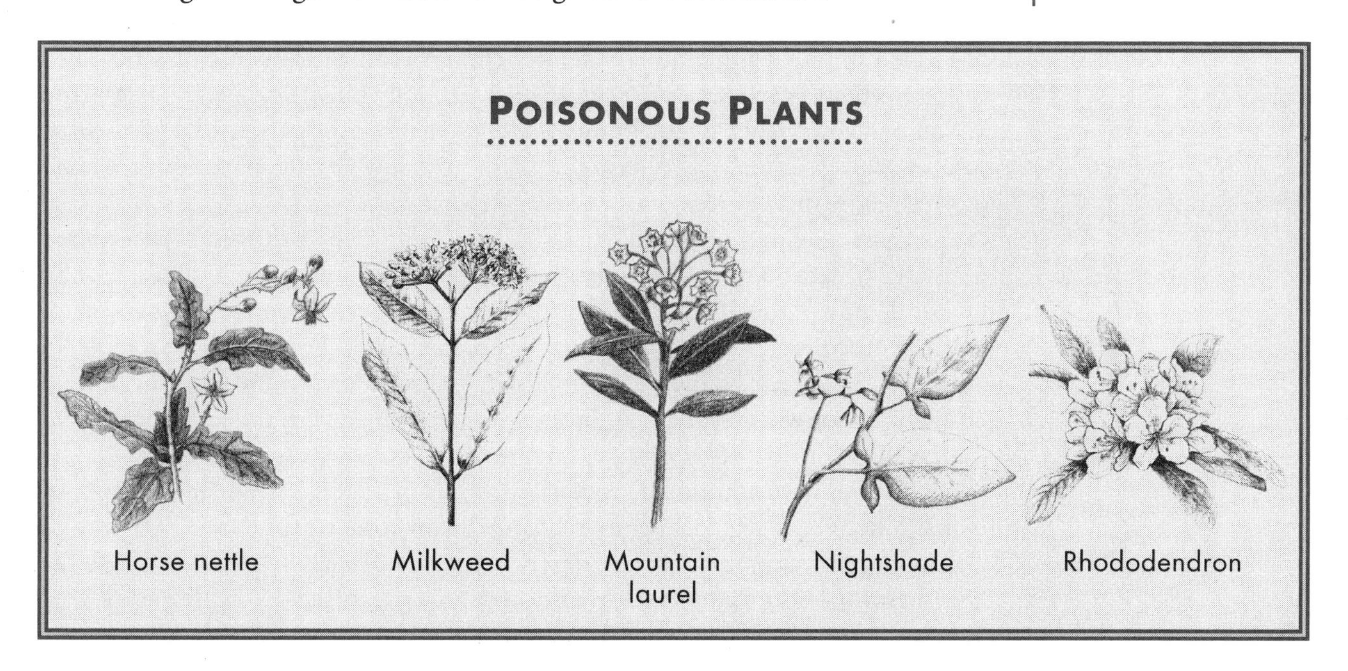

Pneumonia

Also called: Lung sickness.
Symptoms/effects: Coughing, runny eyes and nose, fever, loss of appetite, fast breathing, fever; may cause death.
Cause: Various bacteria and viruses (usually after exposure to drafts and dampness), parasites, allergies.
Prevention: Provide dry, draft-free housing with good ventilation. Do not heat housing.
Treatment: Contact your veterinarian about antibiotic treatment.
Human health risk: None.

Ringworm

Also called: Dermatophytosis.
Symptoms: Circular hairless patch, usually on the head, ears, or neck, sometimes on an udder.
Cause: Various fungi in the soil.
Prevention: Avoid contact with infected animals.
Treatment: Scrub area with soapy water, and coat with iodine or fungicide (be careful not to get any in the goat's eyes).
Human health risk: Can infect humans. Wash your hands after handling infected animal.

Scours

Also called: Diarrhea.
Symptoms/effects: Watery, bad-smelling diarrhea, loss of appetite, loss of energy; can cause death.
Cause: Various bacteria.
Prevention: Feed kids properly. Disinfect containers after each feeding, especially when feeding milk. Keep housing clean.
Treatment: Substitute electrolyte fluid for milk ration for 2 days (see page 204). Call your veterinarian if diarrhea does not clear up right away.

Scrapie

A neurologic disease called scrapie is not a significant problem in goats, but the USDA has imposed regulation related to scrapie because the disease may be transmitted from goats to sheep or sheep to goats. Exactly how the disease passes from one animal to another is not known, and its incubation period of 1 to 7 years makes tracking difficult. Scrapie is similar to mad cow disease in cattle and has no known treatment or cure.

The first cases in the United States were transmitted by British sheep imported in 1947. The federal government has been trying to control scrapie since 1952. In the last decade of the 20th century, only seven cases were reported in goats, all of which were acquired from contact with sheep. In an attempt to eradicate scrapie, in 2001 the USDA enacted measures requiring herd certification and identification of individual goats within each herd, hoping to trace any outbreaks back to the herd of origin.

Most states have additional regulations, some of which exceed federal requirements. Since these regulations may change from time to time, the best way to remain current, as well as locate your state regulatory agency, is to contact the USDA Animal and Plant Health Inspection Service office by calling 866-873-2824 (toll-free) or visiting their Web site at www.aphis.usda.gov/vs/scrapie.htm.

Tetanus

Also called: *Clostridium tetani* toxemia, lockjaw.
Symptoms/effects: Stiff muscles, spasms, flared nostrils, wide-open eyes; results in death.
Cause: *Clostridium tetani* bacteria entering a wound.
Prevention: Vaccinate kids with ½ mL of tetanus toxoid at 4 weeks of age (or before disbudding). Repeat in 30 days and annually thereafter.

Treatment: No effective cure.
Human health risk: You cannot get tetanus from your goats, but you can get it from the same sources that they do. Talk to your family doctor about immunization.

Ticks

Symptoms: Rubbing, scratching, loss of hair, loss of weight.
Cause: Browsing in wooded areas.
Prevention: Check goats daily during tick season.
Treatment: Dust or spray with pesticide approved for livestock (for milk goats, use a pesticide approved for dairies). Treat Angoras after shearing; repeat in 15 days. To remove an attached tick, grasp the tick carefully with tweezers near the point of attachment and lift firmly. Wrap the tick in a tissue and flush it down the toilet.
Human health risk: Ticks transmit diseases such as Rocky Mountain Spotted Fever and Lyme disease. During tick season, check your body for ticks after handling your goats.

Urinary Stones

Also called: Calculosis, bladder stones, kidney stones, urinary calculi, urolithiasis, water belly.
Symptoms/effects: Difficulty urinating, kicking at abdomen, loss of appetite; can lead to death.
Cause: Sandlike crystals (calculi) in the urinary tract due to dietary imbalance or drinking too little water.
Prevention: Keep a wether's diet low in concentrate and the leaves from beets, mustard, and Swiss chard, and keep it high in free-choice hay, salt, and clean water. Feed him grass hay, never a legume hay.
Treatment: Surgery.

Worms

Also called: Internal parasites.
Symptoms: Paleness around eyes, loss of weight or failure to gain weight, loss of energy, weakness, poor appetite, diarrhea, coughing, rough coat, reduced milk production, strange-tasting milk.

Taking a Fecal Sample

Place a self-sealing plastic bag, inside out, over your hand. With the covered hand, grasp a handful of fresh droppings. With your other hand, turn the bag right side out to enclose the droppings. Seal the bag and take it to your veterinarian as soon as possible.

Your vet will examine the sample through a microscope. Ask if you can take a look, too. If your goats have worms, the vet will tell you what kind they are and recommend the proper treatment.

Take a fecal sample in fall around breeding time, and then again in spring after kidding. On the basis of these samples, your vet may recommend a routine worming program.

Cause: Eating worm eggs from manure, as may happen when feed is thrown onto the ground or pasture is grazed without frequent rotation. Worm egg numbers can get very high after a particularly mild winter or in areas with a mild climate.
Prevention: Keep feed and water free of manure. Move goats to fresh pasture often. Isolate and worm new animals.
Treatment: Take a fecal sample to your vet and obtain a wormer.

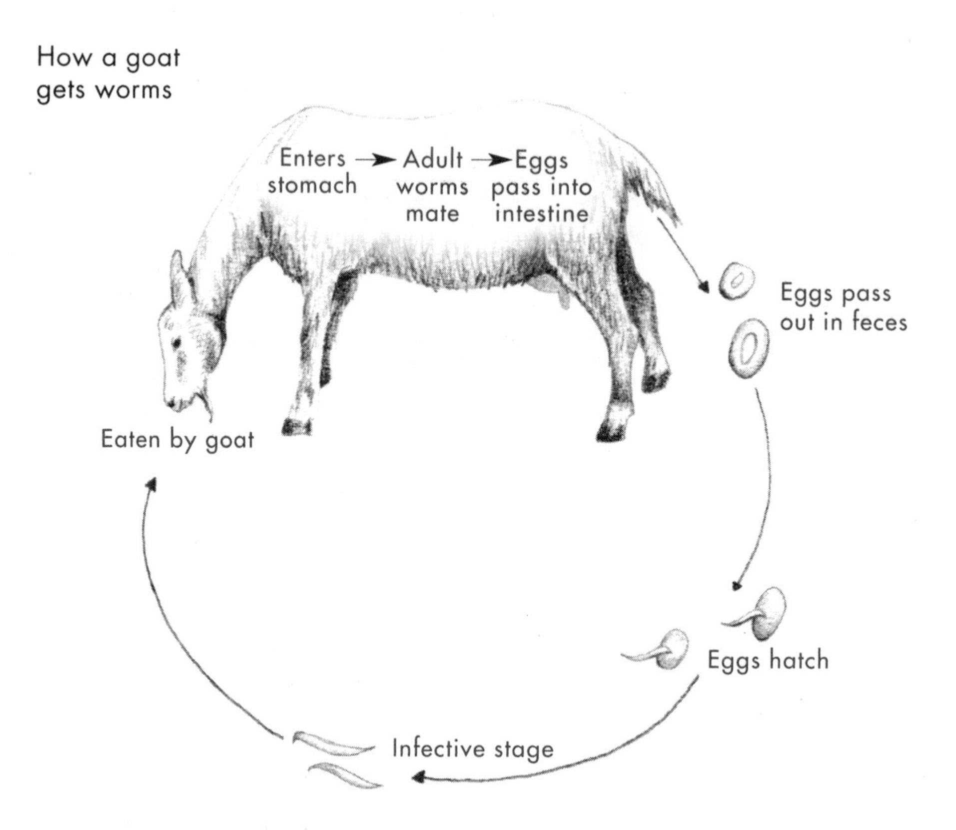

Wounds

Curiosity can get a goat into all sorts of trouble, including serious injury. If one of your goats is wounded, clean the wound with hydrogen peroxide so you can see how serious it is. Cuts on the udder usually look worse than they are because they bleed a lot.

Stop the bleeding by pressing a folded clean towel, cloth, or disposable diaper against the wound; if possible, tape it on tight. If the bleeding does not stop, call your vet. Keep the goat quiet and continue to apply pressure until the vet can take a look.

If the wound is a simple cut or scratch, clip away the hair around the wound. Wash the area with warm, soapy water and rinse with clean water. Pour hydrogen peroxide over the cut, dab it with a clean tissue, and coat the area with an antibiotic ointment (such as Neosporin). Clean the cut daily and coat it with iodine or antibiotic ointment until it heals completely.

Watch for signs of infection, such as redness, swelling, tenderness, and oozing. If infection occurs, call your vet. A cut is not likely to become infected if you keep it clean while it heals.

First-Aid Kit for Goats

If you assemble a first-aid kit ahead of time, you will be ready to handle most emergencies. Keep the items listed below clean and dry in a large lunch box, tackle box, ammunition case, or any sturdy plastic or metal container with a tight lid. Inside the cover, tape the names and phone numbers of at least three goat-oriented veterinarians. In an emergency, you may not be able to reach your regular vet.

Include the following items in the first-aid kit:

- 1 rectal thermometer, to take temperatures
- 1 quart isopropyl alcohol, to sterilize thermometer
- 6 disposable syringes (3 mL and 5 mL), to give shots
- 6 needles (18 gauge), to go with syringes
- 3 clean towels or diapers, to stop bleeding
- 1 bottle tetanus antitoxin, in case of wounds
- 1 pint hydrogen peroxide, to clean wounds
- 1 tube antibiotic ointment, to dress wounds
- 1 container tamed iodine (such as Betadine), to treat wounds
- 1 quart mineral oil, to treat bloat
- 1 quart propylene glycol, to treat ketosis
- 1 package powdered electrolytes, to treat scouring kids
- 1 jar udder balm, for chapped udders (and hands)
- Worming medication, as recommended by your vet
- Pesticide spray or powder, if needed to treat lice and ticks

Health Records

Keep a health maintenance chart for each goat so you can track its medical history, recognize recurring problems, and keep worming and vaccinations up to date.

Health Maintenance Chart

Date	Name of Goat	Medication Used	Dosage	Remarks

Goat Milk

Milk from a properly cared for doe tastes exactly like milk from a cow. Although most people in the United States drink cow milk, around the world more people drink goat milk than cow milk. Goat milk, like all milk, contains solids suspended or dissolved in water. Goat milk is made up of approximately 87 percent water and 13 percent solids. The solids are:

- Lactose (milk sugar), which gives you energy
- Milk fat, which warms your body and gives milk its creamy smooth texture
- Proteins, which help with growth and muscle development
- Minerals, for your general good health

Milking Equipment

Goats are milked in a milk room or milk parlor, which may be built into a corner of your dairy barn or in a separate building. Some people milk their goats in their garage or laundry room. Wherever your milk room is, it should be easy to clean and big enough to hold a milk stand and a few necessary supplies.

A milk stand, homemade or purchased from a dairy goat supplier, gives you a comfortable place to sit while you milk. At the head of the milk stand is a stanchion that locks the doe's head in place so she can't wander away while you are still milking her. Most people feed a doe her ration of concentrate to keep her from fidgeting during milking, but it's better to train your does to be milked without eating. A doe that's used to eating while she's being milked tends to get restless if she finishes eating before you finish milking.

Keep your equipment scrupulously clean to ensure that your milk is healthful and good tasting. Every time you use any of your dairy equipment, rinse it in lukewarm (not hot) water to melt milk fat clinging to the sides. Then scrub everything with hot water mixed with liquid dish detergent and a splash of household chlorine bleach. Use a stiff plastic brush — not a dishcloth (which won't get your equipment clean) or a scouring pad (which causes scratches where bacteria can hide). Rinse your equipment in clean water, then in dairy acid cleaner (which you can obtain from a farm or dairy supply store), then once more in clear water.

A milk stand may be used for hoof trimming as well as for milking.

You will need the following equipment and supplies for milking.

Equipment. These items are a one-time purchase:

- Spray bottle (for teat dip)
- Strip cup
- Stainless-steel milk pail
- Dairy strainer or funnel
- Milk storage jars
- Pasteurizer
- California Mastitis Test kit
- Milk scale

Supplies. These items must be replaced as you use them:

- Baby wipes
- Teat dip
- Bag Balm or Corn Huskers lotion
- Milk filters
- Chlorine bleach
- Dairy acid cleaner

Milking a Goat

When a doe gives birth, her body begins producing milk for her kids, a process called freshening. If the doe is a milking breed, she may give more milk than her kids need and continue to produce milk long after the kids are weaned. The amount of milk a doe gives increases for the first 4 weeks after she freshens, then levels off for about 15 weeks, after which production gradually decreases and eventually stops until the doe freshens again to start a new lactation cycle.

A doe's milk is produced and stored in her udder. At the bottom of the udder are two teats, each with a hole at the end through which the milk squirts out. The two most important things to remember when you milk a goat are to keep her calm and not pull down on her teats, both of which can be tricky when you're first learning. Keep the doe calm by singing or talking to her and by remaining calm yourself. Not pulling her teats takes practice. The doe will kick the milk pail if you pull her teat, pinch her with a fingernail, or pull a hair on her teat. To avoid pulling a doe's hair during milking, and to keep hair and dirt out of the milk pail, use clippers to trim the long hairs from her udder, flanks, thighs, tail, and the back part of her belly.

To get milk to squirt out the hole, you must squeeze the teat rather than pull it. The first time you try, chances are milk will not squirt out but will instead go back up into the udder. To force the milk downward, apply pressure at the top of the teat with your thumb and index finger. With the rest of your fingers, gently squeeze the teat to move the milk downward. If you are milking a miniature goat, her tiny teats may have room for only your thumb and two fingers.

After you get one squirt out, release the pressure on the teat to let more milk flow in. Since you will be sitting on the milk stand and facing the doe's tail, work the right teat with your left hand and the left teat with your right hand. Get a steady rhythm going by alternating right, left, right, left. Aim the stream into your pail beneath the doe's udder. At first the milk may squirt up the wall, down your sleeve, or into your face, while the doe dances a little jig on the milk stand. Keep at it and before long you will both handle the job like pros.

How to Milk

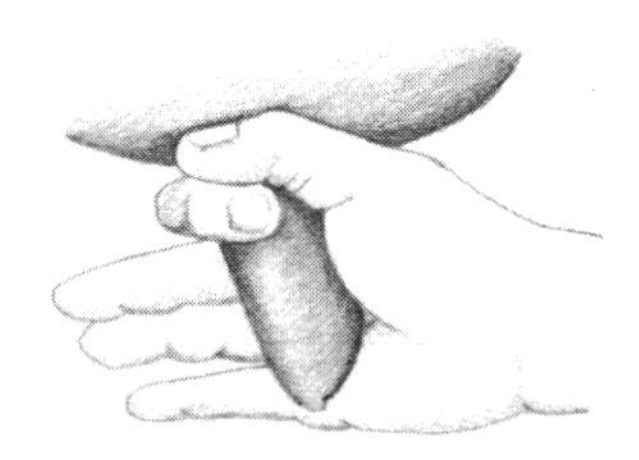

Apply pressure with your thumb and index finger to keep the milk from going back up into the udder.

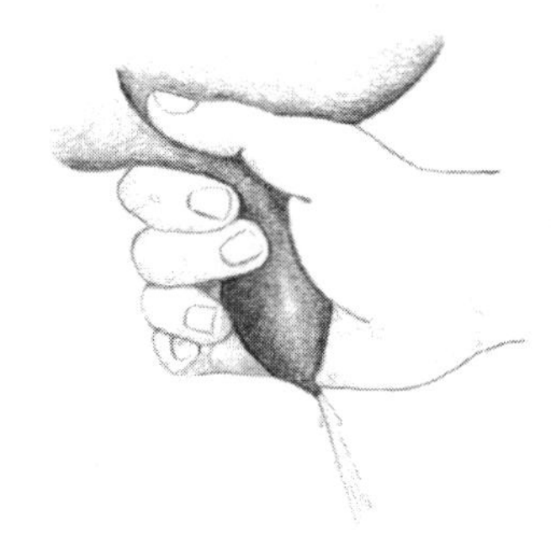

Use your remaining fingers to move the milk downward into the milk pail.

Milking a large goat

Milking a miniature goat

When the flow of milk stops, gently bump and massage the udder. If more milk comes down, keep milking. When the udder is empty, the teats will become soft and flat instead of firm and swollen.

If you milk more than one doe, always milk them in the same order every day, starting with the dominant doe and working your way down to the meekest. Your goats will get used to the routine and will know whose turn is next.

As you take each doe to the milk stand, brush her to remove loose hair and wipe her udder with a fresh baby wipe to remove clinging dirt. While you clean the doe's udder, watch for signs of trouble — wounds, lumps, or unusual warmth or coolness. Squirt the first few drops of milk from each side into a cup or small bowl, called a strip cup because it is used to examine the first squirt or stripping. Check the stripping to see whether it is lumpy or thick, two signs of mastitis.

When you finish milking each doe, spray both teats with teat dip so bacteria can't enter the openings. Use a brand recommended for goats — some dips used for cows are too harsh for a doe's tender udder. In dry or cold weather, prevent chapping by rubbing the teats and udder with Bag Balm (available from a dairy supplier) or Corn Huskers lotion (available from a drugstore).

Milk Output

Exactly how much milk a doe produces in each cycle depends on her age, breed, ancestry, feeding, health and general well-being, and how often you milk. The more often you milk, the more milk the doe will produce. Most goat keepers milk twice a day, as close to 12 hours apart as possible. If milking twice a day gives you more milk than you can use, milk only once a day. Do it every day at about the same time. If you don't milk regularly, your doe's udder will bag up, or swell with milk. Bagging up signals the doe's body that her milk is no longer needed, and the doe dries off.

Milk sold at the grocery store is measured by volume: 1 pint, 1 quart, or ½ gallon. Milk producers measure milk by weight: pounds and tenths of a pound. One pint of water weighs approximately 1 pound, giving rise to the old saying "A pint's a pound the world around." Milk also weighs approximately 1 pound per pint, although its exact weight depends on the amount of milk fat it contains, which varies by goat and by season.

During the peak of production, a good doe in her prime should give at least 8 pounds of milk (about 1 gallon) per day. She will then gradually taper off to about 2 pounds (1 quart) per day by the end of her lactation cycle. During the entire lactation, the average doe will give you about 1,800 pounds (900 quarts). A miniature doe averages one third as much milk as a large doe.

Weighing each doe's output helps you manage your goats properly. A sudden decrease in production may mean the doe is unhealthy, is not getting enough to eat, or is in heat.

Weigh milk by hanging the full pail from a dairy scale. A dairy scale has two indicator arms. Set the arm on the right to zero. With your empty pail hanging from the scale, set the left arm to zero. When you hang a pail of milk on the scale, the left arm automatically deducts the weight of the pail. Use the right arm to weigh other things, such as newborn kids.

If you don't have a scale, you can keep track of each doe's output by volume, although this method is not as accurate as weighing, because fresh milk has foam on top and it's hard to tell where the foam stops and the milk starts.

Keep a record of each doe's milk output, noting not only the amount of milk obtained from each milking but also anything that might affect output, such as changes you've made in the doe's ration, rainy weather that has kept your herd from going out to graze, or the time of day you milked (whether earlier or later than usual). At the end of each month, add up each doe's milk output. At the end of her lactation cycle, add up each doe's total output.

A standard cycle lasts 10 months, or 305 days. To accurately compare the annual output from each doe, or to compare the output of one doe against another, adjust production to a 305-day cycle: Divide the total output by the actual number of days in the cycle, then multiply by 305.

Pasteurizing Milk

Pasteurizing destroys harmful bacteria that may be in milk. Milk is pasteurized by making it hot enough for a long enough time to destroy bacteria. Pasteurize your milk in a home pasteurizer or on top of the stove.

Home pasteurizers, which come with full instructions, are sold through dairy supply catalogs. They are expensive but easy to use, because they automatically control the time and temperature.

Home pasteurizer

If you have only a few quarts of milk to pasteurize at one time, you can do it on top of the stove. (Milk cannot be safely pasteurized in a microwave oven.) To get the right temperature, use a candy thermometer. To keep the milk from scorching, use a double boiler or a clean pail set in a large pot of water.

Heat the milk to 165°F, stirring to distribute the heat evenly. Do not take the thermometer or stirring spoon out of the milk and put it back in during pasteurization, or you will recontaminate the milk. When the milk reaches 165°F, continue heating it for 30 seconds more.

Cool the pasteurized milk quickly for best taste. Set the pan or pail of milk in a basin of ice or cold water and stir until the milk is cool. Pour the milk into clean jars with tight-fitting lids. Store the jars on the bottom shelf of the refrigerator, where the temperature is coolest and milk keeps the longest.

Milk Sensitivities

Milk gets its sweet taste from the complex sugar lactose. For your body to digest lactose, it needs the enzyme lactase to break down the lactose into two simple sugars, glucose and galactose. About 75 percent of all adults are lactase deficient. They cannot digest lactose, and when they ingest it, the result is bloating, cramps, gas,

Making a Milk Compress

Milk protein is good for soothing sunburns and rashes from poison ivy, poison oak, and poison sumac. To make a milk compress, combine 1 cup of cold milk with 4 cups of cold water. Soak a clean cloth in the mixture. Lay the wet cloth on the sunburn or rash for 15 to 20 minutes; resoak the cloth whenever it feels warm. Repeat every 2 to 4 hours.

nausea, and diarrhea. The problem may be resolved by taking a lactase concentrate, available at most drugstores. Fermentation reduces lactose content by as much as 50 percent, so if you have a problem drinking milk because of lactose intolerance, you may not have a problem eating yogurt.

Not all problems associated with milk are caused by lactose deficiency. About 5 percent of the population is allergic to milk protein. In children, symptoms of milk protein allergy are eczema and digestive problems, including diarrhea, vomiting, and colic. In adults, milk protein allergy causes a feeling of being bloated and gassy. Since the protein in goat milk is not the same as the protein in cow milk, someone who is sensitive to cow milk protein may have no trouble drinking goat milk. Besides having a different protein makeup, goat milk has proportionally more small fat globules, making it easier to digest than cow milk and therefore leaving less undigested residue in the stomach to cause gas and cramps.

Whole and Skimmed Milk

After goat milk has been refrigerated for a day or two, its milk fat rises to the surface. Milk fat thinned with a little milk is cream. The milk fat content of goat milk ranges from 2 to 6 percent, depending on genetics, diet, and other factors. An average of 4 percent will give you about 5 tablespoons of milk fat per quart. The milk from Nubians and African Pygmies contains more fat than other milk, and the milk from all does varies in fat content during the lactation cycle. Milk fat content is important in making ice cream, butter, and certain kinds of cheese.

If you are trying to limit the amount of fat in your diet, you may remove the milk fat to create skimmed milk. Store fresh milk in a widemouthed container. In about 2 days, most of the milk fat will rise to the surface, and you can skim it off.

Making Yogurt

Yogurt is easy to make from fresh goat milk and tastes better than any yogurt you can buy. You'll need a candy thermometer. Heat 1 quart of milk to 115°F or, if you have just pasteurized the milk, let it cool to 115°F. Meanwhile, sprinkle one packet of unflavored gelatin over a little cold water. Allow the gelatin to soften for 1 minute. Heat the water and gelatin in a small saucepan or in the microwave oven until the gelatin thoroughly dissolves.

Combine the dissolved gelatin with the warm milk and stir in 2 tablespoons of store-bought unflavored yogurt. Make sure that the yogurt label says "live culture," indicating that it contains the live organisms needed to start fermentation. You can purchase yogurt culture from a cheese-making supplier, but it is expensive.

Place the cultured milk in an electric yogurt maker and turn it on, or place the milk in a glass casserole dish, wrap the dish in towels, and put it in a warm place, such as on a heating pad or an electric hot plate set on "warm." In about 9 hours, your milk will ferment into yogurt. It is ready when it thickens and tastes right to you. The longer you let it ferment, the more tart it will be, so adjust the time according to your taste. Store the yogurt in the refrigerator. Fermentation stops when yogurt is refrigerated.

Making Cheese Spread

Make a delicious cheese spread from fresh yogurt by straining it through a yogurt strainer or through a kitchen strainer or colander lined with cheesecloth (available from a supermarket, kitchen supply store, or cheese-making supply catalog). Place the strainer over a sink or deep bowl.

Remember Little Miss Muffet, who sat on a tuffet eating her curds and whey? When you strain yogurt, the liquid that drains off is whey. The thick part left in the strainer is curds. In 8 to 12 hours, 1 quart of yogurt will give you about 1½ cups of curds, which you can use to make a soft cheese spread to serve on crackers or toast.

For a sweet spread, stir ⅛ teaspoon of salt and honey or maple syrup to taste into the curds. Flavor the spread with bits of chopped pear or apple, drained crushed pineapple, grated lemon or orange zest, or raisins and nuts. For an herbal spread, stir in ⅛ teaspoon of salt, your favorite herbs, cracked pepper, and perhaps a little crushed garlic. Either spread will keep in a covered container in the refrigerator for up to 3 weeks.

Making Soft Cheese

You can make soft cheese in a hurry without making yogurt first. Place 1 quart of milk in a stainless-steel or enamel (not aluminum) pan. Heat the milk to 170°F. Squeeze two lemons and stir the juice into the milk.

Continue stirring gently for 15 minutes. If stringy curds do not form, add a little more lemon juice. Pour the mixture into a strainer or colander lined with cheesecloth. Drain over a bowl for at least 2 hours. Save the whey — combined with a little sugar or honey and chilled, it makes a delicious drink.

The drained curds will give you about ½ cup of mild cheese to use as a cream cheese spread or to serve like cottage cheese. Refrigerated in a covered container, this cheese will keep for up to 1 week.

Making Hard Cheese

You can use lemon juice to make a hard cheese. This time, start with 4 quarts (1 gallon) of raw milk. Stirring constantly, heat the milk to 185°F and cook for 5 minutes. Gradually stir in ½ cup of lemon juice. If curds and whey do not form within 15 minutes, add a little more lemon juice.

Drain the whey. Stir in ½ teaspoon of salt. Press the drained curds into a cheese mold, which you can easily make by carefully drilling small holes into the sides and bottom of a 1-pint plastic freezer container. Set the mold into a strainer or colander, and let the curds drain until the dripping stops.

In about 2 hours, you will have 1½ pounds of mild-tasting hard cheese to grate on top of stew, pasta, or soup or to slice for sandwiches. Wrap the cheese in plastic wrap and keep it in the refrigerator for up to 2 weeks.

These simple recipes are offered to get you started. If you are seriously interested in learning to make cheese, take a class on the subject or get a good beginner's book (see Recommended Reading, page 379).

Selling Goat Milk

In most states, if you wish to sell milk for human consumption, you must have state-approved (and expensive) equipment for storing, processing, and packaging the milk. You might, however, develop a nice little business selling milk for feeding orphaned animals such as deer, puppies, llamas, foals, bear cubs, and many other kinds of livestock and wildlife. Let your local veterinarians, horse stables, zoos, and wildlife parks know you have raw goat milk for sale.

FIBER GOATS

Most goats have two kinds of hair: primary and secondary. Primary hairs are usually straight, and secondary hairs are usually curly. The main coat on most breeds, including dairy goats, contains mostly primary hairs. Goats originating in cold climates have long primary hairs, giving them a shaggy look.

As insulation against cold weather, some goats grow a coat of secondary hairs. Short, downlike secondary hairs are known as cashmere. The long, densely packed secondary hairs of an Angora's coat are called mohair; the primary hairs, called kemp, are undesirable.

Mohair and cashmere are wonderful fibers for spinning, knitting, and weaving. Fabrics made from mohair are exceptionally soft and silky and are warm but lightweight. Although spinning these fibers is time-consuming, many people enjoy the quiet work and the pleasure of transforming a fleece into an article of clothing or other useful item. Even spinners who do not care to knit or weave take great pleasure in the quiet enjoyment of spinning fiber into yarn.

Mohair

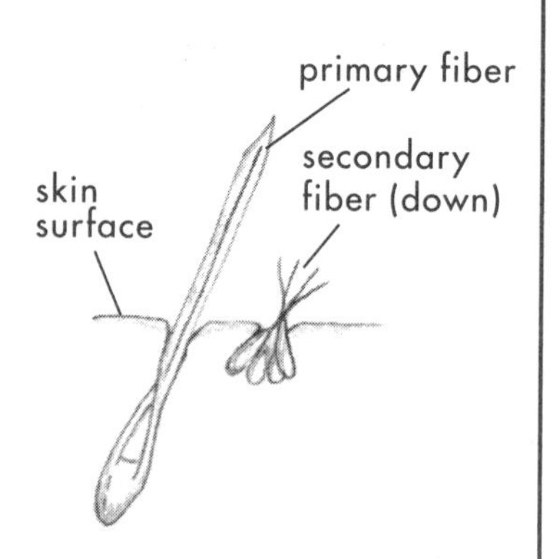

Dairy goat hair

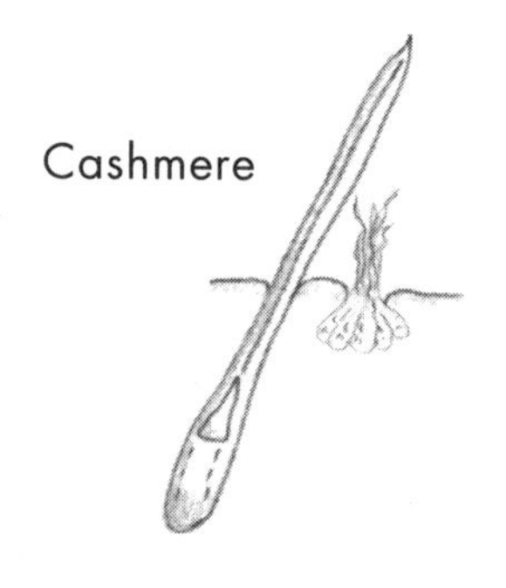
Cashmere

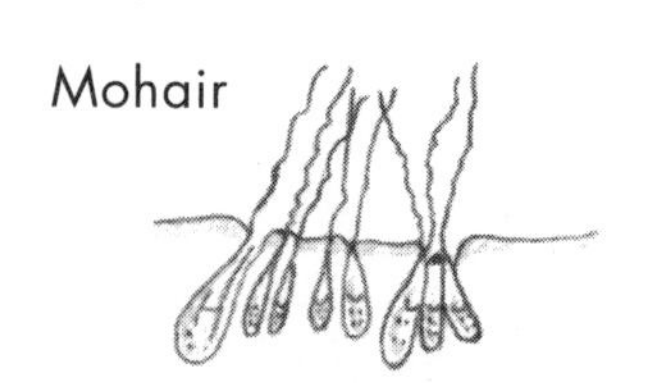
Mohair

Mohair makes a lustrous, luxurious, fuzzy yarn used in products ranging from fine clothing to carpets. Mohair yarn is stronger and warmer than wool, does not shrink like wool, and does not burn as easily. Mohair yarn holds dye well and therefore may be readily dyed into brilliant colors. A mohair fleece should start out creamy white. Colored fibers are undesirable to commercial buyers because they limit the use of dyes. Colored fibers are also an indication that the goat is not pure Angora. Colored fleeces of good quality, however, are valuable to hobbyists who enjoy working with the natural colors.

A fleece consists of all the hair obtained from one goat at one shearing. A shearer doing a less than perfect job may not get the fleece all in one piece; rather, he or she may have to go back and take a second cut. The result shows up as short clumps of hair on the back of the fleece that, if not removed, cause undesirable fuzzy bumps, or noils, when the fleece is spun into yarn.

All the hair from one goat in one year, or all the hair from one herd at one shearing, is called the clip. First clip is the soft, fine hair of kids that have been sheared for the first time. Fineness is determined by the thickness or diameter of individual fibers, measured in microns. One micron equals about 4⁄100,000 of an inch, a value so thin it must be measured in a laboratory. The thinner and finer the hair, the better its quality. First clip produces the finest hair, which may weigh only one fifth as much as adult clip but is worth up to three times more. First clip is nice for making clothing that touches the skin and is soft enough for baby clothing. Coarser, more durable adult mohair is perfect for making cushions and floor mats.

The individual fibers or hairs of a fleece are called staple. The longer the staple, the easier it is to spin. Experienced spinners like to work with staple that's at least 3½ inches long. A group of fibers or hairs clinging together is a lock. Ideal locks are well formed and free of defects, such as urine stains and dry, brittle tips.

How much mohair each goat produces depends on its age, size, sex, genetic background, nutrition, health, and general management. The best coat growth occurs in goats 3 to 6 years old. Thereafter, as a wether ages, its fleece loses character and becomes coarse. As a doe ages, her ability to continue producing both mohair and kids decreases.

On average, one goat will yield 3 to 4 pounds of first-clip mohair, 4 to 5 pounds in the second shearing, and 6½ to 7½ pounds each year thereafter. A well-managed purebred may give you 12 pounds or more. By way of comparison, ¼ pound of mohair will make a scarf and ¾ to 1 pound will make a sweater.

Shearing

Angoras are sheared twice a year: in February or March (just before kidding) and in July or August (after kids are weaned but before does are rebred). Don't be tempted to skip the fall shearing, thinking the retained coat will keep your goats warm during winter. The hair will grow back fast enough to keep them warm.

Goats, like sheep, are usually sheared by traveling crews. If you have only a few goats, it can be hard to get a crew to come. You might arrange to combine your goats with a larger herd of goats or sheep. You might also learn to do your own shearing by taking a class or asking an experienced person to show you how. For only a few goats, shearing with hand shears (or even scissors) is an option.

Manage your goats so their fleeces remain free of dirt, matted or tangled hair, and burrs or other vegetation. Keep your goats indoors for 2 days before shearing, to keep their fleeces dry. Wet mohair is difficult to shear and may turn moldy. Just before shearing, clean the floor of the shearing area so foreign fibers won't stick to your fleeces and reduce their value.

After shearing, the fleece must be skirted, which involves picking out stained locks, short cuts (anything under 2½ inches), matted clumps, and kempy areas. After the fleece has been skirted, roll it inside out and place it in a clean cloth bag, paper bag, or cardboard box. Do not store mohair in a burlap sack or woven plastic bag, and don't tie the fleece with twine; bits of fiber from the burlap or twine may stick to it.

For up to 6 weeks after shearing, or until your goats grow at least 1 inch of new coat, they can get sunburned in sunny weather or catch pneumonia in cold weather. Snug housing and proper nutrition will help your goats rapidly replace their protective coats.

Preparing Mohair for Spinning

Different parts of a fleece handle differently in terms of length and softness. Before spinning, sort the fleece into groups of similar-quality fibers that will result in a uniform yarn. Hold the fleece with one hand and pull out one lock at a time with the other hand. Because bits of dirt and vegetation will fall out, work outdoors or spread newspapers on the floor around you. After handling each lock, place it with locks of similar quality.

Once the fleece is sorted, it must be washed to remove the natural grease. Fill a basin with hot water and add a liquid soap designed for washing wool or use any gentle liquid soap, such as Ivory. Drop about a pound of sorted fleece into the water and gently push down until it is soaked through. Let the fleece soak for a short while, but not long enough for the water to get cold, then gently squeeze out the water and refill the basin for a second washing. Repeat until the water remains clear, then rinse the fleece in hot water and lay the fibers on a towel to dry.

Once the mohair is clean and dry, the fibers must be aligned parallel before they may be spun. Kid fleece needs only teasing, which involves gently pulling apart each lock to separate the fibers. The more carefully you work the fibers apart, the less lumpy your yarn will be. Tease a big pile at once so that you won't have to stop spinning to tease more.

The coarser fleece from adult goats needs the help of some device to get the fibers properly aligned before they may be spun into a smooth yarn. The easiest way to align the fibers is to comb each lock with a tough metal comb, such as that used to comb pets. Grasp a lock at one end and comb out the fibers at the other end, then change sides and comb the uncombed end. A combed fleece is called top.

If you will be doing a lot of spinning, you may want to invest in a set of cards with metal teeth. Fine-toothed cards are used on first-cut mohair, and medium-toothed cards are used on adult fleeces. The method shown on page 279 for carding sheep wool is the same as for carding mohair. Learning to use cards can be frustrating; try to find an experienced person to show you how. A fleece that has been carded is called roving.

Spinning Mohair

To spin teased, combed, or carded mohair, pick up a handful in one hand. With the other hand, pinch a small clump of fibers and pull gently until they form a fan 3 to 4 inches long. This little fan-shaped bunch of fibers is called the draft.

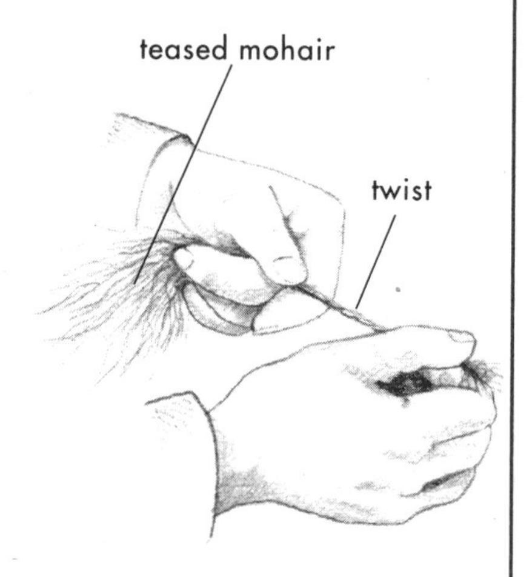

Drafting and twisting

Twirl the pinched end of fibers between your thumb and forefinger to twist them together. Soon the twist will run along the draft toward your other hand. To keep the twist from running into the teased mohair behind the draft, press firmly against it with your finger and thumb, release just enough to pull out more fibers, and continue twisting. Your hands will get farther apart as the yarn grows longer.

When the first bunch of teased mohair is nearly all spun, lay a second bunch over it. Don't wait until the first bunch runs out.

As your hands get farther apart, pretty soon you won't be able to draft and twist anymore. You'll need a spindle to wind the finished yarn onto so that you can keep spinning. A drop spindle consists of a smooth, tapered shaft and a light weight that slips on and off. The weight, called a whorl, keeps the spindle turning like a top, automatically putting in twist while you concentrate on drafting.

Buy an inexpensive wooden drop spindle, or make your own. Whittle down a wooden dowel to taper it, and slip it into a circular wooden base with a hole in the center. The base must be well balanced, and the hole dead in the center, for the spindle to spin evenly. Smooth down all edges with sandpaper, leaving no rough spots for fibers to get caught in.

To get your spindle started, tie on a couple of feet of yarn, as illustrated on the facing page. Fuzz up the end so the yarn will grab your teased mohair. Hold some teased mohair in one hand. Pull out a few fibers and wrap them around the fuzzed-up yarn. With your free hand, give the spindle a gentle spin. Let go of the spindle and draft out a few fibers. Release the draft and give the spindle another gentle turn.

The trick is to keep turning the spindle and drafting fibers with a steady rhythm. The spindle should turn freely at the end of the yarn. Pretty soon, your yarn will be so long that the spindle reaches the floor. Untie the yarn from the top of the shaft and beneath the whorl. Wind the yarn onto the shaft in a cone shape and reattach it as before.

When you have no room to wind on more yarn, remove the cone. Untie the yarn, push up on the whorl, and slip the cone off the shaft. Store the finished cones on a piece of wood or heavy cardboard with long nails or knitting needles stuck through it until you have enough to spin or weave.

If you enjoy spinning, you may eventually wish to purchase a spinning wheel. Spinning on a wheel is quicker than working on a spindle, but a wheel is not as portable as a spindle. Learning to first control twist and draft on a spindle makes learning to spin on a wheel much easier.

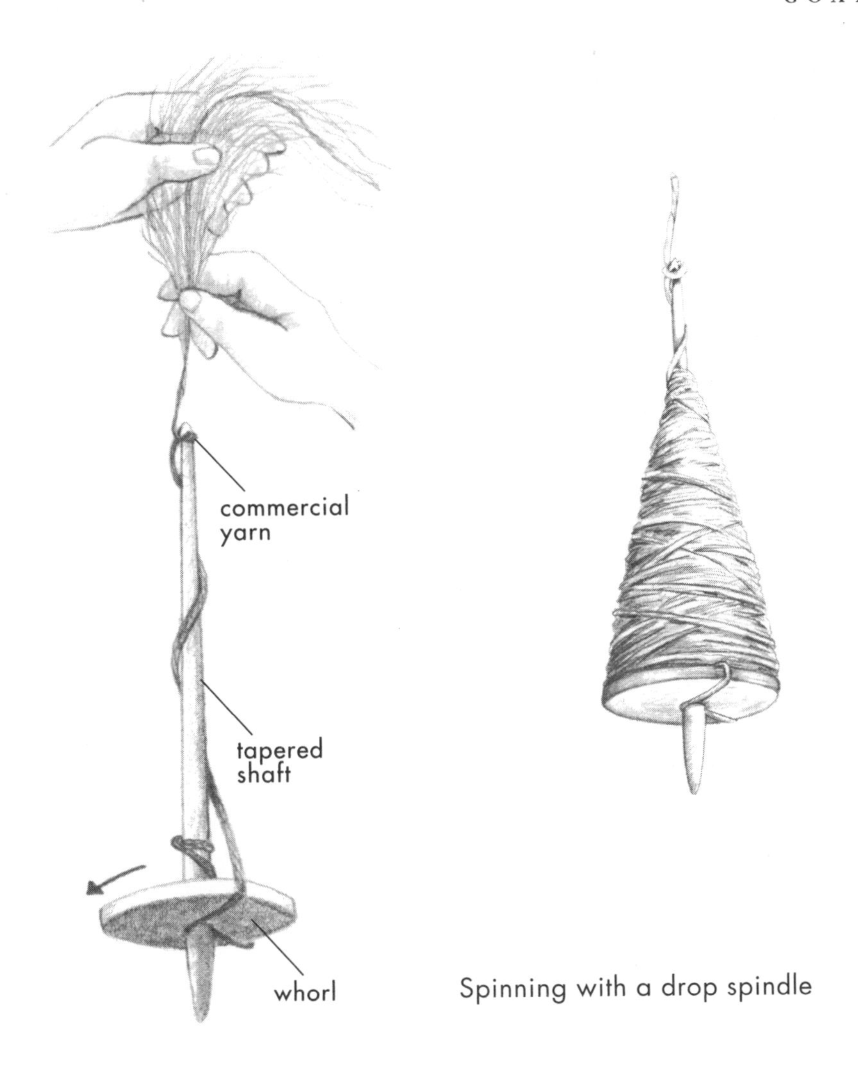

Spinning with a drop spindle

Cashmere

Cashmere, like mohair, is soft, warm, and light. Unlike mohair, cashmere has short fiber and comes in colors other than white. The value of cashmere is based on several properties, including fineness. To qualify as cashmere, the diameter of the fiber must be 19 microns or less; cashmere is one of the world's finest animal fibers. Like mohair, cashmere becomes coarser as a goat grows older. You can get some idea of the fineness of a goat's cashmere by examining its primary hairs. The more primary hairs you find and the coarser they are, the finer the down. The fewer hairs you find and the finer they are, the coarser the down.

Cashmere grows during the time of year when the length of days decreases, between the summer solstice around June 21 and the winter solstice around December 21. When cashmere stops growing, it starts shedding.

Commercially, cashmere is harvested by shearing before shedding starts. If you have only a few goats, you can harvest cashmere the traditional way — by combing it out as it sheds. Since the fibers do not all shed at the same time, comb your goats daily for 2 or 3 weeks until all the down is harvested.

Cashmere that has been harvested by combing has fewer coarse primary hairs than sheared cashmere. Before either may be spun, however, the hairs must be removed, a process called dehairing. Shake out as many hairs as you can, then pick out the rest.

A commercial grade goat averages ⅓ pound of down per year. A good wether may produce as much as 1 pound.

Cashmere may be teased and spun like first-clip mohair. Since the fibers are shorter, draft with your hands closer together and put in more twist.

Income Opportunities

Commercial buyers prefer to purchase large numbers of fleeces at once. If you have only a few goats, you might try finding someone with a large herd who will sell yours for you. Spinners, on the other hand, often buy one or two fleeces at a time. Locate buyers through your county Extension agent, farm store, local spinning and weaving supplier, and yarn shops. If spinning classes are offered in your community, ask the instructor if you may come and exhibit your fleeces.

How much you charge per pound depends on local demand, current commercial prices, and the quality of your fleeces — the finer they are, the more you can ask. Your fleeces will be worth more if they are uniform in length, clean (no weed seeds, dirt, or excess grease), and pure (no stains or off-colored fibers, and low percentage of kemp).

How much you can ask is also determined by how you sell a fleece. The easiest way to sell a fleece is raw or "in the grease," which means it hasn't been washed. You'll get more per pound, however, if you wash the fleece and sell it as scoured. You can charge more yet if you take time to comb your mohair and sell it as top, or card mohair or cashmere and sell it as roving. You will get the best price of all from knitters and weavers by selling well-spun yarn.

Align fiber for spinning by carding the fleece. Carded mohair or cashmere is known as roving.

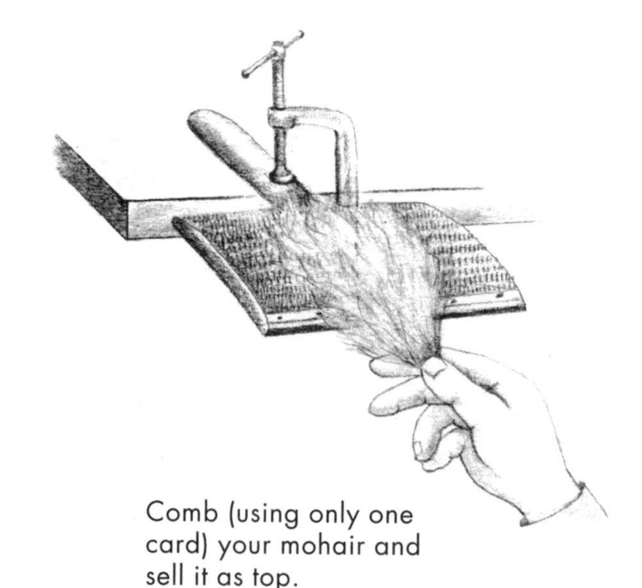

Comb (using only one card) your mohair and sell it as top.

5

Sheep

Introducing Sheep

For thousands of years, people have raised sheep for three critical reasons: milk, meat, and wool. Of course, other barnyard animals are also able to provide humankind with these items; but sheep have many advantages. They are much easier to handle than other farm animals, such as cows, horses, and pigs. Moreover, they require little room, they're fairly easy to care for, and they can be trained to follow, come when called, and stand quietly.

Sheep are also earth-friendly. Land that cannot be used to grow vegetables, fruits, or grains is fine for sheep. They eat weeds, grasses, brush, and other plants that grow on poor land, and their digestive systems are designed to handle parts of food plants such as corn, rice, and wheat that people cannot eat. Many of the world's most popular cheeses are made from sheep milk. Sheep wool, which can be used to make rugs, blankets, clothing, and other materials, is a renewable resource. Sheep manure fertilizes soil. The fat of a sheep raised for meat can be used to make candles and soap, and the pelt of that sheep can be used to make clothing.

Sheep Q & A

Sheep rely on their owners for food, protection from predators, and regular shearing, but they require less special equipment and housing than any other livestock. One or two lambs or ewes can be raised in a backyard with simple fencing and a small shelter.

If you've never owned sheep before, you probably have some questions. Here are some answers to the most common questions prospective sheep owners have.

Are children safe around sheep?

Sheep are among the safest four-legged farm animals for children to handle. Most sheep are small and docile. Rams or bucks can be aggressive at times, but sheep, especially those that are around people every day, are usually very gentle and even-tempered. A sheep can easily become your pet, especially if you raise an orphaned lamb on a bottle.

What do sheep eat?

Sheep don't need fancy food. In summer, they can live on grass; in winter, they can eat hay supplemented with small amounts of grain. Fresh water, salt, and a mineral and vitamin supplement complete their diet.

Are sheep dumb?

Sheep are anything but stupid. They learn very quickly and are among the smartest of all farm animals. Many sheep recognize and respond to their individual names. Sheep are often thought to be stupid because of the way they react to perceived danger. They have no way to defend themselves; if an enemy threatens them, they cannot kick like horses, butt like goats or cattle, or bite like pigs. They can only bunch together and run away. Sometimes, when sheep are frightened, they run headlong into obstacles, which makes them *seem* stupid.

Do sheep stink?

Definitely not. All farm animals have their own distinctive odor. The natural odor of sheep and their manure is not as strong as that of cattle or horses.

How many sheep should I keep?

No sheep should be raised alone. Sheep have a built-in social nature and a flocking instinct. They are happiest when they have companions. Orphaned, or bummer, lambs, however, are often just as happy around humans as they are with other sheep. Orphaned lambs quickly become attached to the person who feeds them.

How old should sheep be when I buy them?

Most people start with weaned lambs, which are 2 to 3 months of age. However, nothing is quite as satisfying as raising an orphaned baby lamb. You may also buy an older ewe that has been bred to lamb in the spring. These ewes may often be purchased at a low cost, and you'll get two sheep for your money.

How much do sheep cost?

Prices for adult ewes vary widely. Shop around at several local farms to get a sense of the average market prices in your area. A good crossbred (a sheep whose parents are of different breeds) will be much less expensive than a purebred or registered animal.

Lamb prices also vary. An orphan lamb often costs little, because a farmer who has a large flock sometimes doesn't have the time to raise an orphan lamb and may be happy to find it a good home. A weaned lamb usually costs more than an orphan.

Choosing a Breed

Before you decide on a breed, evaluate your reasons for wanting sheep. You may wish to raise a sheep simply to have as a pet. A wether (castrated male sheep) makes the best pet. If it was a bummer (bottlefed) lamb, better yet. You may choose a pet sheep on the basis of personality and appearance alone. Flock owners are often eager to sell a black lamb, one whose color is unusual for its breed, or a lamb from a set of twins or triplets whose mother may not have enough milk for all of her lambs. Such animals are often sold at a bargain price.

If you wish to raise sheep for practical or show purposes, you'll need to know which breeds have the characteristics that are most important for your purposes. The climate in your region will help you determine which is the best breed for you. If you live in an area with severe winters, choose a breed that can survive in cold weather. If you live in a wet area, look for a breed that tolerates rainy weather. If you live in a desertlike area, you will want a breed that is adapted to hot, dry climates. Look around and see what breeds of sheep are being raised locally — these breeds may also be the best ones for you.

Sheep come in so many breeds that it would take a whole book to describe them all. This section describes some of the most popular breeds, as well as a few minor ones. If you keep in mind your reasons for owning sheep, these brief descriptions will help you decide which breeds are right for you.

Columbia

Columbias are large animals that produce heavy, dense fleece and fast-growing lambs. Columbias have a calm temperament and are easy to handle. They have an open, white face and are polled.

Corriedale

Corriedales, noted for their long, productive lives, are distributed worldwide. These large, gentle-tempered sheep have been developed as dual-purpose animals, offering both quality wool and quality meat. A strong herding instinct makes them excellent range animals, as well. They have an open, white face and are polled.

Dorset

The Dorset is considered one of the best choices for a first sheep. Dorsets are medium-sized and have a very gentle disposition. A Dorset has very little wool on its face, legs, and belly, which makes lambing easier. Its face is usually open, and it is white on both the face and the legs. Both polled and horned types are available.

Dorsets are a fine choice for both wool and meat. Their lightweight fleece is excellent for handspinning, and they have large, muscular bodies and gain weight fast. Dorset ewes are good mothers, and Dorsets are one of the few breeds that can lamb in late summer or fall.

Hampshire

The Hampshire is among the largest of the meat types, and the lambs grow fast. Its face is partially closed; the wool extends about halfway down. It has a black face and legs and is polled. Hampshires have gentle temperaments that make them popular with children.

Katahdin

The Katahdin breed of sheep is an easy-to-raise meat sheep that has hair instead of wool. It does not require shearing, because it sheds its hair coat once a year. Katahdins can tolerate extremes of weather. Except for the fact that Katahdins do not produce wool, they possess all of the ideal traits for a pet or small flock: They are gentle, with mild temperaments; require no shearing; have few problems with lambing; are excellent mothers; and have a natural resistance to parasites. They have an open, white face and are polled.

Polypay

Large and gentle-tempered, Polypays are a superior lamb-production breed with a high twinning rate, a long breeding season, and good mothering ability. They are also known for having strong flocking instincts, quality meat and wool, and milking ability. They have an open, white face and are polled.

Sheep Terminology

White or black face. These terms describe the color of the wool on the sheep's head and face. Normally, the wool on the lower legs is the same color as that on the face.

Open or closed face. These terms are used to describe how much long wool is on the sheep's face. An open-faced sheep has only short, hairlike wool on its face. A closed-faced sheep has long wool on its face. On a closed-faced sheep, the wool may grow all the way down to the animal's nose. Too much wool around the eyes causes the sheep to become "wool blind." The excess wool must be clipped away so that the animal can see.

Prick or lop ear. Just as a German shepherd's ears stand up and those of a cocker spaniel hang down, a sheep's ears can stand straight up (prick ear) or hang floppily down (lop ear). Some sheep's ears even stand out to the side.

Polled or horned. Polled sheep have no horns, and horned sheep have horns.

Open and black face

Closed and white face

Corriedale

Dorset

Hampshire

Katahdin

Romney

Like other gentle-tempered sheep breeds, Romneys make excellent pets. They are polled and have an open, white face, black points (noses and hooves), and a long, soft fleece that is ideal for handspinning. They also produce good market lambs. Romney ewes are quiet, calm mothers. Romneys are best suited to cool, wet areas.

Suffolk

Suffolks are similar to Hampshires. They, too, are large and have fast-growing lambs. They have an open, black face (unlike the Hampshire's partially closed face) and are polled. Suffolks are usually gentle, but some can be headstrong and difficult for younger children to manage.

Tunis

The Tunis has been around for more than 3,000 years, making it one of the oldest sheep breeds. It is considered a minor breed because relatively few Tunis can be found in the United States. They are medium sized, hardy, docile, and very good mothers. The reddish tan hair that covers their legs and closed faces is an unusual color for sheep. They have long, broad, free-swinging lop ears and are polled. Their medium-heavy fleece is popular for handspinning. The Tunis thrives in a warm climate, and the rams can breed in very hot weather. The ewes often have twins, produce a good supply of milk, and breed for much of their lifespan.

Heritage, Rare, and Minor Breeds

The breeds that have fallen out of favor with industrialized agriculture are referred to as rare, heritage, or minor breeds. Many of these breeds were major breeds just a generation or two ago, but as agriculture has focused on maximum production regardless of an animal's constitution, these old-fashioned breeds have begun to die out. The loss of heritage breeds can have an especially grave impact on homesteaders, who are usually interested in low-input (less work on the part of the farmer) agriculture. These breeds, although not the most productive in an industrialized system, have traits that make them well suited to low-input farming. Some are dual purpose, able to produce both meat and fiber. Others are acclimatized to regional environments, such as hot and humid or dry and cool conditions.

Many perform well on pasture with little or no supplemental feeding. Others resist disease and parasites. Some have such strong mothering skills that the farmer doesn't have to do much work during lambing season.

Interest today in preserving heritage breeds of livestock, including sheep, is increasing. A driving force in this movement is the American Livestock Breeds Conservancy (ALBC). For more information about heritage breeds of sheep, contact the ALBC (see Resources, page 387).

Buying Your Sheep

Try to buy your sheep directly from the person who raised it, because you can ask questions about the sheep's history and see the flock it comes from. Avoid buying a sheep at an auction; you don't usually have the opportunity to talk with the owner, and you may be rushed into making your decision.

What should you look for in the ideal sheep? Above all, your sheep should be healthy. You can gauge a sheep's condition by examining it carefully for the following characteristics:

- ◆ It should seem alert and thriving.
- ◆ It should come from a flock with no major medical problems. If the whole flock is healthy, your sheep has probably been well cared for, is of sturdy family stock, and has not been exposed to diseases or parasites.
- ◆ It should have good conformation. Although you may not find or be able to afford a "perfect" lamb, the one you select should come close to the breed's ideal conformation.
- ◆ The sheep should be the normal size and weight for an animal of that age and breed. A large-boned sheep will have more meat, and the ewes will handle pregnancy and birth more easily.

Qualities of a Good Sheep

- ◆ Good conformation, with no obvious defects
- ◆ Rear legs plump and well muscled
- ◆ Teeth meet dental pad well; jaw not overshot or undershot
- ◆ Strong pasterns (ankle joint just above the hoof)
- ◆ Rams: good fertility (have it checked by a veterinarian)
- ◆ Ewes: good udder with no lumps or damage

Proper sheep conformation

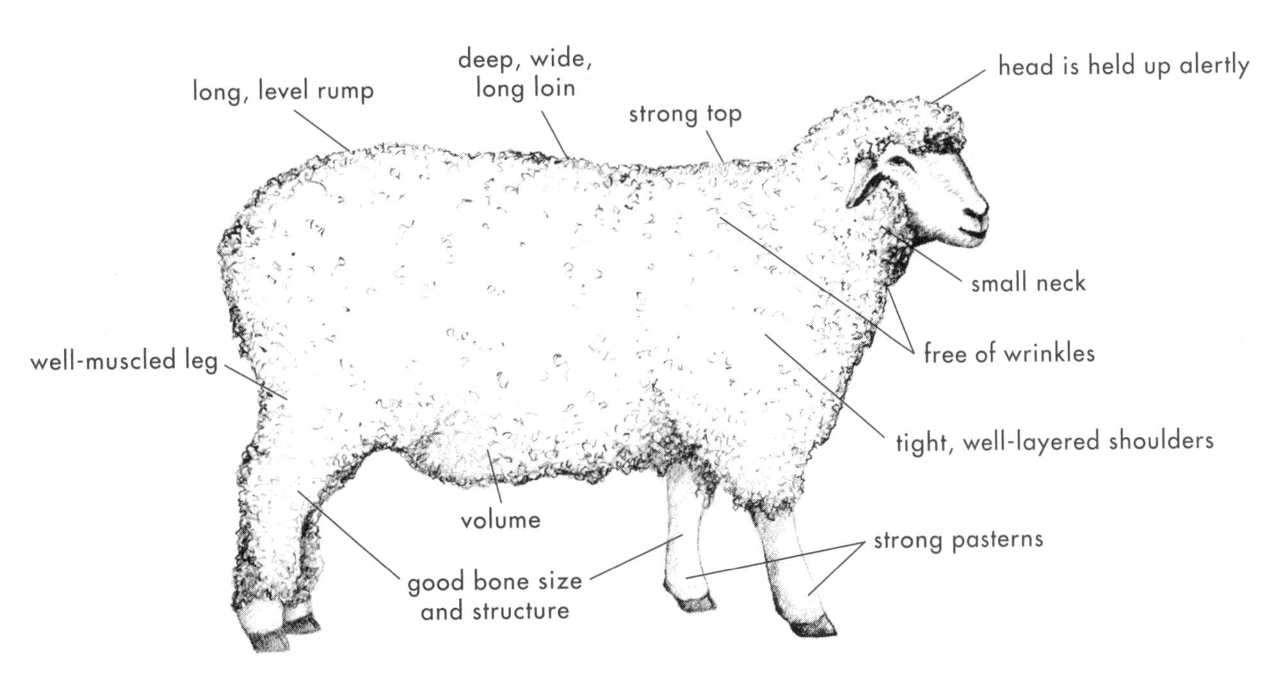

Leg Conformation

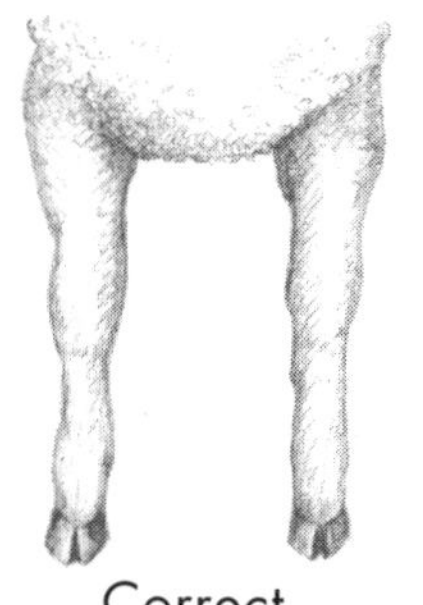

Correct

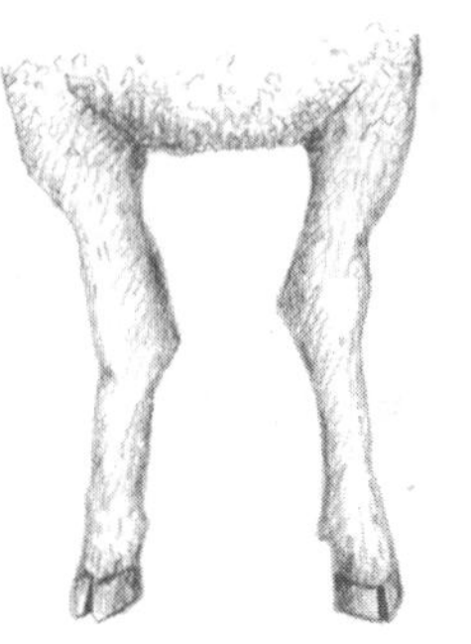

Knock-kneed and splayfooted

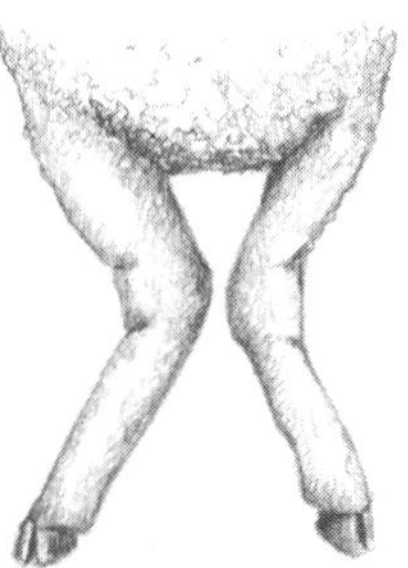

Bent leg

Judging a Sheep's Physical Condition

Examine the sheep's eyes, teeth, feet, and other body parts, including its fleece. Look for signs of good health and conformation, and avoid purchasing a sheep with any of the problems described below.

Eyes. Runny, red, or damaged eyes may mean that the sheep is diseased.

Teeth. Worn or missing teeth will interfere with eating.

Jaw. The lower jaw should not be undershot or overshot (see the illustrations below).

Head and neck. Make sure there are no lumps or swelling under the chin. These may be related to bottle jaw, a disease caused by an untreated worm infestation, or to bacterial infection causing abscesses.

Feet. If the sheep is limping, it may be injured or have foot rot (see pages 254–255). Even if the sheep you're considering seems healthy, notice whether other sheep in the flock are limping. If a sheep in the flock has foot rot or other foot problems, your sheep may also be infected. Also, if the sheep has untrimmed or overgrown hooves, it has not been cared for properly.

Body conformation. Sheep with narrow, shallow bodies tend to be light muscled. Sheep with wide backs and deep bodies are more desirable.

Body condition. Don't choose a ewe that is extremely thin. However, it's normal for a ewe that has just raised lambs to be thin; this does not mean she has a health problem. Don't choose a very fat ewe, either. She may have trouble lambing.

Potbelly. A thin lamb with a potbelly usually has a heavy infestation of worms.

Udder. If you are buying an adult ewe, check the udder. If it is lumpy, she may have had mastitis, which sometimes causes a ewe to have no milk for future lambs.

Tail. A very short tail may mean that the docking (see page 263) was not done properly. This is a serious defect, because the entire area around the tail is weakened and may create problems during lambing.

Good jaw conformation

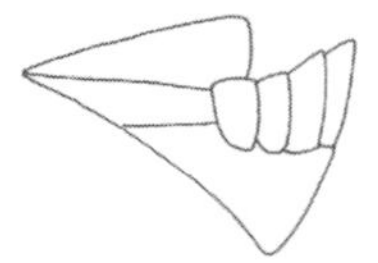

Undershot jaw

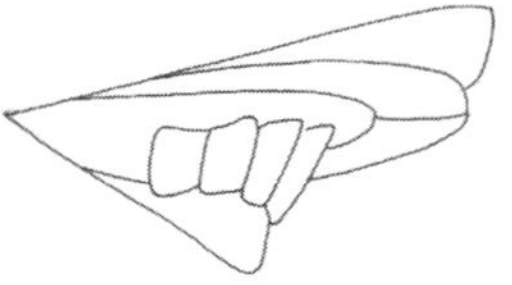

Overshot jaw

Manure. Runny droppings or a messy rear end can mean that the sheep is sick or has worms. However, a lamb that has been grazing on new, lush spring grass will sometimes have runny droppings; this is normal.

Wool covering. Each breed has a certain amount of wool on its face or legs. Your sheep should have the correct amount for its breed. Avoid sheep with excessive wool around the eyes, which can lead to wool blindness.

Fleece. A ragged, unattractive fleece may be an indicator of disease, keds, or lice.

External parasites. Look for signs of external parasites, such as keds or lice.

Judging the Age of a Sheep

The front teeth of sheep are similar to the front teeth of humans, with one big difference: Sheep have front teeth only on the bottom. Where you would expect to find upper teeth, a sheep has a hard gum line called a dental pad.

You can estimate the age of a sheep by examining its teeth. The front teeth of sheep begin to show at 2 to 3 weeks of age, starting with one pair in the center. This pair is followed by three more pairs a few weeks later. The lamb will have eight teeth in all. Sheep lose their baby teeth, and permanent teeth grow in. When the sheep is about 1 year old, the first pair of permanent teeth appears. Each year after that until the sheep is 4 years old, it gains another pair of permanent teeth. When the sheep is 4 years old, all four pairs of baby teeth have been replaced with permanent teeth.

Buying Lambs

- Buy a healthy lamb. Ask your veterinarian to examine the animal for signs of disease.
- Learn to recognize the finer points of conformation. Lambs with exceptional conformation will probably cost more than lambs with average conformation.
- Price does not always reflect quality. The most expensive lamb is not necessarily the best one.
- Most faults don't get better with time. If the lamb has faulty conformation, its defects will become more pronounced with age.
- Don't buy what you cannot see. If you are considering an older lamb with long fleece, you may not be able to see its body form. Learn how to discern desirable traits, such as a thick leg muscle, by feeling the lamb.
- Spend only 80 percent of what you think you can afford. Expenses you haven't planned for will always come up.
- An inexpensive lamb with poor conformation is not a bargain. Keep shopping until you find quality at a price you can afford.

What to Ask the Sheep's Current Owner

Here are some important questions that only the seller can answer. Don't be afraid to ask them.

What vaccinations or treatments has the sheep had, and when?
Young lambs may have received some, but not all, of their primary vaccinations. The sequence and timing of the vaccinations are important. You may be purchasing a lamb in the middle of its immunization schedule. You will need to know what shots were given and when, so that you can get the rest of the shots on schedule and with the same product.

When was the sheep wormed, and with what?
You will need worming information so that you can give scheduled maintenance wormings using the same product the animal has already received. (For more about worming, see pages 250–251.)

What kind of feed is the sheep being fed?
Any sudden change in the kind of feed you give your sheep can make it sick. Digestive upset can also slow down your sheep's growth or even cause its death. If you are not sure what to feed your sheep, continue to give it the same kind and amount of feed it has received up to now.

Was the sheep a twin? Was its mother a twin?
Twinning is a highly desirable genetic trait. The possibility of twinning is mostly determined by the ewe. If the ewe is a twin, she is more likely to give birth to twins. However, even if you are buying a ram, ask if he was a twin. If so, his daughters are also likely to give birth to twins.

If the sheep is an orphan lamb, why is it an orphan?
A lamb may have been pushed away by the mother because she had triplets or quadruplets, and this one was just one too many for her to care for. But if the ewe was simply a poor mother — some ewes are not as interested in mothering as others are — or had no milk, her daughter may have inherited these traits. The female bummer will therefore be less valuable for breeding when she grows up.

Has the flock had a history of medical problems?
Diseases spread quickly in a flock. A sheep from a sick flock may have been exposed to a number of diseases, especially foot problems. Notice whether any sheep in the flock limp or kneel when grazing. If their feet show signs of neglect, such as overgrown hooves, be cautious about buying from the owner of this flock. Along the same lines, don't buy sheep that show signs of having lice, keds, or problems with internal parasites.

Is there a record of this sheep's growth?
Ask for the date of the lamb's birth, its weight at birth, the date of its weaning, and its weight at weaning. You can use this information to figure out the rate of growth. A lamb's rate of growth is more important than its size. Two lambs of equal weight may be 2 or more months apart in age. If you are raising a lamb for

meat, do not purchase a slow-growing runt, because it won't gain enough to be economical.

After you buy your lamb, weigh it frequently and write down the weight on your sheep record.

Breeding Ewes

If you plan to raise a ewe for breeding, consider all of the breed's characteristics. Some breeds are easy to handle, with quiet and easygoing dispositions, whereas others are more high-spirited. Some tend to give birth easily, make superior mothers, and give birth to twins or triplets frequently. Sheep that are themselves twins are more likely to produce more twins. If you are planning to breed sheep, you might therefore wish to buy a twin, because your flock will expand more quickly and you'll have more lambs to sell.

Sheep being raised to produce new breeding stock should be purebred or registered animals from high-quality stock. Try to find out what breeds sell best in your area. For example, do other shepherds want sheep for meat, or for wool? Do they breed sheep for their exotic characteristics?

Consider, too, what breeds are already available in your area, and follow one of two strategies:

- Select a common local breed and try to develop superior animals through intelligent breeding, special care, and good nutrition. If your sheep win many prizes at shows, you will soon have plenty of customers for your lambs.
- Purchase an unusual breed that is scarce in your area. The breed's desirable traits and appearance will attract buyers. Many producers seek out unusual animals to crossbreed with their flocks.

Fencing

Fencing around your sheep pasture should be your first priority. Before you bring your first lamb home, build a fence if you have none, or check existing fences to be sure that they are secure. Fences keep in sheep and keep out predators, such as dogs and coyotes. A dog playing with a sheep in the pasture can kill or seriously injure it, and coyotes prey on sheep.

Of the many types of fence, the best for sheep is a smooth-wire electric or a nonelectric woven-wire fence. Some nonelectric wire fences have six or more tightly stretched wires and heavy posts, but woven-wire "sheep fence" is better. This fencing is designed only to keep sheep in; to keep predators out, put a couple of strands of electric wire on the outside.

The strength of the electric fence is in the shock. Electric fence chargers (energizers) are available at farm supply stores and fence supply dealers. Most of these chargers operate on a few dollars' worth of electricity per month.

Fences permit you to rotate your pasture. Cross-fence, or divide, your large pasture into smaller paddocks by using inner fences. The outer perimeter fence must be permanent and sturdy to keep dogs out of the pasture, but the inner fences can be made of inexpensive woven-wire or portable electric fencing, which is quick and easy to construct.

Sheep Manure

Use sheep manure as fertilizer in your garden. If you don't have a garden, you can sell packaged manure to gardeners in your area. Sheep manure is dry, has little odor, will not burn plants, and is an excellent substitute for commercial fertilizer. Sheep manure also has more nitrogen, more phosphorus, and more potassium than manure of cows or horses.

Shelter

Sheep and lambs need shelter only in bad weather — heat, cold, rain, and snow. The rest of their time will be spent out on pasture, grazing.

When they must seek shelter, sheep are happiest in a south-facing three-sided shed; it's well ventilated and offers adequate protection from the weather. If your sheep shelter is an enclosed shed, it must be well ventilated. You can build or buy a small portable shed that can be moved easily to whichever paddock the lambs and ewes are currently in. Allow 15 to 20 square feet per adult animal.

And whatever sort of shelter you have, keep it clean. If you think it smells bad, so will your sheep — and the dirty shelter will be a potential breeding ground for disease.

Even good fencing can't always provide enough protection for sheep at night. The shelter should be located within a corral or small pen constructed of tightly woven wire or cattle panels that will positively keep dogs and other predators from getting to the sheep. If you place feed in the shed each evening, the sheep will naturally head toward it every night.

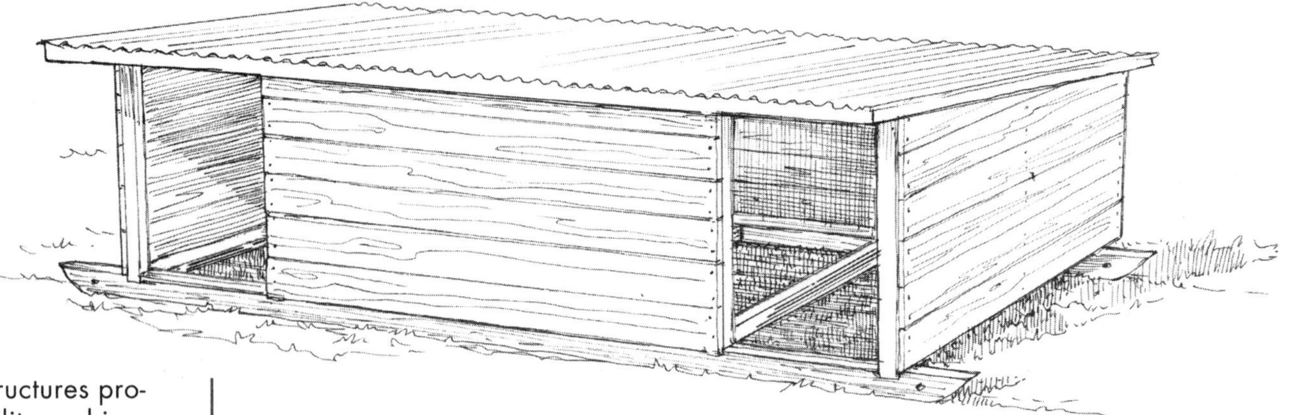

Portable structures provide flexibility and inexpensive shelter for sheep and lambs on pasture. Because sheep use the shelter only in bad weather, up to six ewes and their lambs can share one shelter.

Handling Sheep

To be a successful shepherd, you must learn to take advantage of the sheep's way of thinking. A sheep's most powerful instinct is, at all costs, to avoid being trapped. A cat has claws, a porcupine has quills, and a skunk has scent, but a sheep has only one defense from danger — escape.

This drive to avoid being trapped is why dogs in the pasture can cause so much trouble. When the sheep see the dog, they run to escape. When the sheep begin running, the dog thinks it's fun and begins chasing them. It's a vicious circle.

Sheep will come to you if you don't do anything to startle them or make them think that you're trying to chase them. For instance, if you try to drive sheep into a barn, they will avoid going in if they think you are trying to trap them. But sheep love to eat and will do almost anything to get their favorite treat. They especially love peanuts, apples, and grain. If you show them a bucket of grain and use it to coax them to follow you, you can lead them into the barn easily. In fact, they will be so eager to follow you that they will get pushy.

Allow the sheep time to eat the grain before you try to pen them or catch one. The grain is their reward for coming in, and eating it should be a good experience for them. Once the sheep are indoors or in a pen, use lambing panels or a gate to squeeze them together so they can't run away. Then walk up to the group, catch one sheep, and move it to a clear area where you can handle it.

If you say a sheep's name and offer it a treat at the same time, it will soon come when you call its name. You'll be surprised how quickly it will learn. If you sit quietly on a small stool or box in the middle of the sheep pen, your sheep will be curious about you and begin to approach. Remain still and talk softly; your sheep will soon come close enough to smell your hair and nibble at your clothing.

As your new sheep becomes tame, it will probably want to be scratched and petted. Never attempt to scratch it on the top of its head or nose; sheep don't like it. They prefer to be stroked under the chin and scratched on their chest between their front legs. Some sheep like to have their back or rear scratched.

The time will come when you will need to make your sheep do something it doesn't want to do. By training a sheep properly, you can get it to do such things without becoming frightened or losing its affection for and trust in you.

Although calm adult sheep that are used to people can be trained, it's easier to begin training as early as possible. A lamb that has not been handled much is practically impossible to catch in a pasture. Training should start once the lamb begins to nibble at grass (about 2 weeks of age).

Never hold a sheep by its wool; this hurts the sheep. To make your sheep stand still, place one hand under its chin and the other hand on its hips or dock (tail area) and slightly to the rear. When held this way, it can't go forward or backward, and it can't swivel away from you. With your hand gripped firmly under the chin, walk your sheep forward by giving it a gentle squeeze on the dock. As the sheep starts moving, all you need to do is keep up with it. When you want it to stop, hold it back with the hand under its neck.

Hold your sheep with one hand under its chin and the other on its dock.

Halters

When your lamb is about 1 month old, you should begin to train it with a halter. Your sheep will not like its halter at first, but with practice and repetition, it will soon settle down. After the halter is in place, grip it close to the head and lace your other hand over or behind the sheep's hips, just as you would without a halter. Start leading by pulling on the halter and pushing on its rear. Be patient. A lamb may buck and fuss a bit, but it will soon get used to working with the halter.

Making a Halter

An inexpensive halter can be constructed easily from nylon rope. To make a show halter, which has a short lead strap, you will need a 6-foot length of ⅜-inch-diameter, three-ply rope. If the halter is to be used for training and tying up, you will need a piece of rope 8–10 feet long.

Cut the rope by holding it over a flame at the desired length. Slowly rotate rope over flame. Once the nylon rope has melted apart, and while the melted nylon is still hot enough to stick together, use pliers to squeeze the ends to seal all nylon strands together. Do not use your fingers; the ends of the rope will be hot. Loop the rope together, as shown in the steps below.

To create holes in the rope, twist the ply open

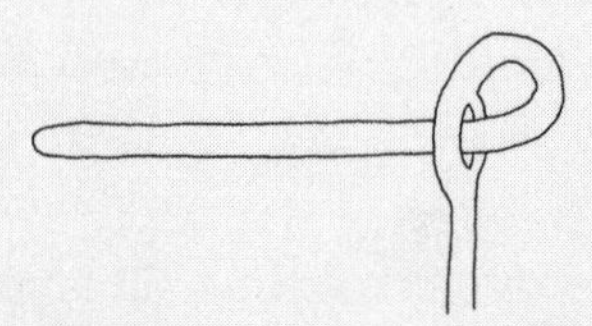

Step 1

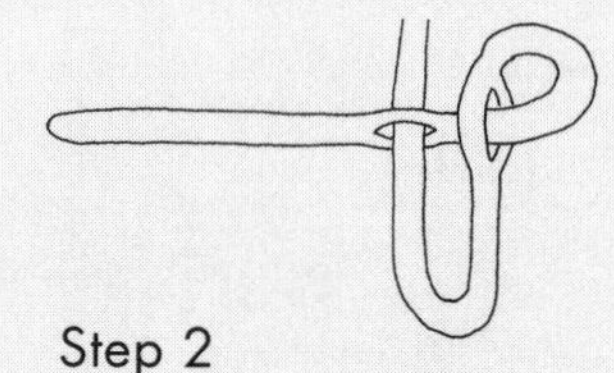

Step 2

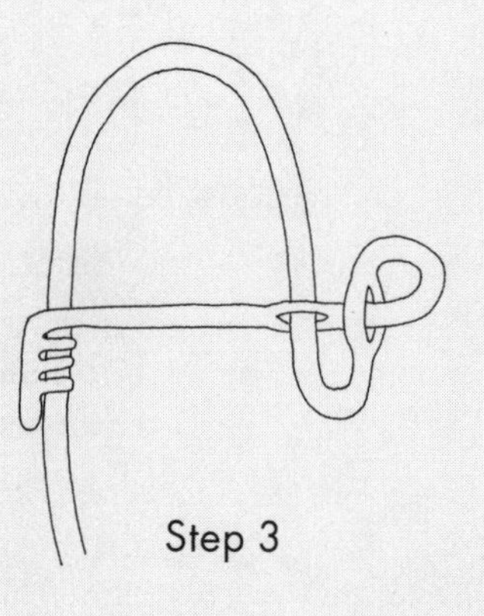

Step 3

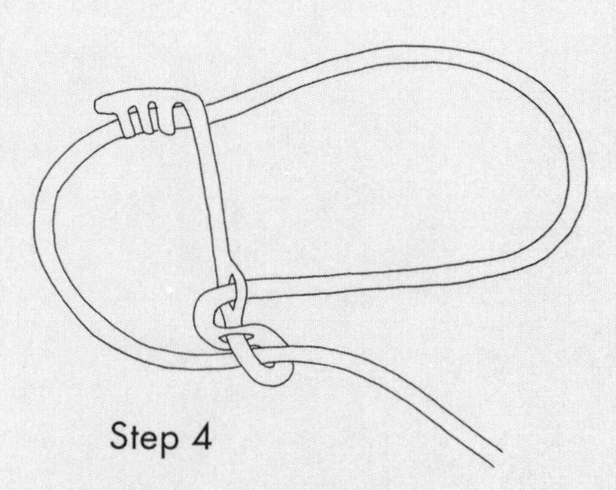

Step 4

The Chair Hold

To give a sheep shots, trim its feet, or shear its wool, take advantage of a reflex all sheep have: Once all four of their feet are off the ground, they can be placed on their rump, and they will sit still.

To get your sheep into this sitting position, slip your left thumb into the sheep's mouth in back of the incisor teeth and place your other hand on the sheep's right hip. Bend the sheep's head sharply over its right shoulder, and swing the sheep toward you. Lower it to the ground as you step back. From this position, you can lower it flat on the ground or set it up on its rump for foot trimming.

Withhold feed for several hours before handling or shearing sheep. This sitting position is uncomfortable for sheep with a full stomach.

Sheep in this sitting position will hold still while you shear, trim, and give shots.

FEEDING SHEEP

Sheep, like goats and cows, are ruminants. This family includes such animals as reindeer, buffalo, elk, moose, deer, cattle, and goats. They have four compartments in the upper part of their digestive systems that work together to break down plant material such as fresh grass and hay. In the three "fore-stomachs" (rumen, reticulum, and omasum), the feeds are mixed together and fermented by microorganisms, and in the one true stomach (abomasum), the fermented feed is finally broken down by enzymes. This large digestive tract with its separate compartments helps ruminants ferment and digest a wide variety of feeds, including flowers, leaves, straw, young twigs, soft bark, apples, grain, and pine needles. (See pages 313–314 for a detailed description of the ruminant digestive system.)

Sheep cannot adapt quickly to big changes in their diet, such as a change from feeding on pasture grass to feeding on grain. The microorganisms that help digest the grass differ from those that digest grain. Therefore, you must make dietary changes slowly, to give the appropriate microorganisms time to grow in numbers. It takes a week or more before a grass-fed animal can digest a large amount of grain.

Bottlefeeding the Orphan Lamb

Feeding the orphan lamb is a lot like feeding a newborn human baby. The key to trouble-free feeding is to offer a small amount of milk four to six times a day. Feeding these small amounts helps prevent digestive problems, such as colic and scours.

1–3 days old. Begin by mixing lamb milk replacer (available at feed stores) with *twice* the water called for on the label. Give approximately 4 ounces per feeding, four to six times a day, using a human baby bottle. Hold the bottle vertically, with the nipple pointing almost straight down, so that the lamb has to stretch upward to nurse. This is the position the lamb would be in if its mother were nursing it, and it prevents the lamb from swallowing air. In the natural nursing position, the lamb's nose and face are somewhat buried under the ewe's hind leg and udder. Some lambs respond to bottlefeeding better if you hold the bottle close to the nipple between your thumb and forefinger, so that your free fingers lie loosely over the lamb's nose. This position provides warmth and closeness that feels natural to the lamb and creates a bond between the two of you.

The act of nursing activates a reflex that causes milk to go directly to the abomasum, bypassing the rumen. If weak lambs are tube fed, this reflex does not occur, and milk can enter the rumen, causing rumenitis.

3–7 days old. At about 3 days of age, begin to increase the amount of milk replacer powder in the formula, until you are mixing it full strength (as per the label instructions) by about the fifth day.

7–14 days old. At this age, the lamb will outgrow the human baby bottle. Get a regular lamb nipple and place it on a 12-ounce beverage bottle. Continue to slowly increase the amount of milk per feeding. At 1 week of age (depending on the size of the lamb), a healthy lamb should be taking about 4 to 6 ounces of formula per feeding. As the lamb gets older and takes more milk at each feeding, reduce the number of feedings to three per day.

About 10 days old. At this time, the lamb will begin to show curiosity and interest in solid foods by nibbling at grass, hay, and grain. Provide a small amount of commercial lamb feed, called creep feed, or finely ground 15 percent protein supplement and water. At this age, the lamb's digestive system begins to develop in response to the roughage it starts to eat. Commercial creep feeds usually contain small doses of medications that help control common lamb illnesses.

15–30 days old. Continue to feed approximately 6 to 8 ounces of milk per feeding, three times daily. During this period, the lamb will begin to eat more creep feed and leafy alfalfa or good grass hay.

Depending on their size, orphan lambs can be gradually weaned from milk if you start to reduce the frequency of feedings beginning at 6 to 8 weeks of age, as long as they are consuming adequate amounts of pasture or creep and hay. If your lamb enjoys and finishes the feed you give it, you can start weaning.

Scours

Bottlefed lambs are more prone to infections and stomach upsets at an early age than are lambs nursing their mothers. Overfeeding may cause scours, or diarrhea. If scouring occurs, reduce the amount of milk you feed the lamb. Using water instead of milk for one feeding every 24 hours helps reduce the diarrhea and ensures that the lamb gets all the water it needs. A few teaspoons of Pepto-Bismol or diarrhea medication for human infants can be used for baby lambs. Scouring lambs may also benefit from electrolyte solutions (see page 204).

Unclean bottles can cause scours. Be sure to sterilize bottles before feedings to eradicate bacteria.

Diarrhea is usually yellow in color and pasty in consistency. Should the diarrhea turn a whitish color, get help from your veterinarian right away. If a vet is not available, begin antibiotic therapy immediately. You can use one of the diarrhea medications for young calves or swine that are available at most feed or animal health stores. The dose of these medications depends on the weight of the animal, as directed on the label.

The Nursing Lamb

The feeding schedule for the nursing lamb is identical to that for the orphan lamb, except that you don't have to bottlefeed. Observe the lamb and check the ewe, especially during the first week, to make sure the ewe is producing enough milk. Lambs that are receiving enough food will be alert and lively. They sleep soundly and stretch when they wake up. Underfed, weak lambs are dull and listless. Ninety percent of lamb deaths during the first week of life are due to chilling and starvation, not disease. The starving lamb will rarely cry out after the first day of hunger, so observe carefully.

Begin feeding creep when the lamb is about 10 days old, just as you would for an orphan lamb. Feed the same type of hay as is being fed to the ewe. The lamb will often follow her example. Mixing some of the ewe's grain mixture with the creep feed will also stimulate the lamb to begin eating solid food. You can offer a small amount of salt and mineral/vitamin supplement, too. At about 2 weeks of age, lambs begin to nibble just about anything within reach.

The age at which nursing lambs are ready for weaning can be anywhere from 6 to 16 weeks, but it's usually about 12 to 14 weeks. Separating the young from their mothers during weaning can be upsetting for everyone, including the shepherd. Leave the lamb in its familiar pen and remove the ewe to new quarters. If you don't have enough space to separate them completely, remove the ewe during the daytime hours and reunite her with her lamb during the night. This results in less nighttime crying. Forcibly weaning a lamb from its mother is not particularly necessary if you don't need a lot of milk. You can milk the ewe in the morning and let the lamb nurse at night. As the ewe's milk supply declines normally, she weans by not allowing the lamb to nurse as long or as often.

Why Wean?

Weaning is necessary if you want the sheep's milk for your own use.

Building a Creep Feeder

Creep feeders are feeding stations for creep. They must be protected by an outer fence that allows entry for lambs but not larger ewes. The creep feeder should be sheltered, should have fresh water provided daily, and should be well bedded with clean hay or straw. If the creep is in a barn, it should be well lit.

You can make your own creep panel from a variety of materials, ranging from welded pipe to 1" x 4" lumber (as shown on page 248). An effective creep gate relies on the perfect combination of height and width. For most breeds, the necessary dimensions are 8 inches wide and 15 inches high.

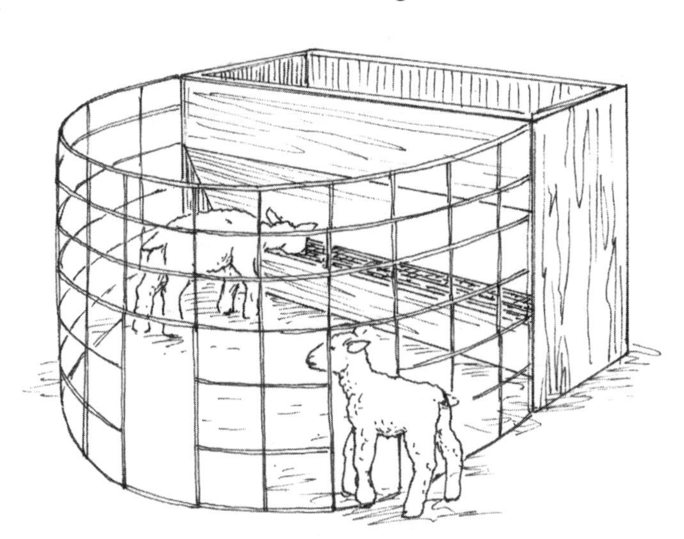

Creep feeders allow lambs to enter and get extra feed, while keeping out larger ewes.

Creep panel

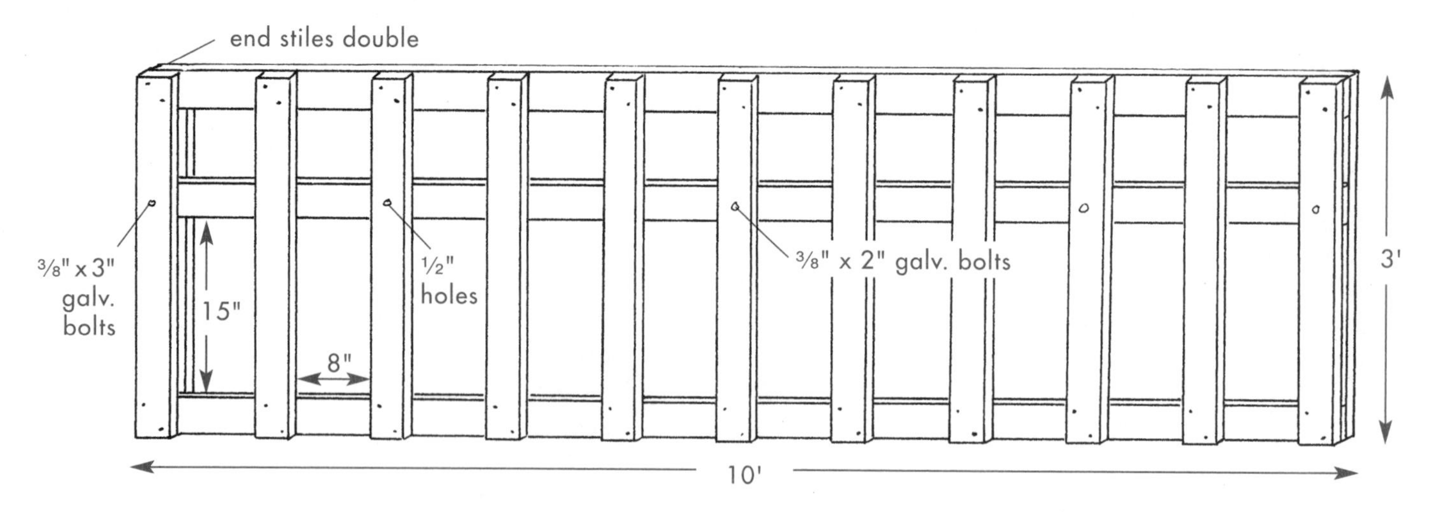

Creep gates take a lot of abuse. If you construct one from wood, do not use nails; fasten all boards with bolts. Fat lambs and big ewes trying to get in will quickly wreck a nailed-together creep gate.

Ewes and Rams

Feed for mature ewes and rams consists of grass pasture, salt, a mineral/vitamin supplement, and water. As long as they have enough forage and water, adult sheep will be properly fed.

Sheep will feed themselves from pasture grass. In periods of drought, when the grass becomes short, or in winter, you must supplement your sheep's diet with hay and/or grain. Feed sheep a green, leafy grass or alfalfa hay. Never throw the hay onto the ground, because it will become dirty and wet and contribute to worm infection. Use a feeder of some type.

A good supply of salt and a mineral and vitamin supplement is important for good health. These supplements are usually placed in sturdy wooden boxes that won't tip over or in a hanging feeder. The salt should be loose or granulated. Never feed sheep salt blocks intended for cattle. The sheep may attempt to chew the block and harm their teeth. The same is true for the mineral/vitamin supplement. Cattle minerals often contain levels of copper that can be toxic to sheep.

Pasture Rotation

Pasture rotation is good for grass growth as well as for sheep health. In a large pasture, use woven-wire fencing or portable electric fencing to divide the larger area into smaller paddocks. While the sheep spend a few days grazing one paddock, the grass in the empty paddocks has time to regenerate. The sheep are confined to smaller pastures, but the grass they eat is always fresh.

Water

From the day you bring your sheep home, you will need to provide them with a constant supply of fresh water. For a few sheep, a washtub works nicely for watering. Do not use a bucket; it is easily tipped over.

Plants Toxic to Sheep

Sheep are curious and love to nibble at strange plants, so you must prevent them from contact with any poisonous plants or shrubs in their grazing area. Your county Extension agent can tell you what poisonous wild plants grow in your area, but you'll need to investigate potentially dangerous ornamental plantings. The most common plants that are poisonous to sheep are American or Japanese yew, lupine, milkweed, nightshade, oak, ragweed, rhododendron, sheep laurel, and tansy ragwort.

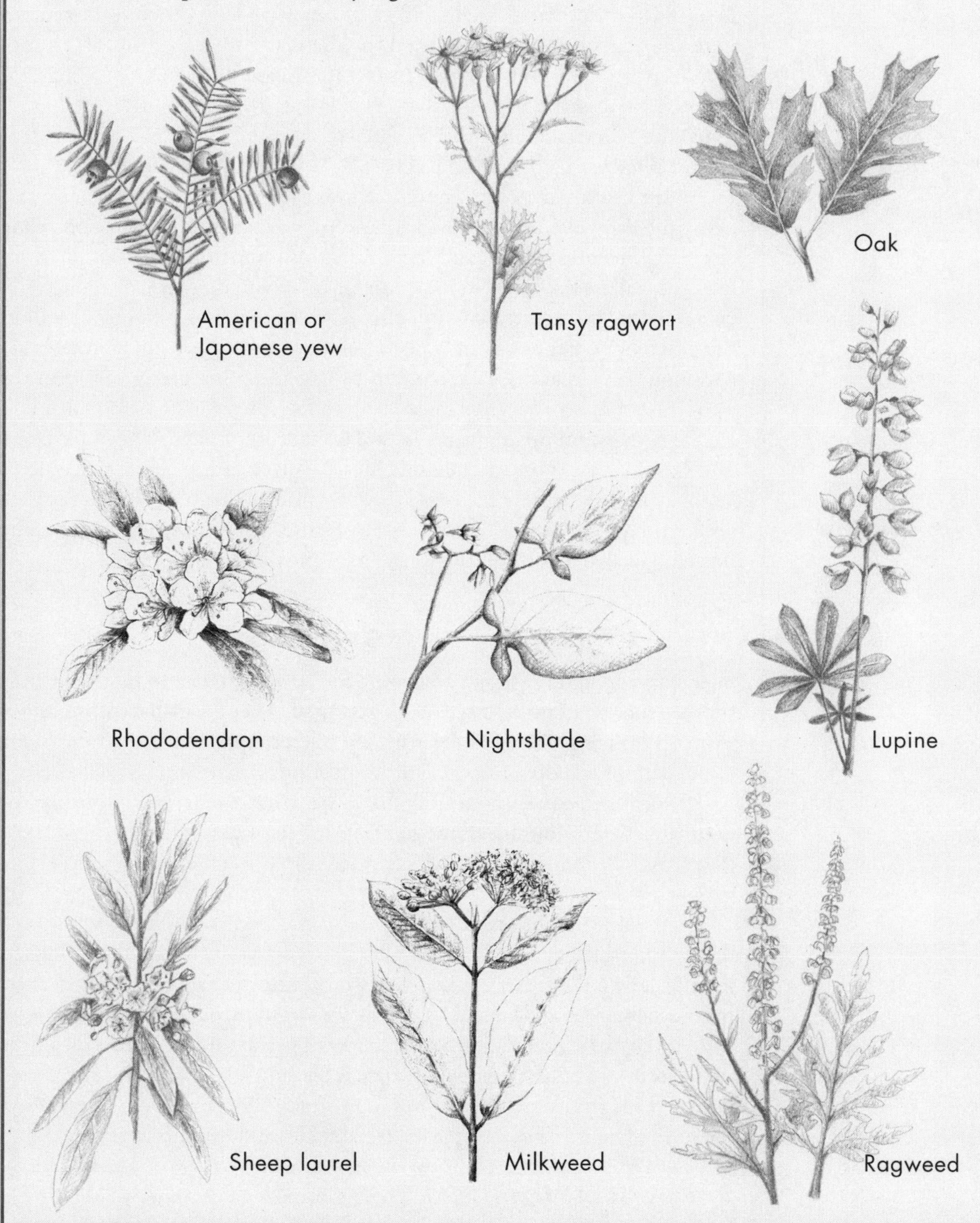

HEALTH CARE, TREATMENT, AND MEDICATION

All sheep need grooming, health care, and medical treatment. Following are the basics for keeping sheep in good health and for dealing with health problems when they arise.

Vaccinations

Lambs should be vaccinated at an early age. Consult your veterinarian, a county Extension agent, or an experienced shepherd about vaccinations for your sheep.

Sheep are commonly vaccinated against diseases that infect the lungs, the digestive system, and the reproductive tract. For example, lambs (as well as ewes and rams) can be vaccinated with Nasalgen-IP, an intranasal vaccine that helps protect against pneumonia. It should be given to lambs during the first 3 to 7 days of life.

All lambs should receive immunizations against enterotoxemia, a clostridial disease caused by overeating. Ewes should be vaccinated or given booster shots for clostridial diseases before lambing. (Take this opportunity to vaccinate your ram, too. You will be less likely to forget any sheep if you do them all at once.) This vaccine not only protects the ewe but also increases the disease-fighting antibodies that she passes on to her newborn lambs when they nurse. Covexin 8 protects against all common clostridial diseases that affect sheep. Another group of vaccines protects ewes from diseases that may cause them to lose their lambs before birth.

Disease control programs can be somewhat complicated. The important thing to remember is that most vaccinations must be given either before breeding or before lambing. Don't wait until the ewes are almost ready to lamb before you vaccinate them.

Keep regular records of your sheep's health history, including dates and vaccinations given. The box on the facing page provides a sample chart.

Worming

Sheep have a high resistance to disease but low resistance to internal parasites. All sheep, especially lambs, need to be wormed. Sheep can pick up many types of worms through grazing. The worms are microscopic and live off the blood of the sheep. Lambs can die of severe worm infection.

If you are raising sheep for the first time, your pasture will probably be uncontaminated by previous use. Cross-fencing and pasture rotation help reduce exposure to worms. Clean feeding facilities further help prevent rapid or serious buildup of stomach worms.

Even with precautions, however, sheep can become infected with worms, and they will need to be treated periodically. This is especially true in mild climates or after a particularly mild winter. Several good worm medicines are available that are safe even for pregnant ewes and small lambs. Obtain worm medicine from your veterinarian, who can give you information on the proper dose and timing. The medicine may come in the form of a bolus (large pill), which is placed at the base of the sheep's tongue by using a bolus gun, available at farm supply stores. Worming medications also come in the form of liquid drenches, pastes, feed blocks, and injections.

Symptoms of Worms. A lamb with worms may be underweight for its age and have a potbelly, prominent hip bones, a scruffy wool coat, runny or loose manure

HEALTH MAINTENANCE RECORD

Name__________ Ear tag no.__________

Birth date__________ Birth weight__________

Color__________ Sex__________

Breed__________ Dam and sire__________

Vaccinations		Worming	
Product	Date	Product	Date

Lambing history__________

Exposure date__________

Lambing date__________

Ear tag nos. of lambs__________

Misc. comments

and manure buildup on its rear, and anemia, which makes it weak. Anemia may be prevented by worming lambs when they are 2 or 3 months old. If you have only one or two sheep on clean pasture, you may need to worm your lambs only once more during the first year. In large flocks that have limited or contaminated pasture, the lambs may have to be wormed monthly. Adult sheep with heavy worm loads will sometimes have a swelling under their throat or chin called bottle jaw. Ewes should be wormed before breeding, before lambing, and before going out on fresh spring pasture.

After Worming. Ideally, you will put your sheep into the last pasture they grazed in immediately after worming and transfer them to a clean pasture 24 hours later. The sheep will expel the worms and eggs in the old pasture and will not contaminate the new pasture as rapidly. This is possible only if you have cross-fenced the pasture. The eggs and larvae of many worm species can survive as long as 3 months in cool, damp weather but may die within a few weeks during hot, dry weather.

Keds

Keds are wingless flies. They cause severe itching and discomfort. If untreated, these parasites cause loss of wool and damage a potentially valuable pelt. Keds are easily controlled by using insect-killing sprays, pour-on formulas, and medicated powder. You can get rid of keds completely if you treat them with an effective pesticide. You may have to apply the pesticide twice, the second application about

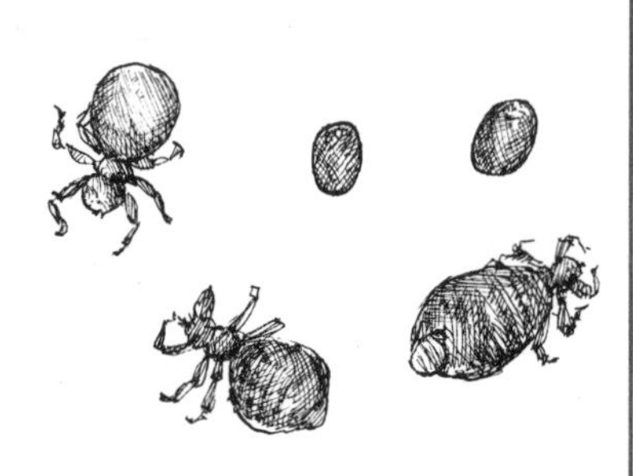

Sheep keds and eggs

3 weeks after the first. Your flock will stay free from keds unless you bring in an infected animal or let your sheep come into contact with infected animals.

The best time to treat sheep for keds is immediately after shearing; however, be sure to wait until any shearing cuts have healed. Treat the entire flock, including lambs and rams. Move the sheared wool at least 50 feet away from the flock, since any keds in it will crawl out in search of the animals.

As with any pesticide, read and follow the directions carefully for proper dose and method of application, and pay attention to warnings about use on pregnant ewes or lambs and on lambs that will be sold for meat. Products containing the active ingredient permethrin (synthetic pyrethrum) are highly effective and safe.

Lice

Less common than keds, lice are external parasites that bite the sheep and cause severe itching. Lice are so small that they can barely be seen with the naked eye. Two types of lice affect sheep — those that suck blood and those that bite. Both cause intense itching. Sheep with lice will constantly rub and scratch against fences, posts, feeders, and almost any other object around the farm. A farm infested with sheep lice will have bits of wool hanging on just about everything the sheep can rub against. To treat lice, follow the same procedure as you would to treat keds.

Other Health Problems

Abscesses. Abscesses caused by bacteria are common in the skin of sheep. The most serious bacterial abscesses are those caused by *Corynebacterium pseudotuberculosis (C. ovis),* which can result in caseous lymphadenitis (pseudotuberculosis).

Symptoms/effects: Abscesses form firm nodules under the skin due to infection of lymph nodes. Severely affected animals may exhibit gradual loss of condition (wasting). Affected animals raised for meat may be condemned at slaughter; that is, their meat will be discarded.

Prevention: The causative bacteria are often present on the skin and in the environment, and enter the skin through wounds. In sheep, these wounds are most often incurred during shearing. Be very careful while shearing, and disinfect any wounds that occur. Disinfect all instruments after shearing an affected animal to help prevent spreading the bacteria to other animals. Culling affected animals is often advised.

Treatment: Antibiotic treatment is not effective. Surgery should be attempted only for very valuable animals.

Copper Toxicity. Copper is an important mineral in the diet of sheep and other animals. High levels can be dangerous, however.

Symptoms/effects: Excess copper builds up in the body, causing changes in the liver, but symptoms are not seen until the animal suddenly develops red blood cell breakdown and dies. Most affected animals are found dead with no previous signs, although yellow-colored gums and conjunctiva and passage of red-colored urine are occasionally seen just before death.

Prevention: Do not feed horse or pig feeds, which are high in copper, to sheep.

Johne's Disease. Johne's disease is also called paratuberculosis. It is a bacterial disease caused by *Mycobacterium johnei (M. paratuberculosis).*

Symptoms/effects: Sheep develop progressive loss of condition. Cattle with Johne's disease often have chronic scouring, but this is much less common in sheep.

Causes of Wasting in Sheep

Severe loss of body condition is known as wasting. Unless you regularly handle your adult sheep, it can be difficult to see when an animal is wasting away. Be sure to regularly check the condition of your flock.

Cause	Diagnosis	Treatment
Inadequate feed	Evaluate your feeding program	Increase amounts of high quality forage and concentrates fed
Bad teeth	Check teeth	As for inadequate feed
Internal parasites	Fecal examination	Appropriate wormer
Johne's Disease	Fecal examination, blood or skin test, postmortem examination	None
Scrapie	Biopsy testing of conjunctiva, postmortem examination	None
OPP	Blood test or postmortem examination	None

Prevention: Purchase sheep from healthy herds. If infection occurs in your flock, talk with your veterinarian about testing all your animals. Any positive animals should be culled.

Orf. Orf is also called contagious ecthyma or sore mouth. It is a viral disease that is readily passed to susceptible animals.

Symptoms/effects: Orf affects lambs, causing sores around the lips and muzzle. Severely affected lambs may have difficulty nursing and eating and can lose condition.

Treatment: The sores will heal by themselves in 1 to 4 weeks. Give extra milk or feed to lambs that are not eating enough to maintain condition.

Prevention: The virus that causes orf can survive for years in some environments. A vaccine is available, but it should be used only in infected flocks and only under a veterinarian's supervision.

Human health hazard: This virus can cause skin sores on humans. Wear gloves when handling affected lambs.

Pneumonia. Pneumonia in lambs is most often caused by bacteria. Pneumonia in adult sheep can also be caused by bacteria, but is more commonly caused by the virus that causes ovine progressive pneumonia (OPP). Pneumonia due to inhalation of food or medications can occur in both lambs and adults.

Symptoms/effects: Difficulty in breathing is common to all pneumonias. Bacterial pneumonia and inhalation pneumonia are often accompanied by fever. Ovine progressive pneumonia causes progressive breathing difficulties and severe loss of condition that may occur over many months or even years.

Treatment: Antibiotic treatment is often effective for lambs with bacterial pneumonia. Inhalation pneumonia often also involves bacteria, but it is more difficult to treat with antibiotics, due to the severe lung damage caused by the inhaled substances. There is no treatment for OPP.

Prevention: Buy healthy animals from a well-maintained facility. Maintenance of a healthy environment is the most important aspect of preventing pneumonia in lambs. Be careful when bottlefeeding and giving medications to avoid inhalation pneumonia. If there is OPP in your flock, blood testing and culling positive animals is the only preventive measure available.

Scrapie. Scrapie is a disease of the nervous system caused by infectious agents known as prions (see also scrapie in goats, page 216).

Symptoms/effects: Affected sheep slowly develop signs of intense itchiness and often scrape themselves against objects (hence the name), rubbing off their wool. Infection of the nervous system causes staggering and trembling. There is severe progressive loss of condition.

Treatment: There is no treatment for scrapie.

Prevention: Scrapie is a reportable disease. Affected animals must be destroyed, and regulatory officials will quarantine the herd and repeat testing until the herd is certified to be free of scrapie. This may seem excessive, but these are necessary measures to control a serious infectious and untreatable disease.

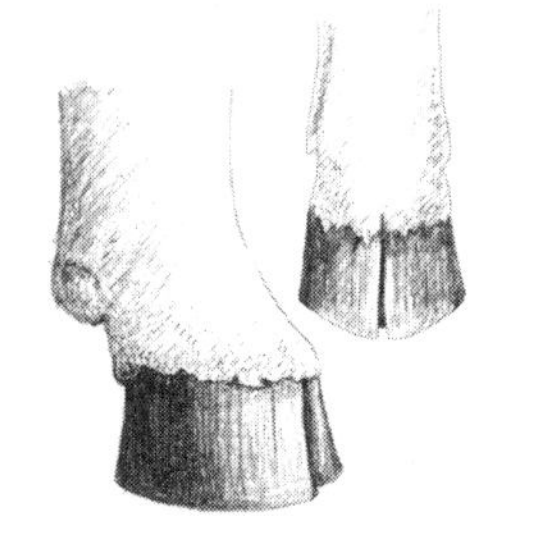

Properly trimmed hooves

HOOF CARE

Wild sheep and those that graze on mountainous pastures wear down their hooves by traveling over rocky ground. Domestic sheep walk primarily on soft soils. Consequently, their hooves become long and overgrown, which makes walking painful. Sheep need to have their hooves trimmed so that they can walk properly and to help prevent hoof diseases, such as scald and foot rot. Before buying an adult sheep, be sure to look at its feet. Ask when they were last trimmed. If they show any need for trimming, the owner should be able to demonstrate the procedure for you. If the owner can't show you how, don't buy the sheep — this is a sign that the sheep's hooves have probably not been properly cared for.

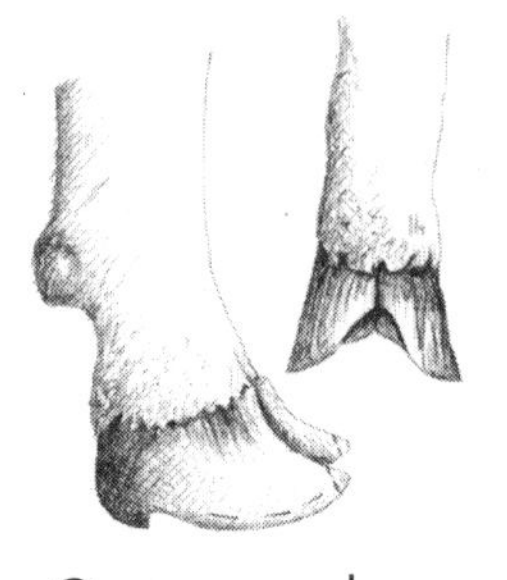

Overgrown hooves

How often to trim your sheep's feet depends on your pasture, paths, and barn floor. Many shepherds must trim twice a year, whereas others need to trim only once a year. If you notice limping sheep or sheep on their knees, check their feet. While you take care of other needs, such as worming or shearing, make a habit of trimming hooves as well.

When you trim hooves, wear leather gloves. Set the sheep or lamb on its rump (as shown on page 245). Trim the rear feet first, and then the front feet. A kicking sheep that is not held firmly or properly can injure you with the sharp edges of the freshly trimmed hoof.

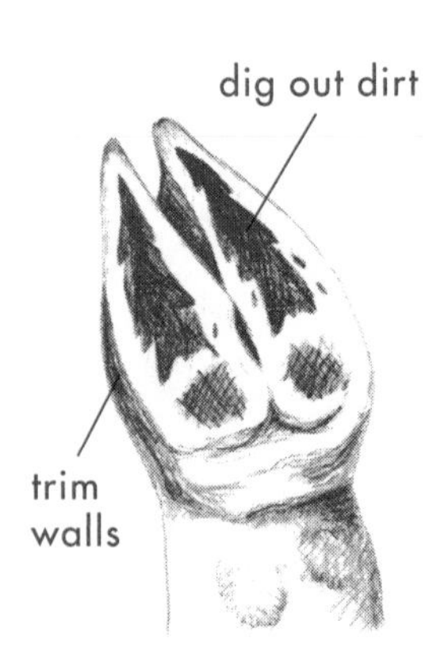

Trim hoof walls and remove dirt between toes

Many styles of hoof shears are available, but few are as easy and safe to use as Felco No. 2 pruning shears. Many farm and garden stores sell them. If you have many sheep to trim, the investment is well worth the money.

Foot Rot

Foot rot, a bacterial disease that causes hard-to-cure infections, is one of the worst diseases that can infect your sheep. Active outbreaks of foot rot occur during warm,

Top 10 Tips for Health and Happiness

1. Have regular feeding times. When you are supplementing your sheep's diet with grain or hay, feed at about the same times every day. Feedings are usually given once in the morning and once in the evening.

2. Stick to the regular diet. Sheep cannot adapt to sudden changes in the type or amount of feed. Such changes can make them sick. If you are going to change feed, do so gradually. Start adding the new feed and reducing the old feed over a period of about 10 days.

3. Provide plenty of fresh water. Water is an important part of your sheep's good health, and they will drink more if it is fresh. If manure gets into the water, empty the container and clean it well before refilling it.

4. Lock down the grain supply. Sheep love grain and will figure out how to get into the grain storage area, if possible. Keep grain in a tightly lidded container that cannot be reached or knocked over by the sheep. Overeating grain can cause severe stomach upset and possibly death.

5. Keep dogs away. Never allow your dog or any other dog to play with your sheep. Sheep are afraid of dogs, and a dog that is barking at them through the fence or running in the pasture causes sheep severe stress, which can lead to overexertion, heat stress, and heart failure.

6. Get rid of keds and lice. If your sheep have keds or lice, treat them immediately.

7. Practice good hoof care. Hoof trimming usually needs to be done once or twice a year. Sore, overgrown hooves are not comfortable. (For more on hoof care, see the facing page.)

8. Treat for worms. Sheep and lambs pick up worms from the pasture and should be treated at least twice a year (more often if you have many sheep on a small pasture). Consult your veterinarian or Extension agent for instructions on the type and use of deworming medications. (For more about worming, see pages 250–251.)

9. Provide shade. Sheep need shade in the summer. An open-sided shed, shade trees, or a canopy roof can keep them cool.

10. Train. If you're not raising your sheep for meat, work with them. Even if you don't plan to take your animal to livestock shows, you will both be happier if you have done some constructive training. Start working with your lamb or sheep as early as possible. Once you have earned its confidence, it can be trained to follow, come when called, and stand still. Peanuts and small bits of apple make great rewards and treats.

wet weather. Foot rot is a painful disease; the bottom of the sheep's hooves literally rots off, exposing the soft tissues beneath.

Before buying a sheep, look at the rest of the flock. If you see any limping sheep or sheep with severely overgrown, misshapen hooves, go somewhere else. Foot rot can be introduced into a clean flock only by an infected animal or a carrier animal (one that shows no symptoms but spreads the infection to others). Once foot rot is introduced into your flock, it is almost impossible to get rid of.

Some sheep raisers disinfect their sheep's feet after they've taken them to shows, where they have been in contact with other animals. To disinfect, you must walk your sheep through a germicidal footbath.

Possible Causes of Lameness

- ◆ Overgrown, untrimmed hooves
- ◆ Mud, stone, or other matter stuck between the toes
- ◆ Plugged toe gland
- ◆ Abnormal foot development (inherited defect)
- ◆ Foot abscess
- ◆ Foot scald
- ◆ Foot rot
- ◆ Thorns, punctures, bruises, or other injuries

Plugged Toe Glands

Sheep have a deep gland between the two toes of each foot. Look for a small opening at the top of the front of the hoof. The gland secretes a waxy substance that has a faint odor and is said to scent the grass to reinforce the herding instinct. When the gland becomes plugged with mud, the secretion is trapped, and swelling and lameness occur.

To unplug a toe gland, squeeze it so that the plug pops out. Then disinfect the gland with Listerine or hydrogen peroxide.

Foot Scald

Foot scald is a skin infection that occurs between the toes. The first sign of scald is limping. Usually, the front feet are the first ones that get sore. On sheep with foot scald, the skin between the toes is moist, hairless, and red from irritation. Scald infections are more common after a long, wet period. The infection can be treated with any of the disinfectants used to treat foot rot. Move the affected animals to a clean, dry area.

Breeding

The normal breeding season for sheep is late summer and fall, from about mid-August to mid-November. Some areas have a breeding season in the late spring, during May and June. Check with local sheep raisers or your Extension agent to determine the best time for breeding in your region. A ewe gives birth 147 to 153 days (about 5 months) after she is bred.

Feeding Requirements

Proper feeding plays an important role in breeding. Before breeding, you must increase the energy value of the ewe's feed to be sure she is well nourished (but not to fatten her). This process, called flushing, increases the number of eggs released by the ewe at breeding, improving her fertility and the chance of twinning.

You can flush your ewe by grazing her on a new lush pasture or by feeding her a small amount of grain (or a combination of the two). About 3 weeks before you plan to breed the ewe, begin feeding her ¼ pound of grain a day. Gradually increase this amount to 1 pound per day. Feed 1 pound a day for 17 days, breed the ewe, then gradually taper off the grain ration. If your pasture or forage is of low quality (as is often the case at breeding time), continue to feed a small amount of grain for 3 weeks after the ewe is bred. Be sure that the sheep pasture is free of clover; it decreases a ewe's fertility.

When the ewe is about 30 days away from lambing, begin the grain ration again, increasing slowly to 1½ to 2 pounds per day, depending on the ewe's size. The ewe needs these additional calories for the growing lamb or lambs she carries. And throughout the sheep's pregnancy, be sure to provide a constant supply of salt, a mineral and vitamin supplement, and fresh water. Provision of selenium in deficient areas is particularly important to ensure the birth of healthy lambs. Talk with your veterinarian or Extension agent about the selenium needs in your area.

Failure to meet these nutritional demands may result in weak lambs, low birth weights, or a sick ewe that could die. Pay particular attention to the ewe during the last 2 weeks of pregnancy. If she acts weak or listless, call your veterinarian for help immediately.

A pregnant ewe, particularly one that is fat or sluggish, needs plenty of exercise, especially in the last month before lambing. Feed her at some distance from the barn so that she'll have to move around to get something to eat.

Preparing the Ewe

Some producers breed ewes to give birth at 1 year of age, while others hold them for more than a year and breed them to lamb as 2-year-olds. As a rule of thumb, ewe lambs should weigh 85 to 100 pounds, or at least 65 percent of their mature weight, before they are bred.

Young ewes need more feeding and management than older ewes. They are growing while pregnant and continue to grow as they are producing milk. Furthermore, because their maternal instinct is not as well developed, the chance that they might reject the lamb is greater.

If your ewe has long fleece or a lot of manure around her rear end, 4 to 6 weeks before lambing, tag or crotch her by shearing the wool from the udder area and around the dock. Crotching reduces the risk that a lamb will suck on a dirty tag of wool before it finds a teat.

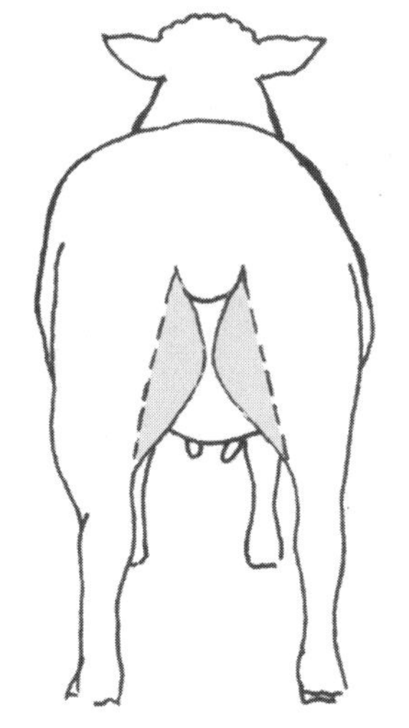

The shaded area indicates the area for crotching.

Preparing the Ram

If you own the ram to which the ewe will be bred, feed him just as you feed the ewe. He'll be working hard, too, and if he is too focused on the ewes to graze properly, the grain ration will ensure that his nutritional needs are met.

Rams may become aggressive during breeding season, so stay on your toes. The docile ram that came up to you all summer to get his back scratched can suddenly become dangerous and may butt you in defense of "his" breeding group.

High daytime temperatures can cause infertility in a ram, and he won't become fertile again for 45 days. Your veterinarian can obtain a sample of semen from the ram to determine if hot weather has damaged the sperm. Even when the ram appears to be breeding normally, if the sperm is damaged, fewer ewes will become pregnant. To prevent hot-weather infertility, confine the ram in a cool place during the day.

When you are ready to have your ewes bred, turn the ram in with them at night, and let nature take its course.

Preparing for the New Arrivals

While you are eagerly awaiting the birth of new lambs, prepare your lambing pens, also known as jugs. A jug is a small pen about 4 by 6 feet square and about 36 inches high. It can be made from a variety of materials and does not have to be elaborate.

Gather the following items:

- **Hand shears.** These are used for tagging.
- **Antiseptic ointment.** You'll need such ointment for lubricating and disinfecting your hands if you have to assist in a delivery.

- **Paper towels and old, clean bath towels.** Towels are used for drying the newborn lamb.
- **Hot water bottle or hair dryer.** These items are used to rescue chilled lambs. If the jug has a source of electricity and you plan to use a hair dryer for warming chilled lambs, make sure you have an extension cord, too.
- **Solution of tamed iodine, such as Betadine, or chlorhexidine (Nolvasan).** This is used to disinfect the umbilical cord. Just before birth, pour the iodine or chlorhexidine solution into a small, widemouthed bottle, so that it is half full.
- **Bucket.** One of the first things a ewe will do after lambing is take a big drink of water (and she will prefer warm water).
- **Livestock molasses.** Mix some livestock molasses into the water for ewes that have just delivered a lamb to give them a boost of energy. It is available at the feed store.
- **Lamb bottle and nipples.** A baby bottle with a nipple that has a slightly enlarged hole is better for the newborn lamb than a standard lamb nipple. Enlarge the hole by making two small slits in the shape of a cross on the tip with a sharp knife. When new, standard nipples tend to be a bit stiff for week-old lambs; dip them in boiling water once or twice to make them easier to use. (With age and use, a nipple tends to become soft and loose. Avoid using an old nipple that the lamb can pull off the bottle and possibly swallow.)
- **Frozen colostrum.** If possible, have on hand a few ounces of colostrum from a heavy milking ewe or a ewe that has lost her lamb. Colostrum is great for use in emergencies and can be kept in the freezer for 1 year or more. Do not thaw colostrum in the microwave; this will destroy the antibodies that protect the lamb from disease. Instead, thaw colostrum slowly at room temperature or in warm water.

Purpose of the Jug

- Aids in mothering up
- Protects the lamb from being trampled by other ewes in the barn
- Prevents drafts
- Keeps the lamb from getting lost
- Prevents the ewe from going outside and exposing her lamb to harsh elements
- Provides you a better opportunity to observe your ewe and lamb during the critical first days of the lamb's life

Lambing pens, or jugs, can be built of any material you have handy. These pens are of the open-slat variety; the pen described at right has solid plywood panels.

Building a Jug

The following plans originated with the Norseman Sheep Company many years ago, and the design remains unchanged. Although these jugs cost slightly more to construct than do jugs built of scrap lumber and old pallets, their durability makes them more economical over time. Because these jugs are solid at floor level, they prevent drafts. In addition, the ewe can readily look out over the solid panels, so she doesn't feel trapped or confined.

The jugs measure 4 by 6 feet and are 3 feet high. They have holes at the four corners where they can be laced together with wire or twine. When more space is required, such as for triplets, four side panels can be laced together to produce a 6-by-6-foot jug. Conversely, if you wish to confine a single lamb because of sickness or injury, for example, four end panels can be used to make a 4-by-4-foot jug.

The jugs can be set up side by side, and as many can be placed in a row as space permits.

1. Cut a 2-by-4-foot piece from the end of each sheet of plywood. Cut one of the remaining sheets of plywood down the center to make two pieces that are each 2 by 6 feet long.

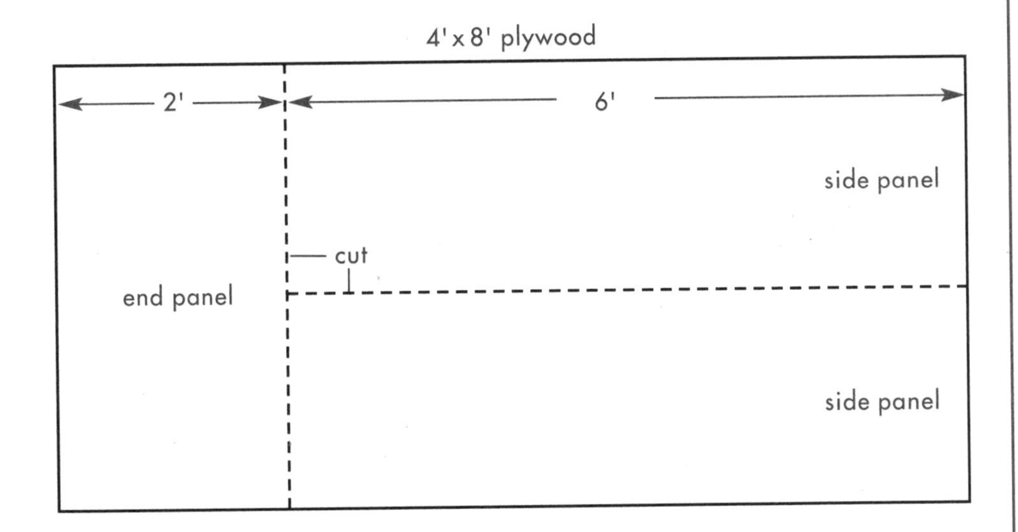

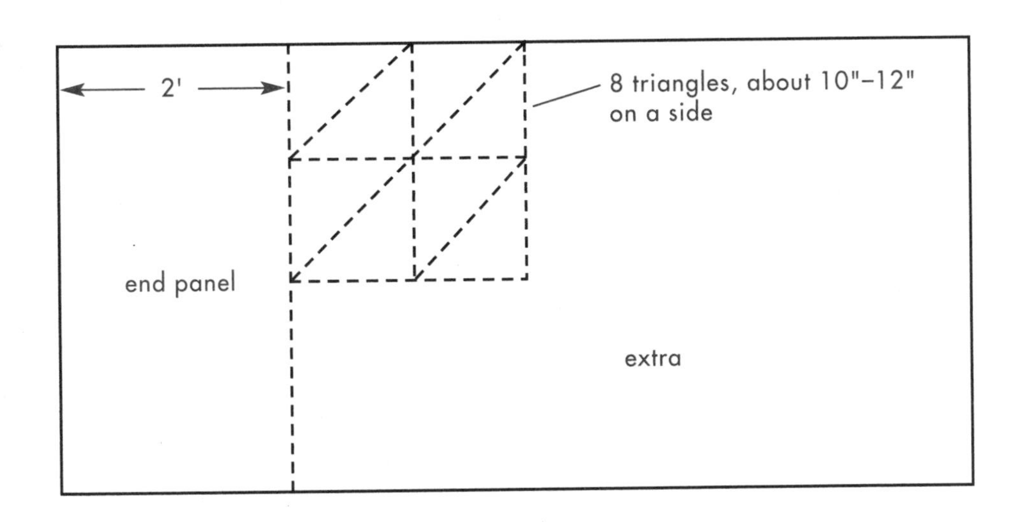

Materials for One Jug

- Two 4' x 8' sheets of ½-inch exterior sheathing grade plywood
- 74' of 2 x 2 lumber, for frames
- 16-penny nails
- 1½-inch drywall screws
- Twine or wire to fasten the panels together

Features of a Good Jug

- Inexpensive to construct
- Solid sides to prevent drafts
- Easily set up or moved
- Adequate space for single lambs, twins, or triplets
- Folds or stacks for easy storage
- Durable, to provide many years of use

2. From the 2 x 2s, make two frames for the ends, each measuring 3 feet by 4 feet. Use one 16-penny nail to fasten the frames together at each corner. The plywood sides and corner braces that you will be adding in steps 4 and 5 will strengthen them.

3. In the same way, construct two frames for the side panels, each measuring 3 feet by 6 feet.

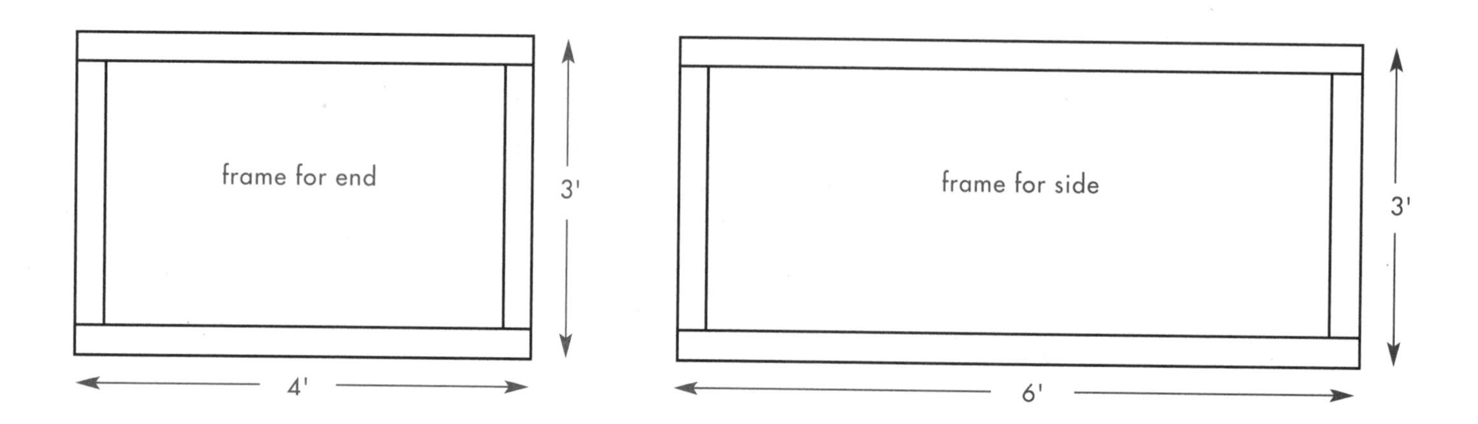

4. Screw the plywood panels to the frames. Fasten the 2-by-4-foot pieces to the smaller frames and the 2-by-6-foot pieces to the larger frames.

5. From the remaining plywood, cut eight triangular pieces about 10 to 12 inches on a side. Screw these to the top corners of the panels.

6. Drill ½-inch holes through the plywood at each corner. Run wire or twine through the holes so that you can fasten the panels together at the corners. Use one end panel as the door. Tie it on one side, so that it can swing open. Put a loop of twine around the free side to keep it closed.

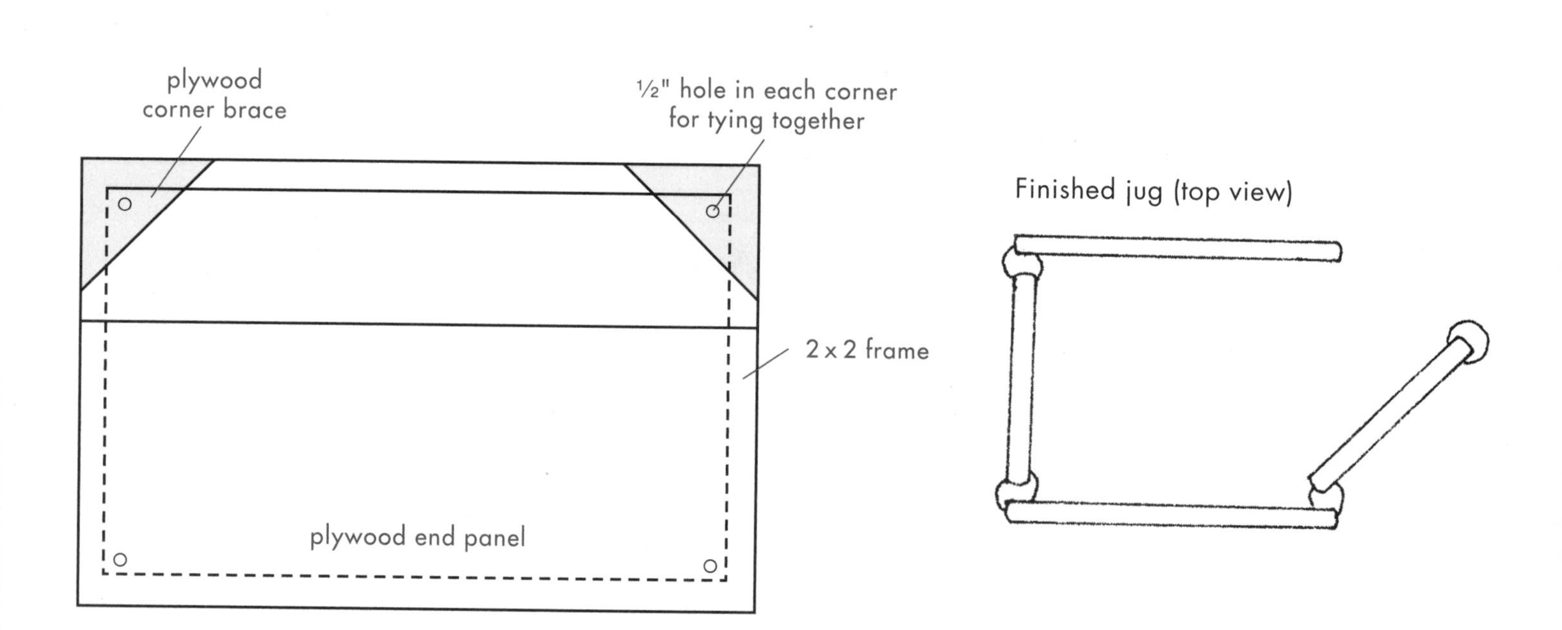

Lambing Time

About a week or two before you think the ewe will lamb, start checking the development of her udder — a process called bagging the ewe — which often gets larger as lambing time nears. However, udder size is not always a dependable sign. Some ewes develop large udders up to 3 weeks before lambing, whereas others have almost no udder development until after the lamb is born.

If your ewe goes off her feed or shows any signs of illness close to lambing time, consult your veterinarian immediately. Sudden signs of illness before lambing could indicate pregnancy toxemia. Signs of illness just before or after lambing could also be the result of a calcium deficiency. The symptoms of both of these conditions are similar, but either disorder is considered an emergency. Call your veterinarian at once; he or she can tell the difference and treat the illness immediately.

First Stage: Dropping of the Lamb

About 24 hours before birth, the ewe's abdomen will appear to droop. This dropping of the lamb produces noticeable triangular-shaped hollows just in front of the hipbone.

Second Stage: Early Labor

About 6 to 12 hours before birth, the ewe may appear uncomfortable and begin to paw at the ground before lying down. During this stage, she is restless and may stand up and lie down frequently. The ewe will be unhappy if you attempt to pen her up at this time, because she wants to choose her own nest. Be patient; observe, but do not bother her. She may occasionally roll over on her side, grunt, then roll back to a normal resting position. She may go to many different locations in the barn or pasture before choosing her final birthing spot.

Third Stage: Real Labor

About 1 to 3 hours before birth, the ewe will act more uncomfortable and may get up and lie down more frequently. As the contractions begin, she may roll a bit and paddle slightly with her feet, as though she were swimming. As the contractions increase, she will often point her nose in the air and let out a grunt. Shortly after, the water bag will emerge and soon break. Don't worry if the ewe gets up and walks around with the water bag hanging out; she is just trying to find a more comfortable position. The lamb will begin to appear soon thereafter. If it hasn't started to emerge after 30 minutes of hard labor (indicated by heavy straining and grunting), get help.

Nine out of 10 deliveries happen without problems. In normal births, the lamb's front feet and head appear first. However, several abnormal lambing positions can occur, which are further complicated if the ewe is carrying twins or triplets. Lambing and kidding are very similar, and assisting an expectant mother involves the same techniques whether you are working with a goat or a ewe; see When a Doe Needs Help on pages 196–197.

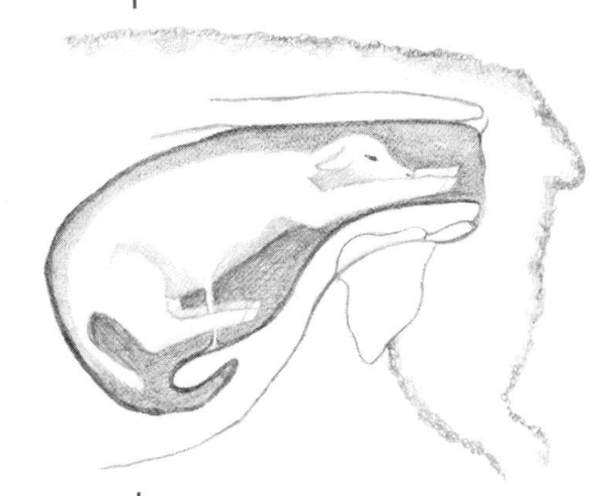

This cross-section view shows a lamb in the normal position for birth

After the Lamb Is Born

The first 10 or 20 minutes after birth are critical for the new lamb, especially if the weather is cold. You can do three things to help:

1. See that no membranes cover the lamb's face or nose.
2. Make sure the lamb is breathing. If it is not, rub it briskly with dry towels to stimulate it to breathe.
3. Place the lamb at the ewe's head and allow her to clean off the lamb.

Try not to disturb the ewe for at least 30 minutes after she gives birth. However, if the lambing was difficult, she may be tired and not attend to the lamb. If that's the case, dry the lamb with a clean towel. Then return the lamb to the ewe and allow her to bond with it with as little interference as possible. Too much activity on your part may cause the ewe to reject the lamb. Be patient, and let nature work.

Once the ewe has clearly accepted the lamb by licking it and making low, soft sounds, place the ewe and the lamb in the jug. Carry the lamb slowly toward the jug; the ewe will usually follow, as long as she can see the lamb or hear it call out. Then "snip, dip, and strip."

Snip. Use scissors to snip off the umbilical cord about 2 inches from the lamb's body.

Dip. Hold a small widemouthed bottle half full of a tamed iodine solution, such as Betadine, or chlorhexidine solution (Nolvasan) tight against the abdomen and dip the umbilical cord in the iodine until the cord is saturated. Use fresh iodine or chlorhexidine for each kid.

Strip. The ewe's teats have a waxy plug that must be removed before the lamb can nurse easily. Strip milk from both of the ewe's teats, gently forcing out the plug. One squirt of milk from each teat will ensure that the milk can flow freely.

In severely selenium-deficient areas, such as parts of the East Coast and the Pacific Northwest, newborn lambs may benefit from a selenium injection soon after birth. Consult your veterinarian or Extension agent regarding the selenium status of the soil in your area.

Under normal conditions, keep the ewe and the lamb in the jug for no more than 3 days. Ewes with very small lambs, twins, or triplets may benefit from an extra day or two in the jug.

Nursing and Colostrum

A lamb usually can stand about 20 to 30 minutes after birth. As soon as it begins to walk, it will be hungry. Most lambs (with experienced mothers) will find the teats and begin nursing naturally within the first hour. A lamb that has difficulty nursing because it is weak or chilled or because it has an uncooperative mother will need help. Milk the ewe and use a bottle to feed about 2 ounces of the colostrum, or first milk, to the lamb with a bottle. Colostrum contains antibodies that prevent disease, as well as laxatives, minerals, and sugars to give the lamb needed energy.

Feeding the Ewe

Shortly after the lamb arrives, the ewe will be hungry. Feed her a good-quality hay and provide a bucket of fresh warm water to which you've added 1 or 2 tablespoons of molasses.

Shortly after lambing, the udder in the normal ewe will fill with milk; this process is called dropping or letdown. If she is a good milking ewe, she will provide more milk than the lamb can drink, and the udder will become too full. If you feed her extra grain, the additional protein it provides will be converted to more milk, which could result in a grossly overfull, painful udder. To avoid this excess, do not feed grain for approximately 3 days after lambing. At that point, begin feeding grain. Gradually increase the ration until you are offering 1½ to 2 pounds daily. Feed the grain twice a day, so that the ewe doesn't overeat.

Caution

Do not use a 5-gallon bucket for the ewe's drinking water in the jug. The lamb could accidentally fall into such a tall bucket and drown. Instead, use a 10- or 12-quart mop bucket.

Ear Tagging

If you have more than two or three ewes, you may want to identify their lambs by applying ear tags. Some tags are self-clinching; others require making a small hole, just like piercing human ears. Apply small metal ear tags at your convenience. Some shepherds tag lambs shortly after birth, while they are doing the "snip, dip, and strip" procedure. Others tag lambs just before releasing them from the jug. Be sure to record the tag number, ewe number, and lamb's birth weight.

Docking

All lambs should have their tails docked at 2 or 3 days of age. Docking causes less pain and trauma at this early stage than it does when the lambs get older. Sheep tails are cut short for health reasons, not just for appearance. Long tails are woolly and may accumulate large amounts of manure, which attracts flies and maggots. In addition, tails interfere with breeding, lambing, and shearing.

Docking should be done before the lamb leaves the lambing pen, while it is easy to catch. Seek the help of an experienced person until you're familiar with the procedure. Several methods are used for docking. Some people favor use of the Burdizzo emasculator and knife. After docking, apply an antiseptic and, in warm weather, a fly repellent to the wound. Some people prefer the Elastrator, which places a small, strong rubber ring around the tail. After the Elastrator band has been on for a couple of weeks, the tail drops off. Any method causes slight but temporary discomfort.

Castration

Ram lambs should be castrated unless you plan to keep them or sell them for breeding. If you plan to sell the lambs at auction or to a packinghouse, you'll find that they're not worth as much if they have not been castrated. The meat from older and heavier uncastrated lambs can have a strong taste.

Male lambs are castrated also to prevent them from breeding with other lambs when they are penned together. Lambs can breed at an early age. If your ewe lambs are bred too early, serious lambing problems can occur.

Male lambs should be castrated early, but not in the lambing pen. Castration cannot be done until both testicles have dropped into the scrotum. The testicles usually drop by 10 days of age.

Seek the help of an experienced person until you learn the procedure. You can use the Elastrator to castrate your lambs. With the Elastrator, pliers are used to stretch a strong rubber band around the scrotum. After several days, the testicles die and drop off because the blood supply has been cut off. This method causes no loss of blood and little shock to the animal and results in only a slight risk of infection.

Problems with Newborn Lambs

Most lambs are born without problems, but if there's trouble, the ewe and her lamb will need your help. This section describes the most common problems after birthing and offers tips on how to handle them.

When the Ewe Rejects Her Lamb

The rejected twin. It is fairly common for a ewe to reject one of her twins. The ewe may be nervous, inexperienced, or confused and accidentally lose her lamb's scent. You can try to force the ewe to accept the lamb, bottlefeed it, or graft it onto a ewe with more milk or one that has just lost her lamb.

If the mother ewe is very young or very old, she may not have enough milk for twins, and it's best to relieve her of the rejected one. When ewes with inadequate milk attempt to raise twins, both twins suffer from lack of milk. One solution to this problem is to partially "bum" them by supplementing the ewe's milk with milk replacer.

The forgotten twin. Occasionally, when a ewe delivers the first twin, the bonding period is interrupted by the arrival of the second lamb, which is dropped several feet away. Similarly, the bonding of second or third lambs is sometimes interrupted when the first lamb comes around to be nursed. If the ewe is distracted, she may forget about the other lamb. After birth, check the entire lambing area. You may be surprised to find that what appears to be a single lamb has a sibling on the other side of the barn. Similarly, a ewe with twins she has acknowledged and accepted won't necessarily miss one of them if it wanders off or gets left behind. To prevent this, keep the ewe and the lambs in the jug a few extra days, or until the lambs are strong enough to keep up with the ewe by themselves.

Difficult labor. Difficult labor is one of the most frequent causes of rejected lambs. The ewe may be so tired afterward that she simply doesn't care. However, an exhausted ewe will often recover her mothering instinct after several hours of rest. In this case, keep the ewe and the lamb in the jug together, but bottlefeed the lamb until the mother wholeheartedly accepts or rejects it.

Painful udder. A good milking ewe may have an overly full and painful udder at lambing, making her reluctant to nurse the lamb. Milk out the udder to relieve the excess pressure. Bottlefeed a few ounces of the colostrum to the lamb and freeze the rest for emergency use.

Lamb with sharp teeth. Check the lamb's mouth. If it has sharp teeth, use an emery board to file them down a little.

The chilled lamb. The ewe may abandon a lamb because she thinks it is dead. When treating a weak or chilled lamb, don't keep it away from its mother too long, or she may reject it when you take it back (especially if you have had it in hot water, which washes off much of the lamb's natural scent).

Strong medicine odors. Use care when treating the lamb's navel with iodine. Iodine has a strong odor that sheep don't like. If you use too much, the ewe may reject the lamb.

The wandering lamb. A ewe may be tired after delivery of a large, robust lamb. While the ewe is resting, a strong lamb will sometimes get up and wander off in search of her before the ewe has had a chance to bond to it. Getting mother and lamb into the jug promptly will prevent this problem.

The stolen lamb. On rare occasions, a ewe will give birth to a lamb, only to have another ewe — usually an older ewe who is close to lambing herself — come by and adopt it. Sometimes the foster mother and the lamb will develop a strong bond. The only problem with this is that the foster mother will have no milk left for her own lambs. In this case, the lamb must be bottlefed.

Trading lambs. Lamb trading occurs frequently when too many ewes and lambs are in one group or when jugs are not used. The lamb from one ewe will bond to another and vice versa. Everyone seems happy and no damage is done, except that the ear tag numbers don't match and your record keeping can become confused.

Ewes with No Milk

Sometimes ewes seem to have no milk. If the udder is hot or lumpy, mastitis may be the problem; call your veterinarian.

If the udder seems normal (soft and full), the ewe may simply be experiencing a delay in letdown; sometimes letdown takes as long as 3 hours. Keep milking the teats every 15 minutes until the milk begins to flow. In the meantime, the lamb must be fed colostrum. If you have another ewe that has just lambed, milk 4 to 6 ounces of colostrum from her and feed it to the lamb. If you have thawed-out frozen colostrum on hand, use it. You can give the lamb an additional feeding of diluted canned milk or lamb milk replacer 2 or 3 hours later, but don't use milk replacer intended for other animals.

If the ewe still does not have milk after 3 hours, keep the lamb with the ewe, but begin feeding it as if it were an orphan. (See pages 245–246 to learn how to feed orphan lambs.)

A Chilled Lamb

A chilled lamb will appear dull, listless, and weak. It may not be able to stand or suck. The inside of the mouth of a severely chilled lamb will be cool to the touch. If the lamb has these symptoms, check its temperature rectally with a thermometer. A lamb's normal body temperature is 103.1°F.

A slightly chilled lamb (temperature of 100 to 102°F) can be warmed by placing a hot water bottle against its belly or by directing hot air from an electric hair dryer over it. If the lamb's rectal temperature is less than 100°F, the lamb needs immediate attention from a veterinarian. If a veterinarian is not immediately available, immerse the lamb up to its neck in warm water that is comfortable to the touch, then gradually heat the water to about 110 to 115°F. When the lamb's body temperature reaches 100°F, its mouth and tongue will again feel warm. It may take several hours for the lamb to warm to this temperature. Keep it in the warm water until its temperature is 102°F. Then remove it from the water, rub it with a towel, and dry it thoroughly with the hair dryer.

After warming, try to bottlefeed the lamb. Return it to the mother as quickly as possible and make sure that she will accept it. Keep the towels used to dry the lamb; they contain the birth fluids that help the mother identify her newborn. You may need to rub the lamb again with the towels to put its scent back.

Baby lambs can tolerate cold temperatures well, but only after they are completely dry and have received colostrum. To keep a weak lamb from becoming chilled in cold or wet weather, wrap it in a homemade plastic or fabric lamb coat (see the box on page 266) just before you turn the ewe and lamb out of the jug.

Newborn Lamb Coats

In cold or wet weather, a coat can help a lamb conserve energy and can prevent hypothermia. Put the coat on after the lamb is dried off or before you turn it out of the jug, depending on the weather. A coat can also save the life of a newborn lamb that is chilled or facilitate the grafting of an orphan lamb to a new mother ewe.

Woven-plastic feed sacks make good lamb coats; the plastic not only conserves body heat but also serves as a raincoat. Cotton duck is also suitable. Use pajama elastic for the leg fittings; it's soft and will not chafe the lamb if the coat is fitted properly. Because lambs grow so fast, you may want to make coats in several sizes.

1. **Belly band:** Stitch elastic to point A on one edge.
2. **Neck fastening:** Stitch elastic to point B on the opposite edge.
3. **Leg fastening:** Stitch the ends of the 4-inch pieces of elastic on the inside of the coat at points C, D, E, and F.

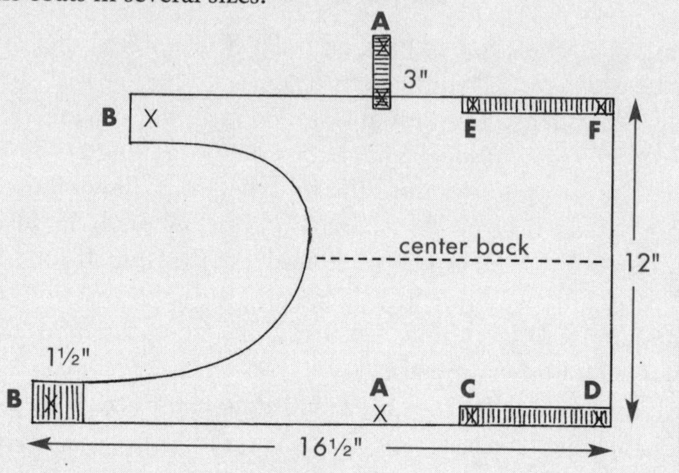

Grafting

Grafting is the process of bonding an orphan lamb to another ewe. Sometimes grafting is simple, and sometimes it isn't.

Grafting is always more successful if done in the first 20 hours after the ewe has lambed, and the sooner, the better. The orphan to be grafted should not be more than 1 day old.

Don't try to graft an older lamb to a ewe that has lost a twin or that has an excess of milk and only a single, small lamb. An older lamb may consume more than its share of milk, causing the younger lamb to become malnourished from starvation.

Bonding between a newborn lamb and a mother ewe is based on the lamb's smell and voice, but not its appearance. For grafting to work, then, you must trick the ewe into thinking that the lamb smells like her own. Thankfully, it's not difficult to fool a sheep. The following grafting techniques will help:

- If the ewe has a single lamb, rub both it and the orphan lamb with molasses water, so that the ewe will lick and accept them both.
- Spray a little bit of vanilla or a nonscented room deodorant on a cloth and rub it on the ewe's nose and the lamb's rear; sometimes this makes the ewe believe that the lamb is hers.
- If a ewe has just dropped a single lamb, take some of the watery fluid from the birth sac of that lamb and rub it all over the orphan you are attempting to graft to her. If the orphan is young and you can rub it with birth fluid within a few minutes of delivery, acceptance of the orphan lamb is almost guaranteed.

- Use an adoption or fostering coat to transfer the smell of a familiar lamb to another lamb. Put the coat (which is like a cotton tube sock) on the ewe's single live or dead wet lamb. The coat absorbs the lamb's smell within 1 hour. Turn the coat inside out and put it on the lamb you want to graft to the ewe. You can purchase a fostering coat or make your own from any stretchy material, such as sweatshirt sleeves.
- Feed the rejected lamb colostrum while you wait for the accepted lamb to have its first bowel movement. This will consist of dark black, tarry material. Take a little bit of the feces and smear it on the rear of the rejected lamb. The feces make both lambs smell the same, and if all goes well, the mother won't notice the difference.
- Another grafting method that has a high success rate is to place the ewe in a stanchion — a piece of equipment that fits loosely around the ewe's neck and keeps her from moving forward or backward. The ewe will be free to eat hay and drink water, but she cannot see, smell, or butt the lamb. The lamb to be grafted is put into the pen with her, where it can nurse at will. You may need to hold the ewe for the first few times, to make sure that the lamb has had its colostrum (and the lamb knows where to find the milk). A stubborn ewe may need up to 5 days to completely accept the lamb.

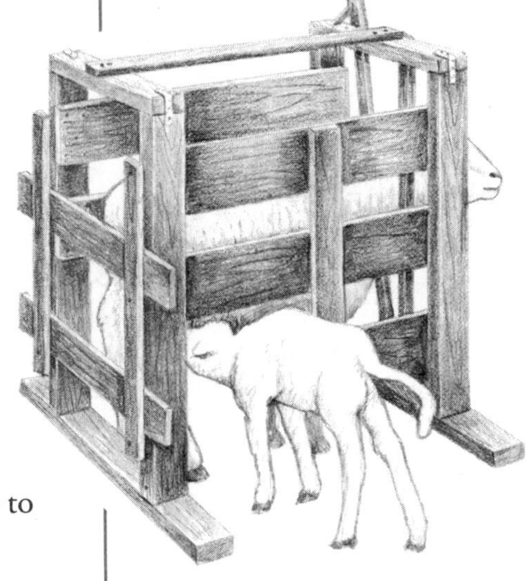

A stanchion holds a ewe still so her lamb can nurse

Caring for the Ewe After Weaning

The ewe's milk production peaks about 4 weeks after lambing, then begins to decline. (For information about weaning, see page 247.) If weaning occurs abruptly, the ewe can develop udder problems. To avoid problems, reduce the ewe's grain ration about 3 weeks before weaning and eliminate it altogether about 2 weeks before weaning. The quality of the hay should be reduced about 3 weeks before weaning; this will also help cut the ewe's milk production. If the ewe is still making a full bag, milk it out partially once a day for a few days to reduce the risk of mastitis (udder infection). Milking more often than that only stimulates continued milk production. Keep the ewe in a clean, dry area for a few days after weaning until the teats close and the udder stops swelling.

Making Sheep-Milk Cheese

You've probably eaten sheep cheese, even if you didn't know it. Many European gourmet cheeses, such as Roquefort, Romano, and Pecorino, are most often made from sheep milk. Sheep milk is ideally suited for cheese making because it contains almost double the solids of cow milk and is high in proteins and minerals; you can produce more cheese with less milk. It also contains a higher percentage of butterfat than does cow milk.

Collecting enough sheep milk to make cheese takes quite some time for one person with just a few sheep. You can collect, chill, and freeze the milk until you have enough to make cheese.

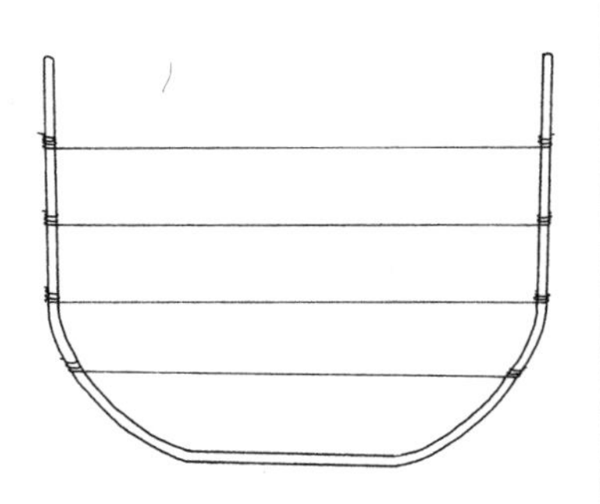

Cheese knife

Cheese press

Cheese making does not require a lot of special equipment. You'll need:

- 6-quart stainless steel (*not* aluminum) kettle
- Dairy thermometer
- Cheese knife
- Fine muslin bag, or cheesecloth and puree sieve or fine colander
- Cheese press (see below)
- Dishpan with rack

You probably already have most of these items in your kitchen. If you don't have a cheese knife or press, they are easy enough to make.

To make a cheese knife. Bend a firm rustproof wire into a U shape that will fit into the kettle in which you will make your cheese. At intervals of 2 inches, anchor rustproof wire from one arm of the U to the other by twisting firmly.

To make a cheese press. Bore ½-inch holes through an 8-inch-square hardwood board at 1-inch intervals; the board should resemble a checkerboard when you're done. Sand the board, rub it with vegetable oil, and wipe it clean. To use, put the cheese bag on this board and top it with another board of the same size. On the top board, place a weight, such as a gallon jug of water, or use C-clamps to apply pressure between the boards and squeeze out moisture.

SHEEP CHEESE RECIPE

This recipe yields a versatile, low-fat cream cheese that makes a great dip or spread when seasoned with parsley, chopped onion, pressed garlic, pepper, or other herbs. When sweetened, it makes a delicious filling for cake.

1 gallon pasteurized whole sheep milk
¼ cup cold water
½ rennet tablet
½ cup fresh commercial buttermilk
1–1½ teaspoons salt

1. Pasteurize the sheep milk by heating it in a 6-quart stainless-steel kettle to 155°F and keeping it at that temperature for 30 minutes.
2. Cool the milk to 85°F.
3. Pour the water into a small bowl. Dissolve the rennet tablet in the water.
4. Add the rennet tablet mixture and the buttermilk to the cooled sheep milk. Stir gently for 10 minutes or longer. Stop stirring when you notice a slight thickening or setting. If you stir too long, you will get a mushy product instead of a firm curd.
5. Keep the mixture at 80 to 85°F. Don't let it get any hotter, or the rennet will be destroyed. The best way to hold this temperature is to set your cheese kettle in a large pan of warm water to which you can add hot water from time to time as it cools. Let the mixture stand until whey, a watery-looking liquid, covers the surface and the curd breaks clean from the sides of the kettle (like gelatin) when it is tipped.
6. Cut the curd into 1-inch cubes by running a long, thin knife through it in both directions, right to the bottom of the pot. Cut the strips horizontally by inserting the cheese knife and drawing it across the kettle.
7. Place a bowl underneath a clean muslin bag or fine colander lined with cheesecloth. Pour or ladle the mixture into the bag or colander. Allow it to drain until nearly all of the whey has been caught in the bowl. Use a cheese press to

squeeze out the rest of the whey. If you don't have a cheese press, place a dish on top of the bag and weight it down with a jar filled with water.
8. Keep the whey in the refrigerator until the cream rises and becomes firm enough to skim off. The cream will have a butterlike consistency. Work it back into the cheese, mixing thoroughly. (Save the thin whey to use as the liquid in bread baking, or feed it back to the sheep.)
9. Once the cheese feels firm, work in the salt.

Culling and Butchering

If you breed sheep, eventually you'll have to cull the flock to improve your stock. Keep your best ewe lambs to replace less productive older ewes. Butcher all ram lambs, unless you intend to raise a breeding ram for your own use or to sell. Unneeded lambs should be butchered at 5 to 6 months of age or at 35 to 45 pounds.

When evaluating older ewes for culling, consider age, productivity, and general health. Cull the following types of older ewe:

- Ewes with defective udders
- Ewes with a broken mouth (teeth missing)
- Limping sheep that do not respond to regular trimming and footbaths
- Ewes with insufficient milk and slow-growing lambs.

Lamb meat is delicate and tender; the meat from older animals, called mutton, can be ground and used like ground beef. However, mutton is easier to digest than beef is, which makes it a good meat for people who have gastrointestinal difficulties.

You can either take your sheep to a custom packing plant to be slaughtered and butchered or do it yourself. If you want to do it yourself, consult a good manual (see Recommended Reading, page 379).

If you're going to work with packers, you'll have to give them some instructions. Consider these guidelines:

- Cut off the lower part of hind legs for soup bones.
- For mutton, have both hind legs smoked for hams.
- For lambs, the hind legs can be left whole, as in the traditional leg of lamb, or cut into sirloin roasts and leg chops.
- The loin, from either mutton or lamb, can be cut as tenderloin into boneless cutlets or as a loin roast.
- Package riblets, spare ribs, and breast meat into 2-pound packages. Riblets, which are sometimes referred to as short ribs, are almost inedible when prepared by most cooking methods, but when prepared in a pressure cooker for about 45 minutes, with 1 inch of water, barbecue sauce, curry sauce, or your favorite marinade in the bottom to start, they are a real delicacy. For lambs, the spareribs and breast can be barbecued or braised. For mutton, these cuts are pretty much waste products.
- For mutton, have the rest boned, trimmed of fat, and ground. Double-wrap in 1-pound packages.
- For lamb, the rack, or rib area, can be cut into lamb chops or left as rack roast. The shoulder can be cut into roasts or chops, and the neck and shank can be used as soup bones.
- If you want kabob meat, make sure to have it cut from the sirloin or loin.

Wool and Shearing

Shearing is usually done once a year, in the spring, so that the sheep are free of their heavy wool coats for the hot summer months. Shearing is important for keeping your sheep healthy and comfortable, and this annual "haircut" has the benefit of producing salable wool for the shepherd.

The first shearing you observe may be quite a surprising scene — a sheep must be held in awkward positions while its fleece is removed with sharp clippers or shears. However, the sheep feels a lot better after shearing, and shorn sheep are easier to keep free of external parasites.

Wool is unique in being the only fiber in the world that retains its warmth when it is wet. It used to be the material of choice for sailors' clothing. You can spin your own wool, or you can sell your fleeces to spinners who will create wool yarn for making wool goods.

If you earn a reputation among spinners for having prime-quality fleeces, your fleeces will be in demand far ahead of shearing time. Several characteristics determine the quality of a fleece, including wool type, strength, and lack of vegetation. The type of wool — silky, curly, or coarse, for instance — is a characteristic of the breed. The strength of the wool is determined by health and nutrition. But even if your sheep has a strong fleece of a desirable type, it will be less valuable if it is not kept clean. To keep a fleece clean, use a sheep coat to protect it from vegetation, such as seeds, hay, burrs, and bedding. (For directions on how to make sheep coats, see pages 276–277.)

Preparing for Shearing Day

While sheep don't enjoy the process of shearing, later they'll be happy to be rid of their heavy winter coat of wool. To reduce the stress for both sheep and shearer:

- Don't give sheep food or water for about 12 hours (or overnight) before shearing. A sheep with an empty stomach is more comfortable during shearing.
- Be sure your sheep is dry. If rain is predicted, keep the sheep indoors the night before.
- Get your sheep into a pen where each one can be caught easily. A shearer must not be expected to chase or catch your sheep.
- Prepare a good shearing floor. A smooth plywood platform is better than a dirt floor or grassy area. Spread a tarp or an old wool rug on the platform to make it less slippery.
- When shearing is done indoors, provide sufficient light.
- Have cold water or another beverage available for the shearer.
- Arrange a skirting table, which is a raised table with a slatted top. When you skirt the fleece (remove a strip about 3 inches wide from the edges of the fleece), the skirtings (the trimmed-off edges) will fall through the table to the floor. Sanded lath, long dowels, or 1-by-2-inch lumber set on edge makes a tabletop suitable for skirting.
- Have first-aid supplies on hand in case a sheep or the shearer is severely cut. Don't worry about little nicks on sheep. They may look bad, but the lanolin in sheep's wool helps them heal quickly. In hot weather, spray each shorn sheep with a fly repellent before releasing it to prevent infection in the tiny cuts.

> ### Crimp
>
> When you look closely at a handful of raw wool, you'll see zigzag waves in the individual strands. These waves are called the crimp, and they are what makes the wool elastic or stretchy. Good crimp acts like a spring; when wool fibers are stretched, they spring back to their original shape. As a rule of thumb, the finer the wool strands or fibers, the finer the crimp. In fine wool, the crimp can barely be seen. In coarse wool, the crimp may range from ½ to 3 inches between the "zig" and the "zag."

- Handle the fleece carefully, so that it doesn't become contaminated with manure, straw, or dirt. Sweep the floor after each sheep.
- Shearing is a good time to trim hooves, check udders, deworm, treat for parasites (after the wool is removed), and check the teeth of oldsters. Have all the necessary supplies on hand.

Skirting

After each sheep is shorn, set the fleece sheared side down on the skirting table. (This position will shake off any second cuts.) Then sweep the floor of the shearing area so that the shearer can get going on another sheep.

Skirt the fleece, allowing the cuttings to fall through the slats of the skirting table onto the floor. Roll up the fleece and secure it with paper twine (available from sheep supply stores) or place it in a paper feed sack (turned inside out) or a cardboard box.

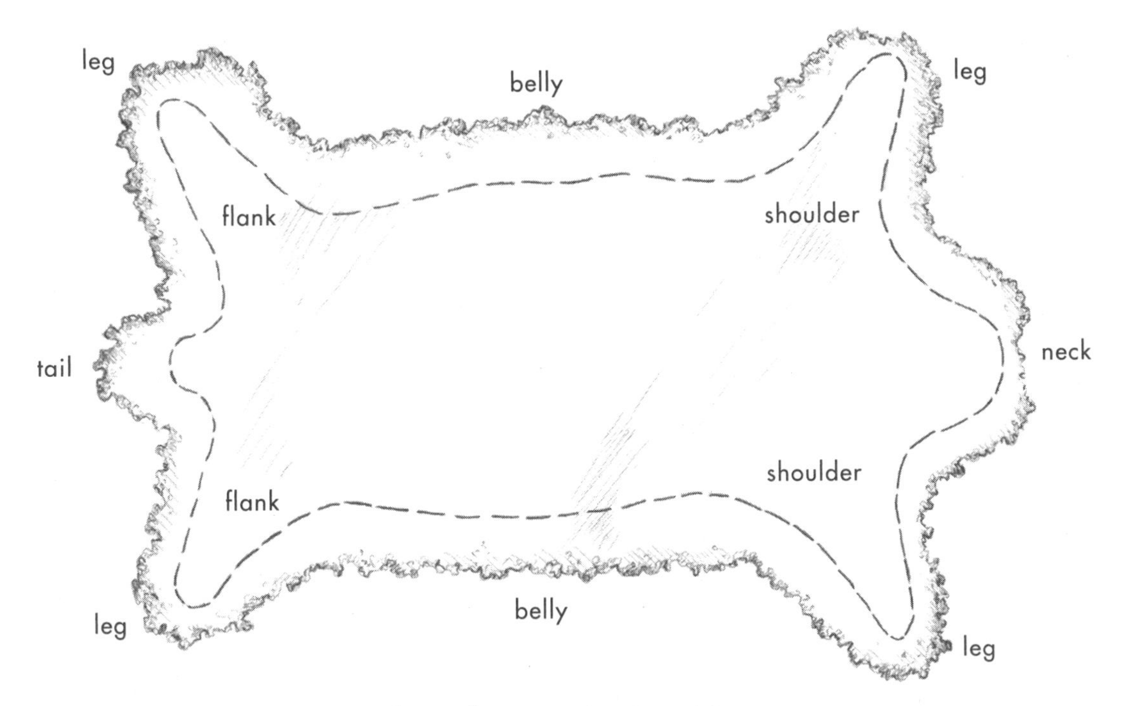

Skirt a fleece at the dotted line.

How to Shear a Sheep in 20 Steps

The real trick to shearing isn't learning the correct pattern of shearing strokes, which lessens the time involved, but immobilizing sheep by various holds that give them no leverage to struggle. A helpless sheep is a quiet sheep. Rendering sheep helpless cannot be done by force alone, for forcible holding makes them struggle more. Try to stay relaxed while you work.

1. Slip your left thumb into the sheep's mouth, in back of the incisor teeth, and place your other hand on the sheep's right hip.

2. Bend the sheep's head sharply over its right shoulder, and swing the sheep toward you.

3. Lower the sheep to the ground as you step back. From this position, you can lower it flat on the ground or set it up on its rump for foot trimming.

4. Start by shearing the brisket, and then shear up into the left shoulder area. Place one knee behind the sheep's back and your other foot in front.

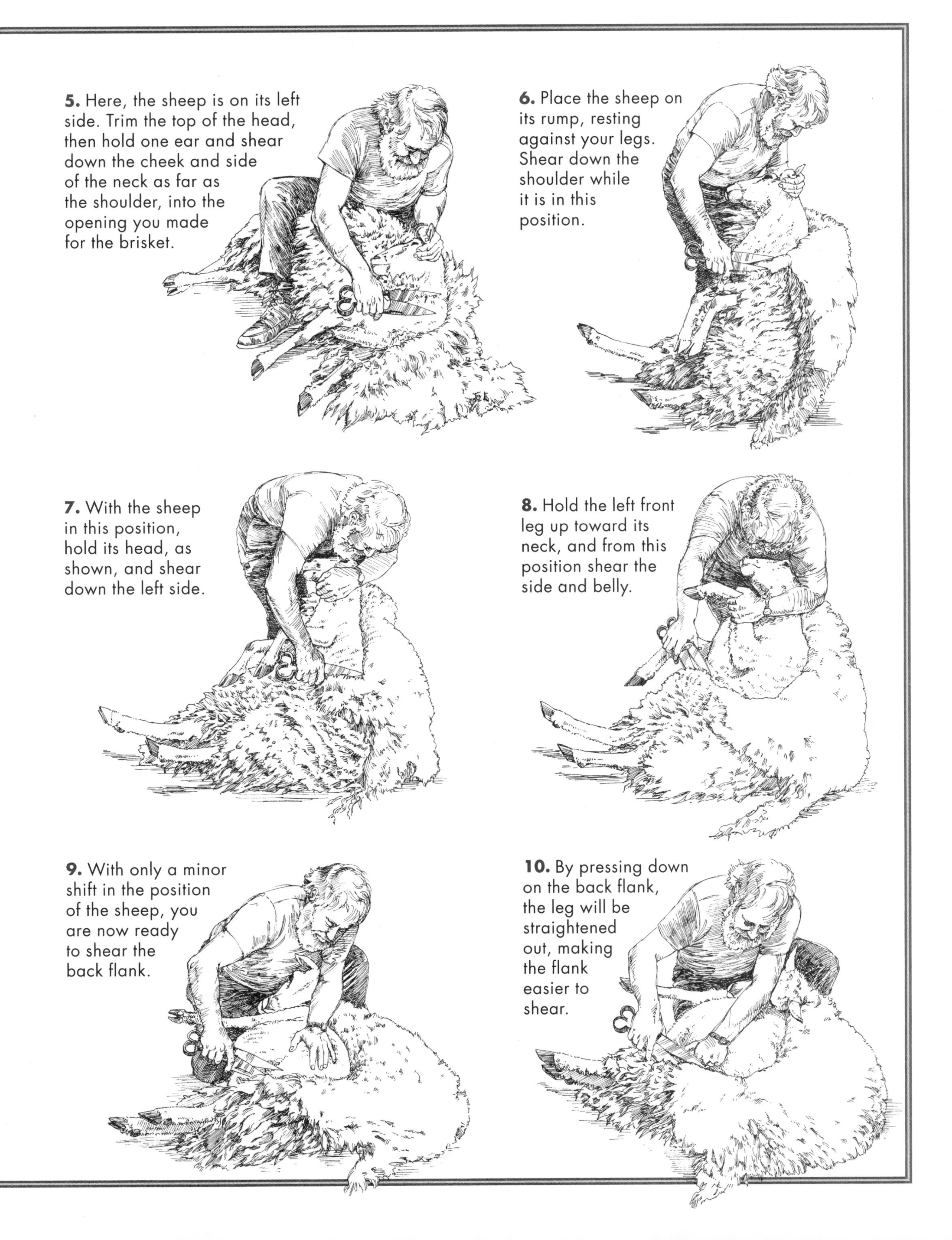

5. Here, the sheep is on its left side. Trim the top of the head, then hold one ear and shear down the cheek and side of the neck as far as the shoulder, into the opening you made for the brisket.

6. Place the sheep on its rump, resting against your legs. Shear down the shoulder while it is in this position.

7. With the sheep in this position, hold its head, as shown, and shear down the left side.

8. Hold the left front leg up toward its neck, and from this position shear the side and belly.

9. With only a minor shift in the position of the sheep, you are now ready to shear the back flank.

10. By pressing down on the back flank, the leg will be straightened out, making the flank easier to shear.

Shearing in 20 Steps, continued

11. From this position, the sheep is shorn along its backbone and a few inches beyond, if possible.

12. By holding up the left leg, it is possible to trim the area around the crotch.

13. The job is half done. The shearer's feet are so close to the sheep's belly that it cannot get up.

14. Holding one ear, start down the right side of the neck. Hold the ear firmly but not tightly, so that you don't hurt it.

15. Hold the sheep with your left hand under the chin and around the neck, and shear the right shoulder.

16. Pull the sheep up against you to expose the right side, so that you can shear down that side.

Second Cuts

Second cuts are short lengths of wool created when the shearer runs the clipper over a place where the sheep has already been sheared. This makes the sheep look smooth and lovely, but the short snips that result are difficult to remove from the fleece when preparing it for handspinning. If they are not removed, they make unsightly little lumps in the yarn and weaken it. When shearing for handspinners, avoid making second cuts.

Clippers versus Shears

Electric clippers make short work of shearing a sheep. However, they're expensive, and in the hands of an inexperienced shearer are apt to result in cut hands and cut sheep. For small flock owners who want to sheer their own sheep, hand shears have a number of advantages:

- No source of electricity is needed.
- The clippers make less noise to annoy the sheep.
- The blades are lightweight and easy to carry.
- The blades are easier to sharpen than are power clipper blades.
- Hand shears are less expensive.

Shearing Your Own Sheep

Your size and weight will give you neither a significant advantage nor a disadvantage when shearing; once you get the sheep into the basic shearing positions, it won't have leverage to scramble to its feet and struggle.

If you have more than one sheep, a great advantage to doing the shearing yourself is that you don't have to shear all the sheep at the same time, as happens if you hire a shearer. If you have ewes with small lambs, shearing can be an upsetting time. The sheep look and smell different after shearing, and all the lambs cry as they try to find their mothers. If you shear only one mother each day, this confusion doesn't happen.

Another advantage of shearing your own sheep is that you gain familiarity with each animal. This offers a good opportunity to check the teeth, udders, and general health of each sheep. In addition, you'll have time to handle the fleece in a way that is in keeping with its intended use. For instance, if you will be using the fleece yourself, you may want to sort and bag the parts separately: The belly wool, skirtings, and tags can be used for felt making, and the prime fleece can be used for spinning or sold to local handspinners. If the wool is going to be handspun, be especially careful to keep it free of vegetation and to avoid second cuts. It will be much more valuable.

Making a Sheep Coat

Sheep coats (also called blankets, covers, or rugs) are used to keep fleeces clean, especially before showing. In areas with severe winters, coats help sheep stay warmer, and the energy that might have gone toward keeping them warm is expended on increased wool growth and heavier lambs. Many owners of small flocks who depend on the sale of choice fleeces use coats on all their sheep.

To keep costs down, make sheep coats from woven-plastic feed sacks, which allow air circulation and hold up well. The edges must be well hemmed so that they don't fray. The easiest material to work with, however, is #10 cotton duck. Prewash it before constructing the coat to prevent shrinkage.

Use the suggested dimensions, or measure your sheep and make a custom-fitted garment. For the best protection, the coat should extend 3 inches below the belly. If your material isn't wide enough, make a seam along the center at the backbone. To custom fit, determine the coat length by measuring from the center of your sheep's breast to the end of the back thigh. Make the hind leg loops of soft pajama elastic, which is about 1 inch wide and not as stiff as most elastic. Although the garment itself should last more than one season, the elastic will probably have to be replaced each season. Check the sheep often to be sure the elastic isn't rubbing its leg and causing injury.

To get the right fit, first make a rough model out of an old sheet and try it on your sheep. The front edge should be as close to the head as possible for maximum protection against hay and weed seeds. If you make a center back seam, you can get a closer fit by shaping the seam to conform to the curve of the sheep's back.

1. Match seam A–B with seam A–B on the other side. Sew, overlapping ½ to 3 inches, depending on the size of the sheep.

2. Make a ½-inch hem on outside edges (shown by dotted line)

3. At points C and D on both sides, stitch soft elastic, 24 to 27 inches long, depending on the size of the sheep.

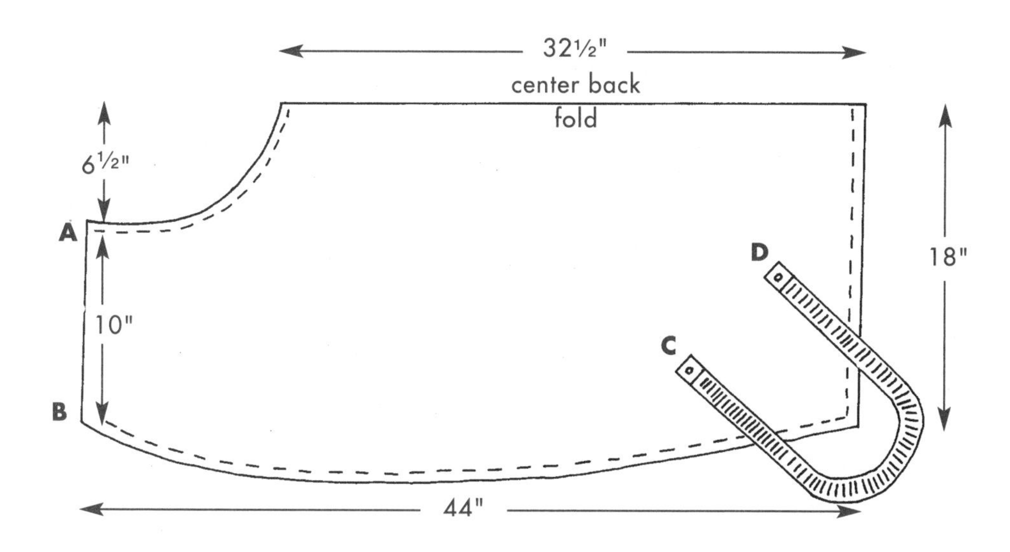

Using this sheep coat pattern, cut two pieces of material to size and stitch together as directed.

Wool Crafts

Many people raise sheep for the single purpose of supplying themselves with wool. You can profit from wool even if you are not a spinner or felter. One way to make money from fleece is to sell it to a wool pool, which combines fleece from many sources and then processes it. However, selling good handspinning fleeces directly to individual spinners can bring in considerably more money. Handspinning fleeces must be free of debris (such as seeds or straw), sheared carefully, and skirted heavily. They should come from a good wool breed and have a staple length (the length of each wool fiber) of at least 3 inches. White wool should be exceptionally white and have no dark fibers. Brown, gray, or black wool can be of any shade, or white with an even sprinkling of dark fibers.

You might also consider providing a custom washing service for your spinner customers. Careful scouring of a handspinning fleece could double its value, and this process does not require expensive equipment.

Washing Fleece

Fleece is washed by soaking it in very hot water and plenty of soap or detergent, without rubbing or moving it about in the water. If the fleece will be used for handspinning, choose only the best and cleanest parts. For felting, any quality of wool will suffice. Detergent cuts grease better than soap and also rinses out more easily. If you are washing a large quantity of fleece (20 pounds, for instance), you will need about 10 cups of detergent for a 20-gallon laundry tub filled to within 5 inches of the top with the hottest tap water possible. Do not use your bathtub for this process, because the water will cool down too fast and the grease that the hot water lifts from the fibers will be redeposited on them.

In washing fleece, the goal is to wash as much wool as possible at a time while getting it free of grease and gumminess, in as little time and with as little effort as possible.

1. Pull the skirted fleece apart and shake out as much of the seeds and dirt as possible.

2. Fill the tub with hot water. Add the detergent (about 1 cup for every 2 gallons of water). Stir well to mix.

3. Push into the tub as much fleece as will fit. Cover the tub to contain the heat, and let the fleece soak for 1 to 2 hours.

4. Remove the fleece from the tub, gently squeezing out the water. Place the fleece in mesh bags or old pillowcases. Place the bags in the washing machine, and run them through the spin cycle.

5. After washing, the fleece is ready for rinsing. If you expose the fleece to an extreme change in temperature, it will felt; the rinse water, therefore, shouldn't be hotter than the wash water was when you removed the fleece from it, but it also shouldn't be cold, because the heat is needed to remove grease and gumminess from the fleece. Take the temperature of the wash water in the tub and then dump it; refill the tub with clean rinse water of the same temperature as the wash water.

6. Immerse half of the washed wool in the rinse water. Squish the wool up and down several times, remove it, and and put it into clean mesh bags or pillowcases. Run the spin cycle in the washing machine. If the wool is still greasy or dirty, you may have to rinse it two or three times.

7. Repeat step 6 with the other half of the washed wool.

8. Fluff the wool and place it on wire racks to dry. (A sweater drying rack or chicken wire stretched across a wooden frame will make a good wool drying rack.)

9. When the wool is dry, put it into a cloth bag (such as a pillowcase) or cardboard container, seal well, and store until use.

Carding

Carding is a mechanical version of teasing; it employes combs to pull fibers apart, remove short fibers, and align the fibers in one direction. Hand cards can be used for small batches of wool, but anyone with two or more sheep would probably be happier with a drum carder.

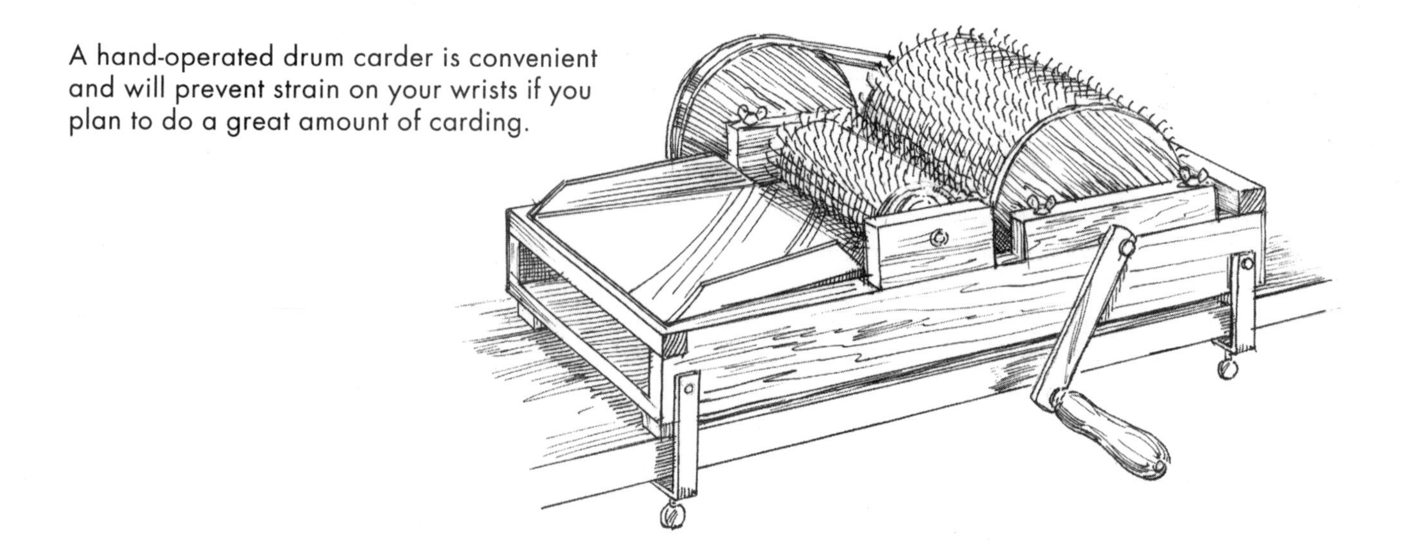

A hand-operated drum carder is convenient and will prevent strain on your wrists if you plan to do a great amount of carding.

Hand carding in four steps

1. Spread the wool on the left card, with the shorn ends at the top of the card.

2. Take the right card and lay it in the center of the left card, with the handles in opposite directions, and draw the right card away from you. Repeat this step several times, until the fibers begin to align.

3. When the fibers are well aligned, lay the right card on your knee, and with the handles in the same direction, brush away from yourself. This deposits the wool on the right card. Switch paddles and repeat this step several times.

4. Roll short or medium wools off the card, or fold over longer wools. The fibers are now ready to spin.

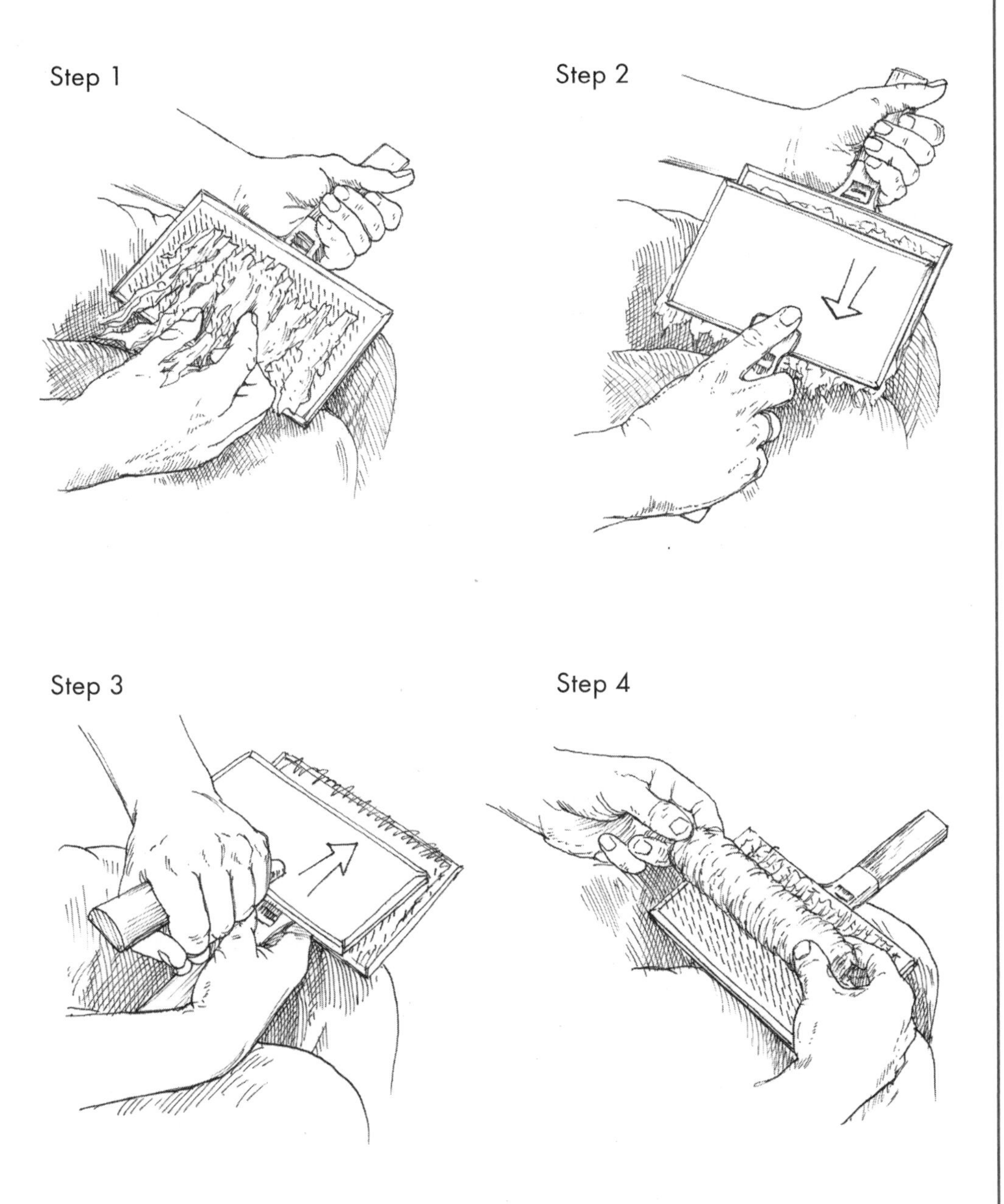

Dyeing Wool

Wool for felting or spinning can easily be dyed at home with grocery-store ingredients. Wash the wool before you begin, and wet the wool before placing it in whichever dye bath you use.

Kool-Aid dyeing. Dyeing with Kool-Aid yields surprisingly bright, lasting colors. In a large jar, dissolve sugarless Kool-Aid in ½ gallon of water. Use 4 packets of the lightest colors and 2½ packets of the vivid colors. Heat 1 quart of water in a large enamel or stainless-steel pan. Add the half-gallon of dye matter and ½ cup of white vinegar. Put in some wet fleece — be sure to leave room for the water to simmer. Heat the water until it barely simmers, and keep it at that temperature for 30 minutes. Remove the pan from the heat and let it cool to lukewarm. Remove the wool from the dye. Rinse it several times in warm water, until the water shows no trace of color. Squeeze out the final rinse water by hand or with the spin cycle of your washer. Put the fleece on a wire rack to dry.

Commercial fabric dyes. Rit, Tintex, and other packaged dyes cost a little less than Kool-Aid and give more predictable colors. Craft shops sell other brands of fabric dyes in a great range of colors. Follow the instructions on the package.

Spinning Wool

Handspinning sheep wool is essentially the same as spinning mohair. See page 228 for instructions. You'll need a good supply of teased or carded grease (raw) or washed wool.

Knitting from Fleece

Even if you haven't learned to spin, you can still knit from your own fleeces, but you'll first have to prepare the wool. While the preparation of the wool may seem tedious, it probably takes about the same amount of time as carding the wool and spinning yarn. The wool for fleece knitting must have a staple length of at least 4 inches.

1. Make bundles of fleece about 4 inches in diameter. Tie each bundle together firmly (but not too tightly) with a strip cut from panty hose.

2. Soak the bundles in hot water and detergent for about 1 hour. Use quite a bit of hot water and fleece and a fairly large pan for the soaking so that the water will not cool down too rapidly.

3. Place the bundles in mesh bags or pillowcases, put them into your washing machine, and use the spin cycle to remove the water from the bundles.

4. Rinse the bundles in warm water, and spin them in the washing machine again.

5. Remove the nylon ties carefully and place the bundles on a rack to dry.

6. When the wool is dry, gently pull the wool fibers out of the bundles into a continuous strand. Try to keep the strand a consistent size as you pull. The wool bundles will remain fairly undisturbed if you have handled them carefully.

7. Knit with this strand as you would with a length of spun yarn.

Wool Felting

The first fabrics made by humans were felted rather than spun or woven. Wool and hair can be felted because each fiber has overlapping scales. When fleece is subjected to a combination of moisture, heat, lubrication, pressure, and movement, the fibers form a fabric up to 1 inch thick. This fabric can easily be made into small items, such as mittens, caps, and slippers. Theoretically, the size of what you can make from felted material has no limit, if you work outdoors and have plenty of helping hands. In fact, some nomadic peoples of Asia use felt tents, or yurts.

You can use washed wool or grease wool, either teased by hand or carded with hand cards, a drum carder, or a carding slope. Short, fine, crimpy wools felt better than long, coarse, lustrous wools, but any wool can be used to make felt.

Felting Equipment

- Clean fleece
- Sheet of heavy plastic (such as a shower curtain)
- Hot and/or cold water
- Liquid soap
- Washing soda (optional)
- An old-fashioned washboard or handmade felting board (optional)

1. Card or tease the wool clumps.

2. Place a sheet of heavy plastic on a table or other hard surface that will not be damaged by hot and cold water.

3. Take a thin layer of teased wool and place it on the plastic. The wool should cover a bigger area than you need for the finished project, because the felting process will shrink the piece by as much as one third.

4. Cover the first layer with a second layer, this time with the wool fibers lying in the opposite direction.

5. Keep adding layers, alternating the direction of the fibers with each layer, until the stack is about 1 inch thick when you press down on it. Try to keep the thickness as even as possible across the entire piece so that the completed felt has no lumps or thin places.

6. Fill a pint jar with very hot water and add about 1 tablespoon of liquid soap. One tablespoon of washing soda will help the process.

7. Pour a small amount of the soapy liquid onto the layers of wool and press gently with your hands to remove the excess water. It shouldn't be too sloppy wet.

8. Move your hands in a circular motion over the surface of the wool, pressing gently at first. After a few minutes, carefully turn the wool piece over and rub the other side. Add more soapy water as needed. The soap not only helps the fibers felt together but also acts as lubricant that makes the surface easier to rub. Some people alternate hot water with cold. Others wait until the wool piece is felted enough to hang together well, then they dunk it alternately in hot water and cold water several times. The piece of material is felted when you pinch the surface and cannot pull fibers from it.

9. Once your piece has begun to felt, you can lay out a design on the surface with pieces of different colored fleece and dyed wool yarn. Continue to rub the material after applying these ornaments, and the pattern will become felted into the finished piece.

10. There are several methods for hardening and shrinking your felt piece, making it firmer and less prone to future shrinkage. You can roll the felt piece tightly and wind it up with a string until the next day, or you can roll it up and pound it with your fist, a stick, or a hammer. Walking and stomping on the rolled felt is an ancient method of hardening it. Rubbing it on an old washboard or felting board is a good way to harden small items.

Felt Projects

- **Boot liners.** Make flat felt pieces large enough that they can be trimmed into boot insoles. Put them into your rubber boots for cold weather.
- **Slipper soles.** Make boot liners, but overcast the edges by hand or by using a zigzag sewing machine stitch. Crochet or knit a slipper top onto the sole.
- **Caps.** Many styles of caps and hats can be made by felting. A round shape is best. When the material starts to felt, begin the actual shaping. Make the piece a little thicker in the middle than at the edge, and then put the piece over a kitchen bowl to form the cap shape. Work the center portion gently onto the bowl, rubbing to stretch and shape it. At the same time, keep working the edges to felt and shrink it. When you are done, trim the edges with scissors.
- **Mittens.** Trace around your fingers and thumb on a piece of paper. Cut ½ to ¾ inch outside the tracing line and use as a pattern to cut four pieces of felt. Stitch one piece on top of another to form the mittens. Once you have some experience in mitten making, try to make each mitten in one piece.

Making a Felting Board

To make a felting board, you'll need the following:

- Waterproof glue or contact cement
- One 18-by-18-inch piece of plywood
- Twenty-four 18-inch-long pieces of ½-inch, ⅝-inch, or ¾-inch half-round molding, or an 18-by-18-inch piece of ribbed rubber floor runner

Glue the molding pieces onto the plywood, spacing them evenly over the entire board, to create a washboardlike surface. Or glue the rubber floor runner to the board.

Felting board

Sheep Calendar

This checklist will help you remember what needs to be done to care for your sheep throughout the year. The page numbers tell where in the chapter you can find more information about each task. The calendar starts with May because that is when most new shepherds begin their flocks.

(Adapted from *The Sheep Calendar,* a product of the much-missed Norseman Sheep Company, formerly of Kansas.)

May

- Check your fences before bringing new sheep home (page 241).
- Check your pasture for toxic plants (page 249).
- Give new sheep hay before turning them onto more lush pasture than they are used to.
- Check for keds and treat your sheep if you see even one ked (pages 251–252).

June

- Practice pasture rotation (page 248).
- Provide plenty of water, salt, and mineral/vitamin mix (page 248).
- Watch for fly strikes (clusters of tiny eggs) and maggots.
- Keep rear ends trimmed to discourage flies.

July

- Provide shade for your sheep in hot weather.
- Provide 1 to 2 gallons of cool, fresh water daily for each adult sheep.
- Check the feet of limping sheep; trim hooves, if necessary (page 254).
- Worm your sheep and record date of worming (pages 250–251).

If you are breeding your sheep:

- Put your ram into a shaded pasture next to the ewe.
- Shear the ram's scrotum to keep him cool.
- Start flushing the ewe before you plan to turn the ram in for breeding (pages 256–257).
- Do not let your ewe eat clover; it decreases her fertility.

August

- Continue to provide shade.
- Worm now, if you didn't do it in July.
- Buy grain for winter and store it in rodent-proof containers that sheep cannot break into.

If you are breeding your sheep:

- For early lambs, keep the ram in a shady place during the day, and turn him in with the ewe in the evening.
- Give the ram ¼ to ½ pound of grain daily.

- Keep the ram with your ewe for at least 6 weeks so that they have at least two chances for mating (page 257). If mating is not likely to be observed, put a marking harness on the ram.
- Mark on your calendar the date you turned the ram in with the ewe, so that you will know the earliest date to expect the lambing (page 256). Also mark the date if you observe any mating take place.

September

- Locate a place to purchase your winter supply of hay.
- Keep water, salt, and mineral/vitamin supplement always available to sheep.
- Make a list of repairs needed on shelter, fencing, and equipment, and start the repairs before cold weather.
- If you are worming, be sure to read the label for any precautions about timing before slaughter.
- Get sizable lambs to market. Shear them first, or ask for their pelts.
- Clean out the place where you plan to store winter hay; save the manure you collect and spread it on the vegetable garden.
- If you have apples, feed a few (but not too many at one time) to sheep. Set some windfalls aside for winter.

If you are breeding your sheep:

- For late lambs, flush the ewe and then turn in the ram (pages 256–257).
- Reduce grain gradually after flushing.
- Record dates when you see breeding take place.

October

- Clean out sheep sheds and barn, and spread sheep manure on garden.
- Get in the winter hay.
- Clean out and check your waterers; winterize your faucets.

If you are breeding your sheep:

- If you have seen no signs of breeding, consider borrowing a ram. Breeding time may be running short now for most breeds.
- Check over your lambing supplies (pages 257–258). Order supplies by mail, if necessary.
- Make lamb jugs (pages 259–260).

November

- Keep grain in rodent-proof containers, and take steps to get rid of rodents.
- Order antibiotics and store them in a refrigerator for emergencies.
- If any of your sheep is limping, check hooves and trim, if necessary (page 254).

If you have bred your sheep:

- Check your lambing supplies (pages 257–258). Order ear tags, if you plan to use them (page 263).

- Put the ram into a separate area from the pregnant ewe, so that he doesn't injure her.
- If the ram is run-down, feed him well.
- Add a small amount of stock molasses to the pregnant ewe's drinking water.
- Make lamb jugs if you haven't already (pages 259–260).

December

- Put molasses into ewe's drinking water — it helps keep the water from freezing.

If you have bred your sheep:

- Crotch the ewes to prepare them for lambing (page 257). Remove dirty tags from their udder and legs.
- Begin checking the udder of each pregnant ewe; if a ewe's udder is hard and lumpy, she may have no milk, and you should be prepared to bottlefeed her lamb (pages 245–246).
- Four weeks before lambing time, begin feeding ewes ¼ to ½ pound of grain daily and some hay (page 256).
- If a ewe is listless, she may have pregnancy toxemia; call your veterinarian (page 261).
- Give calcium supplement to your pregnant ewes.

January

If you have bred your sheep:

- Watch pregnant ewes carefully for signs of labor (page 261).
- Be sure pregnant ewes are getting exercise.
- If a ewe refuses to eat, she may have pregnancy toxemia or be about to lamb.
- Crotch ewes, if you haven't yet done so (page 257).
- Add molasses to pregnant ewes' drinking water.
- After a lambing, put the ewe and the lamb into the jug (page 262). Strip the ewe's teats, and "snip and dip" the lamb's umbilical cord (page 262). Give the ewe warm molasses water.
- Dock lambs' tails when they are 2 to 3 days old, and castrate male lambs when they are about 10 days old (page 263).

February

- Make sure salt is available.

If you have lambs:

- Watch lambs to be sure that they are having normal bowel movements.
- Watch twin lambs to be sure that one isn't growing more rapidly than the other; if that happens, supplement the feed for the slower-growing lamb.
- If a ewe loses a lamb, encourage her to adopt an orphan lamb (pages 266–267).
- Ear tag lambs (page 263).

- Check each ewe's feet, and trim and treat them, if necessary, before turning her out of the pen (page 254).
- Give plenty of fresh, clean water.
- Continue feeding grain to nursing ewes (page 263).
- Prepare the lamb creep (pages 247–248).

March

- Worm all sheep and restock worming supplies (page 250).
- Start shearing, if weather permits (pages 270–271). Keep mothers with their lambs as much as possible, to avoid confusion. Do not shear wet sheep. Keep fleeces clean.
- At shearing time or 10 days later, treat for keds (pages 251–252).
- Trim hooves at shearing time (page 254).
- Clean out the barn or shed, and put old hay and manure onto the vegetable garden or onto an area of the pasture that needs to be fertilized.

April

- Place your feeder on well-drained ground to avoid hoof trouble.

If you have lambs:

- Keep fresh water, salt, and feed in the lamb creep.
- Before turning ewes onto new pasture, let lambs in first, so that they get the best of the grass.
- Worm lambs when they weigh about 40 pounds. You may want to wean them at this time.
- Before lambs are weaned, decrease the ewe's grain ration and feed her only hay.

6

Dairy Cows & Beef Cattle

Introducing Cattle

Raising cattle, a milk cow, or even just a calf or two can be a profitable and satisfying experience. Cattle are an efficient way to produce food, since they can graze on land that won't grow crops. They can eat roughages that humans can't. They can mow the hillside behind your house that is too steep for a garden or survive on a back 40 that has too much brush, rocks, or swamp to grow any crop other than grass. Cattle provide us with meat or milk while keeping weeds trimmed, which is a good measure for fire control and yields a neater landscape. If you have enough acreage to raise a few extra animals to sell, cattle can turn otherwise useless land into a salable product. Moreover, getting set up to raise cattle often does not involve much expense, except for the initial fencing to keep them where you want them. And if you don't want to bother with purchasing hay and grain, you can use weaned calves to harvest your grass during the growing season, then send them for butchering when the grass is gone — thus making seasonal use of pasture and creating a "harvest" of meat.

Raising cattle can also be a soul-satisfying experience. They are fascinating and entertaining animals; working with cattle is never boring. It can be physically challenging at times, as when delivering a calf in a difficult birth or trying to catch an elusive animal. But for those of us who enjoy raising cattle, the chores of caring for them are not really work. Our interaction with these animals is part of our enjoyment of life.

Finally, when you raise cattle, you participate in one of the oldest known human activities. Humans have lived with cattle since prehistoric times. Wild cattle were the main meat in the diet of Stone Age people. Our prehistoric ancestors began to domesticate cattle about 10,000 years ago, so that they'd have a supply of meat for food and hides for leather clothing. Later, humans discovered that they could hitch these animals to a cart or a plow. In fact, oxen were used for transportation before horses were; cattle were domesticated sooner than horses, probably because they were easier to catch.

What Do I Need to Raise Cattle?

You can raise a steer in your backyard in a corral or on a small acreage, or you can raise a herd of cattle on a large pasture, on crop stubble after harvest, or on steep rocky hillsides. Cattle can be fed hay and grain or can forage for themselves. Economics and individual circumstances will dictate your methods. If you have pasture, all you'll need is proper fencing to keep the animals in, so they won't trample your flower beds or visit the neighbors.

You will need a reliable source of water and a pen to corral the animals when they need to be handled for vaccinations or other management procedures. If you have a milk cow, you may want a little run-in shed or at least a roof, so you can milk her out of the weather if it's raining or snowing. Most of the time, cattle don't need shelter; their heavy hair coat insulates them against wind, rain, and cold. In hot climates, however, shade in summer is important. A simple roof with one or two walls can provide shade in summer and protection from wind and storm in winter.

The novice cattle raiser may also need advice from time to time from a veterinarian, cattle breeder or dairyman, or your county Extension Service agent. Don't be afraid to contact an experienced person to ask questions or to request help.

Choosing the Right Kind of Animal

Your choice of calf will depend on how much space you have and what your goals are. Do you want to raise a steer to butcher or sell for beef or a heifer that will grow up to be a cow? A calf can be raised in a small area, even in your backyard, if your town's ordinances permit livestock. But if your goal is to have a cow that someday will have a calf of her own, she'll need more room.

If you are raising a calf to sell, you should probably raise a steer. Steers weigh more than heifers of the same age and bring more money per pound when sold. However, heifers are more flexible — you can raise them as beef or keep them as breeding or dairy animals. Dairy heifers are worth more money than beef cattle when they mature. If you want to eventually raise a small herd of cattle, choose a heifer to start with. Her calves will become your herd.

Bulls. When a male calf is born, he is considered a bull, because he still has male reproductive organs. Most bulls are castrated and become steers. Only the best males are kept as bulls for breeding. A commercial herd may include no bull calves. A rancher may buy all his bulls from a purebred breeder and not raise any of his own.

Steers. A bull becomes a steer when he is castrated. The steer may still have a small scrotum, or his scrotum may be entirely gone, depending on the method used to castrate him. Beef calves are sold as steers.

Heifers. A heifer is a young female animal. Between her hind legs, she has an udder with teats on it that will grow as she matures. A bull or steer calf also has small teats, just as a boy has nipples, but they don't become large. A heifer's vulva is located under her tail, below the rectal opening. The female animal is called a heifer until she is older than 2 years and has had a calf. After this, she is called a cow.

Bull

Steer

Heifer

Cow

Choosing a Dairy Breed

Why raise dairy cattle? Perhaps you want to keep a heifer as a family milk cow or start your own dairy herd. Or maybe you want to raise a dairy heifer to sell. A good young milk cow is worth more than a beef cow; a dairy cow can make more money producing milk than a beef cow can make by producing beef calves.

You can be successful with any of the dairy breeds, but you may want to choose one that is popular in your area, especially if you want to sell your heifers.

Ayrshire. These cattle are red and white. The red can be any shade and is sometimes dark brown. The spots are usually jagged at the edges. Cows are medium sized, weighing 1,200 pounds. Bulls weigh 1,800 pounds. Cows are noted for their good udders, long lives, and hardiness. They manage well without pampering. They give rich white milk.

Brown Swiss. Brown Swiss are light or dark brown or gray. Cows weigh 1,400 pounds and bulls weigh 1,900. Brown Swiss are noted for their sturdy ruggedness and long lives; they give milk with high butterfat and protein content.

Guernsey. Guernseys are fawn and white, with yellow skin. The cows weigh 1,100 to 1,200 pounds. Bulls weigh 1,700 pounds. Guernsey cows have good dispositions and few problems with calving. Their milk is yellow in color and rich in butterfat. Heifers mature early and breed quickly.

Holstein. Holstein cattle are black and white or red and white. They are large: Cows weigh 1,500 pounds and bulls weigh 2,000 pounds. A Holstein calf weighs about 90 pounds at birth. The cows produce large volumes of milk that is low in butterfat. Holsteins are the most numerous dairy breed in the United States.

Jersey. Jerseys are fawn colored or cream, mouse gray, brown, or black, with or without white markings. The tail, muzzle, and tongue are usually black. They are small cattle: Cows weigh 900 to 1,000 pounds, and bulls weigh 1,500. Jerseys calve easily and mature quickly and are noted for their fertility. Jerseys produce more milk per pound of body weight than any other breed, and their milk is the richest in butterfat.

Milking Shorthorn. These cattle are red, red and white, white, or roan (a mix of red and white hair). Cows are large, weighing 1,400 to 1,600 pounds; bulls weigh 2,000 pounds or more. They are hardy, noted for long lives and easy calving. Their milk is richer than that of Holsteins but is not as high in butterfat as that of Jerseys or Guernseys.

Milk Factory

A top-producing dairy cow gives enough milk in one day to supply an average family for a month. The average milk cow produces 6 gallons a day (96 glasses of milk). A world-record dairy cow can produce 60,000 pounds of milk per year — that's 120,000 glasses of milk!

Choosing a Beef Breed

Beef breeds in the United States are descendants of cattle imported from the British Isles, European countries, or India. Many modern breeds are mixes of these imported breeds. The first cattle came here from Spain, but they were soon outnumbered by British cattle.

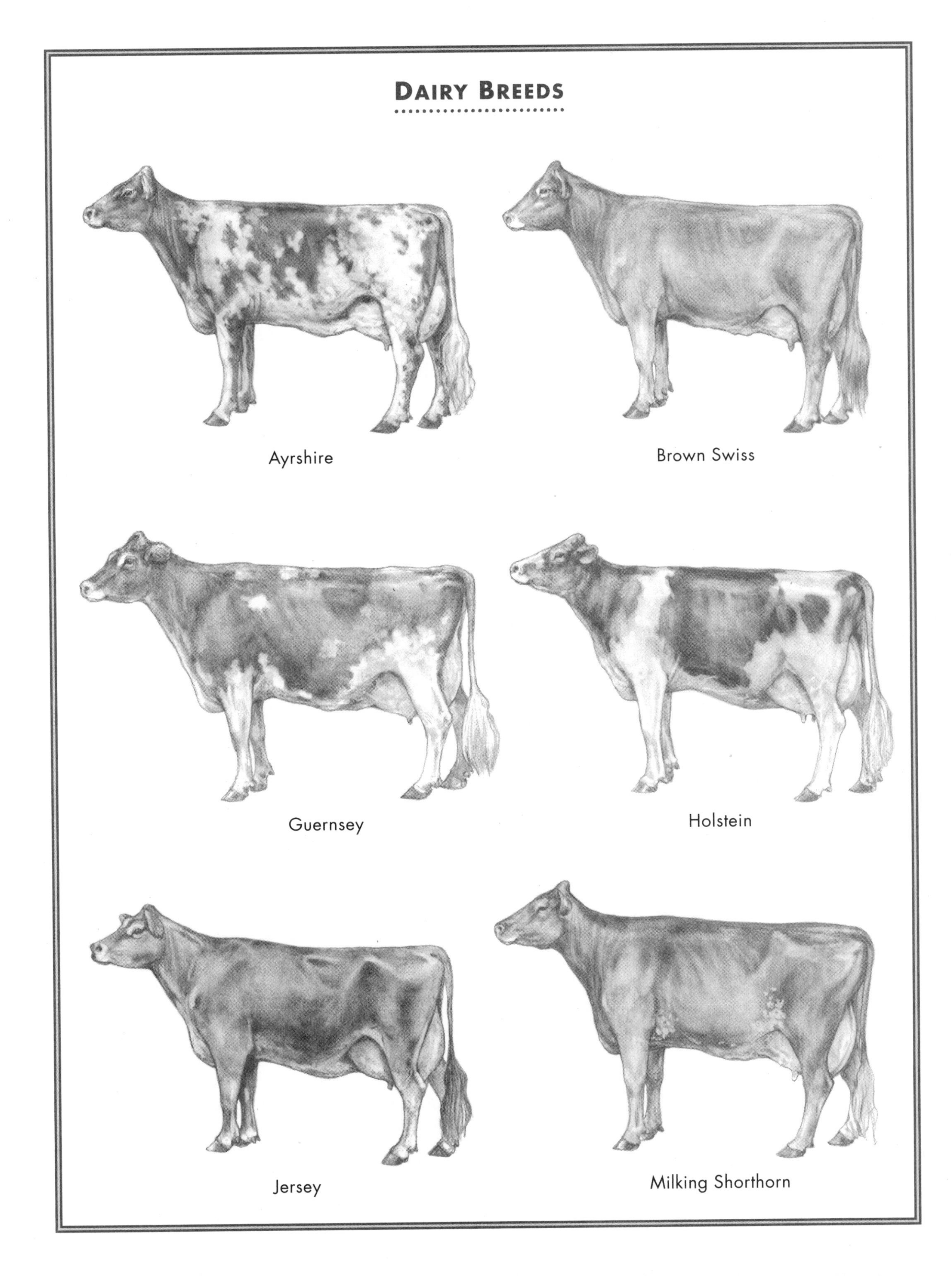
DAIRY BREEDS
Ayrshire
Brown Swiss
Guernsey
Holstein
Jersey
Milking Shorthorn

Parts of a beef animal

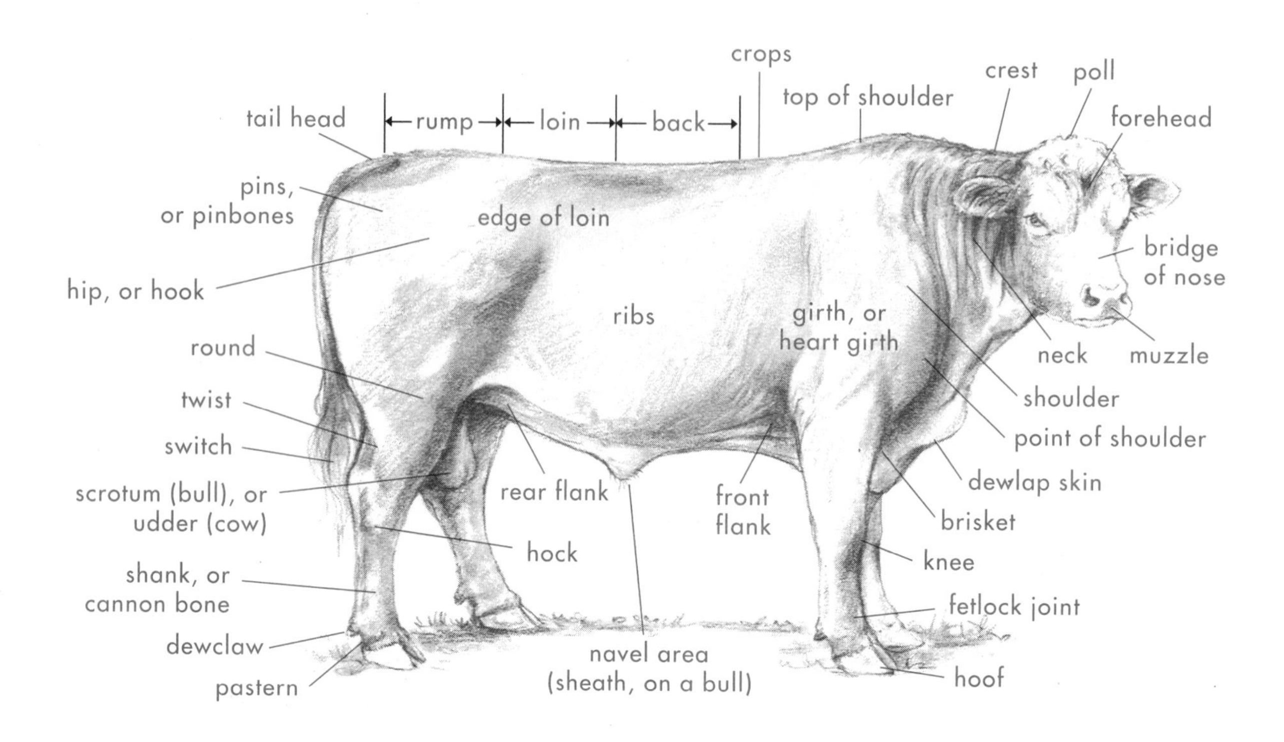

British Breeds

British breeds are those that originated in England, Scotland, and Ireland.

Angus. Angus cattle are black or red and genetically polled (always born hornless). Smaller and finer boned than Herefords, Angus are known for ease in calving, since they give birth to small calves. This characteristic makes them popular for crossbreeding with larger, heavy-muscled cattle. Angus are noted for early maturity, marbling of meat, and motherliness. Angus cows are aggressive in protecting their calves and give more milk than Herefords.

Dexter. Dexters are probably the smallest cattle in the world and are used for milk and beef. The average cow weighs less than 750 pounds and is only 36 to 42 inches tall at the shoulder. Bulls weigh less than 1,000 pounds and are 38 to 44 inches tall. Dexter cattle are quiet and easy to handle, and the cows give rich milk.

Galloway. Galloways are hardy and have a heavy winter coat. Galloway cows live a long time and often produce calves until 15 or 20 years of age. The calves are born easily because they are small, but they grow fast. Most Galloways are black, but some are red, brown, white (with black ears, muzzle, feet, and teats), or belted (black with white midsection). All Galloways are polled.

Hereford. The Hereford is well known for its red body and white face. The feet, belly, flanks, crest (top of neck), and tail switch are white. Other characteristics of the Hereford are large frame and good bone (heavier bones than those of many breeds). The Hereford has a mellow disposition compared with that of Angus or some Continental breeds.

The Polled Hereford is identical to the Hereford except that it has no horns.

Scotch Highland. These cattle are small, with long shaggy hair and impressive horns. These hardy cattle do well in cold weather. Their shaggy coats provide protection from insects, and their long forelocks protect their eyes from flies.

Shorthorn. Shorthorn cows have good udders and give a lot of milk. They have few problems with calving. Even though the calves are born small, they grow big quickly. Shorthorn cattle can be red, roan, white, or red-and-white spotted.

Continental (European) Breeds

Many European beef breeds are raised in the United States. They have become especially popular over the past 30 years, adding size and muscle (and sometimes more milk) to crossbred herds in this country.

Charolais. Charolais are white, thick-muscled cattle. Cattlemen in the United States like Charolais for crossbreeding because they are larger than British breeds.

Chianina. Chianina are the largest cattle in the world. They are white in color. Due to their size and color, they make impressive oxen.

Gelbvieh. Gelbvieh cattle are light tan to golden in color. The calves grow fast, and the heifers mature more quickly than heifers of most other Continental breeds.

Limousin. The Limousin is a red, well-muscled breed. Cattlemen like Limousins' moderate size and abundance of lean muscle. The small calves are born easily and grow fast.

Salers. Salers cattle are horned and dark red. They are popular in the United States for crossbreeding because of their good milking ability, fertility, ease of calving, and hardiness.

Simmental. Simmentals are yellow-brown cattle with white markings. They are known for rapid growth and milk production.

Tarentaise. Tarentaise are a breed of red cattle with dark ears, nose, and feet. They are moderate-sized animals that are used for milk and meat. They reach maturity early and have good fertility.

Other Continental Breeds. Many other Continental breeds are available in the United States today, including Maine Anjou, Pinzgauer, Piedmontese, Braunvieh, Normandy, and Romagnola. In general, Continental cattle are larger, leaner, and slower to mature than the British breeds.

Breeds from Other Places

Some of the beef cattle breeds in the United States originated in places other than the British Isles and Europe.

Brahman. Brahman cattle, which originated in India, are easily recognized by the large hump over the neck and shoulders, loose floppy skin on the dewlap and under the belly, large droopy ears, and horns that curve up and back. They come in a variety of colors. Brahmans do well in the southern part of North America, because they can withstand heat and are resistant to ticks and other hot-climate insects. They are large cattle, but the calves are small at birth and grow rapidly, because the cows give lots of rich milk.

Murray Grey. The Murray Grey is a silver-gray breed from Australia. Murray Greys are gaining popularity in the United States because of their moderate size,

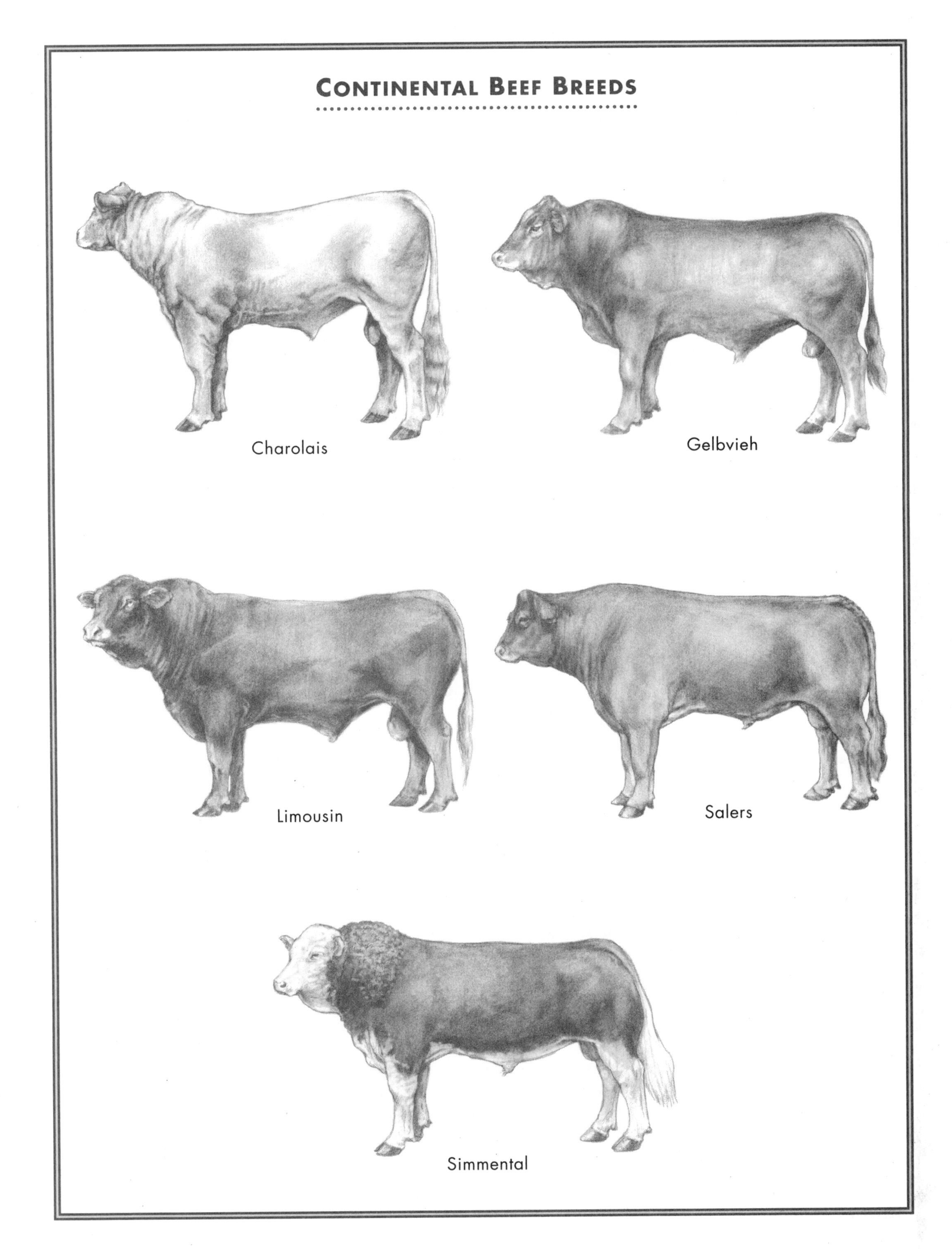
CONTINENTAL BEEF BREEDS
Charolais
Gelbvieh
Limousin
Salers
Simmental

gentle disposition, and fast-growing calves. The calves are small at birth but often grow to 700 pounds by weaning.

Texas Longhorn. The Texas Longhorn is descended from wild cattle left by early Spanish settlers in the Southwest. Longhorns are moderate sized and are well known for ease of calving, hardiness, long life, and good fertility.

Crossbred Cattle

Crossbreeding is a useful tool for the beef producer. There are nearly 100 cattle breeds in the United States. Cattlemen often cross them to create unique cow herds with the traits they want. For example, cattle are raised in a wide variety of environments, from lush green pastures to dry deserts and steep mountains. Each farmer or rancher tries to create a type of cow that will do well and raise good calves in his or her situation. No single breed offers all the traits that are important to beef production.

Hybrid Vigor. The most effective genetic advantage in cattle breeding is hybrid vigor, which results when two animals that are different mate. Hybrid vigor — displayed by the offspring of such parents — increases fertility, milk production, and life span of cows and the robustness and health of young calves. With careful crossbreeding, the rancher can develop crossbred cows that will do better than the parent breeds. Good crossbred females make the best beef cows.

Composites. Composite cattle are created from different breeds that have been blended into a uniform type of crossbred. Several composites have been created in the past 20 years, and new ones are being formed all the time. Nearly every breed registered today began as a composite. Brangus, Santa Gertrudis, and Beefmaster are examples of successful composites. More recent blends are the Hays Converter (a Canadian breed made up of several beef breeds combined with Holstein, a dairy breed) and the RX3 (a blend of Hereford, Red Angus, and Holstein).

Grazing Cattle

Beef cattle can graze on land that won't grow crops. More than 90 percent of the 810 million acres of cow pasture in the United States is too rough and steep, too dry, too wet, or too high to grow food crops. Raising beef cattle is a good way to use these lands to create food.

Selecting and Buying a New Animal

Whether you're purchasing a dairy calf, a beef calf, or a breeding heifer, you'll want to decide carefully where to buy the animal and which qualities and characteristics will best meet your needs. Consider local climate conditions as well. Before purchasing an animal, however, be sure that you have adequate housing and resources to meet its needs.

Selecting a Dairy Heifer

When you have identified your goals, determined which breed you want, and decided on a purebred or grade (unregistered, not purebred) calf, you are ready to select your heifer. Use pedigree information and consider physical appearance, age, and health when making your choice.

Pedigree Information. Use pedigree information to choose a good heifer. A pedigree is a record of the calf's ancestry. It gives genetic data and information about how well the calf's ancestors milked. Information about the sire and dam can help predict how the heifer will milk as a cow. If this is the first time you've looked at dairy cow pedigrees and performance information, have someone explain them.

Pedigree information is important, but common sense and good judgment are just as critical when selecting a heifer. Careful consideration should always be given to the physical appearance of the calf herself.

Physical Appearance. Look closely at the way a heifer is built and the functional traits that determine whether she will develop into a good cow. She should have outstanding breed character, which means that she closely resembles her ideal breed type.

The heifer should be alert and have good length of body; a deep, wide rib cage; a long, graceful neck; sharp withers; and a straight back (not humped or swayed) with wide, strong loins. Her rump should be level and square, not tipped up at the rear or slanted downward. If her rump is tipped up and her tail head is too high, she will have trouble calving.

Her hind legs should be straight, not too close together at the hocks or splayfooted, and set squarely under her body. Her front legs should be straight. She should walk smoothly without throwing her feet out to the side or swinging them inward. Avoid calves with coarse, flat-topped shoulders or low, saggy backs. The calf should be well balanced and well proportioned in all of her body parts, not short bodied, shallow bodied, or too short legged.

The udder should have well-spaced teats that are neither too long nor too fat. It's difficult to predict the future shape and size of a calf's udder. Seeing what her mother's udder looks like, and maybe seeing a photo of her sire's mother, can be helpful.

Age. A newborn calf is cheaper but a riskier purchase than an older one. Newborn calves are more likely than started calves — calves that are several months old — to get sick, unless you buy them from a well-managed dairy and take good care of them through their first weeks of life. The future conformation of a heifer is more difficult to predict in the newborn calf than in an older animal.

Take Your Time

Don't be in a big hurry to buy a calf. Look at several so that you can make a wise choice. Ask questions; find out all you can about the calf and her parents.

A started calf is easier to judge and less likely to get sick. Caring for a calf that is several months old and already weaned is easier, because it needs no nursing bottles and is past the critical age for scours (diarrhea). However, a started calf will also be more expensive than a newborn.

Health. Be sure the animal is healthy before you take it home. It should look bright and perky, be lively and energetic, and have a glossy hair coat and a sparkle to its eyes. Its bowel movements should be firm but soft, not hard or excessively runny.

If a calf or a heifer is dull or slow moving, has a dull or rough hair coat, has foul-smelling manure, or has droopy ears, it is sick. Also beware of an animal that stands with its back humped up or has a cough or a runny, snotty nose. If you are in doubt about the health of an animal, have someone else look at it. An experienced person is always a help when you are selecting a first calf or heifer.

Purebred versus Grade. If you are going to start a breeding herd, you'll have to decide whether to have registered purebreds or grade cattle. Grade cattle can take advantage of the benefits of crossbreeding. If you want a purebred heifer, select a breed being raised in your area, so that you can use bulls from someone's herd close to home. But don't select an animal just because it's registered. Registration papers won't guarantee high production or good conformation. You are better off buying a good grade heifer than a poor registered one.

If you are buying a purebred dairy heifer or a calf that is already registered, join the breed association. Belonging to a breed association has advantages. The association can give you educational material and information and help you market your heifer later if you decide to sell her. Register or transfer the registration of your new heifer to your name. Be sure the color markings or the ear tattoo on the registration certificate is correct. For the Ayrshire, Guernsey, and Holstein breeds, you may use photos of both sides of your calf or sketches of the markings. For the solid-color breeds such as Jersey and Brown Swiss, an ear tattoo is required.

Where to Make the Purchase. Some dairies sell calves at auctions. An auction is the riskiest place to buy a young calf. Even though the calf may have been healthy when taken to the auction, it may get sick after you take it home. Some calves become sick because they are taken from their mothers and sold before they have had a chance to nurse enough colostrum. Thus, they don't have antibodies against diseases.

A sale yard is also a good place to pick up diseases. Cattle come and go, and they spend time in pens before being sold. Some of the cattle brought to a sale may be sick or coming down with an illness. Even if most of the cattle that go through the pens are healthy, germs may contaminate those pens. Don't buy a calf at an auction if you have other options.

Ask for Help

Whether you are looking for a heifer or a market beef steer, many people are available to help you find one. Your county Extension agent can tell you which breeds and crossbreeds are being raised in your area and help you contact the stockmen who are raising the kind you would like to purchase.

Dairy Calves

A dairy heifer will probably be a young calf when you get her, younger than a weaned beef calf would be. A young dairy calf is both easier and harder to handle than a weaned beef calf — she's smaller at first but requires a lot of work to raise properly. A young calf is more susceptible to various diseases. An older calf has already developed some immunities. But if you take care of it well, a young calf will usually stay healthy and become a sassy character that looks forward to seeing you. You will be her substitute mother for a while.

What Does "Purebred" Mean?

The word *purebred* refers to an individual animal whose ancestors are all of a single breed (for example, a purebred Angus). A registered purebred has a registration number, recorded in the herd book of its breed association. The association gives the owner a certificate stating that the animal is the offspring of certain registered parents. However, not all purebreds are registered.

Most dairy farms have many newborn calves to sell in the spring. A baby calf can be inexpensive compared with a big weaned beef calf, but a good dairy heifer will not be as cheap as a newborn bull calf. Bull calves are cheaper because most dairy people don't want to take the time to raise them.

Check local dairies to find out which ones will let you purchase a new calf. If you buy your calf at a dairy, you can ask questions and obtain a lot of information. This is especially important if you are going to raise a dairy heifer. You'll want to ask about the health and conformation of her sire and dam, and you can be sure you buy a calf that has had colostrum and is off to a good start.

Why Do Dairies Sell Young Calves? Dairy cows must have a calf every year to produce their maximum amount of milk. A cow makes much more milk after she freshens (has given birth and begins to produce milk). Her volume of milk is greatest a month or two after calving. From then on, her production gradually declines. Dairy cows are kept at maximum production by being bred every year to have new calves and then being allowed to "dry up" briefly before the new calves arrive.

Dairies may keep the heifer calves to grow up to be milk cows, but they don't need the bull calves. Some dairies sell all their calves. Others keep their heifers and raise them to sell to other dairies. Bull calves are usually sold as soon as they are born.

Some of the calves at a dairy may be crossbred (half beef), if the dairyman breeds his heifers to a beef bull that sires small calves for easy calving. Crossbred calves can often be purchased cheaply. They make good bucket calves to raise for beef. A crossbred heifer can become a good family milk cow or nurse cow, but she won't give enough milk to be a good dairy cow.

Beef Calves

There are several factors to consider when choosing and purchasing a beef animal. Protect your investment by making careful, informed decisions before you buy. You will want to buy from a good source, pick the right breed, consider health and disposition, and pay a fair price. You can buy a beef calf or heifer at a feeder calf sale in the fall or at a farm or ranch. A local purebred breeder or commercial cattle producer is probably a good source. By buying directly from the person who raised the animal, you can find out the things you need to know, such as what vaccinations the animal has had and when they were given.

Find Out About Prices. Before you head out to purchase a beef calf, find out what beef calves are selling for that week, per pound. You don't want to pay more than market price, especially if you are hoping to make a profit by selling your animal as a yearling.

Picking Out Your Animal. A beef steer doesn't need to be purebred. In fact, the best beef steers are usually crossbred. A good crossbred steer often grows faster and has better feed conversion (more pounds of beef produced on less feed) than most purebreds.

The best steer to raise for butcher or for sale is a fast-growing, well-muscled animal that will reach a market weight of 1,050 to 1,250 pounds by the time he is 14 to 20 months of age.

Frame Score. Beef cattle are categorized by frame score, that is, whether they are small-, medium-, or large-bodied. A small-framed, early-maturing steer will not produce as much meat on his small carcass as larger-framed animals. If you try to get him to grow bigger, he'll just get too fat; he is not genetically capable of attaining a larger size. A large-framed steer will grow too big before he gets fat enough to butcher and will use up more feed than is necessary. The most practical kind of beef steer has a medium frame.

The animal you pick should have a lot of muscle, not a lot of fat. He should have nice smooth lines and should not be swaybacked. He should have a deep body, neither shallow nor potbellied. He should be long and tall, but not overly tall.

Not Too Thin, Not Too Fat

Thinness may indicate that the animal has been sick or is not healthy. However, the animal you select shouldn't be too fat, either. If the calf is already fat, he may not grow as well.

Disposition. The disposition of a calf is just as important as his weight and conformation. Some calves are more placid and easygoing than others. If you are getting only one calf, try to select a mellow one, not a skittish one that will get nervous being by himself. Try to choose a smart and gentle one that will learn to trust people.

Don't choose a wild calf. A wild, snorty calf is a poor risk, even if he is big and beautiful. You may have trouble gentling him, and he may try to go through fences. He could also be dangerous — he might knock you down or kick you. A wild calf won't gain weight as well as a placid calf. Rate of gain (pounds gained per day) is almost always better with a gentle calf.

Local Conditions. When selecting a calf, try to make an objective decision about breeds. Evaluate cattle in your region and find out which breeds are best suited to local conditions such as temperature, humidity, and precipitation. For example, a woolly breed such as the Scotch Highland may not do well in a hot climate, and a Brahman won't do well in the cold.

Selecting a Beef Heifer for Breeding

When selecting a heifer for starting a beef herd, evaluate the animal's conformation and disposition. You'll also want to know her mother's performance in producing calves and her sire's performance in producing offspring. The heifer will be a lot like her parents in this regard.

Performance Records for Purebreds. If you are buying a purebred heifer, use the breed's performance records to help you in your selection. Most successful breeders keep detailed records and use them to identify genetic differences in their cattle. You can use this information to compare such factors as birth weight, weaning and

Figuring Frame Score for Beef Calves

To figure a calf's frame score, measure his height from the ground at the hip when he is standing squarely. Then look up his age on this chart and find the hip height on that age line. Look to the top of the column for the frame score. For instance, a 10-month-old calf that is 45 inches tall at the hips would have a frame score of 4.

Age	Frame Score (hip height in inches)						
(months)	1	2	3	4	5	6	7
5	34	36	38	40	42	44	46
6	35	37	39	41	43	45	47
7	36	38	40	42	44	46	48
8	37	39	41	43	45	47	49
9	38	40	42	44	46	48	50
10	39	41	43	45	47	49	51
11	40	42	44	46	48	50	52
12	41	43	45	47	49	51	53
13	41.5	43.5	45.5	47.5	49.5	51.5	53.5
14	42	44	46	48	50	52	54
15	42.5	44.5	46.5	48.5	50.5	52.5	54.5
16	43	45	47	49	51	53	55
17	43.5	45.5	47.5	49.5	51.5	53.5	55.5
18	44	46	48	50	52	54	56

yearling weights, milk production, and fertility. You can review performance records to see how a heifer you are considering compares with other animals in the herd or breed. Have a breeder explain the records so that you can understand how to use them to compare animals.

Most breed associations and breeders use expected progeny differences in evaluating their cattle. When you select a heifer as a future cow, you are also selecting the possible genetic traits of all her offspring; the expected progeny difference helps predict the possible characteristics of a heifer's calves. The cattle are compared with one another to come up with a score that shows how they rank in the herd or in the breed as a whole on any specific trait, such as weaning weight.

Choosing a Heifer Without Performance Records. If the calf you choose is not a purebred, a performance record won't be available to help you in your selection. However, the stockman who raised her can answer important questions about the heifer's dam and sire. Was the dam a good mother? Did she calve easily without any help? How many calves has she had? Is she fertile, always breeding early in the season and never skipping a year or calving late? Does she have a good udder?

Remember that a good crossbred heifer will usually outperform most purebreds in all traits because of her added hybrid vigor — but she must have good parents to do this. Find out as much as you can about her sire and dam.

Conformation and Frame. Cattlemen judge a cow as much by the way she is built as by any other factors. Conformation is as important as performance records. Your heifer should have good feet and legs. You don't want a cow that will eventually become crippled because of poorly formed feet and legs.

The cow must have a feminine head and neck and a long body. A long body gives a cow more room for carrying a calf. You'll want her calves to be long, because that means they'll be better beef animals. She should have a deep body, not narrow or shallow. She should have muscling, but not the bulging muscles of a steer or bull. If a heifer looks like a steer, she is not a good choice for breeding. She should move freely and athletically as she walks. She should have style — all of her parts come together to create a good-looking animal.

The heifer should have a good frame, not too small or too large. A really small cow won't raise big calves, whereas a really big heifer may grow into an enormous cow and may take too long to reach breeding age. Huge beef cows use too much feed and are often not as fertile as they should be. A good medium-sized cow is more productive, and she'll generally wean a calf that is larger in relation to her own body size.

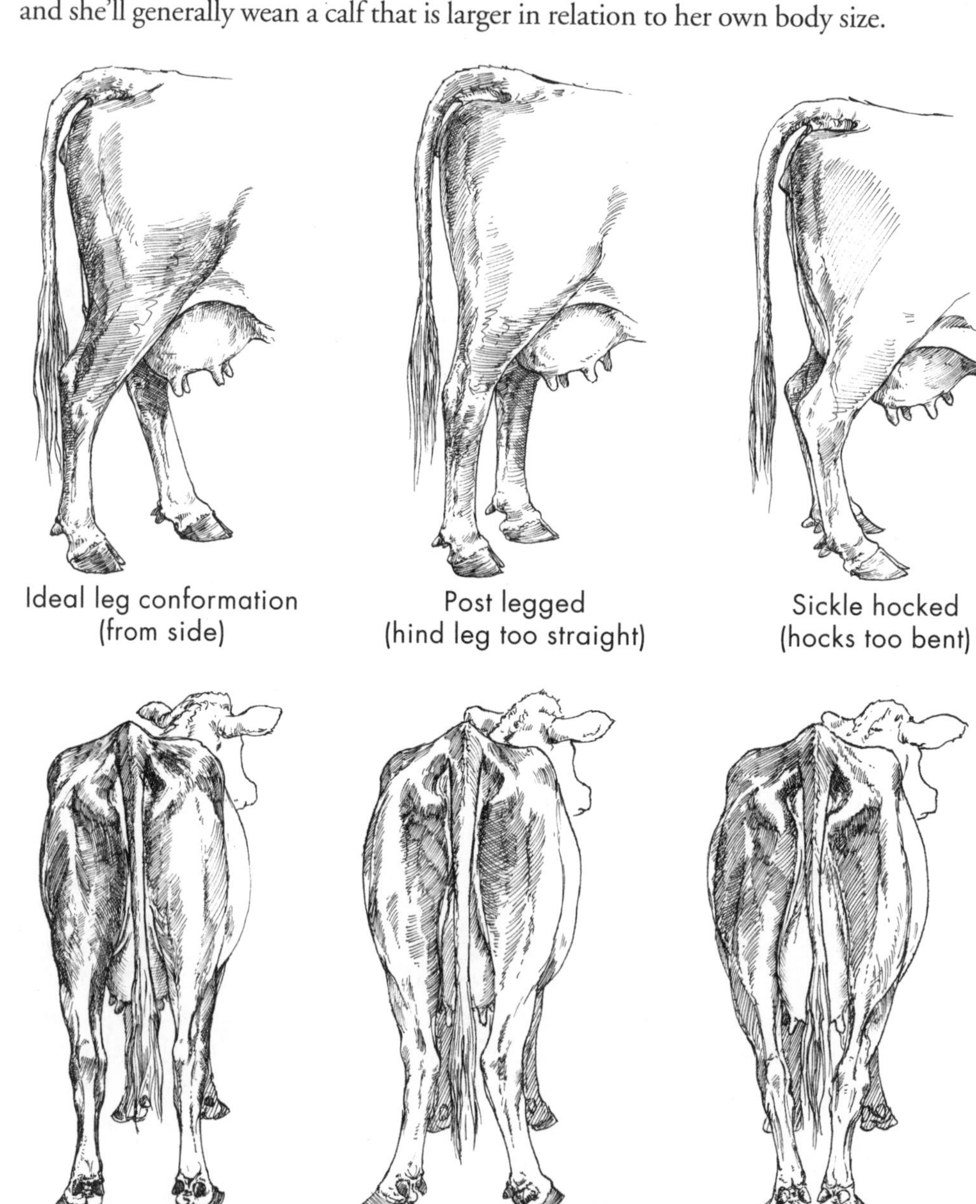

Ideal leg conformation (from side)

Post legged (hind leg too straight)

Sickle hocked (hocks too bent)

Ideal, straight leg conformation

Cow hocked (hocks too close together)

Feet too close together

Udder. You'll want your heifer to give lots of milk when she grows up. To give lots of milk, she must have a good udder. A poor udder may become saggy and injured; big teats or long teats may give a young beef calf trouble nursing.

Udder shape and size are inherited. When selecting a heifer, try to find out what the udders of her dam and her sire's dam (her father's mother) were like. When selecting a heifer calf from someone's herd, ask to see her dam and ask about her sire's dam. Sometimes you can look at beef heifer calves before they are weaned, which allows you to examine their dams.

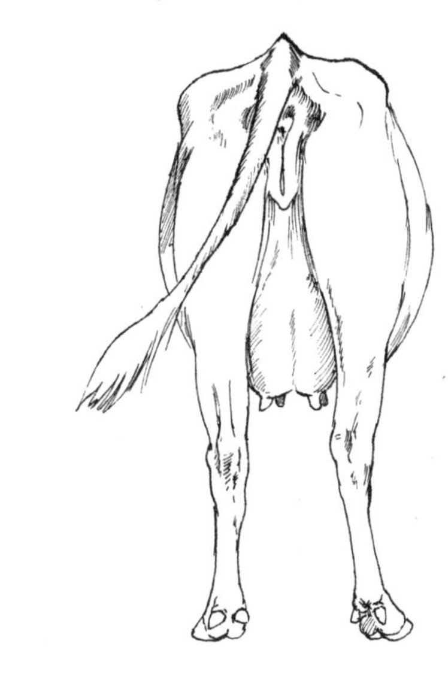

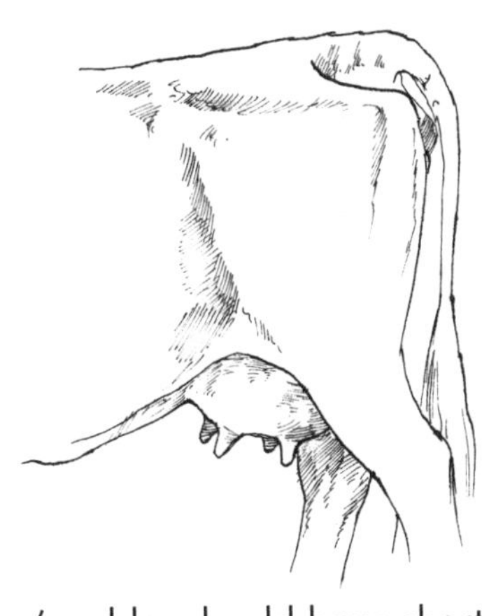

A cow's udder should have short, small teats and well-balanced quarters.

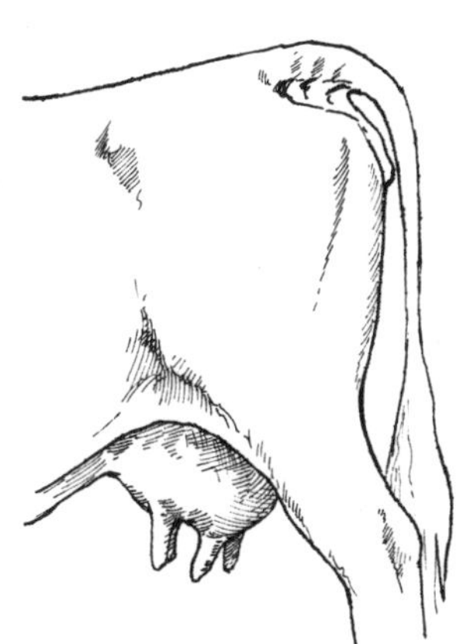

Teats too long

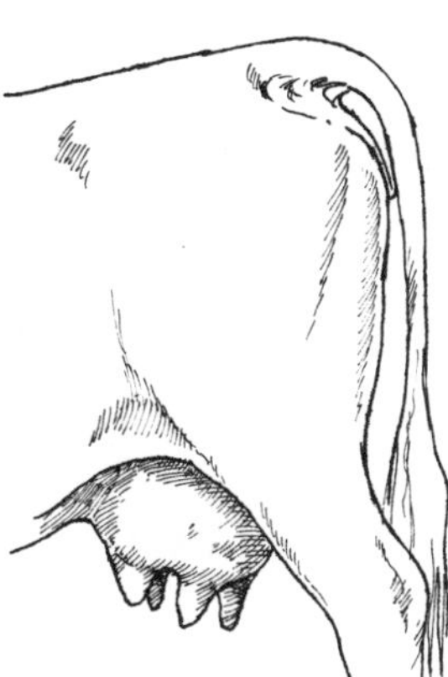

Teats too fat

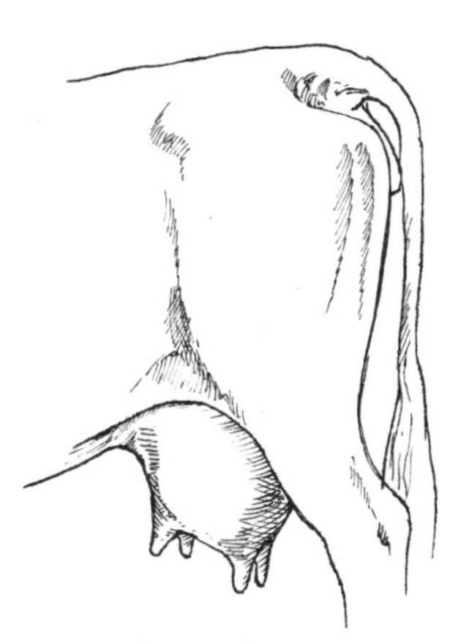

Unbalanced quarters or saggy udder

Getting Started with a Purebred Heifer

Most breed associations have regional managers who work with cattle breeders to promote their breed, help people get started in the purebred business, and help them improve their herds. If you contact your breed association, someone will direct you to local breeders who can sell you a heifer. They can also tell you how to become involved in state and national breed activities. Once you buy your registered heifer, give the seller all the information needed to transfer the registration to you.

Disposition. A cow's disposition is created partly by heredity. She inherits from her parents a tendency toward being nervous or placid, flighty or calm, smart or stupid, kind or mean. Just like humans, some cattle are smarter than others, and some are more emotional. But disposition is also influenced by how the heifer is handled or trained. A timid, nervous heifer that is smart will often gentle down with patient handling. On the other hand, some wild and nervous cattle can be frustrating, and dangerous, because they never learn to trust you.

Getting Your New Animal Home

Make arrangements with someone who has a trailer or a pickup truck with a rack to haul your animal home. If you are buying the animal from a farmer, rancher, or dairyman, he may be able to haul it for you; ask what he would charge.

If you will be unloading into a pen or a pasture, a trailer often works best, because it is low to the ground and the calf or cow can step out of it easily. An animal transported on a truck must be unloaded at a loading chute. Even a pickup truck with a rack is often too high for an animal to jump out of without risk of injury, unless the truck can be backed up to a bank or you have a ramp.

Remember that most beef calves have lived with their mothers in large pastures. Some may have seen people close up only during vaccinations or medical treatment, which are scary and painful experiences. Therefore, your calf may be frightened by you, and it may even try to run over you if you get in its way as it comes out of the truck or trailer.

When you get home, make sure the trailer is backed far enough into the pen that the animal has no choice but to enter the pen. The gate of the pen should be swung tight against the trailer. A scared animal may try to bolt through even a small opening. Don't stand in a place where the animal might run over you. If you are unloading from a pickup truck, make a ramp of sturdy boards.

Watch for Signs of Sickness

Pay close attention to your calf after you bring it home, to detect signs of illness early. If it gets pneumonia, the calf may die. (See pages 329–330 for signs of illness.)

If your calf doesn't feel well, don't wait to see whether it gets better or worse. Get advice immediately from a person experienced at raising cattle, or call a veterinarian.

Understanding Your Animal's Behavior

Wild cattle were safer in herds. If wolves approached, the cows would bellow and form a tight group. That's why yearlings and young cattle generally travel in a group. If one goes to water, they all do; if the leader decides it's time to graze, they all go. They are not just copycats; they do this for protection.

Your new calf or heifer is probably lonely and scared, unless it has another animal for a buddy. Try not to frighten the calf; it may become so upset and frantic that it crashes into the fence. To understand your new animal, try to think as it does. Cattle are herd animals and are happiest when they are in a family group with other cattle. Your calf or cow has probably not spent much time with people.

If your calf was weaned before you bought it, it has already gone through the emotional panic of losing its mother. It will miss the other calves it was with, but it won't be quite so desperate to get out of your pen to find its mother. But if the calf is still going through weaning when you bring it home, it will have several days of stressful adjustment. It may pace the fence and bawl, and it may show little interest in feed or water. A calf being weaned is more susceptible to illness, because stress hinders the immune system. In cold, rainy, or windy weather, a weaning calf may be particularly likely to get pneumonia.

Getting Acquainted

As you begin to get acquainted with your new animal, give it time and space. Don't try to get too close. Until it gets to know you, it may react explosively if it feels cornered.

Speak softly and move slowly around the calf. As you approach the pen to feed the animal, let it know you are there. If its attention is diverted and then it suddenly sees you, it may run off. Talking softly or humming a little tune can help gentle a frightened calf.

When you are in the pen with your animal, don't look directly at it. It will relax more if you act as if you aren't paying attention to it. If you come too close, approach too quickly, and look directly at it, it will see you as a predator. Instead, ignore the animal, but talk softly as you go by. Pretty soon, it will come eagerly to meet you when you bring his feed.

If you bring home a cow or two for your pasture, don't just turn them out and ignore them. Spend some time walking around in the pasture and let them get used to you. They will let you get closer once they know you are nothing to fear.

Making Friends

Cattle like to be petted and scratched, especially in places that are hard for them to reach. Most love to be scratched under the chin, behind the ears, and at the base of the tail. But don't rub the top of the head or the front of the face. Rubbing these spots will encourage a cow or a calf to bunt at you.

Handling and Gentling Cattle

Some cattle are not timid and will be curious about you from the beginning. Use this curiosity to your advantage. If you are patient and quiet, the cattle will come closer to you.

The Flight Zone. Cattle have a certain personal space in which they feel secure. As long as you don't enter that zone, they feel safe. But if you get too close, they'll get nervous or scared and run off.

Different cattle have different-sized zones. A wild or timid calf has a large zone; a gentle or curious animal has a much smaller one. As your cattle get to know you well, their flight zones will disappear.

Use Feeding to Your Advantage. When cows and calves begin to associate you with food, most will lose their fear and come right up to you. They may need a few more days before they will let you touch them, but they will soon stand beside you and eat.

Cattle are good at making associations between things. If you have a special call for feeding time, they'll come to you every time they hear it.

Don't Spoil Your Animal. Don't make the mistake of spoiling a cow or calf. It should trust you, but it must also respect you. Remember that cattle are social animals and accustomed to life in a group, in which they boss other cattle around or get bossed. Your calves and cows will think of you as one of the herd. You must be the dominant herd member; they must accept you as the boss cow. Otherwise, they will try to be too pushy.

If a calf or cow starts pushing you or bunting at you when you are feeding or petting it, discipline it with a swat. Pushing and bunting is the cattle version of play. A calf or cow will naturally want to "play fight" with you, as cattle do with one another.

If you spoil your animal by letting it do whatever it pleases, you will regret it later. Carry a small stick with you when you go to feed your calf; if it gets sassy, rap it on the nose. This swat will remind it that you are the boss.

Be Careful. Although calves and cows are not likely to attack a person (unless a cow is defending a new calf), they can accidentally hurt you because of their size and weight. Always keep an escape route in mind when trying to corner or work with a cow or a calf. Leave enough room to dodge aside if one backs into you or turns around and runs back out of a corner.

Cattle can be dangerous when handled in a confined area, because they may panic. Don't wave your arms, scream, or use a whip. If an animal won't move forward into the catch area, prod it with a blunt stick or twist its tail. Just be careful to not twist it too hard. You can twist it into a loop or push it up to form a sideways S curve. If you have to twist the animal's tail to get it to move, stand to one side so it can't kick you.

Be Gentle. If you yell or chase your cattle, you may scare them badly. Even if they are stubborn or suspicious and won't go into the catch pen or behind the gate or panel on the first try, don't get impatient. If you lose your temper and yell, you'll confuse or scare them and make things worse. They'll be harder to handle next time.

Housing, Facilities, and Equipment

Young calves need shelter from the sun, wind, and rain. Mature animals are hardier and may not need shelter.

The Pen

Before you buy a calf or a cow, prepare the place where you'll keep it. If the animal will be living by itself, build a strong pen to put it into for a few days before you turn it out to pasture. Be sure the fence is constructed so that the animal cannot jump over or crawl through it. A frantic, homesick animal in a new place may try to escape. If the calf you purchase has already been weaned, it won't be so desperate to get back to its mother.

Make sure your pen or pasture has no hazards, such as nails or loose wires, that might injure the animal. Calves are curious, just like little kids, and they often get into trouble. A pole or board on the ground with nails sticking out of it can cause serious injury if stepped on. Wire or nails lying around near your calf's feed may puncture its stomach if it eats them, causing hardware disease, which is often fatal. Don't leave baling twine hanging on a fence or lying on the ground, and watch for

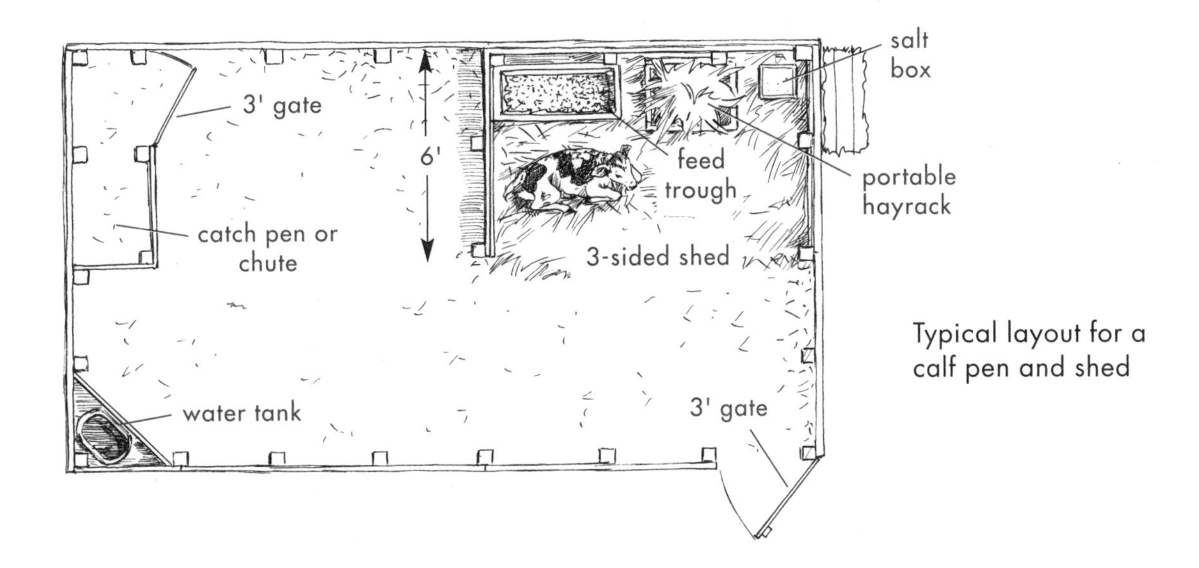

Typical layout for a calf pen and shed

stray garbage: If a calf tries to eat baling twine or chews on a plastic bag that has blown out of your trash, pieces of the material may plug its digestive tract and kill it. Keep electrical wires in the barn out of your calf's reach.

The pen must be dry and have good drainage. If necessary, put sand in the bedding area or a shady spot where the calf sleeps, to make sure the area stays dry.

You can build a pen with sturdy wood posts. The posts should be at least 8 feet long, enough to set deeply into the ground but make a fence at least 5 feet high, and should be set 8 to 10 feet apart. Use poles, boards, wood or metal panels, or strong woven-wire netting as fencing between the posts. Barbed wire or smooth wire won't work, because a calf can get through it if it tries hard enough. Do not use an electric fence to create the pen. You may need to corner the calf in the pen — or it may corner you — and you don't want it or you to get shocked. Don't skimp on materials; a good pen may be expensive to build but will last a long time.

Digging Postholes and Setting Posts

Build pens and erect fencing on solid ground. Posts set in a boggy, wet area will become loose and wobbly. The postholes can be dug with a shovel if the ground is mainly dirt with just a few rocks. If the ground is very rocky, you'll need an iron bar to loosen the rocks as you dig.

Use metal or pressure-treated wood posts to prevent rot. Set the posts in a straight line. A crooked fence will not be as strong as a straight one. Set the corner posts and sight between them to line up your postholes and your posts, or stretch a long string between them to give you an exact line. Set each post at least 2½ feet deep in the ground.

Fill the holes around the posts with dirt and rocks. To set the posts solidly, put in a little material at a time and tamp the dirt firmly with an iron bar or a tamping stick before adding more. (For more information on putting up fencing, consult a good reference book [see Recommended Reading, page 379].)

If the calf will spend all of its time in the pen, it should be large enough to give the calf room for exercise, at least 900 to 1,000 square feet. You can configure the pen however you like, such as 10 by 100 feet, 20 by 50 feet, or 30 by 30 feet — whatever fits the space you have. If you are raising more than one calf, add at least 200 square feet of space to the total area for every additional calf. The pen should offer shade from a building or a tree. A calf needs about 100 square feet (an area 10 feet square) of shade in summer.

You'll need a small catch pen in one corner of your pen, and a place where you can restrain the calf for giving shots and medications. A small enclosed shed or feeding area can be used for cornering and catching the calf. If you make a gated chute at one corner of the pen, you can herd your calf along the fence and into the chute and swing the gate shut behind it. You might include a head catcher or stanchion in the calf's feeding area, which allows you to lock its head in place when it sticks it through to eat. The stanchion will restrain the calf sufficiently for veterinary care.

Pasture Fencing

If you're raising animals on pasture, you'll need to check the fences frequently and carefully to ensure that they are in good shape. And be sure to give fences a good once-over *before* you bring home your first calf. A calf that has never lived alone may be frantic when you first bring it home. It may try to get out of the pasture to rejoin the herd it lived with.

Wire Fence. A good wire fence will hold cattle that are not being crowded or trying hard to get out. The wire must be tight, without slack, so that the cattle won't get into the habit of reaching through it. If they can reach through it to eat grass, they may eventually push through it. Net wire is the best option, because cattle cannot get a nose through it.

When you inspect the fence, replace any missing staples on wood posts or clips on metal posts so that the wire is attached properly. Tighten any sagging wires. Make sure there are no holes through which a calf might be able to get out. If the fence stretches over a dry ditch, a calf or yearling may be able to walk right under the fence. Put a pole across the ditch, under the fence, and secure it so that the calves can't push it away.

Electric Fence. An electric fence will prevent animals from getting through or rubbing against a wire fence. After being shocked a few times, the animals won't touch the fence.

For an electric fence, you'll need a battery-powered or electric fence charger and insulators to attach the wire to the fence posts. Do not allow the electric wire to touch anything metal; it will short out and won't work. It also shouldn't touch wood posts or poles, because it will short out whenever the wood gets wet. Keep all weeds and brush around the electric fence clipped to avoid a fire.

Portable or temporary electric fences can be used to divide a large pasture into several smaller ones for pasture rotation.

Pasture Rotation. Even if you have only a small acreage, the grass will last longer if you practice rotational grazing; that is, if you divide your pasture into two or three portions and graze your cattle on them sequentially. If cattle stay in the same area all the time, they tend to overgraze short, tender grasses and ignore the mature, coarse grass unless nothing else is available to eat. The grasses and plants in a pasture become less healthy if they are overgrazed or undergrazed. To avoid this problem, confine cattle

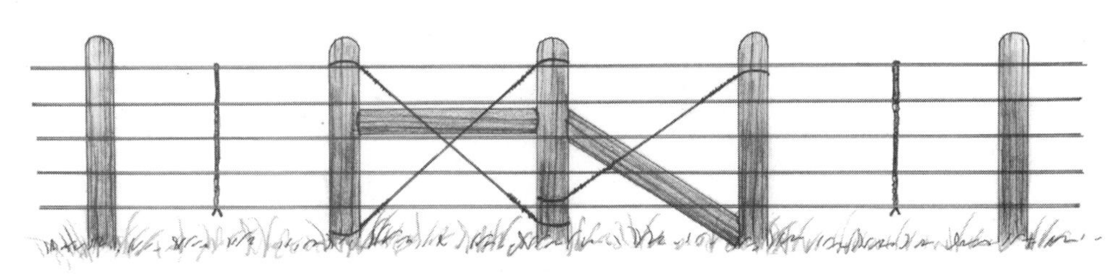

Wire fence with braces and metal stays

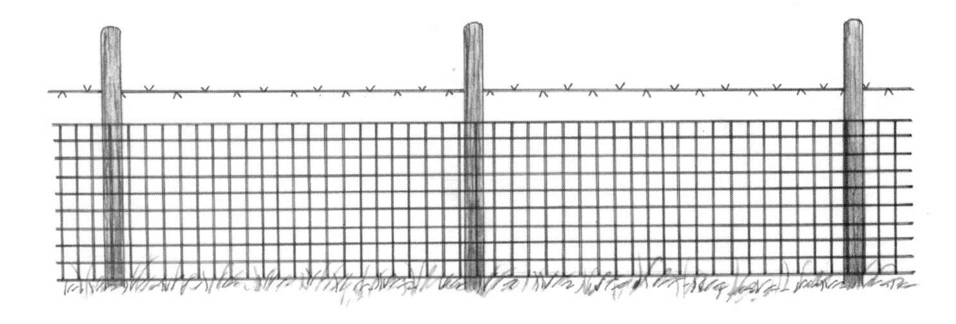

Net wire fence with one barbed wire on top

Corral fence with posts and poles

to one segment of pasture where the grass is at least 4 inches tall and move them to another segment before they graze the first one too close to the ground. The first section will have regrown by the time the cattle get back to it.

Pasture Shelter

Before you bring a calf home, make sure you have a good place to put it. All calves need shelter, but a brand-new calf is especially fragile and needs to be kept warm and dry.

In a mild climate, a calf may need only a small three-sided shed, or a protected fence corner with a roof and some boards or plywood on the sides for a windbreak. Add clean straw or wood chips for your calf to lie in. But if you live in a cold-weather region, you'll need to keep a young calf in a warm barn stall or in your garage or back porch until the calf is several days old and can live in an outdoor shed.

You can make a simple shed by setting a roof on tall, sturdy posts. A freestanding shed with walls on three sides will better protect the animals from bad weather.

The shed should be built on a high, dry spot with good drainage. The roof should slope so that rain or melting snow will run off. Make sure it slopes away from the main pen so that the water doesn't create mud in the pen or flow into the shed.

Before you build a shed, figure out which way the wind usually blows in that spot. Place the shed walls to offer the greatest protection from wind. Two sheets of plywood placed on each side of a fence corner make a nice windbreak; add another sheet of plywood to make a roof.

A simple shed with two or three walls can provide shelter.

If you live in a hot climate, a shed roof will provide shade, but you'll also need airflow to help keep the animals cool. The roof should be high rather than low, and the shelter should have no walls, which would halt air movement.

Straw, bark mulch, or wood chips scattered into a bed in the corner of the shed will give the calf a dry place to sleep. Make sure the bedding area is in a high, dry spot. The calf should always have dry bedding. Moist, dirty bedding contains harmful bacteria and also conducts warmth away from the calf's body, causing it to become chilled and more susceptible to disease. Also, ammonia gases given off by bedding that is wet from urine and manure can irritate and weaken the calf's lungs and allow bacteria to become established, leading to pneumonia.

Very young dairy calves should be kept alone in a pen to prevent spread of disease. Individual calf hutches work nicely. A calf hutch is a 4-by-6-foot or 4-by-8-foot board pen with a roof. Hang a water bucket and feed tub from the wall about 20 inches off the floor or ground; this will keep the calf from stepping in them or getting manure in them. Hutches allow each calf to have its own little barn and a yard next to it for exercise and sunshine.

Halters and Ropes

You will need a good halter and rope for restraining your calf or cow so that you can tie the animal to the fence or to the side of the chute if necessary. An inexpensive, adjustable rope halter with lead rope can be purchased at a feed store or through a mail order catalog from a livestock supply company. When putting on a halter, place it on the animal so the adjustable side is at the left. When you pull it tight, the pressure should be mostly on the rope under the chin, rather than behind the ears.

Tying Knots. At some point, you may need to tie up your calf or cow. For your safety and that of your animals, you'll need to know how to make a good knot that will stay tied but can be untied easily, even if the animal has pulled hard on the rope.

◆ *Overhand knot.* The simple overhand knot is the one you make first when tying your shoes. This knot is often the first step in forming more complex knots.

◆ *Bowline knot.* The bowline knot is probably the most useful nonslip knot for working with livestock. It allows you to tie a rope around the animal's neck or body without the danger that it might tighten when the rope is pulled, and it is relatively easy to untie. An easy technique for remembering how to tie a bowline knot is to think of the following story. The first loop is the "rabbit hole," the standing part of the rope is the "tree," and the working end of the rope is the "rabbit." The rabbit comes out of the hole, runs around the tree, and goes back down its hole.

◆ *Double half hitch.* The double half hitch knot is quick and easy to tie, acts like a slipknot, and is a handy way to secure the rope around an animal's leg when you are tying a leg back or to secure the end of the rope when no other knot seems appropriate.

◆ *Square knot.* The square knot is a stronger version of the overhand knot; it consists of two overhand knots, one tied on top of the other. The square knot is perfect for joining two pieces of rope, as when you are joining a broken rope or tying a rope around a gate and gatepost to keep the gate closed. A properly tied square knot will not slip.

◆ *Quick-release knot.* The quick-release knot (also called a reefer's knot, a bowknot, or a manger tie) is useful for tying your calf to a fence post. Like the square knot, it is a good nonslip knot. The quick-release knot has the advantage of being easily untied even after it has been pulled tight, as will happen if your calf pulls back on the rope.

Common Knots

To tie the knots described above, follow these simple step-by-step diagrams.

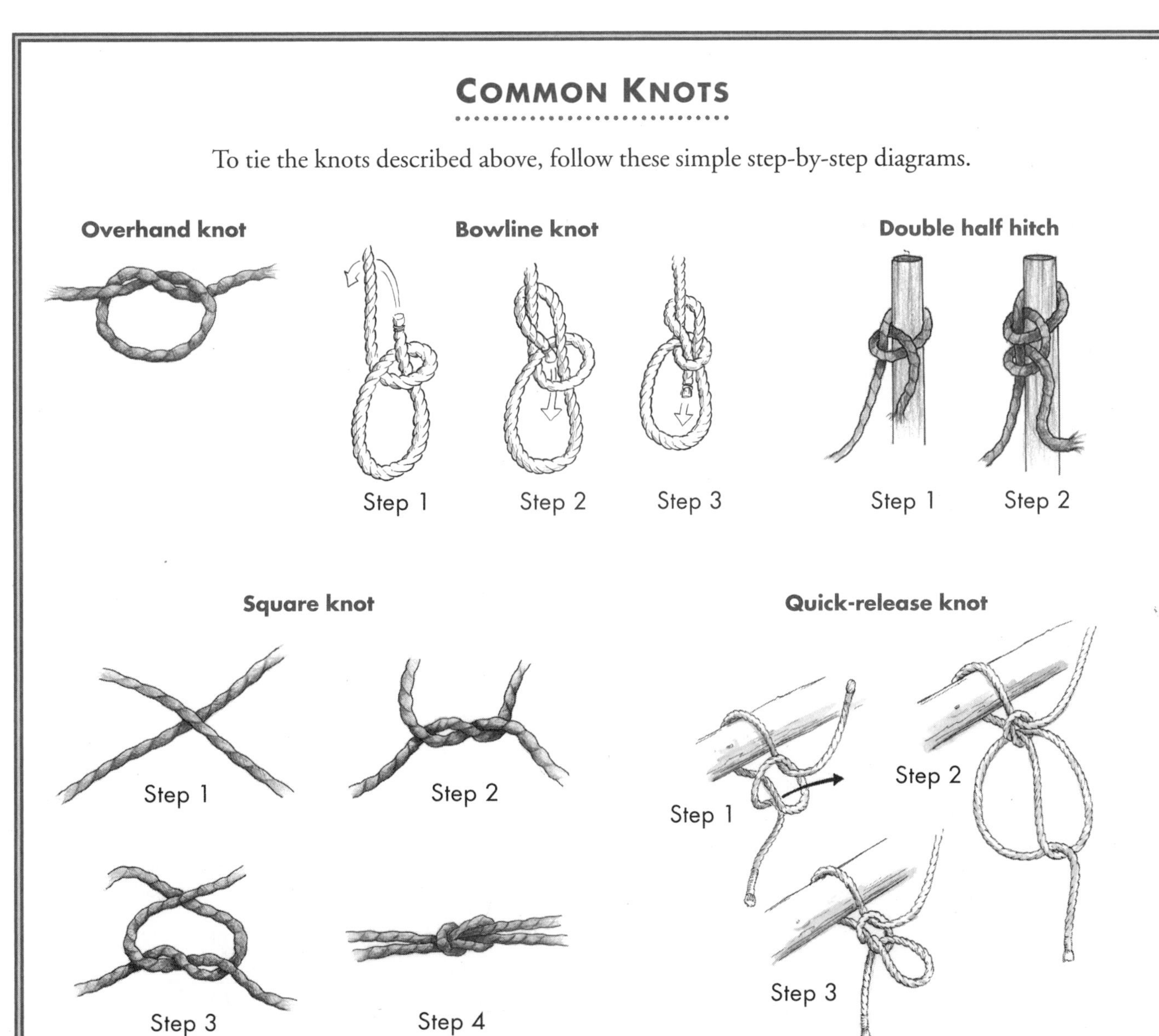

Make Your Own Water Trough

A water trough can be made from anything that will hold water and can be cleaned easily, such as an old bathtub or washtub.

Water Supply

Cattle require a source of fresh water, which can consist of a tub or tank filled twice a day with a garden hose. A water tub for calves should be set up off the ground, but no higher than 20 inches; anything higher will keep them from drinking easily. A tub on the ground will get dirty, because calves may step or poop in it.

You can easily make a stand or frame to hold the tub. Nail a board across the corner of a pen or stall, leaving room for a tub or bucket to fit snugly between it and the walls. You can pull the tub or bucket up out of the corner to rinse and clean it, but the calf can't tip it over.

You must make sure that your calf or cattle have water available at all times. Cattle drink more in hot weather than in cold weather, so check the tub more often in summer. In the winter, keep the water from freezing, even if this means breaking ice every morning and evening. In really cold climates, a rubber tub is a useful water tub, because you can tip it over and pound on it to get the ice out without creating a leak. If you use a hose to water your calf, be sure to drain it thoroughly after each use in cold weather to keep it from freezing.

The water should be kept away from the feed rack or feeding area, to keep cattle from dragging feed into the water. It should be located far from where the cattle bed. If the cattle have to walk some distance to the water, they will be less likely to stand close to it and defecate in it by accident. Keep the water fresh and clean, even if you have to dump and rinse the tub every day; cattle won't drink dirty water. Use a tub or tank that is easy to dump or has a drain hole at the bottom, and rinse it before refilling. The more cattle you have, the larger the tank you will need, or the more often you'll have to fill it.

Hay Feeders

Using a feed rack or manger will ensure that less hay gets wasted. Cattle won't eat hay that has been stepped on, is muddy, or has manure on it. Clean out any hay that collects in the bottom or corners; if it gets wet, hay may become moldy. Moldy feed may make your calf sick, and the mold spores that are released into the air when the calf eats may make it cough. To help prevent this, place the feed area inside a

Make a Feed Trough

You can make an inexpensive feed trough with 2-inch lumber cut into lengths. If several calves will be using the trough, allow 3 square feet per calf. Make the sides of the trough at least 6 inches high. Set the trough no higher than 18 to 20 inches off the ground.

weather-resistant shed. If you feed your calf outdoors, build a roof over its feed manger, hayrack, or grain box or tub.

If you are feeding hay in winter or when pasture gets short, scatter the hay on well-sodded ground. More hay will be wasted if cattle are fed on bare dirt or mud.

Grain Box

To fatten an animal for butchering, you'll have to feed it grain every day. Use a sturdy trough or grain box. The box should be mounted off the ground so that the calf won't step in it, and it must be held securely so that the calf doesn't pull it down. Calves will not eat dirty grain. A rubber tub works well if you have only one calf; these tubs are easy to wash out. Build a roof over the tub or feed box to keep the grain dry.

Clean out any leftover kernels before adding new grain to the box. If birds have pooped in the tub or trough or there's old, fermented, or moldy grain in the corners, the calf may refuse to eat the next batch you put in.

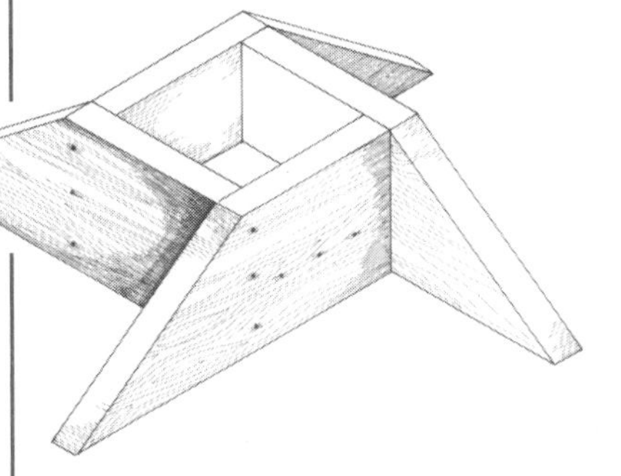

A salt box or grain box can be made of four 1" x 8" boards and a bottom.

Manure Disposal

In a large pasture, manure serves as fertilizer. But in a pen or shed, it needs to be cleaned out. A corral may be easiest to clean with a tractor and blade or loader, whereas a wheelbarrow and manure fork will suffice to clean a shed or bedding area. (A manure fork resembles a pitchfork but has more tines, so that manure and straw can't fall through it easily.) If a calf spends much time in its shed, clean out the manure and soiled bedding regularly so that it doesn't build up.

Manure makes excellent fertilizer. Spread the manure over your pasture or garden, or make a compost pile from manure and old bedding. You may be able to sell manure and compost to gardeners in your neighborhood.

Nutrition

Cattle are ruminants, meaning that they have four stomach compartments and chew their cud. The four compartments of a ruminant's stomach are the rumen, the reticulum, the omasum, and the abomasum or true stomach, which is similar to the human stomach.

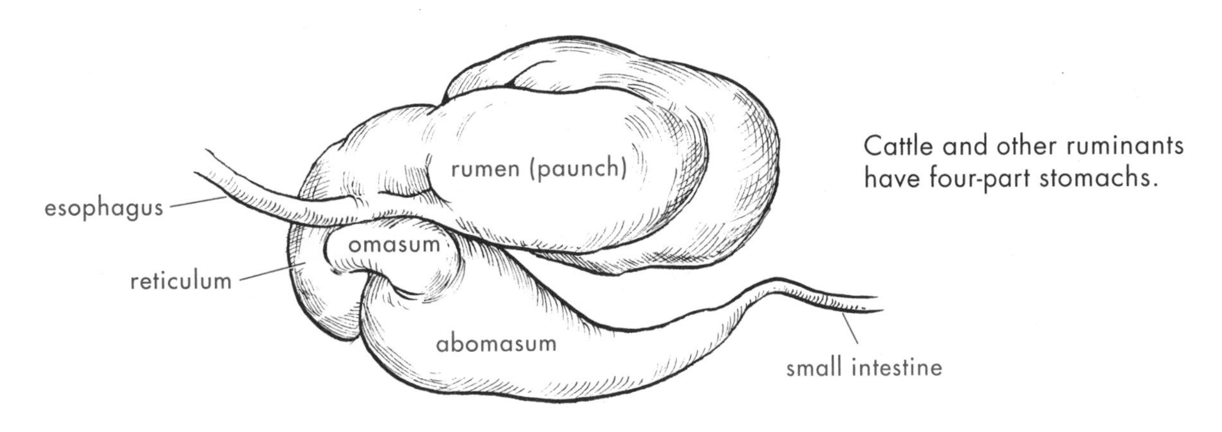

Cattle and other ruminants have four-part stomachs.

When a ruminant eats, it chews food only enough to moisten it for swallowing. After being swallowed, the food goes into the rumen to be softened by digestive juices. After the animal has eaten its fill, it finds a quiet place to chew its cud. It burps up a mass of food along with some liquid, swallows the liquid part, and then chews the mass thoroughly before swallowing it and burping up some more. The rechewed food goes into the omasum, where the liquid is squeezed out, and then goes on into the abomasum. Ruminants developed this way of eating so that they could cram in a lot of feed while grazing in open meadows and then retreat to a safe, secluded place to chew more fully.

Cattle do well on a wide variety of feeds. To some extent, what you feed your animal depends on whether it is a beef or dairy calf or a mature beef or dairy animal. However, the basic elements of good nutrition are the same for all cattle. Make sure that the feeds contain a balance of the basic ingredients for good nutrition: protein, carbohydrates, fats, vitamins, and minerals.

◆ **Protein.** Protein is necessary for growth. Good sources of protein include high-quality legume hay, such as alfalfa or clover; pasture grasses; or high-quality grass hay. (With alfalfa, care must be taken to avoid bloat — see page 335.) Protein supplements include cottonseed meal, soybean meal, and linseed meal. Cattle that are feeding on good hay or pasture don't need supplements.

◆ **Carbohydrates and Fats.** Carbohydrates and fats provide energy and are used for body maintenance and weight gain. Barley, wheat, corn, milo (grain sorghum), oats, and grain by-products, such as mill run and molasses, contain a high proportion of carbohydrates and a small amount of fat. Extra fat can be fed using high-fat products designed for ruminants, such as calf manna.

◆ **Vitamins.** Vitamins are necessary for health and growth. Green pasture, alfalfa hay, and good grass hay contain carotene, which the animal's body converts into vitamin A. Overly mature, dry hay may be deficient in carotene. The other vitamins your calf needs will be in its feed or created in its gut, except for vitamin D, which the calf's body synthesizes from sunshine. Your calf will get enough vitamin D unless it spends all of its time indoors.

◆ **Minerals.** Minerals occur naturally in roughages and grain. Cattle don't normally need mineral supplements beyond those found in ordinary feeds. However, if the soil in which their feed was grown is deficient in iodine or selenium, they may require supplements of these minerals.

Salt is important for proper body functions and for stimulating the appetite. It is the only mineral not found in grass or hay. Always provide salt for your cattle, either in a block or as loose salt in a salt box. Trace mineral salt may be used if feeds in your geographic region are deficient in certain minerals. Trace minerals include cobalt, copper, iodine, iron, manganese, selenium, and zinc. Your veterinarian or county Extension agent can help you figure out which kind of salt to use and whether it should include trace minerals.

Be Careful with Supplements

Check with your veterinarian or county Extension agent before you add any supplements to your cattle's diet. Some supplements are harmful if overfed.

Roughages

Roughages (feeds high in fiber but low in energy, such as hay or pasture) are the most natural feeds for cattle. Cattle do well on them but don't grow or fatten as fast as they do on grain. If you are raising a beef animal and don't need it to grow quickly, or you are raising a weaned heifer to keep as a cow, feed mostly roughages (sometimes called forages) and little or no grain. Whereas grain enables a beef calf to reach market weight faster, a grain-fed heifer may become too fat. Some people feed a combination of grains and roughages.

If you don't allow cattle to feed on pasture, you'll have to feed them hay. The type of hay you feed will depend on what's available in your area. Alfalfa, clover, and timothy are common, but any number of other grass and legume hays are suitable.

A calf needs about 3 pounds of hay daily for each 100 pounds of body weight. For example, a 500-pound calf needs about 15 pounds of good hay each day. If you wish, you can replace part of the hay ration with grain.

Alfalfa hay is richer in vitamin A, vitamin E, protein, and calcium than grass hay, but be careful not to overfeed your calf on it. Alfalfa hay can cause digestive problems and bloat. If the rumen becomes too full of gas (called "bloat"), it expands like a balloon and puts pressure on the animal's lungs and large internal blood vessels, causing it to die. Feeding a mix of grass hay and alfalfa hay is safer.

First-cutting alfalfa often has a little grass in it and can be an ideal hay for your calf. Second- or third-cutting alfalfa is generally richer and more likely to bloat the calf. In addition, alfalfa hay becomes moldy more readily than grass hay if it gets wet or is baled when it is too green.

Selecting Hay

If you have no experience in buying hay, ask your county Extension agent or another knowledgeable person to help you.

Hay Quality

Make sure the hay you buy is not moldy or dusty, because mold and dust may create digestive or respiratory problems. Hay for calves should not be stemmy or coarse. Because the protein and nutrition of hay are mainly in the leaves, stemmy hay is not nutritious, and it's hard for a calf to chew. Adult cattle can handle coarser feed than calves can.

When buying alfalfa, make sure it is green and bright, with lots of leaves and fine stems; it should not be coarse or brown and dry. Alfalfa that is cut early, before it blooms, is finer and more nutritious. Alfalfa that has bloomed has less protein and coarse fiber with larger stems.

Grains

Grains, also called concentrates, include corn, milo, oats, barley, and wheat. In the Pacific Northwest, barley is plentiful and may be used instead of milo or corn. Wheat is usually too expensive to feed to cattle. Corn is high in energy and is commonly used for calf feed when it is available. Oats also make good feed, as does dried beet pulp with molasses.

Pasture Management

Pasture should contain several types of nutritious grasses. Cattle won't do well in a weed patch. If you are a pasture novice, have your county Extension agent look at your pasture.

In the early spring, pen up your cattle and feed them hay for a while to let the pasture grow. Otherwise, your cattle will eat the new green grass as soon as it starts to grow, and it won't become tall enough to provide sufficient feed for the summer. Some pasture plants become coarse as they mature, and your calf will not eat them. Weedy areas may also be a problem. You can improve the pasture by mowing or clipping weeds so that they don't go to seed and spread. If the pasture has bare spots, you can seed them by hand-scattering a pasture mix when the ground is wet.

If you live in a rainy area, your pasture will grow just fine without much help. But in dry climates, pasture must be watered with a ditch or sprinklers so it won't dry out by late summer.

Lush green grass has as much protein and vitamin A as good alfalfa hay. For a growing calf or heifer, good pasture is hard to beat, but keep a close watch on your grass. Dry pasture is poor feed, because it loses its nutrients. If the grass gets short or dry, feed your cattle some good hay to supplement the pasture.

Feeding the Dairy Calf

Whether you purchase a calf from a dairy or the calf is born on your farm, you'll be responsible for feeding it. (If it is born on your farm, you'll be milking its mother.) For the first few days of a calf's life, split the daily feeding into three parts and feed every 8 hours. You can feed the calf early in the morning when you get up, again in the middle of the day, and at night just before you go to bed. Once the calf is 1 week old, you can begin feeding twice a day (every 12 hours, morning and evening), which will be easier.

Keep Feeding Equipment Clean

Always carefully wash your bottle or nipple bucket after each feeding; otherwise, bacteria will grow on it and may make the calf sick. Nipple buckets must be taken apart and cleaned. Use a bottle brush to thoroughly clean bottles.

Colostrum. Your calf should have adequate colostrum. Some dairymen let the calf stay with her mother until she has nursed once or twice. Others prefer to take the calf away before she has nursed, put her into a clean pen, and feed her from a bottle. The colostrum from the cow is milked out and saved to feed to calves.

When taking a newborn calf home from a dairy, ask to buy a gallon or two of fresh colostrum to take with you. Store the colostrum in very clean containers in your refrigerator and feed it as long as it lasts.

Bottlefeeding a Calf. Teaching a calf to drink from a bottle is easier if she has never nursed from her mother. A hungry newborn calf will eagerly suck a bottle for her first meal. But the calf that has already nursed from her mother is spoiled, preferring the taste and feel of the udder. These calves can be stubborn, and it takes patience to get them to nurse from a bottle.

If the calf was with her mother for a while before you got her, she knows how to nurse from a cow but not from a bottle. You'll need to quickly teach her to nurse from a bottle — you don't want her to go hungry for long. A young calf should be fed several times a day.

A newborn dairy calf is more easily fed from a bottle if she has never nursed from her mother.

If a young calf's first few feedings with a bottle are colostrum instead of milk replacer, she'll be more willing to suck the bottle, because she'll like the taste better. Colostrum is also the best food for her at this time.

The keys to teaching a stubborn calf to suck a bottle are persistence and the use of real milk (preferably colostrum). The milk should be warm; young calves hate cold milk. Heat the milk so that it feels pleasantly warm on your skin but not hot. If it is too hot, it will burn the calf's tender mouth and she won't suck.

To feed the calf, back her into a corner so that she can't get away from you or wiggle around too much. Straddle her neck and use your legs to hold her still, leaving both hands free to handle her head and the bottle.

Use a nipple that flows freely when the calf is sucking, so that she won't get discouraged by having to work too hard. However, the milk shouldn't flow so fast that it chokes her. Hold the bottle so the milk will flow to the nipple. The calf shouldn't be sucking air. Don't let her pull the nipple off the bottle.

Don't Overheat Milk

Never overheat milk or milk replacer. Overheating damages the proteins.

Transitioning to Milk Replacer. If you get some colostrum for your young calf, divide it into several feedings to get through the first day or two while you are teaching the calf to nurse from a bottle. If you cannot obtain colostrum, use whole milk, preferably raw milk from a dairy rather than pasteurized milk from a store. The calf will like the taste of whole milk better than that of milk replacer.

Before you run out of colostrum or milk, start mixing it with milk replacer (if that's what you'll be using) to gradually adjust the calf to the taste of what she'll drink from then on. If you switch suddenly to milk replacer, she may dislike the taste of the new stuff and be stubborn about accepting it.

You can buy milk replacers at feed stores. Of the many kinds and brands, some are better than others. Read the label on the bag to find out what a certain milk replacer contains. You can also ask a dairyman or your county Extension agent to recommend a good brand.

Protein and fat content. The National Research Council recommends using a milk replacer with a minimum of 22 percent protein and 10 percent fat. But calves will do better if the milk replacer contains 15 to 20 percent fat; they will grow faster and be less apt to get scours from inadequate nutrition.

Fiber content. Check the fiber level in your milk replacer. Low fiber content (0.5 percent or less) means the replacer has more high-quality milk products and less filler.

Protein sources. Check the protein sources in a milk replacer. Are they milk-based or vegetable proteins? Milk protein is the highest quality and best for the calf, because the newborn calf has a simple stomach. Her rumen, for digesting roughages and fiber, is not working yet. She can digest and use protein from milk or milk by-products more easily and efficiently than she can use protein from plants.

Mixing milk replacer. Follow the directions on the bag. The powder is mixed with warm water and fed like milk. The recommended amount varies by brand.

The powder mixes better if you put the warm water into your container first and then add the replacer to the water and stir well, until it is all dissolved. It won't mix quickly if the water is cool or lukewarm. Start with water that is a little hotter than you want it to be when you feed the calf, so that it will be just the right temperature by the time you mix in the powder and take it out to feed her.

Storing milk replacer. Keep milk replacers dry and clean. The powder will spoil if it gets damp. Close the bag immediately after measuring out the correct amount. Keep it in a container with a tight cover. The quality may be reduced and the replacer may become contaminated with germs if the bag is left open and exposed to light, moisture, flies, and mice.

Don't Overfeed

Don't feed greedy calves as much as they want; stick to recommended amounts. Make changes gradually. A sudden switch in the quantity or quality of milk can cause digestive problems.

How Much Should I Feed? It's just as bad to overfeed a calf as to underfeed her. Too much milk can upset her digestion and give her diarrhea. Feed your calf according to her size: A big calf needs more milk than a little one. Weigh or measure the milk to make sure you are not overfeeding the calf.

Feed 1 pound (about 1 pint) of milk daily for each 10 pounds of body weight. Thus, a calf that weighs 90 pounds should get 9 pints daily — 4½ pints (just over 2 quarts) in the morning and again in the evening — or about a gallon a day.

Feed at the same time each day on a regular schedule so as to not upset the calf's digestive system. If the calf gets diarrhea from being overfed, immediately halve the amount of milk for the next feeding. Then gradually increase it to the recommended amount for her size. As she grows, you can increase the amount of milk, but don't feed more than 12 pounds (1½ gallons) of milk daily.

Feeding from a Nipple Bucket. Once your calf learns how to suck a bottle, you can feed her from a nipple bucket. A nipple bucket can be hung from a fence or a stall wall; you don't have to hang on to it while the calf nurses. Nipple buckets can save you time if you are feeding a group of calves. If you use a nipple bucket, hang it a little higher than the calf's head, so that she can reach it easily.

Don't enlarge the nipple hole on a nipple bucket. Some people widen it so the milk flows faster, decreasing the time the calf takes to drink the milk. But this is not a good idea. If the milk runs too fast, the calf may inhale some of it because she can't swallow it fast enough. Milk in the lungs can lead to aspiration pneumonia, which can't be cured with antibiotics and will kill the calf.

Feeding from a Pail. You can teach your calf to drink from a pail instead of a nipple bucket. Put fresh warm milk into a clean pail and back the calf into a corner. Straddle her neck and put two fingers into her mouth. While she is sucking your fingers, gently push her head down so that her mouth goes into the milk. Spread your fingers so that milk goes into the calf's mouth as she sucks. After several swallows, remove your fingers. Repeat this procedure until she figures out that she can suck up the milk. A pail is easier to wash than a nipple bucket or a bottle.

Feeding from a nipple bucket saves time, because you don't have to hold it while the calf drinks.

A calf needs help to drink from a pail.

Dairy Calf Feeding Program

Age	Ration
Birth to 3 days	Colostrum
4 days to 3 weeks	Whole milk or replacer; grain mix or starter
3 to 8 weeks	Whole milk or replacer; grain mix or starter, with access to good roughage
8 weeks to 4 months	2 to 5 pounds of calf ration (grain mix) with access to good roughage; calves can be weaned as early as 8 weeks but do better if weaned a little later
4 to 12 months	3 to 5 pounds of calf ration, with access to good roughage

Getting Your Calf Started on Solid Feed. A growing calf needs concentrates and roughages. Concentrates are feeds that are low in fiber and high in energy, such as grain. A roughage is a feed that is high in fiber (bulkiness) and low in energy. Hay, grass, corn silage, straw, and cornstalks are roughages. A calf needs roughage to help develop her digestive system so that her rumen can begin to function properly.

Teach your calf to eat grain or calf starter pellets as soon as possible. Put some into her mouth after each feeding of milk until she learns to like it. You can then feed it in a tub or feed box. About 1 cup (¼ pound) of grain is all a young calf can eat each day at first. Increase the amount gradually until she is eating about 2 pounds of grain daily by the time she is 3 months old.

Your calf should have hay as soon as she will start to nibble on it. Calves have small mouths and cannot handle coarse hay, but they will nibble on tender leafy hay. Fine alfalfa, clover, or grass hays, or a mix of these, are all nutritious. Don't give your calf much hay at one time, because she will waste it; she won't be able to eat all of it, and baby calves won't eat hay that has been tromped or lain on. Give her just a little bit of fresh hay once or twice a day.

Good green pasture is excellent feed for a calf, as long as she is getting some milk (or milk substitute) and grain. She may also need a little alfalfa hay. Make sure she has fresh clean water every day and access to trace mineral salt. Calves need water, even though they get fluid with their milk replacer. Water is especially important in hot weather.

When to Wean. The age at which you wean your calf will depend on several things, including feed sources, the health of your heifer, and how long she has been eating enough solid food. A dairy heifer can be weaned from milk when she is as young as 8 weeks. But a young calf weaned too early, before she is eating enough grain and hay, won't do well. It's better for your calf to stay on a nursing program longer.

Avoid Scours

Scours (diarrhea) in baby calves is not always due to infection. It can also be caused by nutritional problems, such as overfeeding or the use of poor-quality milk replacers.

Don't Wean Too Soon

It doesn't hurt to keep a calf on milk or milk replacer for quite a while, but it does hurt to wean a calf too soon. Use your best judgment to decide when you think your calf is ready to be weaned.

Dairy calves can be weaned young if they have been eating a high-quality dry starter ration that contains milk products. The grain starter that contains milk products can be offered as early as the first week of life, and grain should be offered by 3 weeks of age. If the calf will eat enough of this starter, she won't need milk, and early weaning can reduce costs. But don't wean her until she is eating about 2 pounds per day of a grain concentrate mix.

Get the calf to eat dry feed as soon as possible. She won't consume much dry feed at first, but she should learn how to eat it. Give her all the grain starter she will finish daily. By the time she is 3 to 4 months old, she may eat as much as 4 to 5 pounds of grain daily. She should also be given some good hay starting at a young age.

If a calf is fed a complete starter that contains both grain and roughages, she can eat as much as she wants; leave feed available at all times. She won't need hay until she is about 3 months old. Grain starter or complete starter can be fed to a calf until she is 4 months of age to help her through the weaning process. Calves should eat at least 1 pound of starter daily for every 100 pounds of body weight before they are weaned. Use a weight tape to estimate your calf's weight.

Calves being fed a complete starter don't need hay. Before you discontinue the starter, give some hay for at least 2 weeks. Make sure your calf is eating the hay well. Calves being fed a grain starter should also be given hay, and they should be eating hay well for at least 1 week before weaning.

If you can get calves to eat hay at an early age, the rumen will start functioning. Baby beef calves begin eating hay or grass at just a few days of age, following their mothers' example. But because the dairy calf doesn't have her mother to show her how to eat, you have to encourage her to eat hay and grain. Put a little grain or leafy alfalfa hay into her mouth after every milk feeding, until she learns to like it.

The Weaning Process. Weaning is easier if the process is gradual. Start by decreasing the amount in her twice-daily bottle or bucket feedings. Cut back to about three quarters of the amount you've been giving. Feed this reduced amount for a few days, and encourage her to eat more grain, feeding it right after she finishes her bottle or bucket. She'll then be interested in the grain and not as upset with you for shortchanging her on milk. Then go to one feeding of milk a day, giving grain at her other feeding time. Then stop the milk feedings. Give grain at the time of day when you used to offer the milk.

Don't Spoil Your Calf

Make a pet of your calf if you wish, but take care not to spoil her. Don't let her get away with bad behavior, such as butting, kicking, or dragging you around when you are trying to lead her. She must respect you and learn to behave well around people. If she is handled gently but firmly, she will grow up with a good attitude and be nice to work with.

At a young age, put a halter on your calf and teach her to lead and tie up. A well-trained heifer is easy to work with for the rest of her life. Your young heifer will be even easier to train than a weaned calf, because she is smaller and not wild or scared.

It takes a while for the rumen to enlarge so that the calf can eat enough solid food to give her the nutrition she needs. Right after weaning, young calves still don't have much rumen capacity. They may eat just a small amount of hay compared with the amount of grain they can handle. The amount of hay consumed will increase as the rumen develops further.

If you have more than one calf, the calves can live together *after* weaning. Keeping dairy heifers with other calves *before* weaning is not a good idea. Even though the calves get enough milk from your bottle or bucket feeding, they drink it up quickly and want more. Right after you take the bottle away or the bucket is empty, the calves want to keep sucking. So they turn to one another and suck on one another's ears or udders, causing problems. In cold weather, ears that are wet from being sucked on may freeze, and the ends of the ears will fall off. Sucking on udders can damage the tiny teats and introduce bacteria from a calf's mouth into the teat, causing infection that can ruin that quarter of the udder and make it useless when the heifer grows up and starts to produce milk.

About 2 weeks after they have been weaned, you can put young heifers together in small groups, preferably no more than five to a pen, or out in a good pasture with shelter and shade. If a heifer starts sucking other heifers, she should be kept by herself for a while.

Feeding After Weaning. After your calf is 3 months old, gradually change from feeding starter to a growing ration. A growing ration should contain at least 15 to 18 percent protein. Often, the necessary protein can be supplied by good pasture or alfalfa hay with 4 to 5 pounds of daily grain to supplement it.

Tracking a Heifer's Weight and Size. A dairy heifer is usually bred at 15 months of age, provided she is well grown and big enough. Skeletal size and weight are more important than age. If a heifer is not quite big enough at 15 months , wait until she is of proper size for breeding.

Feeding Hay

Keep some hay in front of your calf all the time in a place she can easily reach. The hay should be fine-stemmed and leafy, with no mold or dust. As the calf grows, hay can become a larger part of its diet. After it is 5 or 6 months old, good pasture can be used.

Parts of a dairy cow

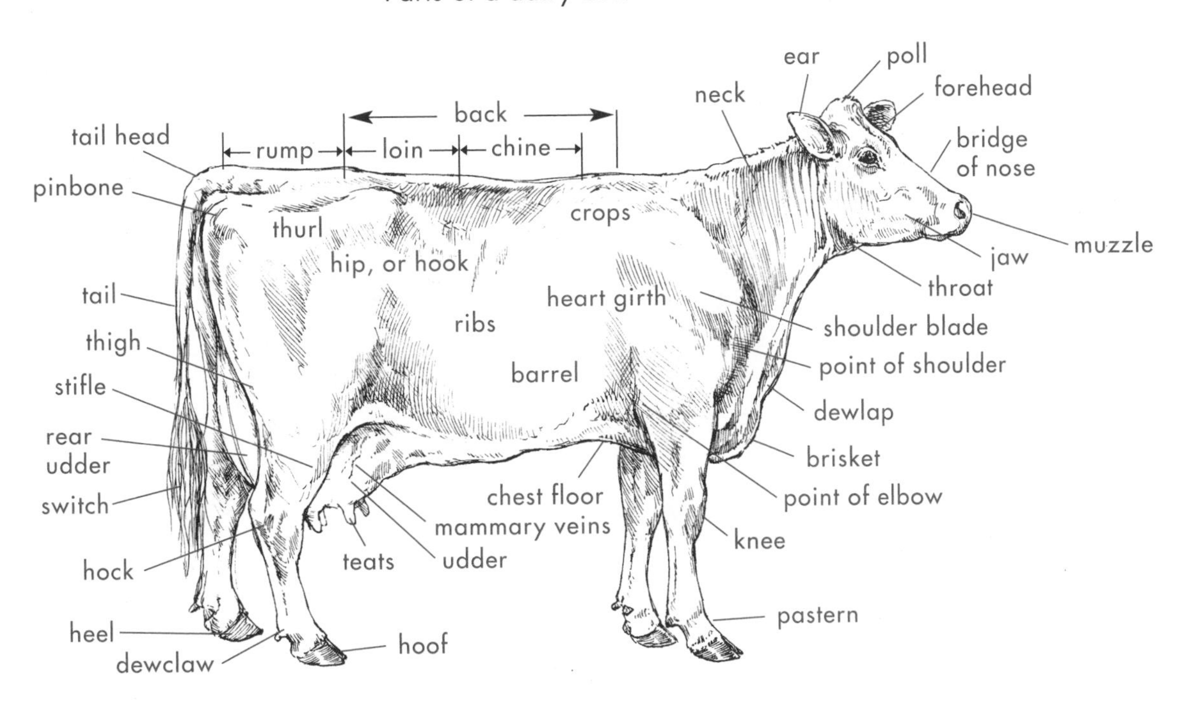

Selling Your Heifer

You may have the opportunity to sell your heifer for a good price. Selling a heifer can be a way to get started in dairying, because the price you get for your heifer will enable you to buy several more calves.

As a guideline for size at breeding, the heifer should weigh about 60 percent of her desired mature weight by the time she is bred. Large-breed heifers (such as Holsteins, Shorthorns, or Brown Swiss) should weigh 800 to 875 pounds at breeding; smaller breeds should weigh less.

Don't overfeed. Even though you want your heifer to grow well, overfeeding her on grain may make her too fat before she gets her full growth. Her skeletal growth may be inadequate with that kind of feeding. If you overfeed her on grain or shortchange her on protein (not enough alfalfa hay or green pasture), she may have adequate weight but inadequate size — her weight is made up of too much fat instead of body growth.

If a large-breed heifer weighs 800 to 850 pounds as early as 11 to 12 months of age, she is too fat. She has reached breeding weight but not breeding size. She does not have enough height and bone growth, especially in the pelvic area, and will have trouble calving if bred at this time. Wait until she is 14 to 15 months old.

Underfeeding (lack of feed or the use of poor-quality feed) is just as bad as overfeeding. An undersized, thin heifer may have difficulty calving, produce less milk, and require more feed than a normal heifer during her milking period, because she is trying to catch up to the size she should be.

You can use a weight tape to measure your heifer's height and weight and see how she is growing. Weight tapes are available at feed stores. If you can't find a weight tape, use a flexible cloth tape measure or a string to determine the distance around her girth. If you use a string, figure out the distance by marking the string and measuring it with a yardstick. To estimate your heifer's weight, make sure she is standing squarely on her feet. Put the tape around her body at the girth, directly behind the front legs. Fit the tape snugly but not tightly, with no slack in it. Then read the measurement that tells her weight.

Use a weight tape to estimate your heifer's weight. Then compare this measurement with the Dairy Heifer Growth Chart on the next page to see whether your heifer's weight is close to what it should be for her age.

Dairy Heifer Growth Chart

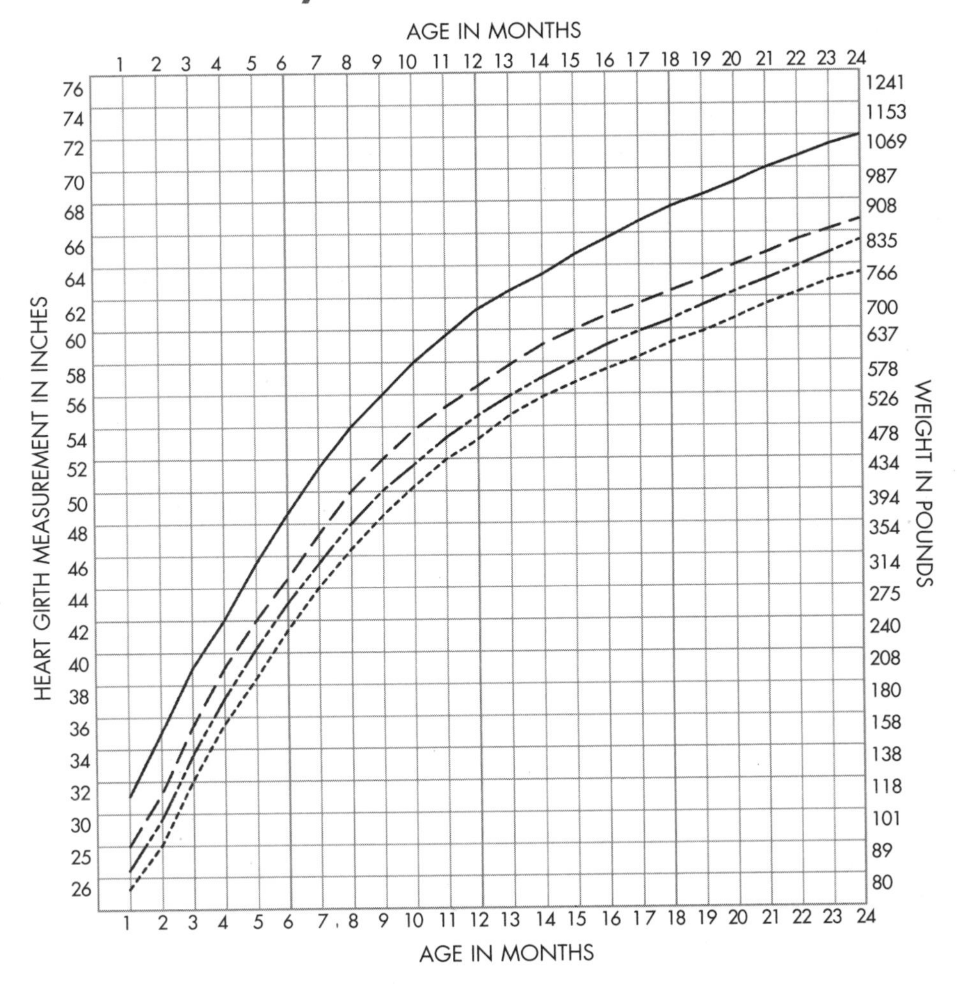

Growth Chart Key

Holstein and Brown Swiss ———

Ayrshire — — —

Guernsey —-—-

Jersey --------

Guideline for Weights and Heights* of Dairy Heifers

Age in Months	Large Breeds (Holstein, et al.) weight in pounds	Large Breeds height in inches	Small Breeds (Jersey, et al.) weight in pounds	Small Breeds height in inches
0	96	29	55	26
2	170	34	115	30
4	270	39	195	34
6	370	44	275	39
8	500	46	385	41
10	600	48	460	43
12	700	50	520	44
14	800	51	575	45
16	900	52	650	46
18	990	53	730	47
20	1,050	54	800	48
22	1,175	55	875	50
24	1,300	56	960	51

**Height is measured from ground to top of withers*

Hay and Grain for a Weaned Calf

When you start feeding your newly purchased calf, give it all the hay it will eat. Then slowly start it on grain, if you wish. Give the calf just a little bit until it learns to eat the grain, and then increase the amount gradually. Too much grain all at once may upset the calf's digestion.

Feeding a Beef Calf

A beef calf purchased from a cattle farmer will already have been weaned by his mother and will be accustomed to eating solid feed. When you bring it home, have feed and water in the pen.

Leave some good hay where the animal can find it easily, but not in a corner or along the fence line, where he'll walk on it every time he goes around his corral looking for a way out. Use really good grass hay or a mix of grass and alfalfa. Don't give rich alfalfa hay at first; it may make the animal sick or bloated. You can gradually adjust it to good alfalfa hay later.

Give water in a water trough or in a bucket that is hooked to the fence or the stall wall so that it can't tip over. If the animal has never drunk from a bucket, you may have to put the bucket next to the feed, or feed the animal next to the water trough for the first day or two, so that it will find the water when it comes to eat the hay.

Basic Nutritional Requirements. Your beef animal will need water, grain, and roughages (hay or pasture). It can be raised on roughages alone but will grow faster and get fatter sooner if you feed it grain.

Figuring a Calf's Rate of Gain. A market steer should weigh 1,000 to 1,300 pounds and have about 0.25 to 0.45 inch of outside fat on the carcass, with the grade "choice" when sold (see the box below). A market beef heifer will finish at a lighter weight (about 900 to 1,000 pounds) than a steer of the same age will.

Most beef animals eat about 7 pounds of feed to gain 1 pound of weight. To some extent, a calf's rate of gain depends on its genetics; some cattle have better rates of gain than others. On average, a growing steer should gain 2 to 3 pounds per day. Some crossbred steers will gain more. Feed your calf 15 to 20 pounds of feed daily.

By knowing the desired weight for a calf you want to butcher or sell and what he weighs when you get it, you can calculate the gain needed to get it ready for butcher or sale. By knowing the total number of days until that time, you can determine the necessary average daily gain.

For example, if you buy a 500-pound steer in November and want to sell him in August (270 days away), you can use the following formula to figure out how much your calf must gain per day to finish at 1,100 pounds: Market weight, minus present weight, divided by number of days until sale, equals the average daily gain. For example, 1,100 pounds, minus 500 pounds, divided by 270 days, equals 2.2 pounds for his necessary daily gain.

Grades of Beef

Marketplace beef cattle carcasses are inspected by the U.S. Department of Agriculture and judged for quality, using several grades to rate the tenderness of the meat. The main thing that determines the grade is the amount of marbling — flecks of fat in the muscle — which makes the meat more tender, tasty, and juicy. The highest grade is "prime," followed by "choice," "select," and "standard." Prime has the most marbling; standard has very little.

Expected Finish Weight (in Pounds) for a Steer

Breed	Angus	Hereford	Shorthorn	Charolais	Simmental	Limousin	Holstein
Angus	1,000	1,050	1,050	1,225	1,225	1,125	1,225
Hereford		1,050	1,075	1,250	1,250	1,150	1,250
Shorthorn			1,050	1,250	1,250	1,150	1,250
Charolais				1,400	1,425	1,325	1,425
Simmental					1,400	1,325	1,425
Limousin						1,200	1,325
Holstein							1,400

If your calf is a cross of two of these breeds, look at the figure where their charts meet. For instance, the top line shows the weights of Angus and Angus crosses. Heifers of the same breeds and crossbreds weigh about 80 percent of these values.

Good average daily gain for a steer is 2½ to 3 pounds. The steer in the example should have no trouble meeting his finish weight. Good average daily gain for a young heifer is 1½ to 2 pounds.

Feeding a Beef Heifer

If your calf is a heifer and you are going to market or butcher her as beef, feed her as you would a steer. Remember, though, that she will not finish out as large as a steer of her same age and breed, and therefore will not need as much feed. Adjust your feeding figures to fit her target finish weight.

If you are raising a heifer to become a breeding animal, feed her adequately for maximum growth so that she will reach puberty in good time for breeding. But don't overfeed her to the point of fatness, since fat will be detrimental to her. If she has too much fat deposited in her udder, she will never milk well. Fat displaces the developing mammary tissue (the milk-producing glands). In addition, too much fat around her internal organs will make her less fertile. She may not breed as quickly as a leaner heifer and may not become pregnant even when bred. If she is too fat during pregnancy, she will have a harder time pushing the calf out. A cow will live longer and stay healthier if she is well fed but not overfed.

Some people feed heifers grain to have them gain weight faster, especially in purebred herds. But most commercial cattlemen, who depend on what their cows will produce without pampering, raise crossbred cattle that grow fast on just hay and grass. A heifer that must have grain or expensive protein supplements to get big enough quickly takes the profit away from the calf she will produce.

Pasture. Good summertime pasture is the ideal feed for a growing heifer. Pasture will also save you money, because you won't have to buy grain and hay to feed the heifer during this season. All you'll have to supply is salt and water.

Cold-Weather Feeding

In cold weather, cattle will need more feed to generate body heat and keep warm. Roughages provide more heat than do grains, because of the fermentation that takes place during digestion. If the weather is cold, increase the ration of grass hay.

Feeding Grain. You shouldn't have to feed your heifer much grain, if any. If she needs grain in order to grow fast enough to meet her breeding weight and mature weight on schedule, she probably won't be a profitable cow. You want a heifer that will grow well and produce good calves without a lot of pampering.

When You Must Feed Grain. You may have to feed your heifer hay or grain for one of several reasons. Perhaps you don't have good pasture, or your area is suffering from drought. Or maybe you don't have space to pasture your heifer all summer. If you have good alfalfa hay to give a heifer on poor pasture, or to add to grass hay, you probably won't need grain. If you don't have alfalfa hay, you can feed grass hay, some grain, and a protein supplement. If you don't have good pasture, your county Extension agent or another experienced person can help you figure out a proper growing ration for the heifer.

With proper care and management after she is weaned, your heifer can continue to grow without getting fat. If she weighs 500 to 600 pounds at weaning and has the genetic potential to weigh 750 to 800 pounds at breeding age (15 months) without being fat, she must gain 150 to 300 pounds in the 160 days between weaning and breeding age. She should be able to gain this much on good hay. But if you have poor hay or if your heifer needs more feed to grow this quickly, add a little grain to her ration. If you add grain, start her gradually, with only a small amount at first.

Rate of Growth for a Heifer. Your heifer calf needs good feed to reach puberty at an early age and be ready to breed by the time she is 15 months old. She needs to reach 65 percent of her mature size by 14 months of age.

The actual desired weight at this age will depend on your calf's breed. Most Angus cattle mature at a lighter weight and are ready to breed at an earlier age than are Herefords or Simmentals, for instance. Know what your heifer should weigh for her age and breed and feed her accordingly.

To be the proper size and maturity for breeding, British-breed heifers should weigh at least 650 to 700 pounds, and Continental breeds should weigh as much as 800 to 850 pounds. Weigh your heifer or use a weight tape (see pages 322–323) to figure out how many days are left before you want to breed her. You can then calculate how much she has to gain to meet that weight. Finally, figure out what her average daily gain should be.

If your heifer isn't gaining weight fast enough, feed her more. If she is getting too much feed, she'll convert the extra feed into fat; in that case, cut back on the grain or eliminate it from her ration.

A growing heifer that is too thin will not breed on schedule. Her heat cycles may start late in the spring. An overfed heifer will convert the extra calories into fat deposits around her internal organs and in her udder instead of into growth. She'll reach her desired weight faster than normal and become fat. Evaluate her growth and fatness and adjust her feed accordingly.

Removing Extra Teats, Castrating, and Dehorning

Removing extra teats on a heifer, castrating a bull, and dehorning should be done while calves are very young. For these procedures, you may want your calf restrained and lying on the ground. To get your calf to the ground easily and gently, without

Flanking a calf

a big struggle, you can flank her: Stand close to the calf. Reach over its back and grab hold of the flank skin with one hand and the front leg (at its knee) with your other hand. Lift the calf off its feet, gently lowering its body to the ground.

To hold the calf still while it's lying on its side being castrated or having extra teats removed, kneel down and hold the calf's front leg (folded at the knee). Put gentle pressure on the neck with your knee so that the calf cannot rise.

Removing Extra Teats

Most heifers have just four teats, but some are born with an extra one or two. An extra teat is of no use to a cow and may cause problems when she is being milked. If your calf has an extra teat, it should be removed.

As soon as the heifer is big enough for you to tell which teat is the extra one (generally at 1 to 3 weeks of age), it can be removed. The extra teat is easier to locate when she is lying down. Examine her udder closely. The four regular teats will be arranged symmetrically, with the two rear teats slightly closer together than the two front ones. An extra teat is usually smaller than the others and located close to the main teats.

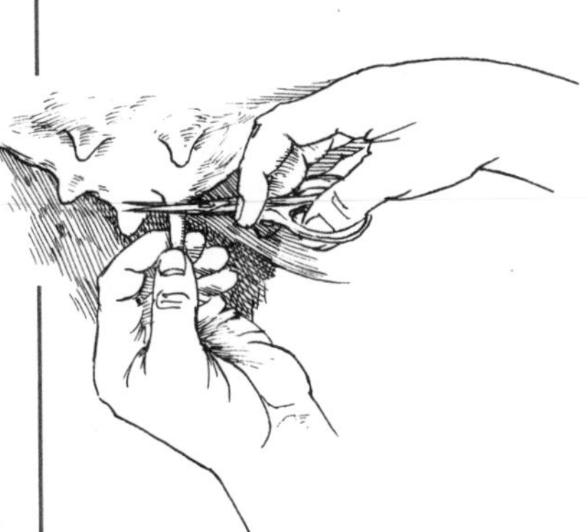

Removing an extra teat with the calf standing

If you are not sure which teat should be taken off, wait until the calf is older and the teat is more developed, then have your veterinarian do the removal. Removal of an extra teat in an older heifer should always be done by a vet, because the wound will bleed more and may need stitches.

The small extra teat on a baby heifer is easy to remove. Flank the heifer and put her onto the ground. Disinfect the teat to be removed and snip it off with sharp, disinfected scissors. Hold the scissors with the handle directed toward the front of the calf and the blades pointed toward her tail end. Make the cut lengthwise with the body. Afterward, dab the wound with iodine.

Another method is to tie the heifer and remove the teat while she is standing, preferably with someone holding her back end so that she can't move around. Pull the extra teat down and snip it off cleanly where it meets the udder. Swab with disinfectant.

A young heifer rarely bleeds when an extra teat is removed. In the summertime, use fly repellent to keep flies away from the small wound.

Before removing a teat, make sure it is truly the extra. You don't want to make a mistake and remove the wrong one.

Castrating

Baby calves should be castrated and dehorned at an early age, the younger the better. All bull calves should be castrated, unless they will be used for breeding.

Castration is harder on a calf the older he gets. As the testicles grow, the blood vessels supplying them also become larger. Bleeding is always a risk when castration is done surgically. This should be performed by an experienced person.

Baby calves are easy to castrate by using elastrator rings (strong rubber rings that resemble a large Cheerio). Using an Elastrator tool, these rings are stretched and placed over the scrotum at the top of the scrotal sac, above the testicles. The tool is removed, leaving the rubber ring to constrict tightly, cutting off all circulation and feeling below it. The testicle tissue dies, and the scrotal sac shrivels up and drops off a few weeks later, leaving a small raw spot that soon heals. Elastrator rings are the easiest and safest way to castrate baby calves, because they cause no bleeding.

To put the rubber ring onto a calf, he should be lying down on his side, with someone holding his head and front leg so he can't wiggle around while another person puts on the ring.

Store Medical Supplies Safely

Make sure all livestock medications and treatments (such as iodine, insecticides, and antibiotics) and needles and syringes are kept in a safe place, out of the reach of children.

Dehorning

If your calves have horns, they should be removed; cattle with horns may hurt one another or you. Dehorning should be done when the calf is young and the horns are small. Older calves or mature cattle with large horns may bleed a lot when their horns are removed, or they may get infections.

There are several methods of dehorning. A caustic paste or stick can be used for young calves a few days old. Electric dehorners are often used on calves up to 3 months of age. A hot dehorning iron is held firmly against the head over the horn for long enough to kill the horn-growing cells at the base of the horn. Have an experienced person show you how to dehorn your calves.

Keeping Your Cattle Healthy

Maintaining proper housing and sanitary conditions, providing adequate diet, and adhering to a vaccination program will go a long way toward keeping cattle healthy. However, even cattle that receive good care may become sick, so you must be able to recognize the signs of disease quickly. This section discusses preventing illness, signs of illness, and common health problems.

Preventing Illness

Take all appropriate measures to prevent illness in your cattle. Otherwise, your cattle may suffer needlessly, and you will incur the expense of treating them. Treating sick animals *can* be expensive, especially when disease spreads around the barnyard, affecting other animals.

Be Careful with Medications

When using any medications or vaccines, always read the labels and follow the directions for dosage, administration, and other factors. If you don't understand the directions, ask your vet for clarification. Store medicines properly. Some need to be refrigerated, whereas others shouldn't be. Some need to be kept out of the sunshine. Some need to be shaken well before use.

Vaccinations. Cattle need vaccinations against such diseases as blackleg, malignant edema, and brucellosis. Talk with your veterinarian about a vaccination program to protect your cattle from the diseases common in your area. A small herd may not be exposed to some diseases that are more likely to affect cattle in larger groups.

For some illnesses, no vaccines are available. Most diseases, however, can be prevented with good care, vaccination, and prevention of exposure to infectious agents.

Keep Your Calf Comfortable. To prevent illness and avoid stress, feed your cattle properly, and make sure they have shelter from the elements in both cold and hot weather. A heat wave in summer may kill a calf if temperature and humidity get too high. During hot weather, use a fan in the stall, or hose down your calf periodically with a misty wet spray from a garden hose.

Sanitation. Good sanitation is important. It is easier to prevent infectious diseases than to cure them. If several calves share a pen or a barn, or if other calves have been there before, bacteria, parasites, and viruses are lurking. To help prevent diseases and infections, thoroughly clean and disinfect your facilities before installing calves or groups of calves. Get rid of all old bedding and scrub the walls and floor with a good disinfectant; your veterinarian or dairyman can recommend one.

Keep pens clean and well bedded. Heifers must not lie on dirty bedding or in mud and manure. Heifer mastitis and blind quarters can result from bacteria entering the teat when heifers lie in dirty places. A blind quarter is a quarter of the udder that does not produce milk because it has been damaged by infection. If your heifer gets an udder infection, consult your vet. Do not let this infection go untreated, or it may ruin your heifer for milking.

A Warning Sign

If you have several animals, one that is off by itself should be checked on. A sick animal often leaves the herd.

Signs of Illness

Get to know your cattle. Pay close attention to them every day so that you can tell whether they are feeling fine or getting sick. If you feed your calf twice a day, you can check for signs of illness at each feeding. If you notice illness early and start treatment promptly, you should be able to catch most problems in time to help the calf recover quickly.

Behavior and Appearance. A healthy animal is bright and alert and has a good appetite. It comes eagerly to its feed or grazes at regular times each day. A sick animal may spend a lot of time lying down. It may seem dull, and its ears may droop instead of standing up alert. It may stop chewing its cud because of pain, fever, or a digestive problem.

A healthy animal usually stretches when it first gets up and shows an interest in its surroundings. It responds with curiosity to sounds and movement and spends some time grooming itself. It walks with a free and easy gait. In contrast, a sick animal may show less interest in things around it. It may stand up slowly or with effort. All of its movements will be slow, and it lacks the spark of vitality and health shown by a normal animal. The more serious the illness, the more indifferent the animal will be to its surroundings, and the more reluctant to move.

An animal that is overly alert or anxious and continually looks around may be in pain or discomfort. Pain may make an animal restless, so that it wanders about, lies down and gets up repeatedly, kicks at its belly or switches its tail, or cranes around to look at its belly.

Respiration Rate. A sick animal with a fever will breathe fast. However, exercise or hot weather will also make a healthy animal breathe faster. Cattle don't have many sweat glands and cool themselves by panting. Check an animal's respiration rate when it is standing quietly or lying down. Its respiration rate should be about 20 breaths per minute (10 to 30 is normal). An easy way to figure the respiration rate is to watch the calf's sides move in and out. Each in-and-out movement counts as one breath. Using the second hand on your watch, count the breaths for 15 seconds, then multiply by 4 to get the number of breaths per minute.

Temperature. The normal body temperature for a cow or calf is 101.5°F. A temperature of more than 102.5°F signals a fever. Learn how to take your calf's temperature by using a rectal animal thermometer, which you can buy from your veterinarian. Tie a string to the ring end to avoid losing it in the animal's rectum.

To take your calf's temperature, restrain the animal in a chute. Shake the thermometer until it reads less than 98°F and moisten it with petroleum jelly so it will slide in easily. Gently lift the calf's tail and insert the thermometer into the rectum. Keep hold of the string, because the animal may poop out the thermometer, along with a bunch of manure. Leave the thermometer in for 2 minutes to get an accurate reading. When you take it out, wipe it with a tissue or paper towel so that you can read it. Disinfect the thermometer after each use.

Other Signs of Illness. Another indication of the health of an animal is whether its eating habits are normal. Does it chew and swallow properly, or does swallowing seem painful? Is it drooling, dribbling food from its mouth, or having trouble with belching and chewing its cud?

The animal's bowel movements and urination can also indicate illness. With some digestive problems, the animal becomes constipated. If it has a gut blockage, the manure may become firm and dry, or it may not be able to excrete at all. Other problems in a sick animal may cause diarrhea. Manure should be moderately firm (not runny and watery) and brown or green.

Urination may become difficult if the urinary tract is blocked, as when a steer develops a bladder stone. The steer may dribble only small amounts of urine, remain in the urinating position for a long time, kick at his belly in pain, or stand stretched. If he shows any of these signs, call your veterinarian.

Pay attention to abnormal posture. Resting a leg may mean lameness. Arching the back with all four legs bunched together is usually a sign of pain. A bloated animal may stand with its front legs uphill to make belching gas easier. A sick animal may lie with its head tucked around toward its flank.

Common Health Problems

Good care can help prevent many health problems. If your animal becomes sick, recognizing the symptoms early and knowing what to do can make all the difference.

A disease may have an infectious or noninfectious cause. Noninfectious diseases include bloat, poisoning, founder, and injury. Infectious diseases are those caused by microorganisms or parasites. The most common microorganisms are bacteria, viruses, and fungi. Infectious diseases may be contagious, such as pinkeye, or noncontagious, such as infection of a wound or abscess.

Scours. Scours, or diarrhea, is the most common disease of young calves and causes the most deaths. Scours may not be a problem in a big weaned beef calf, but it could be life threatening to a young dairy calf. Viral scours tends to affect calves during the first 2 weeks of life, whereas bacterial and parasitic scours can strike from birth up to about 3 months or even up to 1 year. The youngest calves are usually most susceptible and most adversely affected.

Symptoms/effects: The manure is runny and watery. Severe infections, such as those caused by salmonella bacteria, may cause rapid death, sometimes even before scouring is seen.

Causes: Calf scours is a complex problem. Diarrhea can be caused by many things, including bacteria, viruses, and protozoan parasites. Often more than one agent is involved. Overfeeding a calf or using poor-quality milk replacers can also lead to scours. Poor nutrition and a dirty, wet environment can make calves more susceptible to infections that cause scours. Get help from your veterinarian to determine the cause of diarrhea and how to treat it.

Treatment: Treatment includes giving the animal electrolyte fluids and administering antibiotics in liquid or pill form.

A calf with diarrhea must be given fluids. In the early stages, while the calf is still strong, you can give electrolyte fluids with a nursing bottle if the calf will drink it. Oral fluids (by mouth or into the stomach) are adequate and effective because the calf's gut can still absorb them. However, if a sick calf refuses to drink, fluids will have to be administered by stomach tube or esophageal feeder.

As disease progresses, more gut lining is destroyed. The calf becomes weaker, unable to absorb fluid from the digestive tract, and more dehydrated. A calf that becomes this ill requires intravenous fluids given by a veterinarian.

For treating bacterial diarrhea, antibiotic pills are not as effective as a liquid antibiotic. Liquid antibiotic can be squirted into the back of the calf's mouth or added to a fluid-and-electrolyte mix.

If the antibiotic recommended by your vet is available only in pill form, crush the pills or dissolve them and give them in a liquid. Add enough water so that the crushed pills can be added to your fluid mix or squirted into the back of the mouth with a syringe.

Milk need not be withheld from a scouring calf. Rather, the calf should be encouraged to nurse at its regular feedings to keep up its fluid and energy levels. If the calf is too sick to nurse or won't nurse enough, give it regular feedings by an esophageal tube. Administer medicated fluids and electrolytes between feedings.

It is preferable for a calf to nurse instead of taking milk through an esophageal tube. The act of nursing activates a reflex that causes milk to go directly to the abomasum, bypassing the rumen. When milk is given by tubing, this reflex does not occur, and milk enters the rumen, which could cause irritation (rumenitis). Electrolytes and water, however, are not a problem.

Don't Mix Medications with Milk

Do not mix electrolytes with milk, especially if your electrolyte mix contains sodium bicarbonate. Mixing medications with milk or milk replacer can prevent curd formation and can worsen diarrhea. Wait 2 or 3 hours after milk feeding before giving the fluid with electrolytes. Space the fluid treatments between the regular feedings.

A dehydrated calf should be given the fluid-electrolyte mix every 6 to 8 hours, or until the calf feels well enough to nurse from a bottle or nipple bucket again. The liquid antibiotic is needed only once a day, but make sure the calf is nursing or getting fluid three or four times a day to avoid dehydration.

Pneumonia. Pneumonia is the second most common killer of young calves.

Symptoms/effects: Pneumonia can be mild or swift and deadly. If you spot the warning signs early and start treatment quickly, pneumonia will be a lot easier to clear up. An animal coming down with pneumonia usually stops eating and lies around or stands humped up, looking depressed and dull. Its ears may droop. Respiration may be fast or labored and grunting. The calf may cough or may have a snotty or runny nose.

Causes: Pneumonia can be caused by viruses or bacteria. The germs that cause pneumonia are relatively plentiful in the environment. They make a calf sick only if its immunity is poor or its resistance is lowered by stress. Cattle of all ages can develop pneumonia, but young calves 2 weeks to 2 months of age are especially susceptible.

A newborn calf that doesn't nurse soon enough or get enough colostrum (see page 316) will not receive the antibodies needed for adequate immunity. A young calf that has been weakened by a bad case of scours may come down with pneumonia. Severe stress can also make cattle of any age susceptible to pneumonia. Stressful conditions include overcrowding, wet and cold weather, sudden changes in weather from one extreme to another, a long truck haul, or bad weather during weaning. Poor ventilation in sheds and barns can also lead to pneumonia.

Treatment: Confine the animal and take its temperature. If it's greater than 102.5°F, call your veterinarian. A temperature greater than 104°F is serious. Antibiotics should be given immediately, even if the illness is caused by a virus. Antibiotics will fight secondary bacterial invaders, which are the killers. Your vet may leave medication with you, along with instructions for treatment. A sick calf will need good basic care, which includes shelter to keep it warm and dry and sufficient fluids, especially if it isn't eating or nursing enough. Fever can cause dehydration.

For a serious case of pneumonia, try to reduce the pain and fever so that the animal will feel better and start eating again. Dissolve two aspirin tablets in a little warm water and use a syringe to squirt the mixture into the back of the calf's mouth. For a larger animal, other medications can be used; consult your veterinarian.

Don't stop treatment too soon. You may think that you can stop giving medication because the animal seems to be getting better and its fever is down. But if the symptoms return, the animal will be twice as hard to treat. Give antibiotics for at least 2 full days after symptoms have disappeared and the body temperature is normal again. A case of pneumonia usually requires at least 1 full week of care.

Cattle should be vaccinated against the most common viral diseases, such as infectious bovine rhinotracheitis, bovine virus diarrhea, and parainfluenza type 3. These diseases often progress to pneumonia because the virus weakens the calf and allows bacteria to move into the lungs. Discuss an annual vaccination schedule with your veterinarian.

Blackleg and Other Clostridial Diseases. Blackleg is a serious disease caused by bacteria that live in the soil.

Symptoms/effects: Cattle become sick suddenly and usually die.

Prevention: A good vaccine that prevents blackleg is available. Your calf should be vaccinated at about 2 months of age and receive a second dose of vaccine around weaning time.

Blackleg is one of several serious diseases caused by clostridia, a group of bacteria that live in the soil. This family of diseases includes tetanus, red water, malignant edema, and enterotoxemia. Vaccines are available that protect against several of these diseases. Check with your vet to see which vaccine you should use for your calf.

Some clostridial diseases are a problem only in calves. Once calves are vaccinated, they have lifelong immunity. Others, such as red water, can be deadly at any age. If redwater is a problem in your area (as it is in the Northwest), all cattle must be vaccinated every 6 months.

Brucellosis. Brucellosis is also called Bang's disease.

Symptoms/effects: Brucellosis causes abortion in cows.

Prevention: All heifers must be vaccinated against brucellosis between 2 and 10 months of age. Steers do not need this immunization. If your heifer was vaccinated before you bought her, she will have a small metal tag in her ear with a number on it. The same number will be tattooed in her ear. If she doesn't have a tag and a tattoo, ask your vet to vaccinate her.

Bang's Disease in Wildlife

The requirement that all heifers be vaccinated against Bang's disease (brucellosis) has nearly eradicated the disease in cattle in the United States. However, Bang's disease is spread by wildlife, such as elk and bison. For instance, the bison in Yellowstone Park are infected with the organisms that cause brucellosis. Farmers and ranchers in some states must therefore continue to vaccinate their heifers.

Johne's Disease. Johne's disease is also called paratuberculosis. It is caused by a bacterium, *Mycobacterium johne,* also called *Mycobacterium paratuberculosis.*

Symptoms/effects: Johne's disease causes chronic diarrhea and wasting (severe loss of condition) in adult cattle.

Prevention: There is no treatment for Johne's disease. There is also no vaccine. Affected animals must be culled. Other animals must be tested by skin injection or blood or fecal culture. Johne's disease is a reportable disease, and affected herds will likely be quarantined and regularly examined by state regulatory authorities until they are deemed healthy. These measures have drastically reduced the incidence of Johne's disease in cattle.

Leptospirosis. This bacterial disease is spread by mice, rats, and other rodents, wildlife, and infected domestic animals, such as pigs, dogs, and other cattle. Cattle can get "lepto" from contaminated feed or water.

Symptoms/effects: Leptospirosis is a mild disease in cattle, but it can have serious side effects, such as abortions in pregnant cows.

Prevention: Heifers and cows should be vaccinated against leptospirosis at least once a year. Some veterinarians recommend vaccination every 6 months.

Coccidiosis. Coccidiosis is a disease of calves caused by protozoa, tiny one-celled animals that damage the intestinal lining. The protozoa are transmitted in the manure of sick calves and carrier animals. Carriers are not sick but have some protozoans living in their intestines. A calf may pick up the coccidia by eating contaminated feed or water or licking itself after lying on dirty ground or bedding.

Symptoms/effects: Coccidiosis causes severe diarrhea. The loose, watery manure often contains blood.

Treatment: Have a veterinarian examine your calf if it starts passing really loose manure, and especially if it has blood in it or the calf strains after passing the loose bowel movements.

Lumpy Jaw. Lumpy jaw can occur after a calf eats hay or grass with sharp seeds in it. Foxtail or downy brome (cheat grass) seed pods have sharp stickers that can get caught in the mouth and poke into the cheek tissue. A sticker that pokes in deeply may open the way for bacteria to cause infection, and the wound may become an abscess.

Symptoms/effects: An abscess in a calf's mouth may get as large as a tennis ball, causing the cheek to bulge prominently.

Treatment: The abscess must be punctured and drained. Abscesses sometimes break and drain on their own, but they heal faster and with less scar tissue if they are lanced and drained. An experienced person, such as a veterinarian, should treat an abscess. Severe cases may require antibiotic therapy.

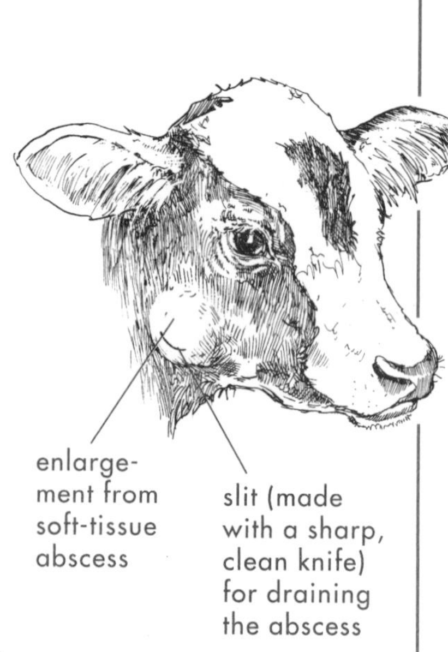

Lancing a cheek abscess

Foot Rot. Foot rot is caused by bacteria that live in the soil. Cattle may get the infection if they have a break in the skin on a foot and walk through wet, muddy areas.

Symptoms/effects: Foot rot causes swelling, heat, and pain in the foot, resulting in severe lameness. The swelling and lameness come on suddenly; the animal may be fine one day and lame the next. The foot may be too sore to walk on.

Treatment: Foot rot heals quickly if it is treated early with proper antibiotics. Consult your veterinarian.

Diphtheria. An infection of the throat and mouth, diphtheria is caused by the same bacteria that cause foot rot. Injury to tissues of the mouth or throat can let the bacteria gain entrance. Cattle are most susceptible through 2 years of age. Emerging teeth or injuries caused by coarse feed or sharp seeds can open the way for infection.

Symptoms/effects: Calves with diphtheria have fever. They may act dull and uninterested in eating. The calf may have a cough and will drool because it has a hard time swallowing. It may have swelling of the cheek tissues, and its breath may smell bad. Swelling at the back of the throat can block the windpipe and make breathing difficult. The calf may die of infection or of obstruction of the air passages unless treated quickly.

Treatment: Diphtheria can be serious. If the symptoms are present, call your vet immediately. Proper antibiotic therapy is important.

Pinkeye. Pinkeye usually appears in summer, because the bacteria that cause it are spread by face flies. Pinkeye often occurs when flies, dust, sunlight, or tall grass irritates eyes. Flies carry the bacteria from one animal to another. Pinkeye is contagious, and it may not be a problem if an animal lives alone, unless other cattle are within the distance that face flies travel.

Symptoms/effects: An animal with pinkeye will hold the eye shut, because it is sensitive to light. The eye will water. After a day or two, a white spot will appear on the cornea (the front of the eyeball). As the disease worsens, the spot grows larger and the cornea becomes cloudy and blue.

Treatment: Mild cases of pinkeye may clear up without treatment, but serious cases may cause blindness. Pinkeye should be treated promptly with an antibiotic powder or spray, which your vet can recommend. Restrain the animal so you can squirt the medication right into its eye. Because the calf's tears will wash medications out of the eye, administer the treatment at least twice a day.

Early detection and diligent treatment should clear up pinkeye within a few days. A long-acting injection of the antibiotic oxytetracycline can also help clear up the infection.

Cancer Eye. Cancer eye is also called ocular squamous cell carcinoma. It is most common in adults over 5 years of age. Herefords and other cattle with unpigmented skin around the eyes are most likely to develop cancer eye.

Symptoms/effects: Cancer eye starts as a sore that does not heal, either on the surface of the eye or in surrounding skin. The sore will continue to get larger, and can involve the entire eye.

Treatment: Surgical removal is the only treatment. Early surgery may save the animal's life.

Bloat. Bloat is a digestive problem that is often caused by highly fermentable feeds. Harmful bacteria that create gas when they multiply can also cause bloat. If too much gas builds up in the rumen, burping may not get rid of it, especially if the gas is frothy. The tight rumen eventually puts so much pressure on the lungs that the animal can't breathe, and pressure on large internal blood vessels interferes with blood circulation.

Symptoms/effects: When viewed from behind, a calf with bloat looks puffed up on its left side, where the rumen is located. As bloat gets worse, both sides puff up, and the calf has trouble breathing.

Prevention: Avoid giving feeds that promote bloat, such as lush alfalfa pasture, rich alfalfa hay, too much grain, or finely chopped hay or grain.

Treatment: Severe bloat must be relieved quickly. Your vet may pass a tube into the animal's stomach to release the gas. If the bloat is frothy, it won't come out easily, and the vet will pour mineral oil through the tube to break up the foam.

If your cattle bloat often, feed them Bloat Guard, which contains an antifoaming agent. It comes in block form. Give Bloat Guard in place of a salt block, and make sure the cattle have used it for several days before you start them on fermentable feed.

Hardware Disease. Hardware disease is also called traumatic reticulopericarditis. Cattle are indiscriminant eaters and will swallow bits of wire and other objects as they eat. Sharp wires can penetrate the reticulum and extend into the heart sac, causing infection.

Symptoms/effects: Affected cattle go off feed, show signs of pain, may have a fever, and may have difficulty breathing.

Prevention: Maintain a safe environment free from wires and other metallic objects. Dairy cattle kept in barns or small enclosures are often given magnets by mouth to try to "bind" metal objects and keep them in the reticulum.

Treatment: Antibiotics can be used but will not be effective in severe cases. Surgery can be considered for valuable animals.

Acidosis. A large increase in a calf's grain ration can cause overproduction of lactic acid in the rumen, resulting in acidosis — too much acid in the calf's body.

Symptoms/effects: If acidosis is not promptly treated, the rumen will stop working. The calf may get fever, diarrhea, or founder (see below), and it may die.

Prevention: Acidosis occurs most often when the grain ration is increased too rapidly. It can also happen if something interferes with the calf's regular feeding schedule, causing it to overeat. For example, if a calf's water has manure in it, the calf won't drink it; then, because it's thirsty, the calf may quit eating. After the calf's water tub is cleaned and it drinks again, the calf may load up on feed.

To prevent acidosis, stick to your feeding schedule. Split the daily grain ration and feed twice a day, so that the calf doesn't eat a large amount of grain at once. When increasing the grain ration, do it gradually, over a couple of weeks.

Treatment: Fast action may be needed to save your calf or prevent founder. Call your vet immediately if your calf shows symptoms of acidosis.

Founder. A calf can develop founder if you feed it too much grain or change its ration suddenly.

Symptoms/effects: In founder, the attachments between the hoof wall and the sole of the foot become sore and may separate, resulting in malformed hooves and severe lameness. Acidosis (see above) is the main cause of founder.

Treatment: Founder is a serious emergency. Contact your vet immediately.

Selenium Deficiency. Selenium deficiency is also called white muscle disease and nutritional myopathy. All cattle need adequate selenium in their diets. Many areas of the world are selenium deficient.

Symptoms/effects: Newborn calves may be too weak to stand or to nurse. Older calves may die suddenly from heart failure, especially after exercise or handling. Heifers and cows may abort.

Prevention: When needed, provide selenium in trace mineral supplements. Selenium salt blocks are not adequate in severely deficient areas. Give a selenium injection to pregnant animals a few months before calving. Newborn calves may benefit from a selenium injection. Talk with your veterinarian or Extension agent about selenium needs in your area.

Treatment: Weak calves may respond to selenium injection, but often the damage is severe and they still die.

Ringworm. This fungal infection occurs most commonly in winter. It is spread from calf to calf directly or by calves rubbing against the same posts.

Symptoms/effects: Ringworm causes hair to fall out in 1- to 2-inch-wide circles. The exposed skin is crusty or scaly.

Treatment: Ringworm generally clears up on its own in spring. Unless you are showing cattle, it may not need to be treated. Ringworm is contagious to other cattle and to people; use gloves or wash thoroughly after handling affected animals.

Warts. Warts are skin growths caused by a virus. They affect calves and yearlings more than adult cattle, since mature cattle have usually developed some resistance to the virus. Warts are unsightly but clear up on their own after a few months.

Lice. Lice are active in winter and can build up in large numbers on cattle.

Symptoms/effects: Calves that behave as if they are itchy, rubbing out the hair over the neck and shoulders, may have lice.

Treatment: Your veterinarian, county Extension agent, or local farm supply store can recommend a product for getting rid of lice. Wear protective clothing, goggles, and rubber gloves when you apply the product. Read the label thoroughly before use. If it is a powder, do not breathe the dust. Apply it on a calm day with no breeze. Easier to use than powder is a pour-on product that needs to be applied only along the animal's back. The treatment will kill the lice but not the louse eggs. It must be repeated at least once to kill lice that have not yet hatched. If your cattle are dairy cows, use only products approved for dairies.

Grubs and Heel Flies. Cattle grubs, also called warbles, sometimes appear under the skin on the backs of cattle in late winter or early spring. The grub is the maggot stage of the heel fly. This fly lays its eggs on the lower part of the legs of cattle during the warm days of early summer. The grub travels through the body to the animal's back.

Symptoms/effects: Grubs look and feel like marbles under the hide. Cattle bothered by heel flies may run wildly with tails in the air and may even crash into fences. They look for shade and try to stand in water holes to escape the flies.

Treatment: If grubs and heel flies are a problem, consult your vet about the best way to get rid of them. Pour-on products are available that treat lice and grubs at the same time.

Flies. Flies bite and suck blood and annoy and irritate cattle.

Symptoms/effects: Cattle trying to escape flies may spend all of their time in the shade. They may therefore not graze or eat as much as they should.

Treatment: Horn flies and face flies can often be controlled by using insecticide ear tags. While in the animal's ear, the tag continuously releases insecticide as it rubs against the hair. The animal rubs the insecticide over its body as it reaches around and scratches itself.

Internal Parasites. Internal worms commonly infest cattle, especially young animals that have not developed resistance to them. These parasites are most often a problem in cattle on pasture.

Symptoms/effects: Animals with worms won't gain weight fast and may lose weight. They may have a rough hair coat, a poor appetite, diarrhea, or a cough. To tell whether your cattle have worms, have your vet examine a sample of manure.

Prevention and treatment: Cattle raised in a clean place will probably not get internal parasites. If a calf is alone in a small pen or hutch and was clean before it went in, you won't have to have the calf dewormed. Worms become a problem once a calf goes out on pasture with other calves or where other cattle have been. Medication to eradicate worms is available.

Lymphoma. Lymphoma is also called lymphosarcoma. In adult cattle, it is caused by the bovine leukemia virus (BLV). Lymphoma also occurs occasionally in calves without viral infection.

Symptoms/effects: Symptoms and effects are variable, depending on what organs are affected. In cattle, skin lumps due to enlarged lymph nodes and loss of condition are common. Lymphoma can affect tissue around the eyes, causing them to bulge out. Lymphoma of the abomasum causes digestive disturbances. Lymphoma of the uterus can cause infertility.

Prevention and treatment: Maintaining a herd free of BLV infection will help. Talk with your veterinarian about blood testing for BLV. Treatment is not effective.

Navel Hernia/Navel Abscess. In some cases, the abdominal wall at the navel area does not close up properly after the birth of the calf, causing a bulge at the navel.

Symptoms/effects: In a hernia, soft tissue passes back and forth through the hole in the abdominal wall. If the swelling at the navel is firm, an abscess is the cause.

A small hernia may go away by itself as the calf grows. However, a large hernia is serious, because a loop of intestine may come through it and strangulate, causing a portion of it to die and killing the calf in the process. A bacterial abscess can spread through the body, causing severe illness and serious joint infections.

Treatment: If your calf has a swelling at the navel, have a veterinarian look at it. An abscess can be treated by lancing, draining, and flushing with antibiotics. If a hernia is present, the vet can tell you whether it will get better on its own or whether stitches are needed to close up the hole.

Giving Injections

Many medications and most vaccines are given to cattle by injection with a syringe and needle. Most injections are given intramuscularly (deep into a big muscle). Some are given subcutaneously (under the skin, between the skin and the muscle), and still others must be given intravenously (directly into a large vein). Intravenous medications should be given only by your veterinarian, but you can learn how to give intramuscular or subcutaneous injections. Have an experienced person show you how to fill a syringe, measure a proper dose, and give an injection.

Safety precaution: After you give an injection, discard the syringe and needle if they are disposable. If they are reusable, boil them before the next use.

Restraining Your Calf or Cow. The calf or cow should always be restrained before you give it a shot. A cow should be confined in a chute; a calf may be tied up and pushed against a fence. A calf that is merely tied to a fence may kick you. Don't stand behind or beside the calf unless it is restrained so it cannot move.

Intramuscular Injections. An intramusculuar injection should be given in the thickest muscle of the neck. Make sure the area where you will put the needle is very clean, free of mud or manure, or the needle will take bacteria into the muscle with it. Wet skin and hair increase the risk that bacteria will enter the muscle with the needle.

Detach the needle from the syringe. Press the area firmly for a moment before putting in the needle. This will desensitize the skin, and the calf will not jump as much when you jab it. Insert the needle with a forceful thrust so it breaks the skin easily.

A new, sharp needle goes in with minimal effort and causes minimal pain for the calf. If the calf jumps, wait until it settles down again before you attach the syringe and give the injection. If the needle starts to ooze blood, you've hit a vein. Take the needle out and try again in a different spot; never give an injection into a vein.

Give Injections in the Neck

If you're raising cattle for beef, give injections into the neck muscle to avoid damage to the best cuts of meat, which are in the rump and buttocks. Sometimes an injection causes a local reaction and a knot in the muscle, or even a small abscess. It's better to have this damage occur in the neck muscle, where it may be trimmed out more easily during butchering.

Subcutaneous Injection. To give a subcutaneous injection, lift a fold of skin on the shoulder or neck, where the skin is loosest, and slip the needle in. Aim it alongside the calf so that it goes under the skin you have pulled up but not into the muscle. With a little practice, you'll find that subcutaneous injections are easy to give.

Giving Oral Medications

Tie your calf or confine it in a chute before you give it pills or liquid medication by mouth. A tied calf can't swing its head away or hit you with its head while trying to avoid the medicine. A large animal should always be restrained in a chute.

Pills. Pills can be given with a balling gun, a long-handled tool that holds the pill as you insert it toward the back of the animal's mouth. When you press the plunger, the balling gun pushes the pill out of its slot, forcing the animal to swallow it. This tool keeps your fingers from being bitten.

Liquids. Giving liquid medication, liquid antibiotics, or pills dissolved in water is easy. Use a big syringe without the needle. Fill the syringe with the proper dose, stick the syringe into the corner of the calf's mouth, and slowly squirt the medication into the back of the mouth. Special syringes known as dose syringes have a tapered nozzle end designed for giving liquids by mouth.

If the dose is large, give the medication a little at a time, allowing the calf to swallow each portion before you squirt in the rest. Keep the calf's head tipped up so that the medication cannot run back out.

Don't Get Bitten!

Be careful when examining the inside of an animal's mouth or giving pills. A calf has no top teeth in front but can still hurt your fingers with the molars if it bites down.

A balling gun is a long-handled tool that makes giving your cow a pill easier.

Butchering a Beef

If you're raising a calf to butcher, you will probably want to let it grow to good size. Some folks like baby beef (from a calf at weaning age), but if you have enough pasture to raise your calf through its second summer, you will get a lot more meat for your money by letting it grow bigger. The ideal age at which to butcher a steer or a heifer is 1½ to 2 years. At that age, the animal is young enough to be tender and is nearly as large as it will get. Butchering at the end of summer or in the fall, before you have to feed hay again during winter, makes the grass-fed beef animal economical to raise.

The breed of the animal can be a factor in determining when it is ready to butcher. Beef animals generally do not marble until they reach puberty (or in the case of a steer, the age at which he would have reached puberty if he had been a bull). Different breeds mature at different ages. Angus and Angus-cross cattle often reach puberty at a younger age (and a smaller weight) than do larger-framed cattle, such as Simmental, Charolais, or Limousin.

An Angus-type beef calf may finish faster and be ready to butcher when it is a yearling or a little older. If you feed it longer, it may not get much bigger, just fatter. A Simmental calf, in contrast, may still be growing and not fill out (carry enough flesh to be in good butchering condition) until it is at least 2 years old.

Thus, the ideal age at which to butcher your beef animal depends on its breed and on whether it is grass fed or grain fed. Cattle will grow faster and finish more quickly on grain, but at greater cost. Whether you feed grain depends on personal preference (some people prefer grain-fed beef to grass-finished beef, and vice versa) and your situation. If you have lots of pasture, raising grass-fed beef is usually most economical.

You can take cattle to a custom packing plant to be slaughtered and butchered, or you can do the butchering yourself. (See Recommended Reading, page 379, for resources on home butchering.)

Steers as Oxen

If you become so attached to your steer that you can't bear to part with him, you might keep him and train him as an ox to help with chores around your place, such as hauling hay, manure, and firewood. An ox is a steer of any breed or cross that has reached maturity and is trained to wear a yoke and do work. A well-trained and properly maintained ox will provide useful hauling power for a dozen years or more. A good book for learning about training and working cattle is *Oxen: A Teamster's Guide,* by Drew Conroy.

Breeding and Calving

Your needs and circumstances will dictate how you choose to breed your heifer. Once she is bred, you must be prepared to see her safely through pregnancy and calving. This section discusses breeding options, care during pregnancy, preparation for calving, potential calving problems and what you should do to help, and caring for the newborn calf.

Age and Weight to Breed a Dairy Heifer

Breed	Age in Months	Approximate Weight
Jersey	13 to 17	550 lbs.
Guernsey	14 to 18	650 lbs.
Ayrshire	15 to 19	700 lbs.
Milking Shorthorn	15 to 19	750 lbs.
Brown Swiss	15 to 19	800 lbs.
Holstein	15 to 19	800 lbs.

Weight of Different Beef Breed Crosses at Puberty

Heifers are assumed to be at least 13 months old. Crosses are from Angus or Hereford cows.

	Percentage of Heifers That Will Be Having Regular Heat Cycles		
Breed	at 600 pounds	at 700 pounds	at 800 pounds
Angus	70	95	100
Angus/Hereford X	45	90	100
Charolais X	10	65	95
Chianina X	10	50	90
Gelbvieh X	30	85	95
Hereford	35	75	95
Limousin X	30	85	90
Maine Anjou X	15	60	95
Shorthorn	75	95	100
Simmental X	25	80	95
Tarentaise X	40	90	100

When to Breed Your Heifer

Your heifer should be bred in the spring when she is about 15 months old so that she will calve the next year as a 2-year-old. Gestation lasts about 9 months. A heifer can be bred after she becomes sexually mature and is having regular heat cycles. Most heifers reach puberty by 12 months of age, but some begin cycling earlier and some start later. A well-grown, healthy dairy heifer can be bred at 14 to 19 months of age. The small breeds, especially Jersey heifers, reach maturity faster than the larger breeds and can be bred younger and at lighter weights.

At breeding age (14 to 19 months), beef heifers should be about 65 percent of their mature weight. British breeds need to weigh at least 650 to 700 pounds at breeding time, and larger-framed breeds (such as Simmental) must weigh 800 to 850 pounds by breeding time, to be 65 percent of their mature weight.

A beef heifer shouldn't calve at younger than 24 months of age. She won't be mature enough yet to have a calf easily, do a good job of raising it, or breed back quickly while she is still growing. Nor do you want her to breed late and calve at more than 24 months of age; if she does calve late, she may do so every year for the rest of her life.

Signs of Heat

A cow or heifer may be bred only when she comes into heat, or estrus. If she is to be bred by artificial insemination (AI), you must be able to determine when she comes into heat, and then have a technician insert a capsule of semen into her uterus at the proper time. Determining when your heifer is in heat can be difficult if no other cattle are around. Signs of heat include increased restlessness, pacing the fence or bawling, or a mucous discharge from the vulva. However, not all heifers show obvious signs.

If a heifer is living with other cattle, telling when she comes into heat is easier. The other cattle will mount her, or she will mount them. The hair over her tail and hips may be ruffled from this activity. The easiest way to tell is to put her with another cow, heifer, or steer for a short time. If you don't have other cattle, take her where she may be left with a bull for 1 to 3 weeks until she is bred.

Importance of Birth Weight for First Calves

The most important factor in breeding your heifer is to choose a good bull that sires calves that are small at birth. Nutrition during pregnancy and whether this calf is your animal's first play some role in the size of the calf, but genetics is the main determinant. To play it safe, don't use a bull that sires calves that are large and heavy at birth.

A heifer having her first calf is smaller than a mature cow. A calf that is too large may die during birth or injure the heifer. A big calf usually needs help being born and may have to be delivered by cesarean section. This is not the way you want your calf to be born.

Distance Is No Problem

The nice thing about breeding your heifer by AI is that you can select an outstanding bull from anywhere in the United States. You have your pick of the best bulls in the breed. These bulls are kept in central locations called bull studs. Get a catalog from one of the major bull studs. Your AI technician can obtain one for you.

Selecting a Sire

Breed your heifer to the best bull available. If a dairy heifer is mated to a good bull of her breed and her calf is a heifer, the calf will be worth more. If you do not plan to sell the calf as a dairy cow, the sire you choose is less important. You may decide to breed her to a beef bull that sires small calves, to make sure she does not have a difficult time with her first calving. A crossbred beef–dairy calf makes a good beef calf if you plan to have it butchered or sell it for beef.

Taking Your Heifer to a Bull

If you bought your heifer from a local cattleman, you might ask him if he would consider putting her with a bull at his place, and what he would charge. Or you may have a neighbor with a bull who is amenable to breeding him with your heifer.

Ask the bull's owner to keep track of the breeding date so you'll have it for your records. With this information, you can predict your heifer's calving date the next spring.

If your heifer is a registered purebred and you want a purebred calf that you can register, she must be bred to a registered bull of the same breed. If she's a crossbred or if you want to raise a crossbred calf, choose a bull of a different breed or a crossbred bull.

Breeding Your Heifer by Artificial Insemination (AI)

Using artificial insemination, a large number of cows can be bred to one bull. The bull's semen is collected, divided into many small portions, and put into tubes called straws. The straws are stored in liquid nitrogen, which keeps them at the very cold temperature of –320 °F. The frozen straws can be shipped anywhere.

If you can tell when your heifer is having heat cycles, you can have her bred by AI. Talk with your local AI technician about ordering semen from a bull of your chosen breed. Several breeding services collect semen from champion bulls all across the United States. Some ranchers and most dairymen use AI instead of buying bulls.

The price of semen varies. Some bulls, especially the most popular champions in their breed, are more expensive. You don't need the most expensive semen; any good bull that sires low-birth-weight calves will be fine. Choose the sire ahead of time and make arrangements with the AI technician so that you can purchase the semen and have your heifer inseminated at the proper time.

Watch your heifer closely to tell when she comes into heat. She will probably be in heat for 12 to 18 hours. Try to spend at least 30 minutes twice a day, morning and evening, watching her for signs of heat. (You may not need to spend this much time if she gives obvious clues.) When you see that she is in heat, call the AI technician.

While your heifer is restrained in a chute, the semen is inserted into her uterus through the vagina. With good luck, she will settle, or become pregnant. If she does not conceive, she will return to heat 17 to 25 days later and can be bred again.

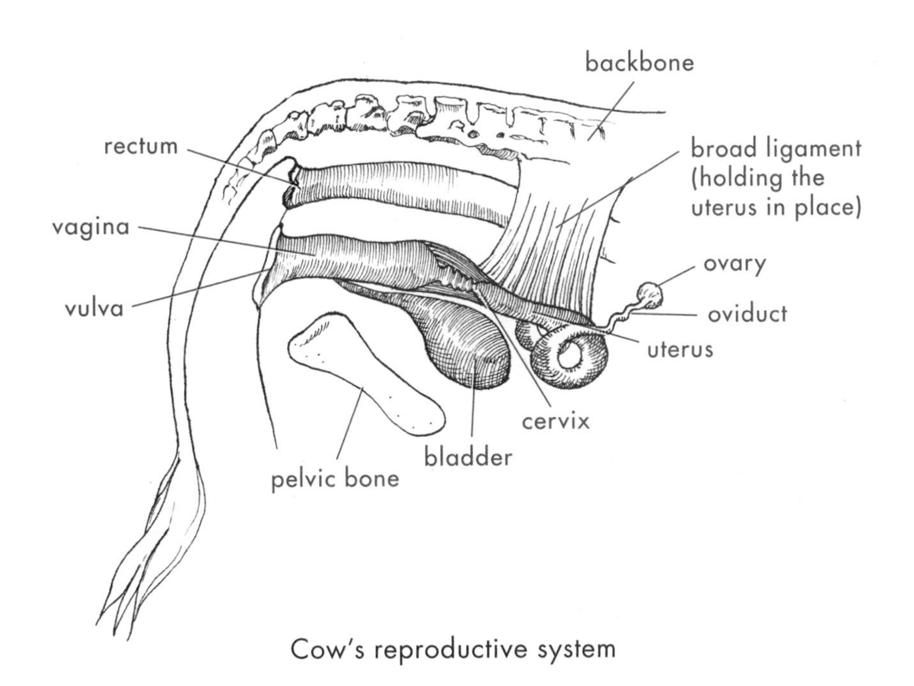

Cow's reproductive system

Keep a Record of Heat Cycles

When your heifer starts cycling, keep a record of her heat periods, which are usually 3 weeks apart, so that you'll know when she should be bred. If you keep good records, you'll be less likely to miss the period during which you want her bred.

Make Sure She's Pregnant

Once your heifer is bred, watch her closely for the next few weeks, especially during the time she would have her next heat period. If she does not come into heat at that time, she's probably pregnant. To make sure, have your vet check her for pregnancy 2 or 3 months after the breeding.

Duration of Pregnancy

The duration of gestation is about 285 days, but a heifer may calve as much as 9 days before or after her due date. Most heifers calve within 3 or 4 days of their due dates.

Size of Fetus During Pregnancy

At 2 months, the fetus is the size of a mouse. By 5 months, it's the size of a large cat. After that, it grows rapidly, becoming calf-sized by 9 months.

Care of the Bred Heifer

After your heifer is pregnant, keep feeding her for good growth so that she'll reach her ideal 2-year-old weight and size by the time she calves. She will need enough feed for her own growth and the growth of the calf inside her. If she is in good condition at calving time, she will produce milk well. If she is thin, she won't be able to milk as well as she should, and she'll have trouble rebreeding on schedule.

Adjust Feed as the Weather Changes. As fall slips into winter, keep close watch on your heifer's pasture. When pasture gets short, start feeding alfalfa hay or a protein supplement. Through the winter, feed a mix of alfalfa and grass hay. If your heifer is gaining weight and growing nicely, she won't need grain. (Most beef heifers don't need grain during pregnancy, unless they start to get thin. This can happen if the weather is severely cold for a long spell or if the hay is of poor quality.)

Feed all the good hay your heifer will eat. Grass hay provides the roughage and nutrients she needs, while alfalfa supplies the extra protein, calcium, and vitamin A necessary for growth and pregnancy.

If the weather gets cold, increase the amount of feed, especially grass hay. Grass hay creates more heat during digestion.

Feed During the Last Trimester. During the last 3 months of pregnancy, your heifer needs more feed. Feed her well, but don't get her fat. If she has a large pasture to move around in, she'll get enough exercise.

Vaccinations During Pregnancy. Your heifer should be on a vaccination schedule in which she receives booster shots for certain diseases once or twice a year. Some vaccines can be given during pregnancy, but others should not. Talk with your vet about the vaccinations your heifer needs.

Also ask the vet about a vaccination to help protect your heifer's calf against scours (infectious diarrhea). If your pens or pastures have held baby calves and have been contaminated with calf diarrhea, vaccinate your heifer against scours. This will create antibodies against many of the diseases that cause scours. She will pass the antibodies on to her calf when he nurses.

Get Ready for Calving

As calving time approaches, your heifer will get large in the belly and clumsy in her movements. She should be in a safe place where she won't slip on ice or get stuck in a ditch.

Make sure you have a good place for your heifer to calve. A shed in her pen or pasture will work if the weather is cold, wet, or windy. If she is confined in a pen or a barn, make sure she has clean bedding. A calf that is born in an unclean place may get an infection.

Pasture Calving. If your heifer is calving in summer and the weather is nice, she can calve at pasture if you check her often. Be sure the pasture is safe and clean, covered with grass rather than dirt or mud. She should have shade if the weather is hot, no gullies or ditches to get stuck in, and strong fences she can't crawl through when she becomes restless during early labor.

The Barn Stall. If your heifer will have her calf in the barn, put lime onto the barn floor before covering it with new bedding. The lime not only helps disinfect the floor but also makes a nonslippery base. The floor must provide good footing so that the calf will be able to stand up and the heifer won't injure herself. She'll be getting up and down during labor, and you don't want her to slip and injure her legs or damage her udder.

The stall should be large and roomy. If the stall is too small, the heifer may lie too close to the wall, and the calf may get jammed into the wall as it emerges.

Make sure the bedding is clean. Never use wet sawdust, moldy straw, or any damp, moldy, dusty material. Many cases of mastitis are caused by dirty bedding. Wet or dirty bedding containing mold or manure will have germs that can invade the uterus or udder of the calving cow or infect the calf's navel.

Things to Have on Hand at Calving Time

- Halter and rope in case you need to tie the heifer
- Strong iodine (7 percent) or chlorhexidine solution in a small wide-mouthed jar, for dipping the calf's navel
- Towels for drying the calf
- Bottle and lamb nipple, in case you need to feed the calf
- Obstetric chains or short, small-diameter (½ inch), smooth nylon rope with a loop at each end, in case you need to pull the calf
- Disposable obstetric gloves (from your vet) and lubricant (obstetric "soap") in a squeeze bottle
- Flashlight for checking on your heifer at night

Prepare Your Dairy Heifer for Milking

Get a dairy heifer used to the milking barn or shed at least 3 or 4 weeks before she is due to calve. Let her go into the barn and get accustomed to putting her head into the stanchion to eat her grain. The milking barn is also a good place to give her supplemental feed if she needs it.

Always be gentle, quiet, and calm when working around your heifer, so that she trusts you and feels secure and at ease. Handling her a lot in the weeks before calving will make training her to be milked easier for both of you.

When you handle your heifer and brush her, touch her udder and teats. Wash them gently a few times so she'll get used to being touched and won't be upset or ticklish. Her udder may be sore when she first calves, because of all the pressure and swelling. But if you have handled her udder a lot, she probably won't try to kick you.

Signs That the Heifer Will Soon Calve

As your heifer approaches calving, her udder gets full. The vulva becomes large and flabby; the muscles are relaxing so that they can stretch when the calf comes through. The area between the heifer's tail head and pinbones becomes loose and sunken. These changes may start several weeks or just a few days before she calves. Her teats may fill with milk, or milk may leak from the teats. Every heifer is a little different, so be alert, observant, and ready. If your heifer is with other cattle, put her into a separate pen so that they can't bother her when she calves.

Stages of the Birth Process

The process of calving has several stages. Knowing what occurs at each stage will help you decide when or if your heifer needs help.

Early Labor. The signs of early labor (first stage) are restlessness and mild discomfort. The heifer has a few early uterine contractions as the uterus prepares to push the calf out. She may kick at her belly or switch her tail.

Contractions become more frequent and more intense as labor progresses. The contractions of early labor usually help turn the calf toward the birth canal.

Early labor may last 2 or 3 hours for a cow but 4 to 6 hours or longer for a heifer. She will get restless and may pace the fence. If she is at pasture, she may go into the bushes or a secluded corner.

Second-Stage Labor. When the cervix is fully open and the calf or the water sac — which often precedes the calf — starts into the birth canal, active (second-stage) labor has begun. The birth should take place in 30 minutes to 2 hours.

The water sac is dark and purplish. When it breaks, dark yellow fluid rushes out. If the sac breaks before it comes out, all you'll see is fluid pouring from the vulva. The water sac should not be confused with the amnion, a white sac full of thick, clear fluid. The amnion protects the calf while it is in the uterus.

Active labor is more intense than early labor. The heifer has strong abdominal contractions. The entrance of the calf into the birth canal stimulates hard straining. Each contraction forces the calf farther along. The calf's feet soon appear at the vulva. Although the calf may safely remain in this position for a couple of hours, it is best if it is born within 1 hour. Give the heifer time to stretch her tissues, however; pulling on the calf too soon may injure her.

Your heifer may get up and down a lot in early labor, but once she starts straining hard, she will probably stay down. Make sure she doesn't lie with her hindquarters against the fence or the stall wall.

Passing the calf's head may take a while, but as long as the heifer is making progress, you won't need to help. After the head emerges, the rest of the calf usually comes easily. Fluid will flow from the calf's mouth and nostrils as its rib cage is squeezed through the cow's pelvis. This fluid was in the calf's air passages while it floated around in the uterus, and it comes out now so it can start to breathe.

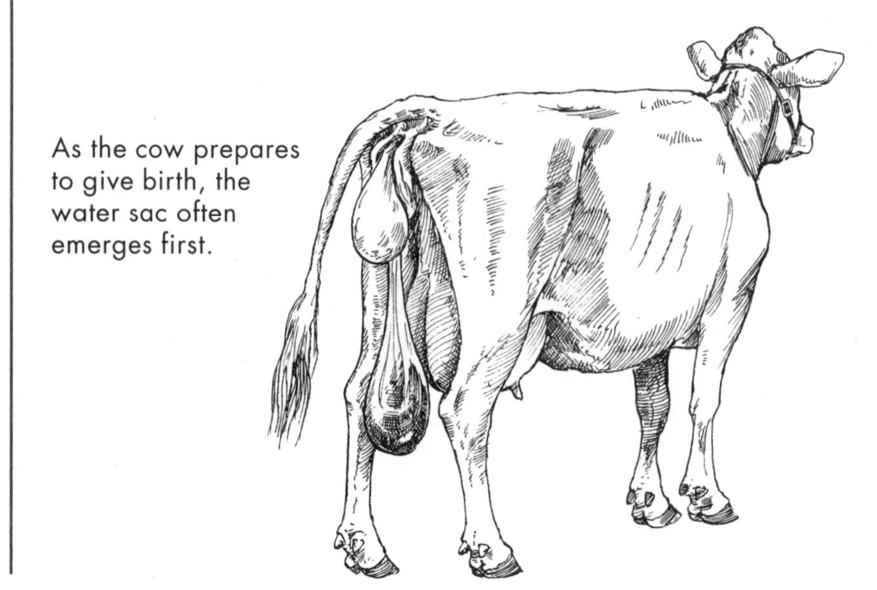

As the cow prepares to give birth, the water sac often emerges first.

When the Calf Is Born

After the calf is born, the cow (she is no longer a heifer after she has calved) may lie still for a few minutes to rest; labor is hard, and she may be tired. But the calf must begin breathing immediately. If it doesn't, or if the sac over its head does not break and is still full of fluids, you must quickly help. Pull the sac away from its nose. Clear the fluid away and make sure the calf starts breathing.

The cow will probably get up and turn around to see her new calf. She should sniff at it and then start to lick it.

Shedding the Afterbirth

When the cow gets up after calving, a lot of red tissue will be hanging from her vulva. This mass is the placenta, which surrounded the calf and attached it to the uterus. The attachments (called "buttons") are dark red dollar-sized objects spaced over the uterus. The afterbirth may take 30 minutes to a few hours to completely detach from her uterus and come out. Never pull on the afterbirth while it is still hanging from the cow. If she doesn't shed the afterbirth for many hours, call the vet. Cows eat their afterbirth so that it won't attract predators; watch for the afterbirth to detach, and remove it from your cow's pen right away so that she won't choke on it.

If the cow takes longer than 10 hours to shed the afterbirth, she may develop a uterine infection. Keep a close watch for pus discharge or illness, which are signs of infection. If your cow won't eat or develops a fever, she'll need immediate treatment. Call the veterinarian.

Helping a Heifer or Cow Calve

You may need to help with the birth. The calf may be positioned wrong in the uterus so that it cannot enter the birth canal or come through it. The calf may be a bit too big to pass through easily, or maybe your heifer has twins. A normal calf should be born within 1 hour of the start of second-stage labor. If a cow is too long at labor and nothing is happening, have her checked by your veterinarian or another experienced person. Definitely call for help if you see only one foot or hind feet coming out.

Your help will most often be needed for birthing bull calves (because they are bigger) and twins.

Be There to Help. Many calving problems can be corrected if someone is there to help. When a cow goes into labor, check on her frequently to make sure the birth is progressing normally. All too often, assistance is given only after the cow or the calf is in critical condition. Be on hand so you can give or get help quickly.

Checking Inside the Cow. During a problem birth, a careful examination inside the cow may be necessary. If your cow is gentle, you can check her. First, tie her up so that she can't move around. If she's lying down and won't get up, check her where she is. Work as cleanly as possible to avoid introducing infection into her. If possible, use a disposable long-sleeved plastic glove that covers your whole arm. If nothing has yet appeared at the vulva , reach into the birth canal to see whether you can feel two feet. If you feel just one foot, or some other abnormality, you'll know why the birth is not progressing.

If no feet have come into the birth canal, reach farther in and examine the cervix. If it is not opening yet, you are interfering too soon. A cervix that is completely open will be 6 to 7 inches wide, and you can easily reach into the uterus.

If the calf is not positioned correctly, the first part of it you touch may be its head, tail, foot, or some other part of its body. In that case, you will need immediate help to reposition the calf so it can be born.

Pulling a Calf. Often, the only problem is that the calf is a little too big and needs a pull. But don't pull on a calf unless it is in perfect position to come out. If the feet have been showing for an hour and you've felt inside the vulva to make sure the nose is right there and the head is advancing properly, pull the calf.

If the calf's nose is showing and the cow's straining starts to push the head out, you can wait. But if she isn't making progress after the feet have been showing for 1 hour, you should help her. First, feel inside the birth canal to see if the head has room to pass through the pelvic opening of the cow. If you cannot get your fingers between the top of the calf's head and the top of the birth canal, the opening may be too small. If that's the case, call a vet.

If you think the head can come through, go ahead and pull on the calf's legs. Having two people working as a team to pull the calf makes the job easier. Pull alternately on one leg and then the other, to ease the calf through the pelvis one shoulder at a time.

You can help deliver a calf by stretching the vulva.

The calf has to come out in an arc. When its feet emerge from the vulva, pull straight out. But after the head comes out, pull slightly downward, more toward the cow's hocks, as its body arches up over the pelvis and then down. If you watch a normal birth, you'll notice that the calf curves around toward the cow's hind legs as she is lying there and it slides out.

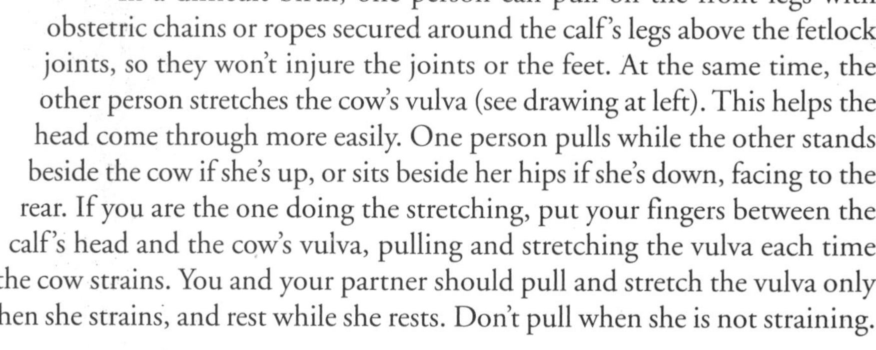

In a difficult birth, one person can pull on the front legs with obstetric chains or ropes secured around the calf's legs above the fetlock joints, so they won't injure the joints or the feet. At the same time, the other person stretches the cow's vulva (see drawing at left). This helps the head come through more easily. One person pulls while the other stands beside the cow if she's up, or sits beside her hips if she's down, facing to the rear. If you are the one doing the stretching, put your fingers between the calf's head and the cow's vulva, pulling and stretching the vulva each time the cow strains. You and your partner should pull and stretch the vulva only when she strains, and rest while she rests. Don't pull when she is not straining.

Hip Lock. Sometimes in a hard birth, you get the calf partly out, only to have it stop at the hips. Don't panic. Remember that it has to come up and over the pelvic bones in an arc. As the calf's body comes out, start pulling downward, toward the cow's hind legs. To avoid hurting its ribs, get it out far enough that its rib cage is free before you pull hard downward. Once its rib cage is out, the calf can start to breathe if the umbilical cord pinches off.

If the cow is standing, pull straight downward and underneath her, pulling the calf between her hind legs. This raises the calf's hips higher, to where the pelvic opening is the widest. If the cow is lying down, pull the calf between her hind legs, toward her belly.

Backward and Breech. A calf coming backward, with hind feet protruding from the vulva, has its heels up. Front legs have the toes pointing down. If the bottoms of the feet are up, the calf may be backward. Before you assume that, reach inside the birth canal to see whether you feel knees (front legs) or hocks (hind legs). The calf may be rotated just a little, sideways or upside down. If the calf is backward, call the vet quickly to help you. He will use a calf puller to get the calf out before it suffocates.

A breech calf is positioned backward, but the legs do not enter the birth canal; he is trying to come rump first. The cow may not start second-stage labor at all. Nothing is in the birth canal to stimulate hard straining, and she seems to be too long in early labor. If you wait too long before checking, the placenta will eventually detach and the calf will die. If you check inside her, all you'll feel is the calf's rump or tail. Call the vet.

Leg Turned Back. Sometimes, one of the calf's front legs will be turned back. One foot will appear but not the other, or sometimes the head and one front foot will show. If you can detect this problem early, before the head is pushed out far, you can push the calf back into the uterus, rearrange it, and get the other leg unbent and coming properly. Otherwise, get help immediately.

When in Doubt, Get Help

Try to recognize problems early and get help before the cow or the calf is in serious trouble. When you are in doubt about a situation or unsure of your ability to handle it, call the vet or an experienced cattleman or dairyman.

Birthing Positions

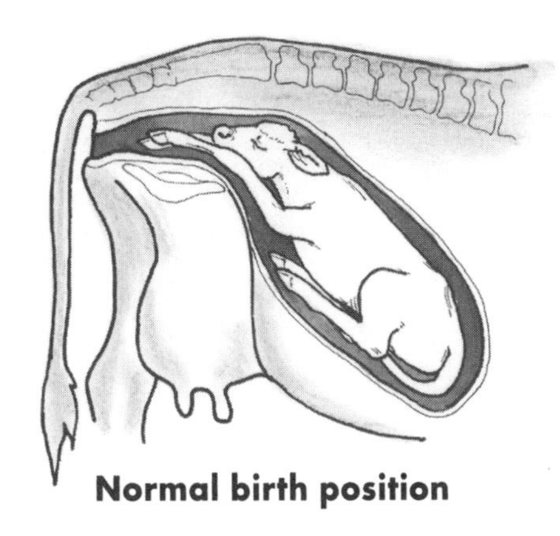

Normal birth position

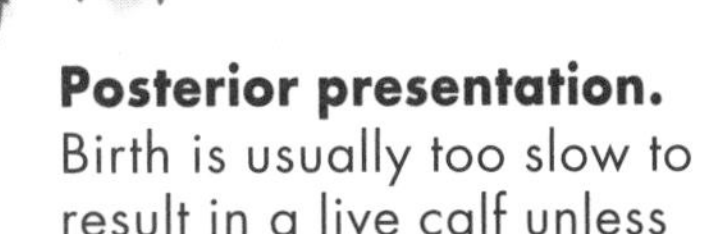

Posterior presentation. Birth is usually too slow to result in a live calf unless assisted.

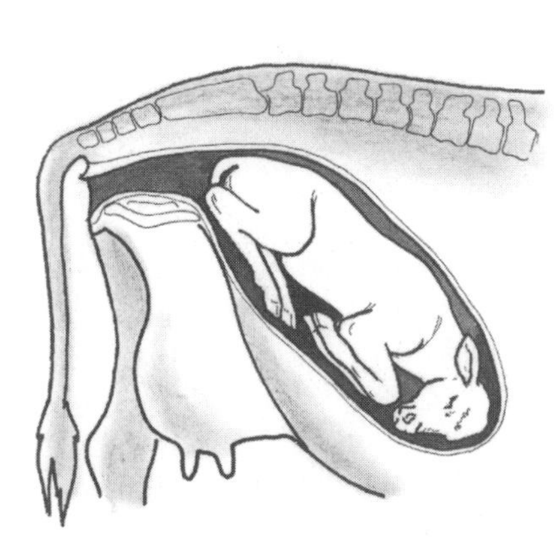

Breech. The calf must be pushed forward far enough so that each hind leg can be tightly flexed at the hock and brought into the birth canal.

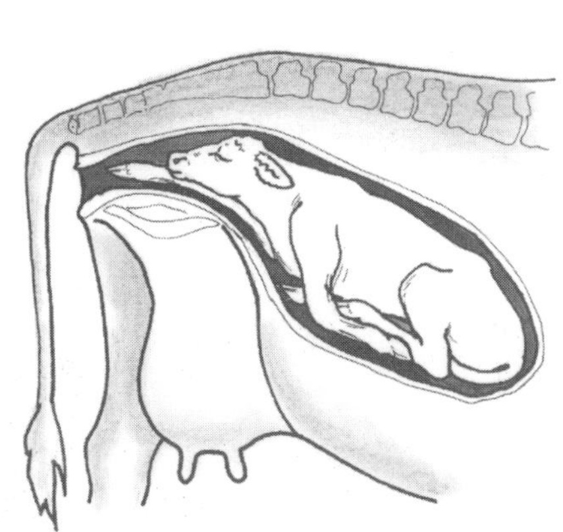

One front leg turned back. The calf must be pushed back and the leg brought into the birth canal.

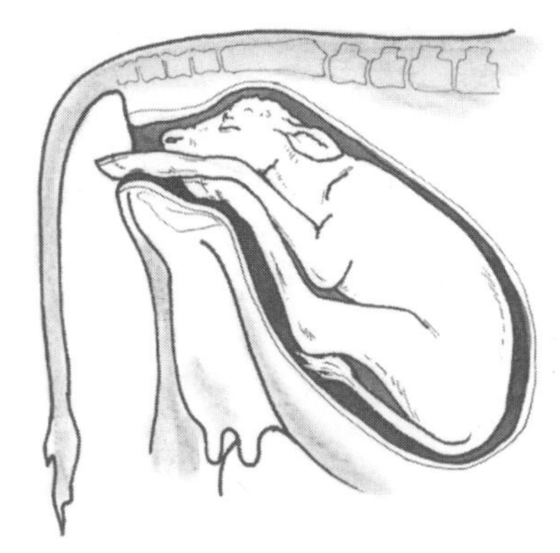

All four feet coming into the birth canal. The calf must be pushed back far enough to push the hind legs back over the pelvic rim and into the uterus.

Get the Calf Breathing

After a difficult delivery, make sure the calf starts breathing as soon as possible. Stimulate it to breathe by sticking a clean piece of straw or hay up one nostril to make it sneeze and cough. If you get no response, give the calf artificial respiration. If it is still alive, you can feel its heartbeat near the rib cage, on the left side, behind its front leg. Blow a full breath of air into one nostril, holding the other nostril and its mouth closed with your hand. Blow until you see the calf's chest rise, then let the air come back out on its own. Blow in another breath, and keep breathing for the calf until it regains consciousness and starts breathing on its own.

Some calves are still encased in the amnion and its fluids after sliding out. If the sac doesn't break, the calf will die because it cannot breathe. This is a very good reason to be on hand when your heifer calves.

Once the calf is born, the cow should get up soon and start licking it. The licking stimulates the calf's circulation and encourages it to try to get up and nurse. If possible, let the cow lick the calf dry. If the weather is cold and the cow is not licking, rub the calf with clean towels.

Disinfect the Navel Stump

To disinfect a newborn's navel stump, use a small jar containing strong iodine or chlorhexidine (Nolvasan) solution.

Navel ill is a serious infection that can kill or cripple a calf. Bacteria that enter through the navel may create an abscess in the navel area or may get into the bloodstream and cause a general infection called septicemia, which can be fatal. Or the bacteria may settle in the joints. Even with diligent treatment, it can be very hard to save a calf once it gets navel ill.

Disinfect the calf's navel stump as soon as the umbilical cord is broken and you have made sure the calf is breathing. Have a small widemouthed jar (such as a baby-food jar) ready ahead of time, containing ½ inch of a strong (7 percent) solution of iodine or chlorhexidine. Immerse the navel stump in the iodine by holding the jar tightly against the calf's belly and making sure the navel cord is thoroughly soaked in the iodine.

The iodine not only kills germs but also acts as an astringent, shrinking the tissues and helping the navel stump dry up quickly and seal off so that bacteria cannot enter the calf. Don't touch the navel cord with your hands unless they are really clean. You need to touch the navel cord only if it drags on the ground when the calf is standing up. In that case, cut it with clean, sharp scissors; leave a 3-inch stump and be careful not to pull on it. Pulling or jerking on the cord can injure the calf internally. As soon as you cut the cord, immediately soak the navel stump in iodine.

The navel stump of a baby heifer usually dries up after just one application of iodine. But a bull calf's stump may take longer. Baby bulls urinate close to the navel, and if they urinate while lying down, as they often do, the navel stump gets wet repeatedly and doesn't dry up quickly. While it is still wet, bacteria can enter it. Reapply the iodine a few hours later, and again if necessary, to dry up the stump within the first 24 hours after birth.

Caution: When dipping the newborn calf's navel in iodine, don't spill any on your hands or the calf. It is a strong chemical and can burn the skin. Never get iodine in your eyes or in the calf's face or eyes. Chlorhexidine (Nolvasan) solution may be substituted for strong iodine. Chlorhexidine is less damaging to the tissue and may be more effective than iodine in preventing navel infections.

Medications for the New Calf. Give the new calf an injection of vitamin A, if recommended by your vet, and any other necessary medications. In some areas, newborn calves need selenium to prevent white muscle disease. You may need to vaccinate newborns against enterotoxemia (a highly fatal gut infection) or tetanus. Discuss your calves' vaccination needs with your vet ahead of time.

Make Sure the Calf Nurses

Colostrum, the cow's first milk, has twice the calories of ordinary milk, with rich, creamy fat that is easily digested and high in energy. Colostrum helps the calf pass its first bowel movements. Most important to the calf are the antibodies in the colostrum that provide immunity to disease. This temporary immunity lasts several weeks, until its immune system starts making its own antibodies. A calf that gets no colostrum or that doesn't nurse until it is several hours old may develop scours or pneumonia.

The best time for the calf to absorb antibodies is during the first 2 hours (and preferably the first 30 minutes) after birth. Make sure the calf is up and nursing within 1 hour of birth. If it can't stand up on its own within that crucial time, help the calf, or milk a little out of your cow's udder and feed it to the calf with a bottle. Before the calf first nurses, wash the cow's udder and teats well with warm water.

If you weren't there when the calf was born, don't assume that it nursed well just because it is with the cow. To make sure it gets an adequate amount of colostrum, try to get the calf to nurse as soon as you arrive on the scene.

Without help, most calves will eventually manage to nurse, but some won't, because they have a hard time getting onto teats. This problem is especially likely if the cow has a low udder or the teats are very full and big. If too much time passes before a calf nurses, it will not absorb enough antibodies because its gut lining will be losing the ability to absorb colostral antibodies after the first couple of hours of life.

A 90-pound calf, such as a Holstein, needs about 3 quarts of colostrum at its first feeding right after birth. A 50-pound calf should receive 1½ quarts.

Colostrum

Good colostrum is thick and creamy. Thin, watery colostrum is low in antibodies. If a cow's colostrum is not good — if it is bloody or she has mastitis — her newborn calf should be fed colostrum from another cow.

If a Calf Can't Nurse

If the newborn calf has trouble nursing or is too weak to stand and nurse after a difficult birth, milk colostrum from its mother to feed it. You'll need about a quart to get it started. Pour the colostrum into a small-necked bottle (one that a lamb nipple will fit) and feed it to the calf. This will usually give it strength to get up and nurse.

If the calf is too weak to suck a bottle for its first nursing, have your veterinarian or another experienced person show you how to use an esophageal feeder.

If the Cow Won't Cooperate

Some cows, particularly first-calf heifers, are slow to mother their calf. Others won't let a calf nurse because their udder is sore. If a heifer kicks at her calf and it cannot nurse, tie her up or restrain her in a stanchion and tie one of her hind legs back so she can't kick. Leave enough slack in the rope so that she can stand comfortably on that leg but cannot kick. You can then help the calf without her striking out at it or you.

Cow tying prevents a cow from kicking her newborn calf.

You may have to put hobbles onto the hind legs of a mother cow if she continues to kick her calf.

A beef calf is left with its mother. If the mother keeps kicking when the calf attempts to nurse, she may have a lot of swelling (called cake) in her udder, which makes nursing painful. You may have to put hobbles onto her hind legs for a few days so that the calf can nurse without being kicked. It may take a few days, a week, or even longer, but eventually the cow will come to accept her calf. Once she starts showing some affection for it and stops kicking at it, the hobbles may be removed.

Leave the calf of a dairy cow with its mother for only a few hours, then put it into a separate pen. A calf that is left with the cow more than 12 hours may bunt at her udder when it nurses. This bunting can bruise the mammary tissue, especially if the heifer has a large, full udder or much swelling. Bruising of the mammary tissue can cause mastitis.

Care of the Cow After Calving

Offer the cow feed and clean, lukewarm water to drink. She should eat and drink soon after calving. If she won't eat or drink or seems dull, she may have a problem; consult your vet. Take her temperature to see if she has a fever.

Heat Stress

If the weather is hot when your cow calves, make sure she and the newborn are in the shade. Heat stress can reduce the immunity the calf receives from colostrum, making it more susceptible to diseases.

Scours

The bacteria that cause scours can lurk on the ground, just waiting for the right conditions, such as wet, muddy weather. Sometimes bacteria are brought into a pen or pasture with purchased calves, or carried from a neighboring farm on the feet of birds, animals, or people.

Preventing Scours

The key to preventing scours is good management — a healthy cow herd, uncontaminated areas for calving, clean bedding for cows, and adequate colostrum for the calf soon after it is born. Vaccinating cows before calving can prevent some but not all types of scours. Following are further preventive measures.

This kind of shed protects calves from the weather while keeping cows out.

Protection from Weather. Young calves need protection from bad weather. If they get wet and cold in a spring storm, they are more susceptible to scours and pneumonia. Make a small shed that the calves can get into but the cows can't. Put clean bedding in regularly.

Keep It Clean. Wash the teats of any cow or heifer you have to assist in calving, clean all your equipment between calves (especially bottles and nipples, and your esophageal feeder tube), and move cows and calves to a clean pasture after they are well bonded.

Separate Sick Calves and Pregnant Cows. Never put sick calves into the same barn stall or pen in which cows will calve. Shelter sick calves in a different shed, and keep pregnant and calving cows separate from those that have already calved. If a pregnant or calving cow is in a pasture with cows and calves and she lies in a bedding area where a calf has had scours, she may get these infective germs on her udder — and her newborn calf will get them as soon as it nurses.

Give Colostrum. Make sure every calf gets colostrum as soon as possible after birth, no more than 2 hours later.

Don't thaw frozen colostrum in a microwave or get it too hot; excessive heat destroys the antibodies. Put the container in hot water to let it thaw. It should feel comfortably warm, but not hot, on your skin before you feed it to a calf.

Keeping Colostrum

If a gentle older cow has a lot of colostrum when she calves, you can milk a quart from her while her own calf is nursing and store it in a plastic container or milk carton. This extra colostrum can be saved for emergencies, such as when a cow has poor colostrum or dies during birth. Extra colostrum will keep for up to 1 week in the refrigerator and for several years in the freezer.

Treating Scours

Scours has different effects on calves at different ages. A month-old calf may recover quickly, whereas the same infection may kill a week-old calf unless you give it intensive treatment.

Give Antibiotics. Give antibiotics as a liquid by mouth rather than shots or pills. Injected antibiotics don't help scours much, because the antibiotic must go directly into the digestive tract, not the muscle, and pills do not dissolve quickly enough in the stomach. Get a good antibiotic from your vet.

Don't Overdo Antibiotics

If you give a calf antibiotics for more than 3 days, it may kill off its good bacteria as well as the bad. You may then have to give the calf a pill or paste containing the proper rumen bacteria (obtained from your vet) after it recovers, to restore proper digestive function.

Replace Fluids. The most important aspect of treating scours is the replacement of the fluids and body salts that the calf is losing, so that it won't become dehydrated. Give a sick calf a quart of warm water by esophageal feeder every 6 to 8 hours until it starts to recover. Have an experienced person show you how to put the feeding tube down the calf's throat. Every time you give your calf warm water, add some electrolyte salts to it. Your veterinarian can provide packets of electrolyte mixtures, or you can use a simple homemade mix.

To make one dose of electrolyte mix for a scouring calf, mix ½ teaspoon of regular table salt (sodium chloride) and ¼ teaspoon of Lite salt (sodium chloride and potassium chloride) into 1 quart of warm water for a small calf or 2 quarts for a large one. If your calf is critically ill, add ½ teaspoon of baking soda (sodium bicarbonate). If the calf is weak, add 1 tablespoon of powdered sugar to the mix. Add a liquid antibiotic, such as neomycin sulfate solution (sometimes called Biosul — you can purchase this from your vet). Use the proper amount for the size of the calf by following the directions on the label.

If you wish, you can add 2 to 4 ounces (¼ to ½ cup) of Kaopectate to this fluid mix. The Kaopectate helps soothe the gut and slow down the diarrhea. Instead of Kaopectate, you can also use a human adult dose of Pepto-Bismol.

Learn to Use an Esophageal Feeder

An esophageal feeder is a handy way to get fluids into a calf that has scours or pneumonia, or into any sick calf that will not nurse. It is also a way to get colostrum into a newborn calf that cannot nurse. The esophageal feeder is a container attached to a tube or stainless-steel probe that goes down the calf's throat and into the stomach. Get an esophageal feeder from your vet, and have him or her show you how to use it.

Detecting Scours Early

To treat calves at the first hint of sickness, watch them closely. Many clues besides a messy hind end can help you spot trouble before a calf becomes critically ill. Often, the calf will act a little dull before it shows diarrhea. It may quit nursing or lie down off by itself while the rest of the calves are running and playing. Feeding time is a good time to check on calves and observe them for signs of illness. Also check the mother's udder. Often, the first sign that a calf doesn't feel well is that it stops nursing. If a cow has a full udder or is only partly nursed out, take a closer look at her calf.

If you catch scours early enough, you can often halt the infection before the calf needs fluids and electrolytes. Neomycin sulfate solution can be put into a syringe, to squirt into the back of the calf's mouth. Use 1 mL per 40 pounds, which means about 2 to 3 mL for a young calf.

CARE OF THE DAIRY COW

After she calves, your dairy heifer will become a dairy cow. Most cows do fine after calving. Just watch your cow closely to make sure all is well. Give her as much good hay as she wants, but only a small amount of grain at first, until the swelling of her udder disappears. If a high-producing cow is fed a lot of grain just before or just after calving, she may have problems.

After calving, heifers often have cake in the udder. Several milkings may be needed to reduce this swelling, and the udder may be sore until the swelling is gone. Milk your heifer at least twice a day, even if the calf is still with her. It cannot drink all of her milk. Milking the cow will help reduce the pressure and swelling in her udder and will relieve a quarter (one of the four teats) the calf may have missed.

Colostrum and Transitional Milk

The calf can be fed for several days with the colostrum you milk from the cow. Milk from your heifer cannot be sold, and people cannot drink it, until it no longer has colostrum in it. You must milk the cow at least twice a day to get the remaining colostrum out of her udder and hasten the production of regular milk. The milk will be ready to be used by people in 4 to 7 days.

True colostrum, the undiluted first milk, is obtained only from the first milking or nursing. The calf needs this first milk immediately after it is born. Later milkings produce transitional milk, a mixture of colostrum and regular milk. It is mostly colostrum at first but becomes more and more diluted by regular milk with time, until no more colostrum is left in the udder.

The colostrum is thicker and richer than regular milk and is usually yellow and sticky. It is waxy when cold. One indication that milk still has colostrum is that it will not go through a strainer quickly.

Milking a Cow

When done correctly, milking is a pleasurable experience for both the cow and the person milking her. You — and your cow — may need a few tries before you feel completely comfortable with each other. Just remember to be gentle and observant.

Short Fingernails, Clean Hands. Before milking a cow, make sure your fingernails aren't long; if you poke the cow's teat with a sharp fingernail, she may kick. Wash your hands as well.

Check and Clean the Udder. The cow's udder is a complex structure that needs good care. Before you start milking, check the udder for problems or injuries. Then check for abnormal milk. If the udder and the milk are fine, wash the udder with a sanitary solution (obtained from your vet) mixed with warm water. Remove any dirt on the teats so it won't get into the milk. Washing the udder with warm water also stimulates the cow to relax and let down her milk. Use a clean paper towel to wash and dry the udder and teats before you milk.

How to Milk. Make yourself comfortable on a stool beside the cow's udder. Hold a clean empty bucket between your legs. Start with two teats — the front teats or the back ones, or two on the same side. Hold one teat in each hand and squeeze one at a time, squeezing the milk down and out through the teat opening. Begin the squeeze at the top of the teat, with your index finger and thumb grip. Finish the squeeze with the lower fingers. By applying pressure with your thumb and index finger, you keep the milk from going back up the teat, so when you squeeze with the rest of your fingers, the milk comes down and out through the hole.

Aim the stream of milk into the bucket; don't let it squirt off to one side. When you become good at milking, you can easily direct the stream and even aim a squirt toward a barn cat waiting to catch the milk in its mouth.

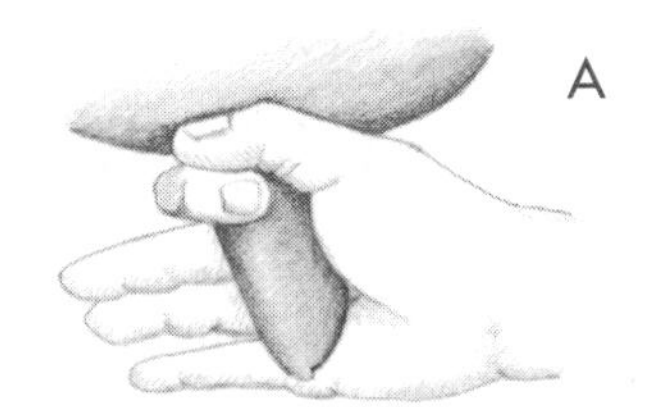

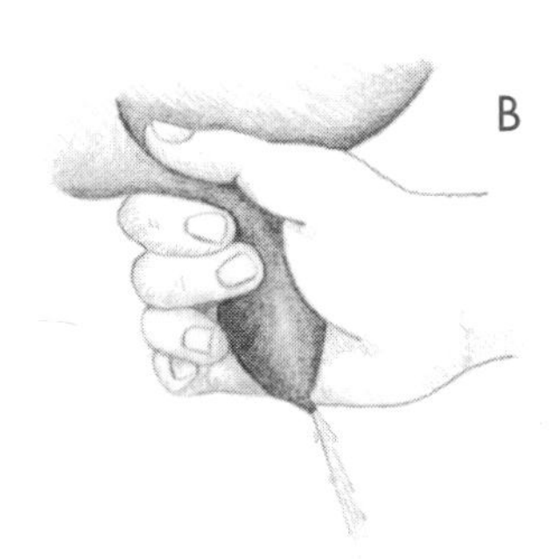

To milk, apply pressure with your thumb and index finger (A) while squeezing with your lower fingers (B).

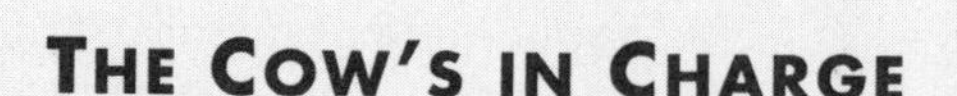

The Cow's in Charge

Cows can let down or hold up their milk. No milk is in the teat until the cow lets it down, which she does by relaxing muscles that keep the teat canal closed. When a cow wants her calf to nurse, or thinks it is milking time, she lets the milk flow down from the udder into the teat. If a cow doesn't want to let down her milk, you won't get much milk out of her. Washing and massaging may encourage a stubborn cow to release her milk.

When milking, try to keep things quiet and calm so the cow can relax. Feed her some grain to eat while you milk. If she gets nervous or upset, she will be tense and won't let down her milk.

After each squirt, release your grip. More milk will flow down into the teat. Keep up a nice rhythm by alternating squirts. When the first two teats are soft and flat and you can't get any more milk from them, milk out the other two quarters. If one quarter seems to have more milk than another, perhaps one of your hands is not yet as strong as the other and you aren't getting quite as much milk out with each squeeze.

Milking Schedule

A cow should be milked twice a day, 12 hours apart. Not sticking to this schedule can be harmful to the cow.

Problem with the First Milking. Sometimes a cow's udder is so sore right after calving that she doesn't want you to touch it. She may kick at her calf as you try to help it nurse, or kick at you. In this case, press your head firmly into her flank as you milk to prevent her from swinging that leg forward. Press your head firmly against the area right in front of the stifle joint. If you press hard every time you feel her tensing up to kick, she won't be able to kick well. If a cow is nervous the first few times you milk her, talk softly to her or hum a little to reassure her and keep her relaxed.

Sometimes a cow will kick if her teats get sore and cracked, as they may do in cold weather. If they get too cracked, they will bleed. To prevent chapping on teats, rub a little ointment (such as Bag Balm) on the teats after you finish each milking.

Sore Muscles. Once you get used to the proper squeezing motion and rhythm, milking is easy. But milking uses muscles in your forearms and hands you may not have used much, and they will get tired. Your arms and hands may ache afterward. The more often you milk, the stronger your arms and your grip will get.

Easy Milkers, Hard Milkers. Some cows are easy to milk. They have teats that are easy to hold on to, and they let their milk down freely. The milk almost flows from the udder into your bucket. Other cows are harder to milk. More effort is needed to squeeze out the milk, and milking takes longer. A cow with short teats may be a difficult milker because getting a good hold is more difficult. More time is needed to milk if you must squeeze with just one or two fingers instead of your whole hand.

Mastitis

Bacteria sometimes enter the udder through the teat canal. Inside the udder, they multiply rapidly because of the warm environment and cause infection, resulting in mastitis. The cow's body sends white blood cells to fight the infection. The infected quarter often feels hot and is swollen. If a quarter has abnormal milk — if it's lumpy, watery, or bloody or has any other abnormality — or heat and swelling, the cow should be treated for mastitis. A cow with mastitis may show no other symptoms.

Testing for Mastitis. You can test milk for mastitis by using the California Mastitis Testing Kit, which measures the amount of somatic cells in the milk. Somatic cells are white blood cells and mammary cells that have been damaged by infection. The somatic cell count in the milk will be high as long as the quarter is infected.

Ask your vet or county Extension agent where you can get a test kit, which comes with complete instructions.

Treating Mastitis. Your vet can provide antibiotic preparations to treat mastitis or you can buy antibiotic tubes at your feed store. Before giving the medication, thoroughly milk out the quarter. To administer the medication, gently insert a syringe tube with a long plastic tip into the teat and squirt the antibiotic up into the quarter. Gently massage the treated quarter to help spread the antibiotic to all parts. Follow directions on the label for proper use and the number of days of treatment. If mastitis does not clear rapidly, call your veterinarian. You may need to have a milk sample cultured to find out which organism is the culprit and what type of antibiotics it is most sensitive to.

If a cow with mastitis shows signs of systemic illness, such as being off feed, acting dully, or having a fever, call your veterinarian.

Not Safe for Humans

Milk from an infected cow should not be consumed by humans. Even after the infection is cleared up, the milk can't be used until it contains no more antibiotic. The label on the medication tells how long the antibiotic persists in the cow's milk. While the cow has mastitis, keep milking her regularly to hasten her recovery. In most cases, the milk from a recovering cow can be fed to calves. Consult your vet for guidance.

Taking Care of Milk

Take care to keep milk clean. Use a clean bucket, a clean strainer, and clean containers. After each use, rinse all equipment in warm water to remove the milk fat. Then scrub everything with hot water and dishwashing soap, using a stiff plastic brush. The brush will clean the equipment much better than a dishcloth. Always use a plastic brush or scouring pad rather than metal. Metal scratches the surface of the equipment and leaves tiny indentations where bacteria can cling. Rinse everything thoroughly in clean water.

If you wish, you can pasteurize your milk to make sure it is perfectly safe to drink. (The milk you buy at the grocery store is pasteurized.) Pasteurized milk is heated to a specific temperature, kept there for a short time, and then quickly cooled. Home pasteurizing units can be purchased from a dairy supply catalog. They consist of metal containers with a heating element in the bottom. Fill the container with water, set a gallon of milk in a covered metal pail into the water, and then plug in the unit. The water heats to the desired temperature for the proper time, and a buzzer sounds when it's done.

Instructions for pasteurizing goat milk without an electric unit are given on page 223. Cow milk may be pasteurized in the same manner.

Feeding the Milking Cow

Proper nutrition is important in determining how much milk your cow will give. She can't produce milk without the building blocks provided by good nutrition.

Hay and Pasture. About two thirds of a dairy cow's total nutrition should come from forages — that is, hay or pasture. Make sure the quality is good. Cows will eat a greater total volume of hay if you feed it fresh several times a day. The more good forage you can get a milk cow to eat, the more milk she will give.

Grain. Even though a dairy cow has a large rumen, she cannot eat enough forage to supply all of her nutritional needs. For top production, the dairy cow needs grain. Your county Extension agent can help you devise a feeding program to create a well-balanced diet for your cow. The amount of grain should be adjusted to fit her needs: more during the peak of her lactation, when she is making the most milk, and less toward the end.

Water. Your cow needs a constant supply of clean water. A milking cow needs 3 to 5 gallons of water, including the moisture in her feed, for every gallon of milk she produces. A cow eating hay needs to drink more than a cow on lush green pasture, which contains a lot of moisture.

Rebreeding

Your cow will need to be rebred about 3 months after she calves. Keep track of her periods of heat when she starts cycling again, so you can have her bred at the proper time. It may be several weeks before she starts having heat cycles. One clue is that milk production usually decreases temporarily on the day the cow is in heat.

After your cow has been bred again, have your vet check her for pregnancy if she does not return to heat. Knowing whether she is pregnant and when she will be due to calve lets you plan the proper time to dry her up before her next calving. If she is not pregnant, you must try to get her rebred.

Managing the Dry Cow

To get ready for the new calf and new milk production, the cow's body needs a rest from making milk. Allow her to go dry for a couple of months before her next calving. The length of the dry period varies depending on the cow's age and condition. She needs at least 45 days of rest to be able to produce a lot of milk during her next lactation and to make enough antibodies in her colostrum for the next calf. A 6- or 7-week dry period is adequate for the average cow. But young cows, such as a first-calf heifer, and high-producing cows generally need 8 weeks (56 to 60 days).

To dry up a cow, simply stop milking her. The transition should be abrupt; don't try to ease the cow into it by partial milking. Cows are designed to stop producing milk when the udder is full and tight, which is what happens under natural conditions when a calf is weaned or dies. The pressure in a cow's udder signals her body to stop making milk. She will be uncomfortable at first, but after a few days the pressure will ease. Her body gradually resorbs the milk left in her udder.

To help a cow dry up, reduce her feed, especially grain. Eliminating grain helps the cow's body adjust to not making milk. Pasture or hay should provide enough nutrition for your cow to go through the dry period without becoming fat. She will need grain only if she needs to gain weight.

Check your cow's udder closely while she is drying up. After the last milking, treat each quarter with an antibiotic recommended by your vet to help prevent mastitis. Watch for heat or swelling in the udder.

Helpful Hints for a Small Dairy

Most of the management concerns of calving time have already been covered. The following hints will be helpful if you have a small dairy.

Feeding Colostrum and Waste Milk. If you have several calves, you can save money on milk replacer by feeding them the extra colostrum and transitional milk from your cows. Every time a cow calves, her milk for the first 5 days can be used to feed all your calves, or to mix with milk replacer. The 8 to 25 gallons of colostrum a cow produces should not be wasted; it will feed calves for quite a while.

If you feed milk from cows with mastitis, feed it only to calves penned individually so that they cannot suck on one another. Mastitis germs can be passed from one calf's mouth to another calf's udder if the calves suck on one another.

Many combinations can be used to feed calves, including mixes of whole milk and milk replacer or sour colostrum and milk replacer. Make any changes in a calf's diet gradually to avoid digestive upsets. Don't suddenly switch a calf from milk replacer to sour colostrum; mix the two. Avoid changing the feed of a young calf during its first 2 weeks of life. After that, changes in diet won't affect it much if they are done gradually.

The Nurse Cow. If you don't need the milk from your cow, she can be a nurse cow. This can increase your income and eliminate the chore of milking or having to find a way to use the extra milk. With a good nurse cow, you can raise four to eight calves each year. A crossbred dairy–beef cow can make a good nurse cow, though she may not be able to feed quite as many calves.

Raising extra calves on a nurse cow. Extra calves can be purchased cheaply and raised with little effort. A nurse cow can raise eight beef steers each year or eight good dairy heifers (two sets of calves each year), if the calves are fed hay and grain before weaning so that they can be weaned at about 4 to 5 months of age. The cow will produce milk for 9 or 10 months before you need to dry her up for her next calf. You can wean the first set of calves and then put four more onto her. Selling 8 to 16 calves a year that are raised on one or two cows is easier than milking the cows and selling milk, and it's more profitable.

To make use of a nurse cow, you'll need a pen or pens next to the milking area for the calves. You can buy your calves just before the cow is due to freshen and feed them on bottles until she calves, or buy them just after she calves.

If the nurse cow produces extra colostrum, freeze it to give to future newborn calves you might buy that have not had adequate colostrum.

Training the nurse calves. Keep the calves separate from the cow except at nursing time. Get the cow accustomed to eating grain in her stanchion, and start the calves on her while she is eating. She probably won't kick her own calf, but she may try to kick off the extra calves. It takes time and patience to persuade some cows to accept all those calves nursing at once. Supervise the nursing, and make sure the calves don't bunt her udder too much or get kicked. If the cow kicks, she may

Sharing the Milk

You can also share a cow's milk with one or two calves. If your dairy cow gives more milk than your family can use, let her raise her own calf and one extra, while you take part of the milk. You can let the calves nurse on one side while you milk the other.

How Many Calves?

A mature dairy cow will easily feed four calves at once, but a first-calf heifer may do best after her first calving with just two at a time.

discourage a timid calf. You may have to hobble her before you let the calves in, until she learns not to kick.

Always put the same calf on the same teat. If each calf has its accustomed place at the udder, you won't have a circus of confusion when you let the calves in with the cow. Be firm; make each calf learn its proper place.

Let the calves nurse until they are clearly done. Don't leave them too long, or they will bunt her udder in their greediness for more milk. Bunting calves can bruise the udder. If a calf is obnoxious and bunts a lot as it nurses, reprimand it with a stick every time it bunts, until it learns not to. If the cow is not hobbled, she will reprimand the unruly calf herself by kicking at it, but she may hit the wrong one.

The hardest job is to get the calves back to their pens after they're done nursing. They won't want to leave the cow. You may have to use a stick to encourage them to go back to the pen. You can put halters on the calves before you let them out of their pens. Leading lessons will make them more manageable when you take them away from the cow and back to their pen. Double the lead rope over the calf's back while it is nursing the cow, to keep it off the ground and prevent it from being stepped on.

Managing Your Herd Through the Seasons

This section looks at managing a herd of cattle and the various things involved in caring for them throughout the year. These guidelines will be useful even if you have just one cow.

Winter

During winter, calves from last spring are already weaned and vaccinated and are growing nicely on good feed. If you have pasture, your calves will do better on it than if confined in a corral for winter. A pen may be knee-deep in mud and manure by spring. Calves get cold if they have to stand in mud to eat. Make sure they have a dry, clean place to sleep.

One of your tasks is to get your pregnant cows and yearling heifers through winter in good shape for calving. Young cows that are still growing need alfalfa or another source of protein; mature cows may get by on good grass hay. Make sure their water supplies are adequate and not frozen. Break ice often, or use a water tank heater.

Delousing. Lice multiply swiftly in cool or cold weather. Delouse all cattle in late fall and again in late winter. Talk with your veterinarian about a good lice control program.

Winter Diseases. Keep close watch for foot rot or weather-related illnesses, such as pneumonia. Your cows and pregnant heifers should already be vaccinated for leptospirosis (and redwater, if it's a problem in your area). If you are vaccinating the cows to protect their newborn calves against scours, give these shots several weeks before calving. Heifers expecting their first calves will need two shots a few weeks apart. Talk with your vet well ahead of time and devise a vaccination schedule.

Spring

Your calves will be born in the spring. If you have several cows or heifers, tag the calves when they are born. Before calving time, buy tags and number them. Also purchase any medications and equipment you might need and have them on hand: iodine, vaccine for newborn calves, obstetric gloves, scour medications, and so on. If a cow calves a week early, you'll be prepared.

Mothers and Calves. If you're raising animals for beef and allowing mothers and calves to stay together, keep each cow–calf pair by themselves for a day or two before they go out with other cattle to let the calf learn who its mother is. Then, when it goes out with other cattle, it won't try to nurse the wrong cow and get kicked.

Keep cows with babies in a different place from the pregnant cows, because they need to be fed differently. Mother cows need more feed so that they can produce milk for their calves.

Vaccinations. Cows must be vaccinated after calving and before rebreeding against leptospirosis, infectious bovine rhinotracheitis, bovine virus diarrhea, and other diseases. Vaccinate at least 3 weeks before rebreeding. Some of the live virus vaccines (such as infectious bovine rhinotracheitis or bovine virus diarrhea) may cause abortions in cows if given while they are pregnant. Also, the cows and heifers need time to build immunities against these diseases before they become pregnant, since some of these diseases can cause abortions. Young calves need to be vaccinated against blackleg, malignant edema, and other clostridial diseases.

Castrating, Dehorning, and Removal of Extra Teats. These important tasks should be taken care of while calves are young. See pages 326–328 for details.

Summer

Your chores will get a little easier in summer if your cows are on pasture. Calving is over, and you aren't feeding hay. You'll still need to watch your cattle closely so that you can treat any health problems that may occur, such as pinkeye or foot rot. Summer is also the time for fly control. You may want to use insecticide ear tags.

Breeding the Cows. If you are breeding your cows by AI, check them for signs of heat twice a day, in the early morning and late evening. For best results, have the technician inseminate each cow about 12 hours after heat is first noticed. Standard practice is to breed a cow in the evening if she was noticed in heat that morning, or early the next morning if she was seen coming into heat in the evening. Keep good records of the dates on which each cow has heat cycles, when she was bred, and the expected calving date.

If you're breeding with a bull, turn him in with the cows 9 months ahead of the date you want the cows to calve: May 1 if you want them to start calving in early February, or the end of May if you want them to calve in March. Keep heifers separate from the cows and breed them to a bull that sires small calves at birth. You can use the same bull for cows and heifers if you know he sires small calves that grow fast.

Remove the Bull from the Herd

After the cows are bred, take the bull out of the herd and keep him in a separate pen. Bulls can be a nuisance and should not be kept with cows year-round. Some bulls are mean and most are obnoxious, wearing out fences in an attempt to get out and fight other bulls or find other cows.

If you have only a few cows and don't want to keep a bull, borrow or lease one for the short time you will need him. Forty-five to 60 days is long enough to leave a bull with the cows if they all calved about the same time and none are really late.

Register Your Calves. If you are raising purebred calves, register them before they are 6 months old. Send the information on each calf to the breed association, along with the registration fees. The older the calf, the more registering him costs, so do it early.

Fall

Fall is the time to wean calves and prepare for winter.

Vaccinations. If you haven't already, vaccinate all heifer calves for brucellosis (Bang's disease). Cows should be revaccinated for leptospirosis and any other diseases your vet recommends. All calves should be revaccinated against blackleg, malignant edema, and any other clostridial diseases in your area and against viral diseases, such as infectious bovine rhinotracheitis, bovine virus diarrhea, and parainfluenza type 3. Calves should be dewormed and deloused.

Sell Steers and Heifers. Decide where to sell your steers and any heifers you don't plan to keep. Do you want to sell them through an auction or directly to a feedlot? Or to a neighbor who is buying weaned calves to put onto pasture? If you aren't sure how to market your calves, talk with your county Extension agent or with local farmers. If you have purebred calves, get advice from your regional breed association representative.

If you sell your calves to a buyer who will be sending them to a feedlot, wean them a few weeks before they are sold, so that they'll be over the stress of weaning.

Weaning a Beef Calf. Weaning is a traumatic experience for a beef calf that has spent all summer by its mother's side, and also for first-time mother cows. Older cows are often not as upset, because they've gone through it before.

The emotional stress of weaning is harder on the calf than is the sudden lack of milk. Separating calves from their mothers and putting them into a corral by themselves causes much anxiety. They frantically miss their mothers and the security of the herd. Their desperation is contagious. The calves mill around and pace the fence, bawling and running and stirring up dust that can irritate their lungs and make them vulnerable to pneumonia. Any frantic activity by one sets off a reaction among the others. They rarely take time to eat.

Weaning is easier on calves if they are in a grassy pasture rather than a dusty corral. If you have just a few cattle, a good way to wean them is by using a net wire fence to divide a small pasture. Put the cows on one side and the calves on the other. The calves will still bawl and pace the fence, but they can see their mothers and be near them, and after a few days they are not so worried. Make sure your fences are strong. Weaning calves will try to crawl through, and so will some of the mothers.

Another way to wean is to leave a "babysitter" cow or heifer with the calves when they are separated from their mothers. She is a calming influence and helps them get through this emotional time.

Buying and Storing Hay. Be sure to buy good-quality hay. Inspect the hay before you buy it; don't purchase rained-on, moldy hay or dry, dusty hay. You can purchase hay during the summer, right from the field, and have it hauled home. If you don't have room to store a lot of hay, you may be able to arrange with the seller to store it for you.

If you store hay at home, stack it in a high, dry spot where the bottom bales won't get wet. Cover the stack with tarps to keep the top bales from spoiling due to rain and melting snow.

Culling Cows. Have your vet test your cows for pregnancy. You should probably cull older cows that are open (not pregnant), that are crippled, or that have poor teeth (which are likely to become thin over winter), a bad udder, or any other serious problem.

Taking Stock of the Year Just Past. Fall is a good time to review the past year and see how the calves grew, how outstanding the heifers are, and what problems and challenges were overcome along the way. You'll feel good if you have saved a calf that might otherwise have died at birth or if you've kept a sick one alive with good doctoring. Seeing the calf now, big and sassy, warms your heart, makes you eager to see next year's babies, and makes you feel ready to face any challenges that may come along.

Appendix

Metric Conversion Chart

Metric (SI*) Conversion Factors

Approximate Conversions to SI Units

Symbol	When You Know	Multiply by	To Find	Symbol
Length				
in	inches	2.54	centimeters	cm
ft	feet	0.3048	meters	m
mi	miles	1.61	kilometers	km
Area				
in^2	square inches	6.452	square centimeters	cm^2
ft^2	square feet	0.0929	square meters	m^2
ac	acres	0.395	hectares	ha
Mass (weight)				
oz	ounces	28.35	grams	g
lb	pounds	0.454	kilograms	kg
Volume				
fl oz	fluid ounces	29.57	milliliters	mL
gal	gallons	3.785	liters	L
Temperature (exact)				
°F	degrees Fahrenheit	5/9 (after subtracting 32)	degrees Celsius	°C

*International System of Measurements

GLOSSARY

Abomasum. Fourth stomach or true stomach of the ruminant animal, in which enzymatic digestion occurs.

Abscess. Boil; localized collection of pus.

Acidosis. Severe digestive upset from change in rumen bacteria.

Acute infection. An infection or disease that has rapid onset and pronounced signs and symptoms.

Afterbirth. Placental tissue that is attached to the uterus during gestation; it is expelled after the birth.

Air cell. Air space usually found in the large end of the egg.

Albumen. The white of the egg.

Anemia. A deficiency in the oxygen-carrying capacity of blood. Can be caused by loss of blood or by certain disease conditions.

Anestrus. The nonbreeding season; the state (for females) of being not in heat.

Angora. A rabbit with a coat about 3 inches long. Raised for wool as well as for meat.

Antibiotic. A drug used to combat bacterial infection.

Antibody. A protein molecule in the blood that fights a specific disease.

Antigen. A "foreign invader," which the body's immune system recognizes as such. Usually a bacteria or virus.

Antiseptic. A chemical used to control bacterial growth.

Artificial insemination (AI). The process in which a technician puts semen from a male animal into the uterus of a female animal to create pregnancy.

Aviary netting. Fencing woven in a honeycomb pattern with ½-inch openings.

Balling gun. A device used to administer a bolus (a large pill).

Bang's disease. *See* Brucellosis.

Bantam. A diminutive chicken about one-fourth the size of a regular chicken. Some bantams are distinct breeds; others are miniatures of large breeds.

Barbicels. Tiny hooks that hold a feather's web together.

Beak. The upper and lower mandibles of chickens, turkeys, pheasants, peafowl, et al.

Bean. A hard protuberance on the upper mandible of waterfowl.

Beard. The feathers bunched under the beaks of some chicken breeds, such as the Antwerp Belgian, the Farcrolle, and the Houdan.

Bedding. Straw, wood shavings, shredded paper, or any other material used to cover the floor of an animal pen to absorb moisture and manure. Also called *litter.*

Bevy. A flock of ducks.

Bill. The upper and lower mandibles of waterfowl.

Bill out. To use the beak to scoop feed out of a trough onto the floor.

Black's disease. A usually fatal disease caused by *Clostridium novyi,* creating acute toxemia; similar to malignant edema and red water.

Blackleg. A serious disease caused by *Clostridium chauvoei,* a soil bacterium, resulting in inflammation of muscles and death.

Bleaching. The fading of color from the beak, shanks, and vent of a yellow-skinned laying hen.

Bloat. An excessive accumulation of gas in the rumen and reticulum, resulting in distension.

Blood spot. Blood in an egg caused by a rupture of small blood vessels, usually at the time of ovulation.

Bloom. The moist, protective coating on a freshly laid egg that dries so fast you rarely see it; also, peak condition in an exhibition bird.

Blowout. Vent damage caused by laying an oversize egg.

Bolus. A large pill for animals; also, regurgitated food that has been chewed (cud).

Bovine. Pertaining to or derived from cattle.

Bovine virus diarrhea (BVD). A viral disease that can cause abortion, diseased calves, or suppression of the immune system.

Break up. To discourage a female bird from being broody.

Breech. The buttocks; a birth in which the fetus is presented "rear" first.

Breed. A group of animals with the same ancestry and characteristics.

Breeder ration. A feed used for the production of hatching eggs.

Brisket. The front of the cow above the legs.

Broiler. A young chicken grown for its tender meat. Also called a *fryer.*

Broken. A color pattern in which blotches of color appear on a white background.

Broken mouth. Having lost teeth.

Brood. To set on a nest of eggs until they hatch. Also, the resulting hatchlings, collectively.

Brooder. A mechanical device used to imitate the warmth and protection a mother bird gives her chicks.

Broody hen. A setting hen.

Browse. Bushy or woody plants; to eat such plants.

Brucellosis. A bacterial disease that causes abortion.

Buck. A mature male goat or rabbit.

Buck rag. A cloth rubbed onto a buck goat and imbued with his odor and kept in a closed container; used by exposing to a doe and observing her reaction to help determine if she's in heat.

Buckling. A young male goat or rabbit.

Bull. An uncastrated male bovine of any age.

Bummer. A lamb that has to be bottlefed by the shepherd. Usually an orphan, though sometimes a lamb whose mother doesn't produce enough milk for multiple lambs.

Bunny. A cutesy term for rabbit. Babies are called bunnies for lack of another term, although some call the babies kits, even pups.

Burdizzo. A castrating device that crushes the spermatic cords to render a male animal sterile.

Calf. A young bovine of either sex, less than a year old.

California Mastitis Test (CMT). A do-it-yourself kit to determine if a female animal has mastitis.

Calve. To give birth to a calf.

Candle. To determine the interior quality of an egg by shining a light through it.

Cannibalism. The bad habit some chickens have of eating one another's flesh or feathers.

Cape. The narrow feathers between a chicken's neck and back.

Capon. A castrated male fowl having an undeveloped comb and wattles and longer hackles, saddle, and tail feathers than the normal male.

Caprine. Pertaining to or derived from a goat.

Caprine arthritis encephalitis (CAE). A serious and widespread type of arthritis, caused by a retrovirus.

Card. To convert loose, clean wool into continuous, untwisted strands. May be done with hand cards or a carding machine.

Carrier. An animal that carries a disease but doesn't show signs of it.

Caruncle. A fleshy outgrowth. Often seen around the eyes and head of certain species of fowl.

Castrate. To remove the testicles of a male animal so that he is permanently incapable of breeding.

Cervix. The opening (usually sealed) between the uterus and the vagina.

Chalazae. White, twisted, ropelike structures that anchor the egg yolk in the center of the egg, by their attachment to the layers of thick albumen.

Chevon. Goat meat.

Chicken wire. Fencing woven in a honeycomb pattern with 1-inch openings.

Clean legged. Having no feathers growing down the shanks.

Clean wool. Usually refers to scoured wool, though handspinners may describe grease wool that has little or no vegetable contamination as clean wool.

Clip. The total annual wool production from a flock.

Cloaca. The cavity just inside a fowl's vent, into which the intestinal and genitourinary tracts empty.

Closed face. In sheep, having heavy wool about the eyes and cheeks.

Closed face

Clostridial diseases. Diseases caused by Clostridia bacteria (including blackleg, Black's disease, tetanus, red water, and enterotoxemia) that produce powerful toxins causing sudden illness.

Clutch. A batch of eggs that are hatched together, either in a nest or in an incubator.

Coccidiosis. An intestinal disease caused by protozoa; usually causes diarrhea.

Coccidiostat. A drug used to prevent coccidiosis.

Cock. A male chicken; also called a *rooster.*

Cockerel. A male bird under 12 months of age.

Colic. An abdominal condition generally characterized by severe pain.

Colostrum. The first milk from a female animal that has just given birth; contains antibodies that give to the newborn animal temporary protection against certain diseases.

Comb. The fleshy prominence on the top of the head of fowl. Also, to remove short fibers from wool and leave long fibers laid out straight and parallel.

Concentrate. Feed that is low in fiber and high in food value; grains and oil meals.

Condition. Degree of health.

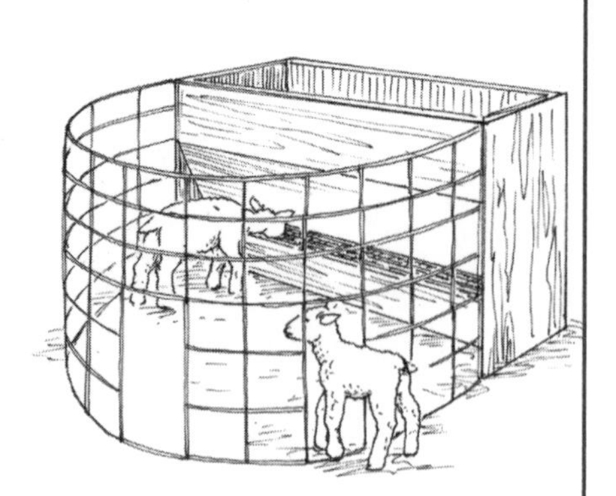

Creep feeder

Conformation. The overall physical attributes of an animal; its shape and design.

Congenital. Acquired before birth (e.g., a birth defect).

Coop. The house or cage in which a chicken lives.

Count. The fineness to which yarn may be spun; a system of grading wool based on how finely it can be spun.

Cow. A bovine female that has had one or more calves.

Creep feeder. An enclosed feeder for supplementing the ration of young animals that excludes larger animals.

Creep-feeding. Providing extra feed (such as grain) to young animals that are still nursing their mothers.

Crest. The elongated feathers on the head of some breeds of ducks and chickens.

Crimp. The "wave" effect in wool fibers.

Crop. An enlargement of the gullet of fowl where food is stored and prepared for digestion.

Crossbreed. The offspring of two different breeds.

Cross-fencing. Fences used to subdivide pastures into smaller paddocks.

Crotch. To trim wool or hair from around the tail and udder.

Cryptosporidiosis. Diarrhea in young animals caused by protozoa; may also cause diarrhea in humans.

Cud. In ruminant animals, a wad of food burped up from the rumen to be rechewed.

Cull. To remove a substandard animal from the herd.

Cycling. Heat cycles in nonpregnant females.

Dam. The female parent.

Degreased wool. Wool that's been cleaned chemically to remove all "grease," or lanolin.

Dehorn. *See* Disbud.

Dewclaw. A horny structure on the lower leg above the hoof.

Dewlap. Loose skin under the neck.

Diphtheria. A bacterial disease in the mouth or the throat.

Disbud. To remove the horn buds from young animals to prevent horn growth.

Disbudding iron. A tool, usually electric, that is heated to burn the horn buds from young animals.

Dished face. Having a concave nose, such as that of the Saanen goat.

Disqualification. A temporary or permanent physical defect serious enough to bar an animal from a show.

Dock. To cut off the tail; the remaining portion of the tail that has been docked.

Doe. A female goat or rabbit.

Doeling. A young female goat or rabbit.

Down. The furlike covering of newly hatched ducklings. Also, the inner layer of soft, light feathers on waterfowl and the fluffy bottom of chicken feathers.

Drake. A male duck.

Drake feather. One of three curly feathers on a drake's tail.

Drakelet. A young male duck.

Drench. To give medication from a bottle.

Dress. To clean meat in preparation for cooking.

Dry. Not producing milk.

Dry period. The time when a female animal is not producing milk.

Duck. Any member of the family Anatidae and specifically a female.

Ducklet. A young female duck.

Duckling. A baby duck.

Dwarf. A rabbit weighing no more than 3 pounds at maturity.

Ear canker. A scabby condition inside the ear caused by mites; usually found in rabbits.

Eclipse molt. A 3- to 4-month period each year, after the breeding season, when the bright plumage of colored adult drakes is replaced with subdued colors similar to those of females.

Edema. Swelling due to excess accumulation of fluid in tissue spaces.

Egg tooth. A small, horny protuberance attached to upper mandible of a hatching bird's beak or bill that it uses to help break open the shell. It falls off several days after hatching.

Elasticity. The ability of wool fibers to return to their original length after being stretched. Good-quality wool has a great deal of elasticity.

Elastrator. The tool used to apply elastrator rings.

Elastrator rings. Castrating rings resembling rubber bands; they are applied to the scrotum so it will atrophy and fall off.

Electrolytes. Important body salts, including sodium, potassium, calcium, and magnesium; need replacing during dehydration.

Endotoxic shock. Shock caused by body systems shutting down in reaction to bacterial poisons.

Endotoxin. The poison created when bacteria multiply in the body.

Enteritis. Inflammation of the intestine.

Enterotoxemia. A bacterial gut infection caused by *Clostridium perfringens,* usually resulting in death; also called pulpy kidney disease and overeating disease.

Escherichia coli. A type of bacterium that has more than 100 strains, some of which cause serious infection.

Esophageal feeder. A tube put down an animal's throat to force-feed fluids from a feeder bag.

Estrous cycle. The time and physiological events that take place in one heat period.

Estrus. The period when a female animal is in heat and will willingly mate with a male animal.

Ewe. A mature female sheep.

Face. To trim wool from around the face of closed-face sheep.

Feather legged. Having feathers growing down the shanks.

Feather out. To grow a full "coat" of feathers.

Feed additive. Anything added to the feed, including preservatives, growth promotants and medications.

Felt. A fabric made of layers of wool pressed and matted tightly together.

Fetus. An animal in the uterus, until birth.

Finish. To mature and fatten enough to butcher (to reach butchering condition).

Fleece. The wool from one sheep or goat.

Flight feathers. The primary feathers on the wing of fowl; sometimes used to denote the primaries and secondaries.

Flight zone. The proximity you can get to an animal before it flees.

Flock. A group of chickens or sheep.

Flush. To feed females more generously 2 to 3 weeks before breeding in order to stimulate the onset of heat and improve the chances of conception.

Foot rot. An infection in the foot causing severe lameness.

Forage. The hay and/or grassy portion of an animal's diet.

Forced-air incubator. A mechanical device for hatching fertile eggs that has a fan to circulate warm air.

Founder. Inflammation of the hooves.

Fowl. A term applied collectively to chickens, ducks, geese, et al, or the market class designation for old laying birds.

Frame size. The measure of hip height; used to determine skeletal size.

Free choice. Being available to be eaten at all times.

Free range. Being allowed to range and forage at will.

Freshen. To give birth and begin to produce milk.

Frizzle. Having feathers that curl rather than lay flat.

Fryer. *See* Broiler.

Furnishing. The long, decorative wool on the head of English Angora rabbits.

Gaggle. A flock of geese.

Gander. A male goose.

Germinal disc. In an egg, the fertility spot from which an embryo grows.

Gestation. The time between breeding and birth.

Giant. A rabbit weighing 12 to 16 pounds or more at maturity.

Gizzard. The muscular stomach of fowl that contains grit for grinding food.

Goose. The female goose, as distinguished from the gander.

Gosling. A young goose of either sex.

Grade. Unregistered; not purebred.

Graft. To have an adult female accept and mother a young animal that isn't her own.

Grease wool. Raw wool that has not been cleaned.

Green. For young waterfowl, to have gone into the first molt.

Grit. The hard, insoluble materials eaten by birds and used by the gizzard to grind up food.

Gut. The digestive tract.

Hackle. A rooster's cape feathers.

Halter. A rope or leather headgear used to control or lead an animal.

Hardware disease. Peritonitis (infection in the abdomen) caused by a sharp foreign object penetrating the gut wall.

Hatch. To come out of the egg; also, a group of birds that come out of their shells at roughly the same time.

Hatchability. The percentage of fertilized eggs that hatch under incubation.

Hatchling. A bird that has just hatched.

Hatchlings

Hay. Dried forage.

Heat. *See* Estrus.

Heifer. A young female bovine that has not calved.

Hemoglobin. The compound in red blood cells that carries oxygen.

Hen. A female chicken more than 12 months of age.

Hen-feathered. In a cock, having round feathers on the hackle and saddle.

Herd. A group of goats or cattle.

Hock. The large joint halfway up the hind leg.

Hoof rot. *See* Foot rot.

Hopper. A food container that is filled from the top and dispenses from the bottom; used for free-choice feeding of grain, grit, and other supplements.

Horn bud. A small bump from which a horn grows.

Hutch. A rabbit cage.

Incubate. To sit on eggs to keep them warm until they hatch.

Incubation period. The number of days it takes eggs to hatch once they are warmed to incubation temperature.

Incubator. A mechanical device for hatching fertile eggs.

Infectious bovine rhinotracheitis (IBR). A respiratory disease caused by a virus; also called red nose.

International unit (IU). A standard unit of potency of a biologic agent such as a vitamin or antibiotic.

Intramuscular (IM). Into a muscle.

Intravenous (IV). Into a vein.

Iodine. A harsh chemical used for disinfecting.

J-clip. A J-shaped metal clip used in hutch construction. Special pliers are required for application.

Johne's disease. A wasting, often fatal form of enteritis.

Jug. A small pen large enough for just one ewe and her offspring.

Junior. *See* Green.

Ked. An external parasite that affects sheep.

Keel. The breastbone or sternum of fowl.

Kemp. Straight, brittle, chalky white mohair fiber.

Ketosis. An overaccumulation of ketones in the body.

Kid. A goat under 1 year of age; also, in goats, to give birth.

Kindle. To give birth to a litter of rabbits.

Knob. A rounded protuberance appearing at the base of the bill (between the eyes) of some species of goose.

Lactation. The period in which an animal is producing milk; the secretion or formation of milk.

Lamb. A newborn or immature sheep, typically under 1 year of age.

Laminitis. *See* Founder.

Lanolin. The naturally occurring "grease" that coats wool.

Lay ration. Feed that is formulated to stimulate high egg production.

Legume. A plant belonging to the pea family (alfalfa, clover, etc.).

Leptospirosis. A bacterial disease that can cause abortion.

Lice. Tiny external parasites on the skin; there are two kinds — biting lice and sucking lice.

Listeriosis. A bacterial disease that can cause abortion.

Litter. Collectively, the offspring of a rabbit, from a single birth. *See also* Bedding.

Liver flukes. Parasites that infest snails and spend part of their life cycle in cattle, damaging the liver and making the host more susceptible to red water and Black's disease.

Loft. The tendency of down to fluff up. The greater the loft, the better the insulating ability of the down.

Lop eared. Having bent or drooping ears.

Lumpy jaw. An abscess in the mouth caused by infection.

Luster. The natural gloss or sheen of a fleece.

Maintenance ration. A feed used for adult fowl that are not in production.

Malocclusion. An abnormal coming together of teeth.

Mammary tissue. Milk-producing tissue in the udder.

Mandible. The upper or lower bony portion of a bird's beak.

Mange. A skin disease caused by mites that feed on the skin.

Marbled. In beef, havings flecks of fat interspersed in muscle.

Mash. A mixture of finely ground grains.

Mastitis. Infection and inflammation in the udder.

Mature. Old enough to reproduce.

Meconium. The dark, sticky first bowel movement of a newborn animal.

Milking bench (or stand). A raised platform, usually with a seat for the milker and a stanchion for the goat's neck, that a goat stands upon to be milked.

Milk letdown. A physiological process that allows milk to be removed from the udder by sucking or mechanical means.

Mites. Very tiny parasites that feed on skin, causing mange or scabies.

Molt. To shed old feathers, fur, or hair and grow a new "coat."

Mount. To rear up over the back of an animal to "ride" it, as a bull does a cow when breeding.

Mutton. Meat from a mature or aged sheep over 1 year old.

Nematode. Roundworm.

Nest box. A place for fowl or rabbits to give birth.

Nolvasan. An all-purpose disinfectant.

Off feed. Not eating as much as normal.

Omasum. One of the four stomach compartments in the ruminant animal.

Open. Nonpregnant.

Open face. In sheep, not having much wool around the eyes and cheeks.

Open face

Overconditioned. Overfed; fat.

Oviduct. The long, glandular tube of female fowl in which egg formation takes place; leads from ovary to the cloaca.

Parainfluenza (P13). A viral respiratory agent that by itself causes a mild disease, but in combination with bacterial infection can be severe.

Parasite. An organism that lives in or on an animal.

Parturition. The birth process.

Pastern. The area between the hoof and the fetlock joint; "ankle."

Pasting. Loose droppings sticking to the anal area.

Pathogen. A harmful invasive microorganism, such as a bacterium or virus.

Peck order. The social rank of chickens.

Pedigree. A paper showing an animal's forebears.

Peritonitis. An infection in the abdominal cavity.

PH. The measure of acidity or alkalinity; on a scale of 1 to 14, 7 is neutral, 1 is most acid, and 14 is most alkaline.

Pigeon-toed. Having toes turning inward instead of pointing straight ahead.

Pinfeathers. New feathers that are just emerging from the skin.

Pinion. The tip of a wing. Also, to cut off the tip of a wing to prevent flight.

Pinkeye. A contagious eye infection spread by face flies.

Pip. The hole a chick makes in its egg's shell when it is ready to hatch; also, the act of making the hole.

Placenta. Afterbirth; attached to the uterus during pregnancy as a buffer and lifeline for the developing animal.

Plumage. All of a bird's feathers, collectively.

Poll. The top of the head.

Polled. Born without horns; naturally hornless.

Post-legged. Having hind legs that are too straight, with not enough angle in the hocks and stifles.

Preen. To clean and organize feathers with the beak or bill.

Primary feather. One of the long feathers at the end of a wing.

Prolapse. Protrusion of an inverted organ such as rectum, vagina, or uterus.

Protozoa. One-celled animals; some can cause disease.

Puberty. The age when an animal matures sexually and can reproduce.

Pullet. A female chicken less than 1 year of age.

Purebred. An animal whose ancestry can be traced back to the establishment of a breed through the records of a registry association.

Quarantine. To keep an animal isolated from other animals to prevent the spread of infections.

Quarter. One of a cow's four teats.

Quill. A primary feather.

Ram. A mature male sheep.

Ram lamb. An immature male sheep.

Ration. The combination of all feed consumed in a day.

Raw milk. Milk as it comes from an animal; unpasteurized milk.

Red water. A deadly bacterial disease of cattle caused by *Clostridium haemolyticum;* animals with liver damaged by flukes are susceptible.

Registered. Having an animal's birth and ancestry recorded by a registry association.

Relative humidity. The percentage of moisture saturation in the air.

Rennet. An enzyme used to curdle milk and make cheese.

Reticulum. The second of the four stomach compartments in the ruminant animal.

Rex

Rex. A rabbit with short, plush fur.

Ringworm. A fungal infection causing scaly patches of skin.

Roaster. A young chicken of either sex, usually 3 to 5 months of age, that is tender-meated with soft, pliable, smooth-textured skin and with a breastbone cartilage somewhat less flexible than that of the broiler-fryer.

Roost. A perch on which fowl rest or sleep; the place where chickens sleep at night. Also, to rest on a roost.

Rooster. A male chicken; also called a *cock.*

Rotational grazing. The use of various pastures in sequence to give each one a chance to regrow before grazing it again.

Roughage. Feed that is high in fiber and low in energy (e.g., hay, pasture).

Rumen. The largest stomach compartment in the ruminant animal, in which roughage is digested with the aid of microorganisms in a fermentation process.

Ruminant. An animal that chews its cud and has four stomach compartments.

Satin. A rabbit with transparent hair shafts that create an extremely lustrous coat.

Scabies. A skin disease caused by a certain type of mite.

Scales. The small, hard, overlapping plates covering a chicken's shanks and toes.

Scours. Persistent diarrhea in young animals.

Scrapie. A usually fatal disease of the nervous system.

Scratch. The habit chickens have of scraping their claws against the ground to dig up tasty things to eat; also, any grain fed to chickens.

Scurs. Horny tissue or rudimentary horns attached to the skin rather than to the skull.

Second cuts. Short lengths of wool resulting from cutting the same spot twice during shearing.

Selenium. A mineral needed in very small amounts in the diet (too much is poisonous).

Set. To keep eggs warm so they will hatch.

Settle. To become pregnant.

Sex. To sort by gender.

Shank. The part of a chicken's leg between the claw and the first joint.

Shear. To clip wool from a sheep or goat.

Sickle-hocked. Having too much angle in the hind legs (weak construction).

Silage. Feed cut and stored green, preserved by fermentation.

Sire. The male parent; to father.

Skirt. To remove the edges of a fleece at shearing.

Snuffles. A highly contagious respiratory disease of rabbits marked by nasal discharge.

Sore hocks. Ulcerated footpads in rabbits.

Splayfooted. Having toes that turn out.

Spur. The sharp points on a rooster's shanks.

Stag. A late-castrated steer or improperly castrated steer that still shows masculine characteristics.

Stanchion. A device for restraining an animal by the neck for feeding or milking.

Standard. The description of an ideal specimen for its breed.

Standing heat. The time during heat when the female animal allows the male animal to mount and breed.

Started. Having survived the first few critical days or weeks of life and begun to develop.

Steer. A male bovine after castration.

Stifle. The large joint high on the hind leg by the flank.

Still-air incubator. A mechanical device for hatching fertile eggs that does not have a fan to circulate air.

Straight run. Not sorted by gender. Usually applied to newly hatched chicks. Also called "unsexed" or "as hatched."

Straw. Dried plant matter (usually oat, wheat, or barley leaves and stems) used as bedding; also, the glass tube semen is stored in for AI.

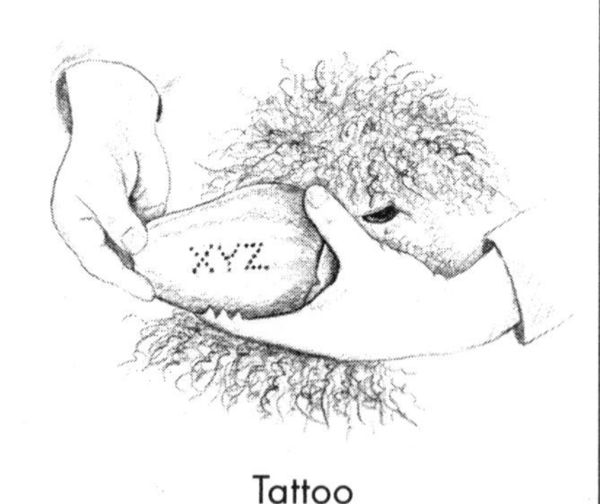

Tattoo

Strip. To remove milk from the udder. Usually refers to removing the last of the milk.

Subcutaneous (SQ). Under the skin.

Supplement. To feed additives that supply something missing in the diet, such as additional protein, vitamins, minerals.

Tag. To cut dung locks from an animal. Also, a lock of wool contaminated by dung and dirt.

Tallow. The extracted fat from sheep and cattle.

Tattoo. The permanent identification of animals produced by placing indelible ink under the skin; to apply a tattoo.

Toxemia. A condition in which bacterial toxins invade the bloodstream and poison the body.

Trace minerals. Minerals needed in the diet in very small amounts.

Trichomoniasis. A venereal disease caused by protozoa.

Trio. Two females and one male of the same breed and variety.

Tuft. A puff of feathers on top of a goose's head.

Udder. Mammary glands and teats.

Udder wash. A dilute chemical solution, usually an iodine compound, for washing udders before milking.

Vaccine. A fluid containing killed or modified live germs, injected into the body to stimulate production of antibodies and immunity.

Vent. The external opening from the cloaca of fowl, through which it emits eggs and droppings.

Vermifuge. Any chemical substance administered to an animal to kill internal parasitic worms.

Vibriosis. A venereal disease of cattle that causes early abortion.

Vulva. The external opening of the vagina.

Warts. Skin growths caused by a virus.

Wattle. A small, fleshy appendage that dangles under the chin of some fowl species.

Wean. To separate a young animal from its mother or stop feeding it milk.

Wether. A castrated male sheep or goat.

Whey. The liquid remaining when the curd is removed from curdled milk as part of the process of making cheese.

White muscle disease. A fatal condition in calves in which heart muscle fibers are replaced with connective tissue; caused by selenium deficiency.

Withdrawal period. The amount of time that must elapse for a drug to be eliminated (through urine, etc.) from an animal's body before it is butchered so there will be no residues in the meat.

Wool block. An illness in rabbits caused by swallowed fur forming a blockage in the digestive tract.

Yearling. A male or female cow between 1 and 2 years of age.

Recommended Reading

The Angora Goat: Its History, Management and Diseases, by Stephanie Mitcham Sexton and Allison Mitcham. Crane Creek Publications, 1999.

Angora Goats the Northern Way, by Susan Black Drummond. Stony Lonesome Farm, 1988.

Basic Butchering of Livestock & Game, by John J. Mettler, Jr. Storey Books, 1986.

Building Small Barns, Sheds & Shelters, by Monte Burch. Storey Books, 1983.

The Chicken Health Handbook, by Gail Damerow. Storey Books, 1994.

Chicken Tractor, by Andy Lee and Patricia Foreman. Good Earth Publications, 1998.

Day Range Poultry, by Andy Lee and Patricia Foreman. Good Earth Publications, 2002.

Earth Ponds, by Tim Matson. Countryman Press, 1991.

The Fairest Fowl: Portraits of Championship Chickens, by Tamara Staples and Ira Glass. Chronicle Books, 2001.

Fences for Pasture & Garden, by Gail Damerow. Storey Books, 1992.

Free-Range Poultry Production & Marketing, by Hermann Beck-Chenoweth. Back Forty Books, 1997.

Goat Health Handbook, by Thomas R. Thedford, D.V.M. Winrock International, 1983.

Goats Produce Too! by Mary Jane Toth. Mary Jane Toth, 1998.

Greener Pastures on Your Side of the Fence, by Bill Murphy. Arriba Publishing, 1999.

A Guide to Canning, Freezing, Curing & Smoking Meat, Fish & Game, by Wilbur F. Eastman Jr. Storey Books, 2002.

Home Cheese Making, by Ricki Carroll. Storey Books, 2002.

Home Sausage Making, by Susan Mahnke Peery and Charles G. Reavis. Storey Books, 2003.

Keeping Livestock Healthy, by N. Bruce Haynes, D.V.M. Storey Books, 2001.

Let It Rot! by Stu Campbell. Storey Books, 1998.

Making Your Small Farm Profitable, by Ron Macher. Storey Books, 1999.

Oxen: A Teamster's Guide, by Drew Conroy. Rural Heritage, 1999.

Pastured Poultry Profit$, by Joel Salatin. Chelsea Green Publishing Co., 1996.

Raising Rare Breeds: Heritage Poultry Breeds Conservancy Guide. Rare Breeds Conservancy, 1994.

Renovating Barns, Sheds & Outbuildings, by Nick Engler. Storey Books, 2001.

Spindle Spinning, by Connie Delaney. Kokovoko Press, 1998.

Storey's Guide to Raising Chickens, by Gail Damerow. Storey Books, 2001.

Tan Your Hide! by Phyllis Hobson. Storey Books, 1977.

Resources

Cooperative State Research, Education, and Extension Service

For more information about barnyard animals, contact the cooperative Extension service. The program is affiliated with each of the nation's land-grant universities. Extension offices are located in each county in the United States. To obtain contact information for the office in your county, visit your state's Extension Web site.

Alabama

Auburn University
109D Duncan Hall
Auburn, AL 36849-5612
Phone: (334) 844-4444
Fax: (334) 844-5544
Web site: www.aces.edu

Tuskegee University
Morrison-Mayberry Hall
Tuskegee, AL 36088
Phone: (334) 727-8808
Fax: 334/724-8812
Web site: www.tusk.edu/academics/cooperative_ext/

Alaska

University of Alaska-Fairbanks
CES Building
P.O. Box 756180
Fairbanks, AK 99775-6180
Phone: (907) 474-7246
Fax: (907) 474-6971
Web site: www.uaf.edu/coop-ext/

Arizona

University of Arizona
Cooperative Extension
Forbes 301, P.O. Box 210066
Tucson, AZ 85721
Phone: (520) 621-7205
Fax: (520) 621-1314
Web site: www.ag.arizona.edu/extension/

Arkansas

University of Arkansas
Cooperative Extension Service
2301 S. University Avenue
Little Rock, AR 72204
Phone: (501) 671-2000
Fax: (501) 671-2209
Web site: www.uaex.edu

University of Arkansas
1890 Cooperative Extension
1200 N. University Drive
Pine Bluff, AR 71611
Phone: (870) 543-8529
Fax: (870) 543-8033
Web site: www.uapb.edu/safhs/page5.html

California

University of California, Davis
Cooperative Extension
DANR Building, Hopkins Road
Davis, CA 95616
Phone: (530) 754-8509
Web site: www.ucanr.org/ce.cfm

Colorado

Colorado State University
Cooperative Extension
1 Administration Building
Fort Collins, CO 80523-4040
Phone: (970) 491-6281
Fax: (970) 491-6208
Web site: www.ext.colostate.edu

Connecticut

Cooperative Extension Program
1376 Storrs Road, Unit 4134
Storrs, CT 02069-4134
Phone: (860) 486-1987
Web site: www.canr.uconn.edu/ces

Delaware

University of Delaware
Cooperative Extension Service
Townsend Hall
Newark, DE 19717
Phone: (302) 831-2506
Web site: ag.udel.edu/extension

District of Columbia

University of the District of Columbia
Cooperative Extension Service
4200 Connecticut Avenue NW
Washington, D.C. 20008
Phone: (202) 274-7115
Web site: www.udc.edu/coes/ces

Florida

Florida Cooperative Extension Service
Box 110210
1038 McCarty Hall
Gainesville, FL 32611-0210
Phone: (352) 392-1761
Fax: (352) 846-0458
Web site: www.ifas.ufl.edu/www/extension/ces.htm

Florida A & M University
Cooperative Extension & Outreach Programs
Perry-Paige Building, Room 215 South
Tallahassee, FL 32307
Phone: (850) 599-3546
Web site: www.famu.edu/acad/colleges/cesta/co-op.html

Georgia

College of Agricultural & Environmental Sciences
Cooperative Extension
The University of Georgia
Room 101, Conner Hall
Athens, GA 30602-7501
Phone: (706) 542-3924
Fax: (706) 542-0803
Web site: www.ces.uga.edu

Fort Valley State University
Cooperative Extension
Fort Valley, GA 31030-3298
Phone: (478) 825-6296
Fax: (478) 825-6299
Web site: www.aginfo.fvsu.edu/ces/overview.htm

Hawaii

Office of Cooperative Extension
3050 Maile Way
Gilmore 203
Honolulu, HI 96822
Phone: (808) 956-8139
Fax: (808) 956-9105
Web site: www.ctahr.hawaii.edu/ctahr2001/Extension/ExtMain.html

Idaho

Extension Administration
P.O. Box 1827
Twin Falls, ID 83303-1827
Phone: (208) 736-3603
Fax: (208) 736-0843
Web site: www.uidaho.edu/extension

Illinois

University of Illinois
Office of Extension and Outreach
214 Mumford Hall (MC-170)
1301 West Gregory Drive
Urbana, IL 61801
Phone: (217) 333-5900
Fax: (217) 244-5403
Web site: www.extension.uiuc.edu/welcome.html

Indiana
Purdue University
Cooperative Extension Service
1140 Agriculture Admin. Bldg.
West Lafayette, IN 47907-1140
Phone: (765) 494-8489
Fax: (765) 494-5876
Web site: www.ces.purdue.edu

Iowa
Iowa State University
2150 Beardshear Hall
Ames, IA 50011-2020
Phone: (515) 294-4576
Fax: (515) 294-4715
Web site: www.extension.iastate.edu

Kansas
Kansas State University
Cooperative Extension Service
123 Umberger Hall
Manhattan, KS 66506-3401
Phone: (785) 532-5820
Web site: www.oznet.ksu.edu

Kentucky
University of Kentucky
306 W. P. Garrigus Building
Lexington, KY 40546-0215
Phone: (859) 257-4302
Fax: (859) 323-1991
Web site: www.ca.uky.edu/ces

Kentucky State University
Cooperative Extension Program Facility
400 E. Main Street
Frankfort, KY 40601
Fax: (502) 227-5933
Web site: www.kysu.edu/landgrant/CEP/cep.htm

Louisiana
Louisiana State University
Cooperative Extension Program
P.O. Box 25100
Baton Rouge, LA 70894-5100
Phone: (225) 578-4141
Web site: www.agctr.lsu.edu/nav/extension.htm

Southern University & A&M College
Cooperative Extension
P.O. Box 10010
Baton Rouge, LA 70813
Phone: (225) 771-2242

Maine
University of Maine
Cooperative Extension
5741 Libby Hall
Orono, ME 04469-5741
Phone: (207) 581-3188
Fax: (207) 581-1387
Web site: www.umext.maine.edu

Maryland
University of Maryland
Maryland Cooperative Extension
1202 Symons Hall
College Park, MD 20742-5565
Phone: (301) 405-2907
Fax: (301) 405-2963
Web site: www.agnr.umd.edu/MCE/

University of Maryland
Eastern Shore
2122 Henson Center
Princess Anne, MD 21853
Phone: (410) 651-6206
Fax: (410) 651-6207
Web site: www.umesde.umes.edu/1890-mce/

Massachusetts
University of Massachusetts
230 Draper Hall
Amherst, MA 01003-9244
Phone: (413) 545-4800
Fax: (413) 545-6555
Web site: www.umass.edu/umext

Michigan
Michigan State University Extension
Agriculture Hall, Room 108
Michigan State University
East Lansing, MI 48824-1039
Phone: (517) 355-2308
Fax: (517) 355-6473
Web site: www.msue.msu.edu

Minnesota
University of Minnesota
Room 240, Coffey Hall
1420 Eckles Avenue
St. Paul, MN 55108-6070
Phone: (612) 624-2703
Fax: (612) 624-1222
Web site: www.extension.umn.edu

Mississippi
Mississippi State University
Extension Service
Box 9601
Mississippi State, MS 39762-9601
Phone: (662) 325-3036
Fax: (662) 325-8407
Web site: www.msucares.com

Alcorn State University
1000 ASU Drive 690
Alcorn State, MS 39096-7500
Phone: (601) 877-6137
Fax: (601) 877-6219
Web site: www.alcorn.edu/academic/academ/ags.htm

Missouri
University of Missouri
Cooperative Extension Service
309 University Hall
Columbia, MO 65211-3020
Phone: (573) 882-7754
Fax: (573) 884-4204
Web site: www.extension.missouri.edu

Lincoln University
Cooperative Research and Extension
900 Chestnut Street
Jefferson City, MO 65102-0029
Phone: (573) 681-5543
Fax: (573) 681-5546
Web site: www.luce.lincolnu.edu

Montana
Montana State University
336 Culbertson Hall
P.O. Box 172230
Bozeman, MT 59717-2230
Phone: (406) 994-1752
Fax: (406) 994-1756
Web site: extn.msu.montana.edu

Nebraska
University of Nebraska
Cooperative Extension
211 Agriculture Hall
Lincoln, NE 68583-0703
Phone: (402) 472-2966
Fax: (402) 472-5557
Web site: extension.unl.edu

Nevada
University of Nevada, Reno Cooperative Extension
Mail Stop 404
Reno, NV 89557-0106
Phone: (775) 784-7070
Fax: (775) 784-7079
Web site: www.unce.unr.edu

New Hampshire
University of New Hampshire
Cooperative Extension
59 College Road
Taylor Hall
Durham, NH 03824-3587
Phone: (603) 862-1520
Fax: (603) 862-1585
Web site: www.ceinfo.unh.edu

New Jersey
Rutgers University
Rutgers Cooperative Extension
Cook College
88 Lipman Drive
New Brunswick, NJ 08901-8525
Phone: (732) 932-9306
Fax: (732) 932-6633
Web site: www.rce.rutgers.edu

New Mexico
New Mexico State University
P.O. Box 30003
Gerald Thomas Hall
MSC 3AE
Las Cruces, NM 88003
Phone: (505) 646-3015
Fax: (505) 646-5975
Web site: www.cahe.nmsu.edu/ces

New York
Cornell Cooperative Extension
365 Roberts Hall
Ithaca, NY 14853-5905
Phone: (607) 255-2237
Fax: (607) 255-0788
Web site: www.cce.cornell.edu

North Carolina
North Carolina State University
Cooperative Extension Service
Box 7602
Raleigh, NC 27695-7602
Phone: (919) 515-2811
Fax: (919) 515-3135
Web site: www.ces.ncsu.edu

North Carolina A&T State University
School of Agriculture and Environmental Sciences
Cooperative Extension
1601 East Market Street
Greensboro, NC 27411
Phone: (336) 334-7979
Fax: (336) 334-7580
Web site: www.ag.ncat.edu/extension

North Dakota
North Dakota State University
Extension Service
315 Morrill Hall, P.O. Box 5437
Fargo, ND 58105-5437
Phone: (701) 231-8944
Fax: (701) 231-8520
Web site: www.ext.nodak.edu

Ohio
Ohio State University
Extension Service Department
3 Agricultural Administration Building
2120 Fyffe Road
Columbus, OH 43210-1084
Phone: (614) 292-4067
Fax: (614) 688-3807
Web site: www.ag.ohio-state.edu

Oklahoma
Oklahoma State University
Cooperative Extension
139 Agriculture Hall
Stillwater, OK 74078
Phone: (405) 744-5398
Fax: (405) 744-5339
Web site: www.dasnr.okstate.edu/oces

Agricultural Research and Extension Programs
P.O. Box 730
Langston, OK 73050
Phone: (405) 466-3836
Fax: (405) 466-3138
Web site: www2.luresext.edu

Oregon
Oregon State University
Extension Administration
101 Ballard Extension Hall
Corvallis, OR 97331-3604
Phone: (541) 737-2713
Fax: (541) 737-4423
Web site: www.osu.orst.edu/extension

Pennsylvania
The Pennsylvania State University
Cooperative Extension
217 Agricultural Administrative Building
University Park, PA 16802
Phone: (814) 863-3438
Fax: (814) 863-7905
Web site: www.extension.psu.edu

Rhode Island
University of Rhode Island
Cooperative Extension Center
3 East Alumni Avenue
Kingston, RI 02881
Phone: (401) 874-2900
Fax: (401) 874-4017
Web site: www.uri.edu/ce

South Carolina
Clemson University
Cooperative Extension Service
Clemson, SC 29634-0310
Phone: (864) 656-3382
Fax: (864) 656-5819
Web site: www.clemson.edu/extension

South Carolina State University
1890 Research and Extension
300 College Street N.E.
P.O. Box 8103
Orangeburg, SC 29117
Phone: (803) 536-8229
Fax: (803) 536-7102
Web site: www.1890.scsu.edu.1890.htm

South Dakota
South Dakota State University
Cooperative Extension Service
AGH 154
P.O. Box 2207D
Brookings, SD 57007
Phone: (605) 688-4792
Fax: (605) 688-6347
Web site: www3.sdstate.edu/CooperativeExtensionb

Tennessee
The University of Tennessee
Agricultural Extension Service
2621 Morgan Circle
121 Morgan Hall
Knoxville, TN 37996-4530
Phone: (865) 974-7114
Fax: (865) 974-1068
Web site: www.utextension.utk.edu

Tennessee State University
Cooperative Extension Program
3500 John A. Merritt Boulevard
Nashville, TN 37209-1561
Phone: (615) 963-5491
Fax: (615) 963-5833
Web site: www.tnstate.edu/cep

Texas
Texas A & M University
Jack K. Williams Administration Building
Room 112
College Station, TX 77843-7101
Phone: (979) 845-7800
Fax: (979) 845-9542
Web site: agextension.tamu.edu

Prairie View A & M University
P.O. Box 3059
Prairie View, TX 77446-0519
Phone: (936) 857-2023
Web site: www.pvamu.edu/gridold/ag_husc/coopext

Utah
Utah State University
4900 Old Main Hill
Logan, UT 84322-4900
Phone: (435) 797-2201
Fax: (435) 797-3268
Web site: www.extension.usu.edu

Vermont
University of Vermont
Extension Systems
601 Main Street
Burlington, VT 05401-3439
Phone: (802) 656-2990
Fax: (802) 656-8642
Web site: www.uvm.edu/extension

Virginia
Virginia Tech
Virginia Cooperative Extension
101 Hutcheson Hall
Blacksburg, VA 24061-0402
Phone: (540) 231-5299
Fax: (540) 231-4370
Web site: www.ext.vt.edu

Washington
Washington State University
Cooperative Extension
P.O. Box 646230
Hulbert Hall 411
Pullman, WA 99164-6230
Phone: (509) 335-2933
Fax: (509) 335-2926
Web site: ext.wsu.edu

West Virginia
West Virginia University
Extension Service
507 Knapp Hall
Morgantown, WV 26506-6031
Phone: (304) 293-4421
Fax: (304) 293-6611
Web site: www.wvu.edu/~extn

Wisconsin
University of Wisconsin
Extension Headquarters
432 N. Lake Street
Madison, WI 53706
Phone: (608) 262-3980
Fax: (608) 265-9317
Web site: www1.uwex.edu/ces

Wyoming
University of Wyoming
Cooperation Extension Service
P.O. Box 3354
Laramie, WY 82071-3354
Phone: (307) 766-5124
Fax: (307) 766-3998
Web site: www.uwyo.edu/ces/ceshome.htm

National Association of State Universities & Land Grant Colleges
1307 New York Avenue, NW
Suite 400
Washington, D.C. 20005-4722
Phone: (202) 478-6040
Fax: (202) 478-6046
Web site: www.nasulgc.org

Helpful Organizations and Web Sites

American Angora Goat Breeders Association
P.O. Box 195
Rocksprings, TX 78880
Maintains the registry of American angora goats.

American Bantam Association
P.O. Box 127
Augusta, NJ 07822
(973) 383-6944
www.bantamclub.com
Represents bantam breeders and their special interests.

American Dairy Goat Association
209 W. Main Street, P.O. Box 865
Spindale, NC 28160
(828) 286-3801
www.adga.org
Promotes dairy goats and dairy goat products, records pedigrees, provides services to dairy goat breeders.

American Egg Board
1460 Renaissance Drive
Park Ridge, IL 60068
(847) 296-7043
www.aeb.org
Promotes egg consumption.

American Harness Goat Association
Carole A. Contreres
15835 Bald Hills Road SE
Yelm, WA
(360) 894-3154
Works to preserve the working harness goat and enhance its integrity and quality.

American Herding Breed Association
277 Central Avenue
Seekonk, MA 02771
(508) 761-4078
www.ahba-herding.org
Promotes appreciation of herding dogs and provides information about breeds, training, behaviors, and herding in general.

American Livestock Breeds Conservancy
P.O. Box 477
Pittsboro, NC 27312
(919) 542-5704
www.albc-usa.org
Encourages the preservation of endangered cattle, goats, horses, asses, sheep, swine, and poultry.

American Meat Goat Association
P.O. Box 676
Sonora, TX 76950
(915) 387-6100
www.ranchmagazine.com/amga
Promotes goat meat and the use of meat goats in agriculture.

American Poultry Association
133 Millville Street
Mendon, MA 01756
www.ampltya.com
Promotes and protects the standard-bred poultry industry; publishes The American Standard of Perfection.

American Rabbit Breeders Association
P.O. Box 426
Bloomington, IL 61702
(309) 664-7500
www.arba.com
Dedicated to the promotion, development, and improvement of the domestic rabbit. Publishes The Standard of Perfection.

Appropriate Technology Transfer for Rural America
P.O. Box 3657
Fayetteville, AR 72702
(800) 346-9140
www.attra.org
ATTRA offers free technical assistance, publications, and resources on sustainable farming production practices, alternative crop and livestock practices, and marketing.

The Cattle Pages
www.cattlepages.com
A comprehensive listing of cattle resources that shows you where to find equipment, supplies, and services and includes a breeder's directory, breed associations, breeds photo gallery, discussion board, and more.

Forage Information System, Oregon State University
www.forages.orst.edu
Oregon State offers a comprehensive Web site about forage plants of all types. The site also lists plant tissue-testing labs.

National Association of State Departments of Agriculture
1156 15th Street NW, Suite 1020
Washington, DC 20005
(202) 296-9680
www.nasda.org
Promotes the agriculture industry while protecting consumers and the environment.

Oklahoma State University, Department of Animal Science
www.ansi.okstate.edu/breeds
A Web site with information about livestock breeds from around the world.

Purina Mills, LLC
Family Livestock Health and Nutrition
www.rabbitnutrition.com
Information on rabbit, goat, and poultry health and nutrition, frequently asked questions, and links.

Rabbit Web
www.rabbitweb.net
Information on rabbit care, health, and breeding; book reviews; discussion board; chat system.

Rare Breeds Canada
Trent University
c/o Environmental and Resource Studies Program
1600 West Bank Drive
Peterborough, ON
Canada K9J 7B8
(705) 748-1011 ext. 1634
www.trentu.ca/rarebreedscanada/
Encourages the preservation of heritage livestock.

Organic Trade Association
P.O. Box 547
Greenfield, MA 01302
(413) 774-7511
www.ota.com
Promotes organic products in the marketplace and protects the integrity of organic standards. OTA can help you identify an organic certifying agency working in your state.

Society for the Preservation of Poultry Antiquities
Route 4, Box 251
Middleburg, PA 17842
www.cyborganic.com/People/feathersite/Poultry/SPPA/SPPA.html
Encourages the preservation of rare exhibition breeds.

United States Animal Health Association
P.O. Box K227
8100 Three Chopt Road
Richmond, VA 23288
(804) 285-3210
www.usaha.org
Works to prevent, control, and eliminate livestock disease. Advises the United States Department of Agriculture.

United States Border Collie Handlers Association
2915 Anderson Lane
Crawford, TX 76638
www.usbcha.org
Serves as the sanctioning body for sheepdog trials throughout the United States and Canada.

United States Department of Agriculture
14th and Independence Avenue SW
Washington, DC 20250
(202) 720-2791
www.usda.gov
For general information.

United States Poultry & Egg Association
1530 Cooledge Road
Tucker, GA 30084
(770) 493-9401
www.poultryegg.org
Promotes research, education, and communication on the United States poultry and egg industries.

USDA Animal and Plant Health Inspection Service
United States Department of Agriculture
12th and Independence Avenue SW
Washington, DC 20250
www.aphis.usda.gov
Functions include veterinary accreditation, biotechnology and trade regulation, wildlife services, and animal health and welfare.

Suppliers

The following sources are listed for your convenience. No endorsement is expressed or implied. An asterisk (*) indicates a mail-order supplier.

All Seasons Rabbitry & Supply*
9805 W. 133rd Street S.
Oktaha, OK 74450
(918) 687-1861
www.allseasons55.homestead.com/allseasons55.html
Rabbits, cages, nest boxes, bedding, vitamins and minerals, tattoo equipment, and other rabbit supplies.

A. S. WebSales*
1918 N. Broadway
Poteau, OK 74953
(800) 451-4660
www.buytack.com
Cattle supplies including branding equipment, halters, feeding and storage, calf chute, and more.

Bass Equipment Company*
P.O. Box 352
Monett, MO 65708
(417) 235-7557
www.bassequipment.com
A full line of rabbit supplies, including housing needs, water and feed equipment, medicine, management tools, and books.

BF Products Inc.*
P.O. Box 61866
Harrisburg, PA 17106-1866
(800) 255-8397
www.bfproducts.com
Fencings and nettings.

Cackle Hatchery*
P.O. Box 529
Lebanon, MO 65536
(417) 532-4581
www.cacklehatchery.com
Chicks, ducklings, goslings, rare breeds, and related books and supplies.

Caprine Supply*
P.O. Box Y
DeSoto, KS 66018
(913) 585-1191
www.caprinesupply.com
A full line of goat supplies, including health, milking, breeding, management, books, and videos.

Cutler's Supply*
1940 Old 51
Applegate, MI 48401
(810) 633-9450
www.cutlersupply.com
General poultry supplies.

Double-R Discount Supply
4000 Dow Road, Suite 8
Melbourne, FL 32934
(321) 259-9465
www.dblrsupply.com
General supplies for poultry and livestock.

Elite Genetics
605 Rossville Road
Waukon, IA 52172
(563) 568-4551
www.elitegenetics.com
A leader in artificial insemination in the sheep industry.

Farmstead Health Supply*
P.O. Box 985
Hillsborough, NC 27278
(919) 643-0300
www.farmsteadhealth.com
Natural parasite control, supplements, minerals, and do-it-yourself test kits for livestock.

First State Veterinary Supply
P.O. Box 190
Parsonsburg, MD 21849
(800) 950-8387
Health supplies for poultry and other farm animals.

G.Q.F. Manufacturing Company*
2343 Louisville Road
P.O. Box 1552
Savannah, GA 31498
(912) 236-0651
www.gqfmfg.com
Wire pens, incubators, brooders, and other poultry supplies.

Hoegger Supply Company*
P.O. Box 331
Fayetteville, GA 30214
(800) 221-GOAT
www.hoeggergoatsupply.com
A complete line of goat supplies, including supplies for disbudding, tattooing, grooming, artificial insemination, cheese and butter making, insect control. Also carts and wagons and packing supplies.

Hoffman Hatchery*
Box 129
Gratz, PA 17030
(717) 365-3694
www.hoffmanhatchery.com
Chicks, ducklings, goslings, and related books and equipment.

Holderread's Waterfowl Farm & Preservation Center*
P.O. Box 492
Corvallis, OR 97339
(541) 929-5338
Rare varieties of ducks and geese.

Lambriar Animal Health Care*
101 Highway Avenue
Mahaska, KS 66955
(800) 344-6337
www.lambriarvetsupplies.com
Cow and cattle veterinary medicine supplies.

Lyon Electric Company*
1690 Brandywine Avenue
Chula Vista, CA 91911
(619) 216-3400
www.lyonelectric.com
Incubators, brooders, parts, and accessories for poultry.

Max-Flex Fence Systems*
U.S. Route 219
Lindside, WV 24951
(800) 356-5458
www.maxflex.com
Various fencing materials for livestock.

Metzer Farms Duck and Goose Hatchery*
26000 Old Stage Road
Gonzales, CA 93926
(800) 424-7755
www.metzerfarms.com
Ducklings, goslings, books, equipment, medicine.

Michigan Farm Systems, Inc.
(989) 224-3839
www.cowmattress.com
Cow supplies, including mattresses, barn curtains, and new and used equipment.

MidStates Livestock Supplies
125 East 10th Avenue
South Hutchinson, KS 67505
(800) 835-9665
www.midstateswoolgrowers.com
General supplies for sheep.

Murray McMurray Hatchery*
P.O. Box 458
191 Closz Drive
Webster City, IA 50595
(800) 456-3280
www.mcmurrayhatchery.com
Chicks, ducklings, goslings, and related supplies and books. Also rabbit equipment and books.

Nasco Farm & Ranch*
901 Janesville Avenue
Fort Atkinson, WI 53538-0901
(800) 558-9595
www.enasco.com
Full line of supplies for livestock.

National Band & Tag*
721 York Street
P.O. Box 72430
Newport, KY 41072
(859) 261-2035
www.nationalband.com
Poultry bands and livestock tags.

Omaha Vaccine Company*
P.O. Box 7228, 3030 "L" Street
Omaha, NE 68107
(800) 367-4444
www.omahavaccine.com
Medications and supplies for livestock.

PBS Animal Health*
P.O. Box 9101
Canton, OH 44711-9101
(800) 321-0235
www.pbsanimalhealth.com
A full line of health supplies for sheep, goats, and dairy and beef cattle.

Parts Department*
45 Lynwood Drive
Trumbull, CT 06611
800-245-8222
www.partsdeptonline.com
A full line of dairy equipment and supplies.

Patrick Green Carders
48793 Chilliwack Lake Road
Chilliwack, British Columbia
Canada V4Z1A6
(877) 898-2273
Wool carding machines of all sizes.

Powell Sheep Company
P.O. Box 183
Ramona, CA 92065
(760) 789-1758
Offers many sizes of sheep coats and spinning wheels.

Premier 1 Supplies*
2031 300th Street
Washington, IA 52353
(800) 282-6631
www.premier1supplies.com
A full line of sheep and goat supplies. Various fencing for livestock.

Sheepman Supply Company*
P.O. Box A
Frederick, MD 21702
www.sheepman.com
(800) 331-9122
A full line of supplies for sheep and other livestock.

Smith Poultry & Game Bird Supply*
14000 West 215th Street
Bucyrus, KS 66013
(913) 879-2587
www.poultrysupplies.com
A full line of poultry supplies.

Springcreek Goat Supplies*
5349W 100S
Larwill, IN 46764
www.springcreekgoatsupplies.com
A full line of goat supplies, including a pasteurizer and artificial insemination materials.

Stromberg's Chicks and Game Birds Unlimited*
P.O. Box 400
Pine River, MN 56474
(800) 720-1134
www.strombergschickens.com
Chicks, full line of poultry supplies, and books.

Tomahawk Live Trap Company*
P.O. Box 323
Tomahawk, WI 54487
(800) 272-8727
www.livetrap.com
A large selection of live traps and animal cages.

Valley Vet Supply*
1118 Pony Express Highway
Marysville, KS 66508
(800) 419-9524
www.valleyvet.com
Livestock equipment and medications.

Vanecek Bunny Farm and Equipment*
51 Sun Valley Drive
Spring Branch, TX 78070
(830) 885-4834
www.bunnyrabbit.com
Rabbits and a full line of rabbit supplies.

Waterford Corporation*
404 North Link Lane
Fort Collins, CO 80524
(800) 525-4952
www.waterfordcorp.com
Electric fencing materials.

Western Ranch Supply*
P.O. Box 1497
Billings, MT 59103
(800) 548-7270
www.truewest.com
Equipment, medicine, and general livestock supplies.

Wiggins & Associates, Inc.*
503 SW Victoria Court
Gresham, OR 97080-9265
(800) 600-0716
www.wigginsinc.com
Veterinary and animal care supplies, including milking equipment, halters and leads, medicines, castrators, dehorners, hoof care, tags, tattoos, semen tanks, spinning wheels, and more.

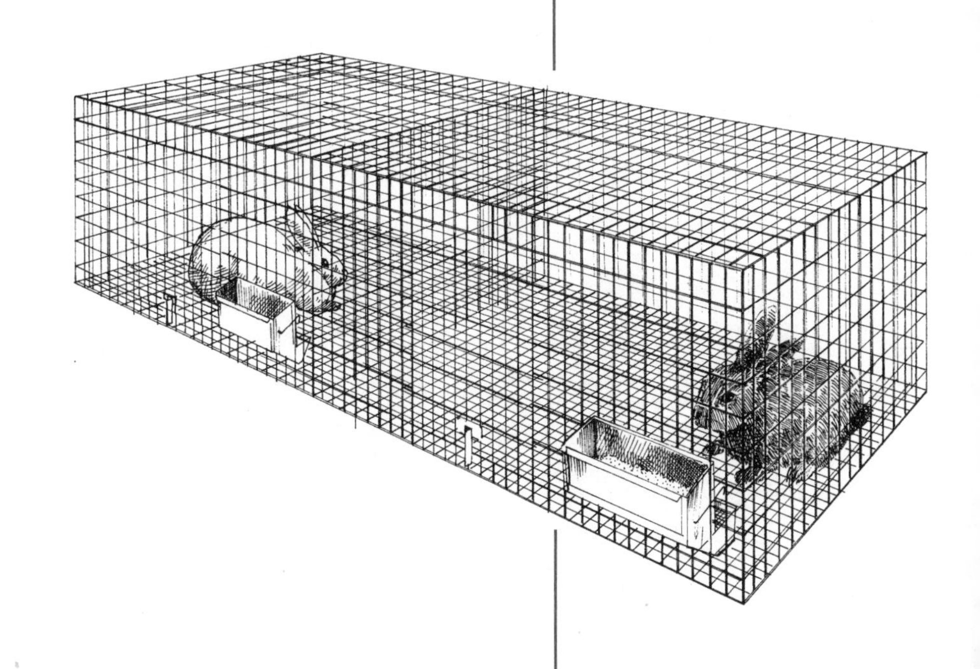

Index

Page references in **bold** indicate a chart; *italicized* references indicate an illustration.

A

B

C

D

H

I

J

K

L

M

N

O

P

R

S

T

U

Illustration Credits

Cathy Baker: 310, 318

Bethany Caskey: 15, 22, 25, 26, 32, 33, 39 (top), 97

Jeff Domm: 309, 311 (double half-hitch, overhand, square knots), 355

Brigita Fuhrmann: 257

Carol Jessop: 116, 125, 127 (top), 128, 130 (bottom), 136, 141 (top), 143, 146, 149, 150, 158, 160, 168, 178, 181, 182, 184, 195, 197, 206, 215, 218, 221–223, 226, 228–230, 235 (box), 237, 238, 243–245, 248, 249, 254, 259, 260, 266–268, 271, 277, 369, 375, 378

Kimberlee Knauf: 80

Elayne Sears: v, 2–5, 7, 8, 10, 14, 17, 18, 23, 24, 28–30, 34, 36, 39 (bottom), 45, 47–51, 55, 65–69, 72, 75, 84–86, 88, 93, 95, 96, 99–101, 103, 105, 106, 118–121, 126, 127 (bottom), 129, 130 (top), 131–134, 137, 139, 140, 141 (bottom), 153, 154, 161,162, 170, 173, 174, 180, 189, 207, 208, 211, 220, 235 (breeds), 236, 242, 247, 252, 258, 261, 272–275, 278, 279, 282, 289, 291–293, 295, 296, 302, 303, 307, 311 (bowline, quick-release knots), 312, 313, 316, 321, 322, 327, 334, 339, 343, 346, 348–350, 352, 353, 366, 370, 373, 376, 381, 382, 385, 386, 389, 393